Chemical Protective Clothing

Permeation and Degradation Compendium

T0173969

Krister Forsberg
Lawrence H. Keith

CRC Press
Taylor & Francis Group
Boca Raton London New York

CRC Press is an imprint of the
Taylor & Francis Group, an **informa** business

CRC Press
Taylor & Francis Group
6000 Broken Sound Parkway NW, Suite 300
Boca Raton, FL 33487-2742

Reissued 2019 by CRC Press

A Library of Congress record exists under LC control number:

Publisher's Note
The publisher has gone to great lengths to ensure the quality of this reprint but points out that some imperfections in the original copies may be apparent.

Disclaimer
The publisher has made every effort to trace copyright holders and welcomes correspondence from those they have been unable to contact.

ISBN 13: 978-0-367-20488-4 (hbk)
ISBN 13: 978-0-367-20489-1 (pbk)
ISBN 13: 978-0-429-26179-4 (ebk)

Visit the Taylor & Francis Web site at http://www.taylorandfrancis.com and the
CRC Press Web site at http://www.crcpress.com

INTRODUCTION

Chemical Protective Clothing Permeation and Degradation Compendium was produced from the electronic publication, *Chemical Protective Clothing Permeation and Degradation Database.* The latter was developed to provide a technical basis for selecting chemical protective clothing (CPC). The objective of this book is the same as that of the database, i.e., to give you the best possible information currently available regarding resistance of CPC to a wide variety of chemicals.

The focus of information in this book is chemical resistance. It is addressed by incorporating all the data in the world that could be found on:

    breakthrough times
    permeation rates
    degradation of CPC materials

To use this information intelligently you should have a basic understanding of chemical hazards, properties of materials from which CPC garments are constructed, and risks associated with chemical exposure. The American Industrial Hygiene Association and a multitude of private entities offer short courses that can help you gain an appreciation of these subjects if you need it.

A Chemical Permeation Index Number has been designed to help people who are not very familiar with the above concepts. It helps them to also use this compendium. This index is an easy numbering system from zero (0) which represents the best selection(s) to five (5) which represents the worst selection(s). It is present with all data having breakthrough times and permeation rates. A thorough discussion of this system is presented later; please become familiar with it.

The database from which this book was derived was begun by Krister Forsberg in the early 1980s when he was working as a Research and Project Manager at the Royal Institute of Technology in Stockholm, Sweden. In 1985, Krister and Larry Keith decided to convert the database into a publishable format so that it could be accessed inexpensively throughout the world. It became one of the world's first "electronic books" under the title of Chemical Permeation and Degradation Database and Selection Guide for Resistant Protective Materials. However, the established publishers of the world didn't know how to produce and market an electronic book, so the publication was rejected by every publisher approached. With no other recourse, Larry and Virginia Keith founded a new publication company, Instant Reference Sources Inc., dedicated to publishing electronic books and databases.

The first edition of the database, published in the summer of 1986, contained about 3,500 test results on about 100 gloves with about 250 chemicals. It ran on IBM and compatible computers using DOS 2.0. The publication has since been updated every year, adding new information but retaining the original format. A key criteria of the publication from its inception has been that every piece of test data must have a documented reference source.

iii

In 1988 the title of the database was shortened and changed to emphasize its chemical permeation index number. The new name was Chemical Protective Clothing Performance Index, and it was published under this title through May of 1992.

The current database, and this book, have been expanded to include over 8,800 test records (which includes over 12,000 tests when degradation tests are included) on over 750 chemicals and almost 300 different types and models of chemical protective clothing. Chemical Protective Clothing Permeation and Degradation Database comes in both IBM-compatible and Macintosh versions. The latter is edited by Mike Blotzer and is a Hypercard version with a feature that allows searches by chemical or by garment.

This book is the world's largest international bound publication of chemical permeation and degradation data. It's electronic version has become the industry standard in its field and is widely used throughout the USA, Canada, Western Europe, and some Pacific Basin countries (notably Australia). The CPC products with test data come primarily from the USA, Europe, and Australia.

We hope that you will find this publication to be a valuable reference for your work. The value of the data, which has been collected, screened, checked and rechecked many times over a ten year period, is in units of many thousands of hours. Translate that to present salary values and you'll find it to be a real bargain. We welcome your comments and suggestions for improvements.

ACKNOWLEDGEMENTS

We thank Mr. Bertil Krakenberger of Apoteksbolaget in Stockholm, Sweden, for his contribution in reviewing chemical nomenclature and the chemical class system.

Larry Keith is a Senior Program Manager and Principal Scientist at Radian Corporation in Austin, Texas. A pioneer in environmental sampling and analysis, method development, and handling of hazardous compounds, Dr. Keith has published over a dozen books and presented and published more than one hundred technical articles. Recent publications have involved electronic book formats and expert systems. Dr. Keith serves on numerous government, academic, publishing, and environmental committees and is past chairman of the ACS Division of Environmental Chemistry. Prior to joining Radian Corporation in 1977, he was a research scientist with the U. S. Environmental Protection Agency.

Krister Forsberg is an industrial hygienist and works for AGA AB, one of the leading gas companies with operations in 30 countries in Europe, the U.S., and Latin America. This database was first developed when Mr. Forsberg was working as a Research and Project Manager at The Royal Institute of Technology in Stockholm, Sweden. Other work has been published in *American Industrial Hygiene Journal, Applied Industrial Hygiene,* and *ASTM's International Symposium on the Performance of Protective Clothing.* He has been an active member of ASTM Committee F-23 "Protective Clothing" since 1982.

TABLE OF CONTENTS

## CHAPTER 1

CHEMICAL PROTECTIVE CLOTHING INFORMATION

Generic Materials With Test Data

The most well known generic names are used in this compendium. However, they consist of both chemical names of the materials as well as trade names (and sometimes trademarks or registered names). These are summarized in Table 1 below:

Table 1. Materials Used In The Manufacture of CPC Garments

Material Name Used in the Compendium	Other Material Names Used in the Literature
Butyl	IIR (rubber)
Chlorobutyl	Chlorinated butyl rubber
CPE	Chlorinated polyethylene or PEC
ECO	Epichlorohydrine rubber
EVAC	Ethylene vinyl acetate or EVA
EVAL	Ethylene-vinyl alcohol or EVOH
Hypalon	Chlorosulphonated ethylene rubber or CSM
Natural Rubber	NR(rubber)
Neoprene	Chloroprene rubber or CR
Nitrile	NBR (rubber)
PE	Polyethylene
Polyurethane or PUR	Polyurethane rubber
PVAL	Polyvinyl alcohol or PVA (Trademark)
PVC	Polyvinyl chloride
Saranex-23 (Trademark)	PE/Polyvinylidene chloride
Teflon (Trademark)	PTFE
Teflon-FEP	PTFE-FEP
Viton (Trademark)	Fluorocarbon rubber or FPM

4H (Laminate of PE/EVAL) is a Trade Mark of Safety4; Saranex is a Trademark of Dow Chemical; PVA is a Trademark of Ansell Edmont; Viton and Teflon are Trademarks of du Pont.

In order to improve the chemical resistance and physical properties of materials, manufacturers often laminate, blend, or support the materials in their products. Examples of laminated materials include butyl/neoprene and PE/EVAL/PE. Examples of blended materials include natural rubber+neoprene+nitrile and also PVC+nitrile.

The supported materials are fabrics of different types of fibers. Flame and heat resistant fibers are

Fiberglass
Nomex (Aramid fiber)

Remaining fibers are used as woven or non-woven fabrics:

Polyester (woven or non-woven fabrics)
Nylon or polyamide (woven fabrics)
Polypropylene or PP (non-woven fabrics)
Polyethylene or PE (non-woven fabrics)

NOTE: "Unknown" materials (those which are made of proprietary materials that manufacturer's keep secret) are not described in the table above. Several of these are listed in the table of suit materials below.

Suit Materials With Test Data

The description of the tested materials in the compendium is limited to generic materials. Most suit materials are supported and many are laminated in several layers. Therefore, a more comprehensive description of the suit materials are listed in Table 2. Many of the products in the table below are Registered or Trademarked; your attention is directed to the fact that many of the product names have these characters. The reader should also be aware that "Teflon" is a trade name applied to Du Pont's brand of FEP, and PTFE. Since each of these resins has a different chemical structure, "Teflon" alone does not identify the exact polymer.

Table 2. Suit Materials and Product Names Tested and Commercially Available

Material Description	Product Name	Manufacturer Name
Butyl-Nylon-Butyl	Butyl Suit	Life-Guard
Butyl-Nylon-Butyl	Acid King	Wheeler Protective Apparel, Inc.
Butyl-Nylon Butyl	306 B/BA	Fryrepel Products, Inc.
Butyl-Nylon-Butyl	Trellchem/Butyl Extra	Trelleborg, Inc.
Butyl-Nylon-Butyl	Type 500/Type 710	Drager
Butyl-Polyester-Neoprene	Betex	MSA, Co.
CPE-Nylon-CPE	Chloropel	ILC Dover, Inc.
CPE-Polyester-CPE	Stasafe CPE	Standard Safety Equipment Co.
Neoprene-Nylon-Neoprene	Neoprene Suit	Life-Guard
PVC-Polyester	Stasafe WG-20	Standard Safety Equipment Co.
PVC-Polyester-PVC	Acid King	Wheeler Protective Apparel, Inc
PVC-Nylon-PVC	306PVC/BA	Fyrepel Products, Inc.
PVC-Nylon-PVC	Trellchem Light Extra	Trelleborg, Inc.
Saranex-Polyethylene	Saranex-23P-Tyvek	Du Pont product
Teflon-Fiberglass-		

Material Description	Product Name	Manufacturer Name
Teflon	Challenge 6000	
Teflon-Nomex-Teflon	Acid King	Wheeler Protective Apparel, Inc
Viton-Nylon-Chlorobutyl	VN-100	Life-Guard
Viton-Nylon-Butyl	Type 500/Type 710	Drager
Viton-Nylon-Neoprene	Vautex	MSA, Co.
Viton-Polyester-Viton	306V/BA	Fyrepel Products, Inc.
Viton/Butyl-Nylon-Butyl	Trellchem Super Extra	Trelleborg, Inc.
Viton/Butyl/Unknown	Trellchem HPS	Trelleborg, Inc.
Unknown plastic-Polyester	Chemrel	Chemron, Inc.
Unknown plastic-Polyester-plastic	Chemrel Max	Chemron, Inc.
Unknown plastic-Polypropylene	Barricade	Du Pont product
Unknown plastic-Polypropylene-plastic	Responder	Life-Guard
Unknown	CPF III	Kappler
Unknown plastic-Polypropylene-plastic	Blue Max	MSA, Co.

dash (-) = Supported; slash (/) = Laminated materials

The National Fire Protection Association (NFPA) has developed a list of minimum requirements under NFPA-1991 that full body suits should meet when tested with both physical and chemical parameters. A list of 17 representative chemicals selected by NFPA are used for chemical permeation tests and the suits in Table 3 meet the NFPA-1991 minimum test requirements. These products are listed in alphabetical order.

Table 3. Suits Meeting NFPA-1991 Minimum Requirements

Barricade (DuPont material)
Blue Max
Challenger 6000
Chemrel Max
Responder
Trellchem HP

Chemical Protective Clothing Tested

Below is information on the name of each manufacturer, tested product, generic material name, and information on the garments. Various clothing types are indicated by letters in the table.

Table 4.   Garment Codes in the Compendium and
Clothing Descriptions

Garment Code	Description of the Clothing Type
g	gloves (supported gloves have no material thickness data)
s	suits (Level A or B hooded or totally-encapsulating)
c	coveralls
v	visors, face shields and other transparent materials
b	boots
m	miscellaneous garments (aprons, booties, etc.)
u	materials under development (not in production)
x	clothing type unknown

A complete summary of information on all of the garments tested appears in Table 5. There are a total of 8,815 test records (which include over 12,000 tests when degradation tests are included) on 280 different models of chemical protective clothing. NIOSH has shown that variations in protective capability offered by a material can vary greatly from manufacturer to manufacturer so this information is very important. It also provides some indication as to the probability of finding test data for any particular manufacturer's product since it shows frequency of the garments referenced.

The use of chemical permeation and chemical degradation data by knowledgeable industrial hygienists and safety professionals in the United States, Australia, New Zealand, and Europe has greatly increased over the last several years. It is recognized as the only reliable way to make decisions regarding selection of the best CPC for specific needs. Many manufacturers of chemical protective clothing now routinely publish this type of information in their brochures. In addition, some users of large amounts of CPC obtain this information through CPC manufacturers (and sometimes independently of them). The latter situation occurs when large companies need this data on proprietary, unusual or complex mixtures of chemicals to protect their employees from unnecessary chemical exposure.

A summary of all the protective clothing garments with counts of the number of test results reported with each appears in three sections of Table 5 below. Section 1 contains information on garments that are currently available. Section 2 contains information on garments that are of unknown availability. Section 3 contains information on garments that are believed to be not available. Although the garments in Section 3 may not be commercially available, the information associated with them is still useful as research data or for helping to make generic material selections when no other data is available. The names for commercially available Edmont products are named ANSELLEDMONT since Ansell has purchased Edmont. However, old Edmont data on products not commercially available still have separated names for those products. A similar situation occurs with MAPA-PIONEER products because MAPA has purchased PIONEER.

North Hand Protection has changed its model numbers for Viton gloves: North F-091 Viton glove is now being manufactured as North F-101 Viton glove and

North F-121 Viton glove is being manufactured as North F-124 Viton glove. Improved process modifications have been made but the formula and chemical resistance has not been altered according to the manufacturer. The former model numbers are still provided with their test data. When new test data is available for these gloves under the new model numbers it will be reported under the new model numbers.

Safety 4 makes gloves, sleeves, booties and aprons of the same material (which is listed as PE/EVAL/PE and is a trademark 4H) as of 1991. The garment designation with the test data is "g" for "gloves" but the test data applies equally for all of the test garments since the exact same material is used for all of Safety 4's products. DuPont's Barricade (a multi-layer of unknown plastic materials) is also used by several suit manufacturers; one example is Kappler's FrontLine• suit.

All Saranex data refer to the product SARANEX 23-P.

In addition to the abbreviations described above for garment type and notations of blended (+) and laminated (/) materials, the following also apply to Table 5 in the 3 sections below:

> n.a. - not available or not applicable
> V - variable lengths under the Nominal Length column;
> Bx - glove box gloves under the Nominal Length column;
> N - not supported under the Supported column;
> Y - supported under the Supported column;
> blank - unknown and/or data is not available

Table 5.   *Garment Information and Frequency of Reported Test Data*

Section 1 — Garments That Are Currently Commercially Available

# of cts	Garm.	Manufacturer/ prod. ident.	Generic material	Length (mm)	Thick (cm)	Sizes	Supp.	Avail.
185	g	ACKWELL 5-109	NATURAL RUBBER	0.15	23	2	N	Y
1	g	AMERICAN SCIENTIFIC	PVC	0.17	25		N	Y
3	g	ANSELL 632	NITRILE	0.36	33	4	N	Y
3	g	ANSELL 9070-5	NATURAL RUBBER	0.25	22	2	N	Y
72	g	ANSELL CHALLENGER	NITRILE	0.38	33	5	N	Y
81	g	ANSELL CHEMI-PRO	NEOPRENE/NATURAL	0.72	33	4	N	Y
1	g	ANSELL CONFORM 4107	NATURAL RUBBER	0.14	25	2	N	Y
46	g	ANSELL CONFORM 4205	NATURAL RUBBER	0.13	23	2	N	Y
11	g	ANSELL CONT ENVIRON	NATURAL RUBBER	0.53	30	5	N	Y
52	g	ANSELL FL 200 254	NATURAL RUBBER	0.51	30	4	N	Y
13	g	ANSELL NEOPRENE 520	NEOPRENE	0.46	30	4	N	Y
68	g	ANSELL NEOPRENE 530	NEOPRENE	0.46	30	4	N	Y
89	g	ANSELL OMNI 276	NATURAL+NEOPRENE	0.56	30	4	N	Y
57	g	ANSELL ORANGE 208	NATURAL RUBBER	0.76	30	4	N	Y
63	g	ANSELL PVL 040	NATURAL RUBBER	0.46	30	5	N	Y
45	g	ANSELL STERILE 832	NATURAL RUBBER	0.23	30	6	N	Y
56	g	ANSELL TECHNICIANS	NATURAL+NEOPRENE	0.43	30	8	N	Y
133	g	ANSELLEDMONT 29-840	NEOPRENE	0.38	28	4	N	Y

# of cts	Garm.	Manufacturer/ prod. ident.	Generic material	Length (mm)	Thick (cm)	Sizes	Supp.	Avail.
10	g	ANSELLEDMONT 29-865	NEOPRENE	0.38	28	4	N	Y
217	g	ANSELLEDMONT 29-870	NEOPRENE	0.38	28	4	N	Y
1	g	ANSELLEDMONT 29-875	NEOPRENE	0.38	28	4	N	Y
4	g	ANSELLEDMONT 3-318	PVC	0.38	28	4	N	Y
21	g	ANSELLEDMONT 30-139	NATURAL RUBBER	0.38	28	4	N	Y
172	g	ANSELLEDMONT 34-100	PVC	0.38	28	4	N	Y
2	g	ANSELLEDMONT 34-500	PVC	0.38	28	4	N	Y
37	g	ANSELLEDMONT 35-125	PE	0.38	28	4	N	Y
121	g	ANSELLEDMONT 36-124	NATURAL RUBBER	0.38	28	4	N	Y
1	g	ANSELLEDMONT 36-755	NATURAL RUBBER	0.38	28	4	N	Y
17	g	ANSELLEDMONT 37-145	NITRILE	0.38	28	4	N	Y
215	g	ANSELLEDMONT 37-155	NITRILE	0.38	28	4	N	Y
146	g	ANSELLEDMONT 37-165	NITRILE	0.38	28	4	N	Y
71	g	ANSELLEDMONT 37-175	NITRILE	0.38	28	4	N	Y
2	g	ANSELLEDMONT 37-185	NITRILE	0.38	28	4	N	Y
119	g	ANSELLEDMONT 392	NATURAL RUBBER	0.48	30	9	N	Y
26	g	ANSELLEDMONT 46-320	NATURAL RUBBER	0.38	28	4	N	Y
1	g	ANSELLEDMONT 46-322	NATURAL RUBBER	0.38	28	4	N	Y
8	g	ANSELLEDMONT 49-125	NITRILE	0.23	28	4	N	Y
23	g	ANSELLEDMONT 49-155	NITRILE	0.38	28	4	N	Y
1	c	ANSELLEDMONT 55-530	PE	0.38	28	4	N	Y
1	g	ANSELLEDMONT HYNIT	NITRILE	1.40	25	5	Y	Y
44	g	ANSELLEDMONT M GRIP	PVC	1.40		4	Y	Y
119	g	ANSELLEDMONT NEOX	NEOPRENE	1.50	V	1	Y	Y
115	g	ANSELLEDMONT PVA	PVAL	1.20			N	Y
87	g	ANSELLEDMONT SNORKE	PVC	1.40	V	4	Y	Y
18	b	BATA HAZMAX	NITRILE+PUR+PVC	2.50		9	Y	Y
32	b	BATA POLYBLEND	PVC+POLYURETHANE	2.00		9	Y	Y
15	b	BATA POLYMAX	PVC+POLYURETHANE	2.00		9	Y	Y
17	b	BATA STANDARD	PVC	2.00		9	Y	Y
15	b	BATA SUPER POLY	PVC+POLYURETHANE	2.00		9	Y	Y
1	g	BEMAC 789 POLARFLEX	PVC	0.68	25	3	Y	Y
52	g	BEST 22R	NITRILE	1.40	30	1	Y	Y
60	g	BEST 32	NEOPRENE	1.50	30	3	Y	Y
40	g	BEST 65NFW	NATURAL RUBBER	1.50	30	1	Y	Y
77	g	BEST 6780	NEOPRENE	1.40	30	4	Y	Y
9	g	BEST 67NFW	NATURAL RUBBER	1.50	35	1	Y	Y
3	g	BEST 723	NEOPRENE	0.38	30	4	N	Y
22	g	BEST 725R	PVC	1.40	30	3	Y	Y
68	g	BEST 727	NITRILE	0.38	33	5	N	Y
2	g	BEST 7712R	PVC	n.a.	30	1	N	Y
21	g	BEST 812	PVC	1.40	30	1	Y	Y
2	g	BEST 814	PVC	1.40	35	1	Y	Y
71	g	BEST 878	BUTYL	0.75	36	3	N	Y
46	g	BRUNSWICK BUTYL STD	BUTYL	1.40	35	1	Y	Y
46	g	BRUNSWICK BUTYL-POL	BUTYL/ECO	0.63	36	5	N	Y
47	g	BRUNSWICK BUTYL-XTR	BUTYL	0.63	36	5	N	Y
34	g	BRUNSWICK NEOPRENE	NEOPRENE	0.88	36	4	N	Y
17	s	CHEMFAB CHALL. 6000	TEFLON	0.35		1	Y	Y
15	v	CHEMFAB CHALL.	TEFLON-FEP	n.a.			N	Y

# of cts	Garm.	Manufacturer/ prod. ident.	Generic material	Length (mm)	Thick (cm)	Sizes	Supp.	Avail.
46	s	CHEMRON CHEMREL	UNKNOWN MATERIAL	0.28		6	Y	Y
31	s	CHEMRON CHEMREL MAX	UNKNOWN MATERIAL	0.28	U	4	Y	Y
19	c	CHEMRON CHEMTUFF	UNKNOWN MATERIAL	0.28		4	Y	Y
1	v	CHEMRON	TEFLON-FEP/PC	n.a.		U	N	Y
2	s	COMASEC ASTRO-PRENE	NEOPRENE	0.28		4	Y	Y
21	g	COMASEC BUTYL	BUTYL	0.50	30	4	N	Y
29	g	COMASEC BUTYL PLUS	BUTYL/NEOPRENE	0.50	30	4	N	Y
46	g	COMASEC COMAPRENE	NEOPRENE	1.50	30	4	Y	Y
46	g	COMASEC COMATRIL	NITRILE	0.55	30	4	N	Y
47	g	COMASEC COMATRIL SU	NITRILE	0.60	30	4	N	Y
24	g	COMASEC DIPCO	HYPALON	0.60	Bx	4	N	Y
48	g	COMASEC FLEXIGUM	NATURAL RUBBER	1.00	28	4	N	Y
50	g	COMASEC FLEXITRIL	NITRILE	0.80	28	4	Y	Y
23	g	COMASEC MULTIMAX	NITRILE+PVC	1.40	V	U	Y	Y
48	g	COMASEC MULTIPLUS	NITRILE+PVC	1.40	V	4	Y	Y
49	g	COMASEC MULTITOP	PVC	1.40	V	4	Y	Y
47	g	COMASEC MULTIPOST	PVC	1.40	V	4	Y	Y
48	g	COMASEC NORMAL	PVC	1.40	V	4	Y	Y
48	g	COMASEC OMNI	PVC	1.30	V	4	Y	Y
17	g	COMASEC SOLVATRIL	PVAL	0.90			Y	Y
4	g	DAYTON SURGICAL	NATURAL RUBBER	0.16		U	N	Y
2	g	DAYTON TRIFLEX	PVC	0.20			N	Y
17	s	DRAGER 500 OR 710	VITON/BUTYL	0.50		2	Y	Y
5	s	DRAGER 500 OR 710	BUTYL	0.52		2	Y	Y
93	s	DU PONT BARRICADE	UNKNOWN MATERIAL	0.58			Y	Y
84	c	DU PONT TYVEK QC	PE	n.a.		U	Y	Y
120	s	DU PONT TYVEK SARAN	SARANEX-23	0.15		U	N	Y
18	s	DU PONT TYVEK SARAN	SARANEX-23 2-PLY	0.15		U	N	Y
5	g	EDMONT CANADA 14112	PVC	n.a.	30	1	Y	Y
25	g	ERISTA BX	BUTYL	0.60	30	7	N	Y
1	g	ERISTA P	NITRILE	0.80	V	3	N	Y
15	g	ERISTA SPECIAL	NITRILE	0.40	28	7	N	Y
29	g	ERISTA VITRIC	VITON/NEOPRENE	0.50	30	7	N	Y
2	s	FAIRPRENE	CHLOROBUTYL	n.a.		U	Y	Y
4	s	FAIRPRENE	NEOPRENE	0.51		U	N	Y
1	s	FYREPEL	BUTYL	n.a.		4	N	Y
1	s	FYREPEL	PVC	n.a.		U	Y	Y
1	s	FYREPEL	VITON	n.a.		U	N	Y
4	g	GRANET 2714	NEOPRENE	1.20	35	1	Y	Y
2	g	GRANET 512-L	PVC	1.30	30	2	Y	Y
2	g	GRANET SAFETY 1012	PVC	n.a.	30	1	Y	Y
27	s	ILC DOVER CLOROPEL	CPE	0.51		2	N	Y
17	b	JORDAN DAVID	PVC	2.30		U	Y	Y
2	g	KACHELE 706	NATURAL RUBBER	0.70	30	U	N	Y
1	g	KACHELE 722	NEOPRENE	0.65	30	U	N	Y
34	s	KAPPLER CPF III	UNKNOWN MATERIAL	n.a.		5	Y	Y
14	g	KID 490	PVC	1.30	V	4	Y	Y
2	g	KID 500	PVC	1.10	30	4	Y	Y
18	g	KID VINYLPRODUKTER	PVC	0.91	33	5	N	Y
16	s	LIFE-GUARD BUTYL	BUTYL	n.a.		5	Y	Y

# of cts	Garm.	Manufacturer/ prod. ident.	Generic material	Length (mm)	Thick (cm)	Sizes	Supp.	Avail.
15	s	LIFE-GUARD NEOPRENE	NEOPRENE	n.a.		5	Y	Y
134	s	LIFE-GUARD RESPONDE	UNKNOWN MATERIAL	n.a.		5	Y	Y
40	s	LIFE-GUARD VC-100	VITON/CHLOROBUTYL	n.a.			Y	Y
1	v	LIFE-GUARD VISOR	PVC	n.a.			N	Y
71	g	MARIGOLD BLACK HEVY	NATURAL RUBBER	0.82	30	4	N	Y
74	g	MARIGOLD BLUE	NITRILE	0.43	32	4	N	Y
68	g	MARIGOLD FEATHERWT	NATURAL+NEOPRENE	0.35	30	4	N	Y
72	g	MARIGOLD GREEN SUPA	NITRILE	0.40	31	4	N	Y
69	g	MARIGOLD MED WT EXT	NATURAL RUBBER	0.45	31	3	N	Y
69	g	MARIGOLD MEDICAL	NATURAL RUBBER	0.28	28	5	N	Y
78	g	MARIGOLD ORANGE SUP	NATURAL RUBBER	0.73	33	4	N	Y
69	g	MARIGOLD RED LT WT	NATURAL RUBBER	0.43	31	3	N	Y
67	g	MARIGOLD R SURGEONS	NATURAL RUBBER	0.28	28	8	N	Y
68	g	MARIGOLD SENSOTECH	NATURAL RUBBER	0.28	28	3	N	Y
71	g	MARIGOLD SUREGRIP	NATURAL RUBBER	0.89	30	4	N	Y
68	g	MARIGOLD YELLOW	NATURAL RUBBER	0.38	30	3	N	Y
5	g	MAPA-PIONEER 258	NATURAL RUBBER	0.60	32	5	N	Y
1	g	MAPA-PIONEER 296	NATURAL RUBBER	0.70	37	4	N	Y
4	g	MAPA-PIONEER 300	NATURAL RUBBER	1.10	31	6	Y	Y
4	g	MAPA-PIONEER 360	NEOPRENE	1.45	38	4	Y	Y
6	g	MAPA-PIONEER 420	NEOPRENE	0.75	32	5	N	Y
3	g	MAPA-PIONEER 370	NITRILE	1.15	34	4	Y	Y
4	g	MAPA-PIONEER 424	NATURAL+NEOPRENE	0.55	32	5	N	Y
5	g	MAPA-PIONEER 490	NITRILE	0.40	31	5	N	Y
3	g	MAPA-PIONEER 492	NITRILE	0.40	31	5	N	Y
3	g	MAPA-PIONEER 493	NITRILE	0.56	39	4	N	Y
4	g	MAPA-PIONEER 730	NATURAL RUBBER	1.00	42	4	N	Y
59	g	MAPA-PIONEER A-14	NITRILE	0.56	35	5	N	Y
16	g	MAPA-PIONEER A-15	NITRILE	0.38	33	5	N	Y
6	g	MAPA-PIONEER AF-18	NITRILE	0.46	33	3	N	Y
41	g	MAPA-PIONEER L-118	NATURAL RUBBER	0.46	30	5	N	Y
7	g	MAPA-PIONEER GF-N-	NEOPRENE	0.38	31	3	N	Y
1	g	MAPA-PIONEER LL-301	NATURAL RUBBER	1.50	30	5	Y	Y
1	g	MAPA-PIONEER N-30	NEOPRENE	0.38	28	4	N	Y
3	g	MAPA-PIONEER N-36	NEOPRENE	0.46	28	3	N	Y
60	g	MAPA-PIONEER N-44	NEOPRENE	0.56	30	3	N	Y
3	g	MAPA-PIONEER N-54	NEOPRENE	0.76	35	4	N	Y
19	g	MAPA-PIONEER N-73	NEOPRENE	0.56	45	4	N	Y
1	g	MAPA-PIONEER NS-35	NEOPRENE	0.55	32	5	N	Y
1	g	MAPA-PIONEER NS-420	NEOPRENE	0.89	32	5	N	Y
2	g	MAPA-PIONEER NS-53	NATURAL+NEOPRENE	0.85	32	5	N	Y
52	g	MAPA-PIONEER TRIONI	NAT+NEOPR+NITRILE	0.46	35	5	N	Y
2	g	MAPA-PIONEER V-10	PVC	0.25	27	4	N	Y
47	g	MAPA-PIONEER V-20	PVC	0.51	27	4	N	Y
5	g	MAPA-PIONEER V-5	PVC	0.13	27	4	N	Y
68	g	MARIGOLD FEATHERLIT	NATURAL RUBBER	0.15	30	4	N	Y
89	g	MARIGOLD NITROSOLVE	NITRILE	0.75	44	3	N	Y
11	c	MOLNLYCKE H D	PE	0.15		4	Y	Y
42	s	MSA BETEX	BUTYL/NEOPRENE	0.48		1	Y	Y
33	s	MSA CHEMPRUF	BUTYL	n.a.		1	Y	Y

# of cts	G a r m.	Manufacturer/ prod. ident.	Generic material	Length (mm)	Thick (cm)	Sizes	Supp.	Avail.
39	s	MSA VAUTEX	VITON/NEOPRENE	0.58		1	Y	Y
5	g	NOLATO 1505	NEOPRENE	0.80	35	2	N	Y
3	g	NORTH 800	PVC	1.20	30	U	Y	Y
2	g	NORTH 8B 1532A	BUTYL	0.38	Bx	2	N	Y
18	g	NORTH B-131	BUTYL	0.31	28	5	N	Y
104	g	NORTH B-161	BUTYL	0.41	28	5	N	Y
181	g	NORTH B-174	BUTYL	0.63	35	5	N	Y
285	g	NORTH F-091	VITON	0.23	28	4	N	Y
23	g	NORTH F-121	VITON	0.31	28	4	N	Y
3	g	NORTH	HYPALON/NEOPRENE	0.88	Bx	1	N	Y
5	g	NORTH LA-102G	NITRILE	0.25	30	5	N	Y
1	g	NORTH LA-111EB	NITRILE	0.28	30	5	N	Y
7	g	NORTH LA-132G	NITRILE	0.33	30	5	N	Y
55	g	NORTH LA-142G	NITRILE	0.36	30	5	N	Y
6	g	NORTH	NEOPRENE	n.a.	Bx	1	N	Y
68	g	NORTH SILVERSHIELD	UNKNOWN MATERIAL	0.08	35	3	N	Y
2	g	NORTH	VITON/BUTYL	0.46		4	N	Y
3	g	NORTH Y-1532	HYPALON	0.38	Bx	1	N	Y
1	g	PEDI	NEOPRENE	0.50	Bx		N	Y
2	g	PERRX-AM TEX	NATURAL RUBBER	0.12			N	Y
7	g	PLAYTEX 827	NATURAL+NEOPRENE	0.40	36	4	N	Y
3	g	PLAYTEX 835	NATURAL+NEOPRENE	0.40	28	4	N	Y
46	g	PLAYTEX ARGUS 123	NATURAL+NEOPRENE	0.91	33	5	N	Y
14	m	PLYMOUTH RUBBER	BUTYL	n.a.			Y	Y
17	b	RAINFAIR	NEOPRENE	1.80		U	Y	Y
17	b	RANGER	NEOPRENE	1.60		U	Y	Y
1	g	REX	NEOPRENE	n.a.			N	Y
363	g	SAFETY4 4H	PE/EVAL/PE	0.07	40	7	N	Y
1	g	SEMPERIT	PVC	0.65	30	4	N	Y
17	b	SERVUS NO 22204	NEOPRENE	1.50		U	Y	Y
17	b	SERVUS NO 73101	NITRILE+PVC	2.40		U	Y	Y
1	m	STD. SAFETY BOOTIE	NEOPRENE	n.a.				Y
22	s	STD. SAFETY	CPE	n.a.		4	Y	Y
18	b	STD. SAFETY GL-20	PVC	n.a.			Y	Y
1	v	STD. SAFETY	PVC	n.a.			N	Y
1	g	STD. SAFETY SD-6100	PVC	n.a.		2	Y	Y
18	s	STD. SAFETY WG-20	PVC	n.a.		4	Y	Y
3	g	SURETY 315R	NITRILE	0.32			N	Y
1	g	TASKALL SPIT FIRE	PVC	n.a.	U	U	N	Y
17	b	TINGLEY	NITRILE+PVC	3.30		U	Y	Y
17	s	TRELLEBORG HPS	VITON/BUTYL/UNKN.	n.a.		U	Y	Y
2	s	TRELLEBORG TRELLCHE	VITON/BUTYL	0.40		3	Y	Y
17	s	TRELLEBORG TRELLCHE	BUTYL	0.40		3	Y	Y
11	s	TRELLEBORG TRELLCHE	PVC	0.40		3	Y	Y
3	g	UNIROYAL 4-15	NITRILE	0.32			N	Y
25	s	WHEELER ACID KING	BUTYL	n.a.		3	Y	Y
26	s	WHEELER ACID KING	PVC	n.a.		3	Y	Y
22	s	WHEELER ACID KING	TEFLON	n.a.		3	Y	Y

There are 7,712 reported chemical permeation tests on 216 different models of chemical protective clothing known or believed to be commercially available as of March, 1992.

Section 2 — Garments That May Or May Not Be Commercially Available

# of cts	G a r m.	Manufacturer/ prod. ident.	Generic material	Length (mm)	Thick (cm)	Sizes	Supp.	Avail.
2	m	ARROWHEAD PRODUCTS	CHLOROBUTYL	0.50			Y	U
15	g	CLEAN ROOM PRODUCTS	TEFLON	0.05			N	U
1	g	DAVIDS GLOVES	NEOPRENE	n.a.	U	U	N	U
1	v	DU PONT	TEFLON-FEP	n.a.		U	N	U
2	g	ELASTYREN	SBR	0.15	23	U	N	U
2	g	GYNO-DAKTARIN	PE	0.03	25	N	U	
3	g	HANDGARDS	PE	0.07	30	U	N	U
37	u	ILC DOVER	CPE	n.a.			N	U
3	g	KURSAAL 85/62	NITRILE	0.40	30	U	N	U
29	g	MARIGOLD R MEDIGLOV	UNKNOWN MATERIAL	0.03	28	4	N	U
4	m	NASA	BUTYL	n.a.			N	U
1	m	PAIGE	BUTYL	n.a.		U	Y	U
1	g	PHARMASEAL LAB 8070	NATURAL RUBBER	0.20			N	U
3	g	PIONEER	NEOPRENE	n.a.				U
2	m	RICH INDUSTRIES	PVC	n.a.			Y	U
6	g	SURETY	NITRILE	n.a.			N	U
1	g	TRAVENOL	NATURAL RUBBER	0.18			N	U
17	u	Unknown	BUTYL	n.a.				U
2	v	Unknown	CR39	1.70	U	U	N	U
4	g	Unknown	EMA	n.a.		U	U	U
1	u	Unknown	EVAC	n.a.				U
48	u	Unknown	NATURAL RUBBER	n.a.				U
7	u	Unknown	NATURAL+NEOPRENE	n.a.				U
60	u	Unknown	NEOPRENE	n.a.				U
27	u	Unknown	NITRILE	n.a.				U
1	u	Unknown	NITRILE+PVC	n.a.				U
23	u	Unknown	PE	n.a.				U
3	u	Unknown	POLYURETHANE	n.a.				U
5	u	Unknown	PVAL	n.a.				U
49	u	Unknown	PVC	n.a.				U
2	u	Unknown	TEFLON	n.a.				U
3	u	Unknown	VITON	n.a.				U
56	u	Unknown	VITON/CHLOROBUTYL	n.a.				U

There are 421 reported chemical permeation tests on 33 different models of chemical protective clothing of uncertain availability as of March, 1992.

Section 3 — Garments That Are Not Commercially Available

1	u	ALLHABO	PE	0.08			N	N
2	g	ANSELL 650	NITRILE	n.a.				N
7	g	ANSELL CANNERS 334	NATURAL RUBBER	n.a.				N
12	g	ANSELL LAMPRECHT	BUTYL	0.50	35		N	N
1	u	ARSIMA	BUTYL	n.a.			N	N
2	b	BATA	BUTYL	1.10			Y	N
1	g	BEST 735-L	NITRILE	n.a.	25	1	Y	N

# of cts	G a r m.	Manufacturer/ prod. ident.	Generic material	L e n g t h (mm)	T h i c k (cm)	S i z e s	S u p p.	A v a i l.
112	s	CHEMFAB CHALL. 5100	TEFLON	0.48		4	Y	N
27	s	CHEMFAB CHALL. 5200	TEFLON	0.28		1	Y	N
12	u	CHEMFAB	TEFLON	0.10			N	N
1	g	COMASEC 7414	NITRILE	n.a.	36	2	Y	N
171	g	EDMONT 15-552	PVAL	1.40	30	2	Y	N
4	g	EDMONT 15-554	PVAL	1.40	35	1	Y	N
2	g	EDMONT 25-250	PVAL	n.a.		1	N	N
68	g	EDMONT 25-545	PVAL	n.a.	U	1	N	N
153	g	EDMONT 25-950	PVAL	n.a.	U	1	N	N
2	g	EDMONT 34-590	PVC	0.15			N	N
3	g	GRANET 2001	NEOPRENE	0.45			N	N
3	g	GRANET 490	NITRILE	0.32	30	5	N	N
5	g	GRANET N11F	NITRILE	n.a.	28	1	Y	N
5	m	ILC DOVER	POLYURETHANE	0.38			N	N
1	g	KURASHIKI	PVAL	n.a.	U	U	N	N
21	s	LIFE-GUARD TEFGUARD	TEFLON	0.28			Y	N
15	s	LIFE-GUARD VNC 200	VITON/CHLOROBUTYL	n.a.		1	Y	N
5	u	MOLNLYCKE	PVDC/PE/PVDC	n.a.			Y	N
29	s	MSA UPC	PVC	0.20			Y	N
3	u	NOLATO	NITRILE	0.35			N	N
5	g	NORTH SF	VITON	n.a.			N	N
7	g	NORTH VITRILE	VITON/NITRILE	0.25			N	N
1	g	PIONEER N-190	NATURAL RUBBER	n.a.			N	N
1	u	Unknown	PVAL	n.a.		U	N	N

There are 682 reported chemical permeation tests on 31 different models of chemical protective clothing that are known or believed to be commercially unavailable as of March 1992.

There are a total of 8,815 reported chemical permeation tests with a total of 280 different models of chemical protective clothing and 758 chemicals and mixtures in this publication. In addition, there are over 3,000 chemical degradation tests reported and the compendium includes about another 20,000 associated data.

## CHAPTER 2

CHEMICAL CLASSES IN THE COMPENDIUM

Table 6 below lists the chemical classes and the corresponding Class Number used. This classification system is the same as the ASTM F1186, "Guide for Classification of Chemicals According to Functional Groups" and has been extended to include mixtures and commercial formulations.

All of the ASTM chemical classes have been included although a few of them do not have test data yet. As chemicals representative of these classes are tested and their data becomes available it will be added. In the meantime, a notation to the effect that no data is available appears when this situation occurs. There are a total of 108 different chemical classes in Table 6.

Table 6. *Chemical Classes in the Compendium*

CLASS No.	CHEMICAL CLASS NAME
* 102	Acids, Carboxylic, Aliphatic and Alicyclic, Unsubstituted
* 103	Acids, Carboxylic, Aliphatic and Alicyclic, Substituted
* 104	Acids, Carboxylic, Aliphatic and Alicyclic, Polybasic
* 105	Acids, Carboxylic, Benzoic (Contains no chemicals or permeation test data)
* 106	Acids, Aromatic, Others (Contains no chemicals or permeation test data)
* 111	Acid Halides, Carboxylic, Aliphatic and Alicyclic
* 112	Acid Halides, Carboxylic, Aromatic
* 121	Aldehydes, Aliphatic and Alicyclic
* 122	Aldehydes, Aromatic
* 132	Amides, Aliphatic and Alicyclic
* 133	Acrylamides
* 134	Amides, Acetanilides (Contains no chemicals or permeation test data)
* 135	Amides, Aromatic, Others
* 137	Amides, Carbamides, and Guanidines
* 141	Amines, Aliphatic and Alicyclic, Primary
* 142	Amines, Aliphatic and Alicyclic, Secondary
* 143	Amines, Aliphatic and Alicyclic, Tertiary
* 145	Amines, Aromatic, Primary
* 146	Amines, Aromatic, Secondary and Tertiary
* 147	Amines, Alkyl-Aryl, Monoamines (Contains no chemicals or permeation test data)
* 148	Amines, Poly, Aliphatic and Alicyclic
* 149	Amines, Poly, Aromatic (Contains no chemicals or permeation test data)
* 161	Anhydrides, Aliphatic and Alicyclic
* 162	Anhydrides, Aromatic
* 170	Azo and Azoxy Compounds
* 211	Isocyanates, Aliphatic and Alicyclic
* 212	Isocyanates, Aromatic
* 221	Esters, Carboxylic, Formates (Contains no chemicals or permeation test data)
* 222	Esters, Carboxylic, Acetates
* 223	Esters, Carboxylic, Acrylates and Methacrylates
* 224	Esters, Carboxylic, Aliphatic, Others
* 225	Esters, Carboxylic, Lactones

* 441   Nitro Compounds, Unsubstituted
* 442   Nitro Compounds, Substituted
* 450   Nitroso Compounds
* 461   Organophosphorus Compounds, Phosphines (Contains no chemicals or permeation test data)
* 462   Organophosphorus Compounds and Derivatives of Phosphorus-based Acids
* 470   Organometallic Compounds
* 480   Organosilicon Compounds
* 501   Sulfur Compounds, Thiols
* 502   Sulfur Compounds, Sulfides and Disulfides
* 503   Sulfur Compounds, Sulfones and Sulfoxides
* 504   Sulfur Compounds, Sulfonic Acids
* 505   Sulfur Compounds, Sulfonyl Chlorides (Contains no chemicals or permeation test data)
* 506   Sulfur Compounds, Sulfonamides (Contains no chemicals or permeation test data)
* 507   Sulfur Compounds, Sulfonates, Sulfates, and Sulfites
* 508   Sulfur Compounds, Thiones
* 510   Nitrates and Nitrites
* 550   Organic Salts (Solutions)
* 800   Multicomponent Mixtures With > 2 Components
* 810   Epoxy Products
* 820   Etching Products
* 830   Lacquer Products
* 840   Pesticides, Mixtures and Formulations
* 850   Cutting Fluids
* 860   Coals, Charcoals, Oils
* 870   Antineoplastic Drugs and Other Pharmaceuticals
* 900   Miscellaneous Unclassified Chemicals

## CHAPTER 3

CHEMICALS AND MIXTURES IN THE COMPENDIUM

Table 7 below contains all of the tested chemicals ordered alpha numerically within each of the chemical classes from Table 6. Synonyms are included with each primary name. In this book the chemicals are listed in alpha-numeric order so you can look them up easily by name.

Table 7. Classes and Chemicals in the Compendium (Arranged by Class)

* 102 Acids, Carboxylic, Aliphatic and Alicyclic, Unsubstituted
 2-ethylhexanoic acid (synonym: butylethylacetic acid)
 acetic acid, 30-70%
 acetic acid, glacial (synonym: ethanoic acid)
 acrylic acid
 butyric acid
 caprylic acid (synonym: octanoic acid)
 formic acid
 lauric acid, 30-70% (synonym: dodecanoic acid, 30-70%)
 lauric acid, >70% (synonym: dodecanoic acid, >70%)
 linoleic acid (synonym: linolic acid)
 methacrylic acid (synonym: 2-methylpropenoic acid)
 oleic acid
 palmitic acid
 propionic acid

* 103 Acids, Carboxylic, Aliphatic and Alicyclic, Substituted
 2-(2,4-dichlorophenoxy)propionic acid, 15% & 2,4-D, 8% & water
 2-(2,4-dichlorophenoxy)propionic acid, 40% & MCPA, 20% & water
 2-(2,4-dichlorophenoxy)propionic acid, formulation
 2-(4-chloro-2-methylphenoxy)propionic acid, formulation
 3-bromopropionic acid
 4-chloro-2-methylphenoxyacetic acid, formulation (synonym: MCPA)
 chloroacetic acid (synonym: monochloroacetic acid)
 lactic acid
 mercaptoacetic acid (synonym: thioglycolic acid)
 trichloroacetic acid (synonym: trichloroethanoic acid)
 trifluoroacetic acid (synonym: perfluoroacetic acid)

* 104 Acids, Carboxylic, Aliphatic and Alicyclic, Polybasic
 citric acid, 30-70%
 citric acid, <30%
 maleic acid
 oxalic acid

* 105 Acids, Carboxylic, Benzoic (Contains no chemicals or permeation test data)

* 106 Acids, Aromatic, Others (Contains no chemicals or permeation test data)

* 111 Acid Halides, Carboxylic, Aliphatic and Alicyclic
 acetyl chloride
 dichloroacetyl chloride

* 112 Acid Halides, Carboxylic, Aromatic

17

benzoyl chloride (synonym: benzoic acid chloride)

* 121 Aldehydes, Aliphatic and Alicyclic
    acetaldehyde (synonym: ethanal)
    acrolein (synonym: acrylaldehyde)
    butyraldehyde
    crotonaldehyde (synonym: 2-butenal)
    crotonaldehyde, <30%
    decanal (synonym: caperaldehyde)
    formaldehyde, 30-70% (synonym: formalin)
    formaldehyde, 37% & methyl alcohol, 10% & water 53%
    formaldehyde, <30%
    glutaraldehyde (synonym: 1,5-pentanedial)
    isobutyraldehyde
    propionaldehyde (synonym: propanal)
    trichloroacetaldehyde (synonym: chloral)

* 122 Aldehydes, Aromatic
    benzaldehyde
    furfural (synonym: 2-furaldehyde)

* 132 Amides, Aliphatic and Alicyclic
    diethylacetamide (synonym: N,N-diethylacetamide)
    dimethyl acetamide (synonyms: DMAC; N,N-dimethyl acetamide)
    dimethyl formamide (synonyms: DMF; N,N-dimethyl formamide)
    N-methyl-2-pyrrolidone (synonym: 1-methyl-2-pyrrolidone)
    N-vinylpyrrolidone (synonym: 1-vinyl-2-pyrrolidone)
    Posistrip LE (see Class 900)

* 133 Acrylamides
    acrylamide (synonym: 2-propenamide)
    acrylamide, 15% & methyl ethyl ketone, 85%
    acrylamide, 88% & N,N-methylenebisacrylamide & water
    N-methylmethacrylamide

* 134 Amides, Acetanilides (Contains no chemicals or permeation test data)

* 135 Amides, Aromatic, Others
    propyzamide, <30% (synonym: KERB 5O R)

* 137 Amides, Carbamides, and Guanidines
    1,3-diphenylguanidine
    thiocarbamide, <30% in ETH

* 141 Amines, Aliphatic and Alicyclic, Primary
    1,3-dimethylbutylamine
    1,3-propanediamine (see Class 148)
    1-(2-aminoethyl)piperazine (see Class 148)
    2-(2-aminoethoxy)ethanol (synonym: diglycolamine)
    3,3'-iminobis(propylamine) (see Class 148)
    3-(dimethylamino)propylamine (see Class 148)
    3-methylaminopropylamine (see Class 148)
    allylamine (synonym: monoallylamine)
    cyclohexylamine
    diethylenetriamine (see Class 148)
    ethanolamine (synonyms: 2-aminoethanol; MEA; monoethanolamine)
    ethylamine (synonym: monoethylamine)
    ethylamine, 30-70% (synonym: monoethylamine, 30-70%)
    ethylenediamine (see Class 148)

isobutylamine
isopropanolamine (synonym: 1-amino-2-propanol)
isopropylamine (synonym: 2-aminopropane)
methylamine (synonym: monomethylamine)
methylamine, 30-70% (synonym: monomethylamine, 30-70%)
n-amylamine (synonyms: 1-aminopentane; n-pentylamine)
n-butylamine (synonym: 1-aminobutane)
nonylamine
propylamine (synonym: monopropylamine)
propylenediamine (see Class 148)
sec-butylamine (synonym: 2-methylpropylamine)
tert-butylamine
tetraethylenepentamine (see Class 148)
triethylenetetraamine (see Class 148)

* 142 Amines, Aliphatic and Alicyclic, Secondary
1,1,1,3,3,3-hexamethyldisilazane (synonyms: bis(trimethylsilyl)amine; HMDS)
2,6-dimethylmorpholine
3,3'-iminobis(propylamine) (see Class 148)
3-methylaminopropylamine (see Class 148)
di-n-amylamine (synonym: di-n-pentylamine)
di-n-butylamine (synonym: dibutylamine)
di-n-propylamine (synonym: dipropylamine)
diallylamine
diethanolamine
diethylamine
diethylenetriamine (see Class 148)
diisobutylamine
diisopropylamine
dimethylamine
ethyl-n-butylamine (synonym: butylethylamine)
ethyleneimine (see Class 274)
morpholine (synonym: tetrahydro-1,4-oxazine)
N-methylethanolamine (synonym: 2-methylaminoethanol)
piperazine (see Class 148)
piperidine (synonym: hexahydropyridine)
Posistrip LE (see Class 900)
tetraethylenepentamine (see Class 148)
triethylenetetraamine (see Class 148)

* 143 Amines, Aliphatic and Alicyclic, Tertiary
2-(diethylamino)ethanol (synonym: N,N-diethylaminoethanol)
2-(dimethylamino)ethanol (synonym: N,N-dimethylethanolamine)
3-(dimethylamino)propylamine (see Class 148)
N,N-dimethylethylamine (synonym: N-ethyldimethylamine)
promethazine hydrochloride (see Class 550)
tri-n-propylamine (synonym: tripropylamine)
triallylamine
triethanolamine (synonym: TEA)
triethanolamine, 30-70%
triethylamine
triethylenediamine, <30%
trimethylamine

* 145 Amines, Aromatic, Primary

4,4'-diaminodiphenylsulfone (synonym: 4,4'-sulfonyldianiline)
4,4'-methylene dianiline, 15% & methyl ethyl ketone, 85%
4,4'-methylenebis(2-chloroaniline), 50% & acetone, 50%
4,4'-methylenedianiline (synonyms: MDA; p,p'-diaminodiphenylmethane)
4,4'-methylenedianiline, 10% & isopropyl alcohol
4,4'-methylenedianiline, 15% & toluene, 85%
4,4'-methylenedianiline, 50% & methyl ethyl ketone, 50%
4,4'-oxidianiline
aniline (synonyms: benzamine; phenylamine)
o-toluidine

* 146 Amines, Aromatic, Secondary and Tertiary
4-nitrodiphenylamine (synonym: p-nitrodiphenylamine)
N,N-dimethylaniline (synonym: DMA)
treflan EC (synonym: trifluralin)

* 147 Amines, Alkyl-Aryl, Monoamines (Contains no chemicals or permeation test data)

* 148 Amines, Poly, Aliphatic and Alicyclic
1,3-propanediamine (synonym: 1,3-diaminopropane)
1,6-hexanediamine (synonym: 1,6-diaminohexane)
1,6-hexanediamine, 30-70%
1,6-hexanediamine, <30%
1-(2-aminoethyl)piperazine (synonym: N-(beta-aminoethyl)piperazine)
3,3'-iminobis(propylamine) (synonym: azaheptamethylenediamine)
3-(dimethylamino)propylamine
3-methylaminopropylamine
diethylenetriamine
ethylenediamine (synonym: 1,2-diaminoethane)
N,N,N',N'-tetramethylethylenediamine (synonym: TEMEDA)
piperazine (synonym: diethylenediamine)
propylenediamine (synonym: 1,2-diaminopropane)
tetraethylenepentamine
triethylenetetraamine (synonym: TETA)
triethylenetetraamine, 50% & methyl ethyl ketone, 50%

* 149 Amines, Poly, Aromatic (Contains no chemicals or permeation test data)

* 161 Anhydrides, Aliphatic and Alicyclic
acetic anhydride (synonym: acetyl oxide)
maleic anhydride
methylnadic anhydride (synonym: methyl-5-norbornene-2,3-dicarboxyl anhydride)
phthalic acid anhydride

* 162 Anhydrides, Aromatic
3,3',4,4'-benzophenonetetracarboxylic dianhydride

* 170 Azo and Azoxy Compounds
pigment yellow 74 (synonym: C.I. Pigment Yellow 74)

* 211 Isocyanates, Aliphatic and Alicyclic
hexamethylene-1,6-diisocyanate (synonym: HMDI)
isophoronediisocyanate (synonym: IDI)
methyl isocyanate (synonym: MIC)

* 212 Isocyanates, Aromatic
methylenebisphenyl-4,4'-diisocyanate (synonym: MDI)
toluene-2,4-diisocyanate (synonym: TDI)
toluene-2,4-diisocyanate, 40% & xylene, 60%

* 221 Esters, Carboxylic, Formates (Contains no chemicals or permeation test data)

* 222 Esters, Carboxylic, Acetates
  1,1,1-TCE, 66% & propyleneglycol monoethylether acetate, 34% (see Class 261)
  1-methoxy-2-propyl acetate (see Class 245)
  2-butoxyethyl acetate (see Class 245)
  2-ethoxyethyl acetate (see Class 245)
  2-methoxyethyl acetate (see Class 245)
  amyl acetate (synonyms: n-amyl acetate; n-pentyl acetate)
  benzyl acetate
  benzyl acetate, 30-70%
  benzyl acetate, <30%
  butyl acetate (synonym: n-butyl acetate)
  butyl acetate, 50% & methyl alcohol, 50%
  butyl acetate, 90% & methyl alcohol, 10%
  ethyl acetate
  ethyl acetate, >70% & ethyl alcohol, <30%
  ethylhydroxynol (synonym: methoxypropanol acetate)
  isoamyl acetate (synonym: isopentyl acetate)
  methyl acetate
  methyl acetate, 50% & ethyl alcohol, 50%
  methyl chloroacetate (see Class 224)
  Photo Resist 1450 (see Class 900)
  propyl acetate
  vinyl acetate

* 223 Esters, Carboxylic, Acrylates and Methacrylates
  1,6-hexanediol diacrylate (synonym: HDDA)
  1,6-hexanediol diacrylate, 50% & 2-ethylhexyl acrylate, 50%
  2-ethylhexyl acrylate, 75% & 1,6-hexanediol diacrylate, 25%
  2-hydroxyethyl acrylate (synonym: ethylene glycol acrylate)
  2-hydroxyethyl methacrylate (synonym: HEMA)
  allyl acrylate (synonym: allyl 2-propenoate)
  butyl acrylate (synonyms: n-butyl 2-propenoate; n-butyl acrylate)
  ethyl acrylate (synonym: ethyl 2-propenoate)
  ethyl methacrylate (synonym: ethyl 2-methylpropenoate)
  glycerolpropoxy acrylate
  glycidyl methacrylate (synonym: methacrylic acid 2,3-epoxypropyl ester)
  isobutyl acrylate (synonym: isobutyl 2-propenoate)
  isopropyl methacrylate (synonym: isopropyl 2-methylpropenoate)
  methyl acrylate (synonym: methyl 2-propenoate)
  methyl methacrylate (synonym: methyl 2-methylpropenoate)
  propyl methacrylate (synonym: n-propyl 2-methylpropenoate)
  trimethylolpropane triacrylate (synonym: TMPTA)
  tripropyleneglycol diacrylate

* 224 Esters, Carboxylic, Aliphatic, Others
  benzyl neocaprate
  cypermethrin (see Class 840)
  glycerol monothioglycolate, > 70% (see Class 501)
  methyl chloroacetate

* 225 Esters, Carboxylic, Lactones
  beta-butyrolactone (synonym: 3-hydroxybutyric acid lactone)
  beta-propiolactone

* 226 Esters, Carboxylic, Benzoates and Phthalates
  butyl benzyl phthalate (synonym: benzyl butyl phthalate)

di(2-ethylhexyl) phthalate (synonym: bis(2-ethylhexyl) phthalate)
di-n-butyl phthalate (synonym: DBP)
di-n-octyl phthalate (synonym: DOP)
diethyl phthalate (synonyms: bis(ethyl) phthalate; ethyl phthalate)
diisooctyl phthalate

* 227 Esters, Carboxylic, Aromatic, Others (Contains no chemicals or permeation test data)

* 231 Esters, Non-Carboxylic, Ortho Esters (Contains no chemicals or permeation test data)

* 232 Esters, Non-Carboxylic, Carbonates
1-bromoethyl ethyl carbonate
diethyl carbonate

* 233 Esters, Non-Carboxylic, Carbamates and Others
ethion 4 (see Class 462)
ethyl parathion, 30-70% (see Class 462)
guthion (see Class 462)
malathion (see Class 462)
methomyl (synonym: lannate)
methyl parathion, 30-70% (see Class 462)
methyl parathion, <30% (see Class 462)
naled (see Class 462)
sevin 50W (synonyms: 1-naphthol methyl carbamate; carbaryl)
sulfallate (synonym: 2-chloro-2-propenyldiethyldithiocarbamate)
tricresyl phosphate (see Class 462)
trimethyl phosphate (see Class 462)

* 241 Ethers, Aliphatic and Alicyclic
1,2-dichloroethyl ether
1-ethoxy-2-propanol (see Class 245)
1-methoxy-2-propyl acetate (see Class 245)
2,2'-dichloroethyl ether
2-(2-aminoethoxy)ethanol (see Class 141)
2-butoxyethanol (see Class 245)
2-butoxyethyl acetate (see Class 245)
2-ethoxyethyl acetate (see Class 245)
2-methoxyethanol (see Class 245)
2-methoxyethyl acetate (see Class 245)
benzyl ether
butyl ether (synonym: di-n-butyl ether)
dimethyl ether
ethyl ether (synonym: diethyl ether)
ethylhydroxynol (see Class 222)
isopropyl ether (synonym: diisopropyl ether)
tert-butyl methyl ether (synonym: 2-methoxy-2-methylpropane)
tetrahydrofuran (synonym: THF)

* 242 Ethers, Aromatic
methyl eugenol

* 243 Ethers, Alkyl-Aryl
2-(2,4-dichlorophenoxy)propionic acid, formulation (see Class 103)
2-(4-chloro-2-methylphenoxy)propionic acid, formulation (see Class 103)
4-chloro-2-methylphenoxyacetic acid, formulation (see Class 103)

* 244 Ethers, Ketals, and Acetals (Contains no chemicals or permeation test data)

* 245 Ethers, Glycols
    1,1,1-TCE, 66% & propyleneglycol monoethylether acetate, 34% (see Class 261)
    1-ethoxy-2-propanol (synonym: ethyl 1-isopropanol ether)
    1-methoxy-2-propanol (synonym: propylene glycol monomethyl ether)
    1-methoxy-2-propyl acetate
    2-(2-ethoxyethoxy)ethanol (synonym: diethyleneglycol mono ethyl ether)
    2-(2-methoxyethoxy)ethanol (synonym: diethylene glycol monomethyl ether)
    2-butoxyethanol (synonyms: butylcellosolve; butylglycol)
    2-butoxyethanol, 50% & naphtha, 50% (<3% aromatics, b.p. 150-200 degrees C)
    2-butoxyethyl acetate (synonyms: butylcellosolve acetate; butylglycol acetate)
    2-ethoxy-1-propanol (synonym: alpha-propylene glycol monoethyl ether)
    2-ethoxyethanol (synonyms: cellosolve; ethylglycol)
    2-ethoxyethyl acetate (synonyms: cellosolve acetate; ethylglycol acetate)
    2-methoxyethanol (synonyms: methylcellosolve; methylglycol)
    2-methoxyethanol, 96% & ninhydrin, 4%
    2-methoxyethyl acetate (synonym: methylcellosolve acetate)
    ethylene glycol dimethyl ether (synonym: 1,2-dimethoxyethane)
    ethylhydroxynol (see Class 222)
    methyltriglycol (synonym: triethylene glycol monomethyl ether)
    Photo Resist 1450 (see Class 900)
    propylene glycol monoethylether acetate
    xylene, 50% & 2-ethoxyethanol, 50% (see Class 292)
    xylene, 75% & 2-butoxyethanol, 25% (see Class 292)

* 246 Ethers, Vinyl
    ethyl vinyl ether (synonym: vinyl ethyl ether)

* 261 Halogen Compounds, Aliphatic and Alicyclic
    1,1,1,2-tetrachloroethane
    1,1,1-TCE, 66% & propyleneglycol monoethylether acetate, 34%
    1,1,1-trichloroethane (synonyms: 1,1,1-TCE; methylchloroform)
    1,1,2,2-tetrachloroethane
    1,1,2-trichloroethane
    1,1-dichloroethane (synonym: ethylidene chloride)
    1,2,3-trichloropropane
    1,2-dichloropropane (synonym: propylene dichloride)
    1-bromo-2-propanol (see Class 315)
    1-chloro-2-propanol (see Class 315)
    2,2'-dichloroethyl ether (see Class 241)
    2,2,2-trichloroethanol (see Class 315)
    2,2,2-trifluoroethanol (see Class 315)
    2-bromoethanol (see Class 315)
    2-chloro-2-nitropropane (see Class 442)
    2-chloroethanol (see Class 315)
    3-bromo-1-propanol (see Class 315)
    3-bromopropionic acid (see Class 103)
    3-chloro-1-propanol (see Class 315)
    bromoacetonitrile (see Class 431)
    bromochloromethane (synonym: chlorobromomethane)
    bromodichloromethane (synonym: dichlorobromomethane)
    carbon tetrachloride (synonym: tetrachloromethane)
    chloral (see Class 121)
    chlordane, >70%
    chloroacetone (see Class 391)
    chloroacetonitrile (see Class 431)

chlorodibromomethane (synonym: dibromochloromethane)
chloroform (synonym: trichloromethane)
chloroform, 50% & toluene, 50%
chloroform, 96% & isopentyl alcohol, 4%
chloroform, 97% & methyl alcohol, 3%
chloroprene (see Class 263)
cypermethrin (see Class 840)
dichloroacetyl chloride (see Class 111)
dichloroethane
dimethylvinyl chloride (see Class 263)
epibromohydrin (see Class 275)
epichlorohydrin (see Class 275)
ethyl bromide (synonym: bromoethane)
ethylene dibromide (synonym: 1,2-dibromoethane)
ethylene dichloride (synonyms: 1,2-dichloroethane; EDC)
freon 113 (TF) (synonym: trichlorotrifluoroethane)
freon TMC (synonym: trichlorotrifluoroethane, 50% & methylene chloride, 50%)
halothane (synonym: 2-bromo-2-chloro-1,1,1-trifluoroethane)
methyl bromide (synonym: bromomethane)
.methyl chloride (synonym: chloromethane)
methyl chloroacetate (see Class 224)
methyl iodide (synonym: iodomethane)
methylene bromide (synonym: dibromomethane)
methylene chloride (synonym: dichloromethane)
methylene chloride, 50% & hexane, 50%
methylene chloride, 50% & toluene, 50%
methylene chloride, 70% & paint stripper D23, 30%
methylene chloride, 90% & isopropyl alcohol, 10%
methylene chloride, >70% & ethanol, <30%
methylene chloride, >70% & phenol, <30%
n-butyl chloride (synonym: 1-chlorobutane)
tribromomethane (synonym: bromoform)
trichloroacetonitrile (see Class 431)
trifluoroacetic acid (see Class 103)
vinyl chloride (see Class 263)
vinylidene chloride (see Class 263)
vinylidene fluoride (see Class 263)

* 262 Halogen Compounds, Allylic and Benzylic
1,3-dichloro-2-butene
1,3-dichloropropene
1,4-dichloro-2-butene (synonym: 2-butylene dichloride)
2,3-dichloro-1-propene
3-chloro-2-methylpropene (synonym: 2-methylallyl chloride)
allyl bromide (synonym: 3-bromopropene)
allyl chloride (synonym: 3-chloropropene)
benzyl chloride (synonyms: alpha-chloromethylbenzene; alpha-chlorotoluene)
hexachlorocyclopentadiene

* 263 Halogen Compounds, Vinylic
chloroprene (synonym: 2-chloro-1,3-butadiene)
cis,trans-1,2-dichlorethylene (synonym: 1,2-dichloroethylene)
cis-1,2-dichloroethylene
dimethylvinyl chloride (synonym: 2-methyl-1-chloropropene)
hexachloro-1,3-butadiene (synonym: HCBD)

tetrachloroethylene (synonym: perchloroethylene)
tetrafluoroethylene (synonym: perfluoroethylene)
trans-1,2-dichloroethylene
trichloroethylene (synonyms: TCE; trichloroethene)
trichloroethylene, 50% & pentane, 50%
vinyl chloride (synonym: chloroethene)
vinyl fluoride (synonym: fluoroethene)
vinylidene chloride (synonym: 1,1-dichloroethene)
vinylidene fluoride (synonym: 1,1-difluoroethene)

* 264 Halogen Compounds, Aromatic
1,2,4,5-tetrachlorobenzene
1,2,4-trichlorobenzene
1,2-dichlorobenzene (synonym: o-dichlorobenzene)
1,3-dichlorobenzene (synonym: m-dichlorobenzene)
1,4-dichlorobenzene (synonym: p-dichlorobenzene)
1-chloronaphthalene (synonym: naphthyl chloride)
4-chlorobenzotrifluoride (synonym: 1-chloro-4-(trifluoromethyl)benzene)
bromobenzene
chlorobenzene
o-chlorotoluene (synonyms: 2-chlorotoluene; 2-methylchlorobenzene)
p-chlorotoluene (synonyms: 4-chlorotoluene; 4-methylchlorobenzene)
PCB & transformer oil
PCB, 1% & naphtha, 99%
PCB, 4% & 1,2,4-trichlorobenzene, 6% & naphtha, 90%
PCB, 50% & naphtha, 50% (synonym: PCB, 50% & mineral spirits, 50%)
polychlorinated biphenyls (synonym: PCBs)
polychlorinated biphenyls, 58% & 1,2,4-trichlorobenzene, 42%

* 271 Heterocyclic Compounds, Nitrogen, Pyridines
beta-picoline (synonym: 3-methylpyridine)
nicotine
pyridine (synonym: azine)

* 274 Heterocyclic Compounds, Nitrogen, Others
1-(2-aminoethyl)piperazine (see Class 148)
2,6-dimethylmorpholine (see Class 142)
9-aminoacridine hydrochloride (see Class 550)
Benlate (synonym: Benomyl)
dichlorotriazine, 80% & toluene, 20%
diquat dibromide (synonym: Reglone)
ethyleneimine (synonym: aziridine)
piperadine (see Class 142)
piperazine (see Class 148)
promethazine hydrochloride (see Class 550)
quinoline (synonym: 1-azanaphthalene)

* 275 Heterocyclic Compounds, Oxygen, Epoxides
1,2-epoxybutane (synonym: 1,2-butylene oxide)
1,2-propylene oxide (synonym: 1,2-epoxypropane)
1,4-butanedioldiglycidyl ether
1,4-butanediolglycidyl ether, 50% & methyl ethyl ketone, 50%
bisphenol A diglycidyl ether (synonyms: DGBA; diglycidyl ether of bisphenol A)
diglycidyl ether with bisphenol A, 50% & methyl ethyl ketone, 50%
epibromohydrin (synonym: 1,2-epoxy-3-bromopropane)
epichlorohydrin (synonym: 1,2-epoxy-3-chloropropane)

epoxytrichloropropane (synonym: trichloroepoxypropane)
ethylene oxide (synonym: oxirane)

* 277 Heterocyclic Compounds, Oxygen, Furans
furan (synonym: furfurane)
furfural (see Class 122)
tetrahydrofuran (see Class 241)

* 278 Heterocyclic Compounds, Oxygen, Others
1,3-dioxane
1,4-dioxane

* 279 Heterocyclic Compounds, Sulfur
thiophene

* 280 Hydrazines
1,1-dimethylhydrazine (synonym: unsymmetrical hydrazine)
hydrazine (synonym: diamine)
hydrazine hydrate
hydrazine, 30-70%
methylhydrazine

* 291 Hydrocarbons, Aliphatic and Alicyclic, Saturated
cyclohexane (synonym: hexamethylene)
cyclopentane (synonym: pentamethylene)
diesel fuel
dodecane (synonym: n-dodecane)
heptane (synonym: n-heptane)
hexane (synonym: n-hexane)
hexane, 50% & acetone, 50%
hexane, 50% & methyl alcohol, 50%
hexane, 50% & methyl ethyl ketone, 50%
hexane, 50% & methylene chloride, 50%
hexane, 50% & toluene, 50%
hexane, 90% & benzene, 10%
hexane, 90% & methyl ethyl ketone, 10%
isooctane
jet fuel with <30% aromatics (synonym: JP-4)
methane
naphtha with 10-15% aromatics (b.p. 50-120 deg C) (synonym: rubber solvent)
naphtha with 10-15% aromatics (b.p. 120-140 deg C) (synonym: Naphtha VMP)
naphtha with 10-15% aromatics (b.p. 150-200 deg C) (synonym: Stoddard solvent)
naphtha with 15-20% aromatics (b.p. 150-200 deg C) (synonym: mineral spirits)
naphtha with 15-20% aromatics (b.p. 180-260 deg C) (synonym: kerosene)
naphtha with <3% aromatics (b.p. 150-200 deg C) (synonym: Rule 66)
naphtha with <3% aromatics (b.p. 150-200 deg C), 50% & 2-butoxyethanol, 50%
naphtha with <3% aromatics (b.p. 150-200 deg C), 95% & 2-butoxyethanol, 5%
naphtha with <3% aromatics (b.p. 180-260 deg C)
octane (synonym: n-octane)
pentane (synonym: n-pentane)
pentane, 50% & trichloroethylene, 50%
petroleum ether with <0.5% aromatics (b.p. 80-110 deg C)
propane

* 292 Hydrocarbons, Aromatic
1,2,4-trimethylbenzene (synonym: pseudocumene)
alpha-methylstyrene (synonyms: 2-phenylpropene; isopropenylbenzene)

benzene
coal tar & benzene, 1:1 (see Class 860)
cumene (synonym: isopropylbenzene)
diethylbenzene
divinylbenzene (synonym: vinylstyrene)
ethylbenzene
gasoline with 40-55% aromatics (b.p. 35-210 deg C)
gasoline, 50% & acetone, 50%
gasoline, 80% & N-methylpyrrolidone, 20%
gasoline, unleaded
m-cresol (see Class 316)
m-xylene (synonym: 1,3-dimethylbenzene)
o-xylene (synonym: 1,2-dimethylbenzene)
p-cresol (see Class 316)
p-tert-butyltoluene (synonym: 4-methyl-t-butylbenzene)
p-xylene (synonym: 1,4-dimethylbenzene)
p-xylene, 50% & toluene, 50%
p-xylene, 75% & toluene, 25%
PCB, 1% & naphtha, 99% (see Class 264)
Photo Resist 1450 (see Class 900)
styrene (synonym: vinylbenzene)
toluene (synonym: methylbenzene)
toluene, 50% & chloroform, 50%
toluene, 50% & dimethylsulfoxide, 50%
toluene, 50% & hexane, 50%
toluene, 50% & isopropyl alcohol, 50%
toluene, 50% & methyl alcohol, 50%
toluene, 50% & methyl ethyl ketone, 50% (see Class 391)
toluene, 50% & methyl isobutyl ketone, 50%
toluene, 50% & methylene chloride, 50%
toluene, 50% & p-xylene, 50%
toluene, 50% & water, 50%
toluene, 75% & naphthalene, 25%
toluene, 75% & p-xylene, 25%
xylene (undefined mixture)
xylene (undefined mixture), 50% & methyl isobutyl ketone, 50%
xylene, 50% & 2-ethoxyethanol, 50%
xylene, 60% & toluene-2,4-diisocyanate, 40%
xylene, 75% & 2-butoxyethanol, 25%

* 293 Hydrocarbons, Polynuclear Aromatic
naphthalene

* 294 Hydrocarbons, Aliphatic and Alicyclic, Unsaturated
1,3-butadiene
4-vinyl-1-cyclohexene
alpha-methylstyrene (see Class 292)
d,l-limonene (synonyms: d,l-p-mentha-1,8-diene; dipentene; limonene)
divinylbenzene (see Class 292)
isoprene (synonym: 2-methyl-1,3-butadiene)
styrene (see Class 292)
turpentine
vinyl acetate (see Class 222)

* 295 Hydrocarbons, Acetylenes (Contains no chemicals or permeation test data)

* 296 Hydrocarbons, Polyenes (Contains no chemicals or permeation test data)

* 300 Peroxides
    2-butanone peroxide
    cumene hydroperoxide
    hydrogen peroxide, 30-70%
    peroxyacetic acid
    t-butyl peroxybenzoate
    t-butylhydroperoxide

* 311 Hydroxy Compounds, Aliphatic and Alicyclic, Primary
    1-ethoxy-2-propanol (see Class 245)
    2-(2-aminoethoxy)ethanol (see Class 141)
    2-(diethylamino)ethanol (see Class 143)
    2-(dimethylamino)ethanol (see Class 143)
    2-butoxyethanol (see Class 245)
    2-ethoxy-1-propanol (see Class 245)
    2-ethyl-1-hexanol (synonym: 2-ethylhexyl alcohol)
    2-mercaptoethanol (see Class 501)
    2-methoxyethanol (see Class 245)
    allyl alcohol (synonym: 2-propenol)
    amyl alcohol (synonyms: 1-pentanol; n-pentanol; pentyl alcohol)
    benzyl alcohol
    butyl alcohol (synonyms: 1-butanol; n-butanol)
    chloroform, 96% & isopentyl alcohol, 4% (see Class 261)
    diethanolamine (see Class 142)
    ethanolamine (see Class 141)
    ethyl alcohol (synonyms: ethanol; grain alcohol)
    ethyl alcohol, 50% & methyl acetate, 50%
    formaldehyde, 37% & methyl alcohol, 10% & water 53% (see Class 121)
    furfuryl alcohol
    isobutyl alcohol (synonyms: 2-methyl-1-propanol; isobutanol)
    methyl alcohol (synonyms: methanol; wood alcohol)
    methyl alcohol, 50% & butyl acetate, 50%
    methyl alcohol, 50% & hexane, 50%
    methyl alcohol, 50% & toluene, 50%
    methyl alcohol, 50% & water, 50%
    methyl alcohol, 90% & butyl acetate, 10%
    methyl alcohol, 95% & 4,4'-methylenedianiline, 5%
    methylene chloride, >70% & ethanol, <30% (see Class 261)
    N-methylethanolamine (see Class 142)
    octyl alcohol (synonyms: 1-octanol; n-octanol)
    propyl alcohol (synonyms: 1-propanol; n-propanol)
    triethanolamine (see Class 143)

* 312 Hydroxy Compounds, Aliphatic and Alicyclic, Secondary
    cyclohexanol (synonym: hexahydrophenol)
    isopropanolamine (see Class 141)
    isopropyl alcohol (synonyms: 2-propanol; isopropanol)
    isopropyl alcohol, 75% & 1-napthylamine, 25%
    lactic acid (see Class 103)
    sec-butyl alcohol (synonyms: 2-butanol; sec-butanol)
    toluene, 50% & isopropyl alcohol, 50% (see Class 292)

* 313 Hydroxy Compounds, Aliphatic and Alicyclic, Tertiary
    citric acid, <30% (see Class 104)
    tert-butyl alcohol (synonyms: 2-methyl-2-propanol; tert-butanol)

* 314 Hydroxy Compounds, Aliphatic and Alicyclic, Polyols
    1,4-butyleneglycol (synonym: 1,4-butanediol)
    diethylene glycol (synonym: 2-hydroxyethyl ether)
    dipropylene glycol (synonym: 2,2'-dihydroxyisopropyl ether)
    ethylene glycol (synonym: 1,2-ethanediol)
    glycerol (synonym: glycerin)
    nitroglycerol (see Class 442)
    propylene glycol (synonym: 1,2-propanediol)

* 315 Hydroxy Compounds, Aliphatic and Alicyclic, Substituted
    1-bromo-2-propanol
    1-chloro-2-propanol
    2,2,2-trichloroethanol
    2,2,2-trifluoroethanol
    2-bromoethanol
    2-chloroethanol (synonym: ethylene chlorohydrin)
    3-bromo-1-propanol
    3-chloro-1-propanol

* 316 Hydroxy Compounds, Aromatic (Phenols)
    creosote
    cresols (isomeric mixture) (synonym: methylphenols (isomeric mixture))
    dinoseb, 48% & xylene, 52%
    hydroquinone (synonym: p-benzenediol)
    m-cresol (synonyms: 3-hydroxytoluene; 3-methylphenol)
    m-cresol, 50% & methyl ethyl ketone, 50%
    nonylphenol
    p-cresol (synonym: 4-methylphenol)
    pentachlorophenol
    phenol (synonym: carbolic acid)
    phenol, 50% & methyl ethyl ketone, 50%
    phenolphthalein
    picric acid (see Class 442)
    tannic acid (synonym: tannin)
    wood creosote
    xylenol

* 317 Hydroxy Compounds, Aromatic, Naphthols (Contains no chemicals or permeation test data)

* 318 Hydroxy Compounds, Aromatic, Others (Contains no chemicals or permeation test data)

* 330 Elements
    bromine
    chlorine (see Class 350)
    chlorine, liquid
    fluorine (see Class 350)
    iodine, solid
    mercury (synonym: quick silver)

* 340 Inorganic Salts
    ammonium acetate (saturated solution) (synonym: acetic acid ammonium salt)
    ammonium carbonate (saturated solution)
    ammonium fluoride <30%
    ammonium fluoride, 30-70%
    ammonium nitrate
    cadmium oxide (solid)

calcium chloride
calcium hydrogen phosphate (synonym: dicalcium phosphate)
cobalt sulfate heptahydrate
ferric chloride (synonym: ferric trichloride)
ferrous chloride
mercuric chloride (saturated solution)
potassium acetate (saturated solution)
potassium chromate (saturated solution)
potassium iodide (saturated solution)
potassium permanganate
sodium carbonate (saturated solution)
sodium chloride (saturated solution)
sodium dichromate, <30% (synonym: sodium bichromate)
sodium fluoride
sodium hypochlorite (saturated solution)
sodium hypochlorite, 30-70%
sodium hypochlorite, < 30%
sodium thiosulfate (synonym: sodium hyposulfate)
sodium thiosulfate (saturated solution)

* 345 Inorganic Cyano Compounds
cyanide salt, 45% solution in water (synonym: sodium or potassium cyanide)
cyanogen bromide
hydrogen cyanide
silver cyanide, < 30%
sodium cyanide (solid)
sodium cyanide, 30-70% (salt solution)
sodium cyanide, <30% (salt solution)
sodium cyanide, >70%

* 350 Inorganic Gases and Vapors
ammonia (synonym: ammonia gas)
arsine (synonym: arsenic trihydride)
chlorine (synonym: chlorine gas)
chlorine dioxide, <30%
chlorine, liquid (see Class 330)
diborane
fluorine
hydrogen chloride (synonym: hydrochloric acid, anhydrous)
hydrogen fluoride (synonym: hydrofluoric acid, anhydrous)
hydrogen sulfide
nitrogen dioxide
nitrogen tetroxide
phosgene (synonym: carbonyl chloride)
phosphine
sulfur dioxide

* 360 Inorganic Acid Halides
antimony pentachloride
bromotrifluoride
phosgene (see Class 350)
phosphorus tribromide
phosphorus trichloride
silicon tetrachloride
sulfur dichloride (synonym: chlorine sulfide)

sulphuryl chloride
titanium tetrachloride

* 365 Inorganic Acid Oxides
nitrogen dioxide (see Class 350)
nitrogen tetroxide (see Class 350)
sulfuric acid, 96% & sulphur dioxide, 65% in H2O (see Class 370)
sulphur trioxide (synonym: sulphuric anhydride)

* 370 Inorganic Acids
aqua regia (hydrochloric acid, 25-37% & nitric acid, 63-75%)
boric acid
chlorosulfonic acid
chromic acid (synonym: chromium trioxide)
chromic acid & sulfuric acid
chromic acid, 30-70%
chromic acid, <30%
fluorosilic acid (synonym: fluosilic acid)
fluorosulfonic acid
hydrobromic acid, 30-70%
hydrochloric acid, 30-70%
hydrochloric acid, <30%
hydrofluoric acid, 30-70%
hydrofluoric acid, <30%
hydrofluoric acid, >70%
nitric acid, 30-70%
nitric acid, <30%
nitric acid, >70%
nitric acid, red fuming
perchloric acid, 30-70%
phosphoric acid, >70%
phosphorus oxychloride (synonym: phosphoryl chloride)
sulfuric acid, 30-70%
sulfuric acid, 96% & sulphur dioxide, 65% in H2O
sulfuric acid, <30%
sulfuric acid, >70%
sulfuric acid, fuming (synonym: oleum)
tetrafluoroboric acid (synonym: fluoroboric acid)

* 380 Inorganic Bases
ammonia, liquid
ammonium hydroxide, 30-70%
ammonium hydroxide, <30%
calcium hydroxide (saturated solution)
nickel subsulfide
potassium hydroxide, 30-70%
sodium hydroxide, 30-70%
sodium hydroxide, >70%

* 391 Ketones, Aliphatic and Alicyclic
4,4'-methylenedianiline, 50% & methyl ethyl ketone, 50% (see Class 145)
4-hydroxy-4-methyl-2-pentanone (synonym: diacetone alcohol)
4-methoxy-4-methyl-2-pentanone
5-methyl-2-hexanone (synonym: isoamyl methyl ketone)
acetone
acetone, 50% & gasoline, 50%

acetone, 50% & hexane, 50%
beta-ionone
beta-tetralone (synonym: tetralone, beta-)
chloroacetone (synonyms: 1-chloro-2-propanone; monochloroacetone)
cyclohexanone
cyclopentanone
diisobutyl ketone (synonyms: 2,6-dimethyl-4-heptanone; DIBK)
isophorone
methyl ethyl ketone (synonyms: 2-butanone; MEK)
methyl ethyl ketone, 50% & 1,4-butanediolglycidyl ether, 50%
methyl ethyl ketone, 50% & diglycidyl ether with bisphenol A, 50%
methyl ethyl ketone, 50% & hexane, 50%
methyl ethyl ketone, 50% & p-cresol, 50%
methyl ethyl ketone, 50% & phenol, 50%
methyl ethyl ketone, 50% & toluene, 50%
methyl ethyl ketone, 50% & triethylenetetraamine, 50%
methyl ethyl ketone, 85% & acrylamide, 15%
methyl ethyl ketone, 90% & hexane, 10%
methyl isobutyl ketone (synonyms: 4-methyl-2-pentanone; MIBK)
methyl isobutyl ketone, 50% & toluene, 50%
methyl isobutyl ketone, 50% & xylenes, 50%
methyl pentyl ketone (synonym: 2-heptanone)
misityl oxide (synonym: isopropylidene acetone)
N-methyl-2-pyrrolidone (see Class 132)

* 392 Ketones, Aromatic
acetophenone
chloroacetophenone (synonym: tear gas)
propiophenone (synonyms: 1-phenyl-1-propanone; ethyl phenyl ketone)

* 410 Quinones
p-benzoquinone (synonym: benzoquinone)

* 431 Nitriles, Aliphatic and Alicyclic
acetonitrile (synonym: methyl cyanide)
acrylonitrile (synonym: propenenitrile)
adiponitrile
bromoacetonitrile
chloroacetonitrile
methylacrylonitrile (synonym: 2-methyl-2-propenenitrile)
propionitrile (synonym: ethyl cyanide)
trichloroacetonitrile (synonym: trichloromethyl cyanide)
valeronitrile (synonym: 1-pentanenitrile)

* 432 Nitriles, Aromatic
benzonitrile (synonym: phenyl cyanide)
benzylnitrile (synonyms: benzyl cyanide; phenylacetonitrile)

* 441 Nitro Compounds, Unsubstituted
1-nitropropane
2,4-dinitrotoluene (synonym: 2,4-DNT)
2-nitropropane
2-nitrotoluene (synonym: o-nitrotoluene)
nitrobenzene
nitroethane
nitromethane

* 442 Nitro Compounds, Substituted
    2-chloro-2-nitropropane
    4-nitrodiphenylamine (see Class 146)
    chloropicrin
    nitroglycerol (synonym: nitroglycerin)
    nitroglycol
    picric acid
    treflan EC (see Class 146)

* 450 Nitroso Compounds
    N-nitrosodiethylamine (synonym: N,N-diethylnitrosamine)
    N-nitrosodimethylamine (synonym: N,N-dimethylnitrosamine)

* 461 Organophosphorus Compounds, Phosphines (Contains no chemicals or permeation test data)

* 462 Organophosphorus Compounds and Derivatives of Phosphorus-based Acids
    ethion 4
    ethyl parathion, 30-70% (synonym: parathion)
    ethyl parathion, <30% (synonym: parathion <30%)
    guthion (synonym: azinphos-methyl)
    hexamethylphosphoramide (synonym: HMPA)
    KVK parathion
    malathion
    methyl parathion, 30-70%
    methyl parathion, <30%
    naled
    Round Up R solution (synonym: N,N-bis(phosphonemethyl)glycine isopropylamine)
    tributyl phosphate (synonym: TBP)
    tricresyl phosphate (synonyms: TCP; tritolyl phosphate)
    trimethyl phosphate
    tris(1,3-dichloropropyl) phosphate

* 470 Organometallic Compounds
    tributyltin oxide (synonym: bis(tributyltin) oxide)

* 480 Organosilicon Compounds
    1,1,1,3,3,3-hexamethyldisilazane (see Class 142)
    chlorotrimethylsilane (synonym: trimethylchlorosilane)
    diethyldichlorosilane (synonym: dichlorodiethylsilane)
    dimethyldichlorosilane (synonym: dichlorodimethylsilane)
    methyltrichlorosilane (synonym: trichloromethylsilane)

* 501 Sulfur Compounds, Thiols
    2-mercaptoethanol (synonyms: 2-hydroxyethyl mercaptan; thioglycol)
    glycerol monothioglycolate, > 70%
    methyl mercaptan (synonym: thiomethanol)

* 502 Sulfur Compounds, Sulfides and Disulfides
    carbon disulfide (see Class 508)
    mustard gas (synonym: bis(2-chloroethyl)sulfide)
    thiophene (see Class 279)

* 503 Sulfur Compounds, Sulfones and Sulfoxides
    dimethyl sulfoxide (synonyms: DMSO; methyl sulfoxide)
    dimethyl sulfoxide, 50% & toluene, 50%
    dimethyl-d6 sulfoxide (synonym: hexadeuterodimethyl sulfoxide)

* 504 Sulfur Compounds, Sulfonic Acids
    chlorosulfonic acid (see Class 370)
    fluorosulfonic acid (see Class 370)
    methanesulfonic acid
    p-toluenesulfonic acid (synonym: 4-methylbenzenesulfonic acid)
    phenolsulfonic acid (synonym: hydroxybenzenesulfonic acid)
    xylene sulfonic acid sodium salt (see Class 550)

* 505 Sulfur Compounds, Sulfonyl Chlorides (Contains no chemicals or permeation test data)

* 506 Sulfur Compounds, Sulfonamides (Contains no chemicals or permeation test data)

* 507 Sulfur Compounds, Sulfonates, Sulfates, and Sulfites
    dimethyl sulfate (synonyms: DMS; methyl sulfate; sulfuric acid dimethyl ester)

* 508 Sulfur Compounds, Thiones
    carbon disulfide (synonym: carbon bisulfide)
    xylene sulfonic acid sodium salt (see Class 550)

* 510 Nitrates and Nitrites
    dynamite (synonym: ethylene glycol dinitrate, 70% & nitroglycerin, 30%)
    isoamyl nitrite (synonyms: 3-methylbutyl nitrite; isopentyl nitrite)
    isobutyl nitrite
    isopropyl nitrite

* 550 Organic Salts (Solutions)
    2,4-D dimethylamine ammonium salt (synonym: 2,4-D amine 96)
    2-(2,4-dichlorophenoxy)propionic acid, formulation (see Class 103)
    2-(4-chloro-2-methylphenoxy)propionic acid, formulation (see Class 103)
    2-hydroxyethyl-N,N,N-trimethylammonium hydroxide (synonym: choline)
    4-chloro-2-methylphenoxyacetic acid, formulation (see Class 103)
    9-aminoacridine hydrochloride
    benzethonium chloride (synonym: hyamine)
    chlorpromazine hydrochloride (see Class 870)
    promethazine hydrochloride
    tetramethylammonium hydroxide
    xylene sulfonic acid sodium salt (synonym: sodium dimethylbenzene sulfonate)

* 800 Multicomponent Mixtures With > 2 Components
    1,1,1-TCE, 73% & dichloromethane, 17% & dodecylbenzene sulfonic acid, 10%
    1,1,1-trichloroethane & ethyl alcohol & turpentine, unknown %
    1,1,1-trichloroethane, 97% & 1,4-dioxane, 3%
    2-ethoxyethyl acetate, 68% & butyl acetate, 7.5% & xylene, 7.5%
    2-ethoxyethyl acetate, 82% & butyl acetate, 9% & xylene, 9%
    acetone & toluene & methylated spirits & conc. ammonia, 2:1:1:1
    acetone, 88% & propylene glycol, 10% & 2,3-diphenylcyclopropene-1-one, 2%
    chromic acid, 30-70% & (sodium fluoride and potassium ferricyanide), 30-70%
    cyclohexylamine, 32% & morpholine, 8% & water, 60%
    dimercaptothiodiazole, 10% in butyldioxothol & methyl ethyl ketone, 1:1
    ethyl acetate, >70% & acetone, ethyl alcohol and methyl alcohol
    gasoline, 7% & ethanol, 60% & methanol, 33% (synonym: gashol)
    methyl ethyl ketone, 30-70% & ethylene glycol acetate, MIBK, isopropyl alcohol
    methyl ethyl ketone, 30-70% & MIBK, isopropyl alcohol, binding material
    methylene chloride, 30-70% & phenol, <30% & formic acid, <30%
    phenol, 50% & chloroform, 48% & isopentyl alcohol, 2%
    toluene, 30-70% & butyl alcohol, butyl acetate, ethyl acetate and methanol
    toluene, 5-20% & butanol, butyl acetate, ethanol, methyl ethyl ketone, xylene

toluene, 50% & MEK, 15% & methanol, 15% & butylglycol, 10% & ethyl acetate, 10%

xylene, 80-92% & 2-methoxyethanol, < 6% & cyclized polyisoprene, 5-15%

* 810 Epoxy Products

1,4-butanediolglycidyl ether, 50% & methyl ethyl ketone, 50% (see Class 275)

DGEBA, 50% & methyl ethyl ketone, 50%

epoxy (accelerator)

epoxy (base & accelerator), & acetone, methoxyethanol, methyl alcohol

epoxy (base)

epoxy (base) & toluene, 30-70% & methyl isobutyl ketone, <30% & xylene, <30%

epoxy (DGEBA), 50% & n-butanol, <30% & methyl ethyl ketone, <30% & xylene, <30%

polyamide (accelerator), 30-70% & isobutyl alcohol, isopropyl alcohol, MIBK

styrene monomer resin

* 820 Etching Products

hydrofluoric acid & ammonium trifluoride (synonym: oxide etch)

hydrofluoric acid & nitric acid & acetic acid (synonym: silicon etch)

phosphoric acid & nitric acid (synonym: aluminum etch)

phosphoric acid & nitric acid & acetic acid (synonym: slope etch)

potassium hydroxide & butyl alcohol & propyl alcohol (synonym: KOH etch)

sulfuric acid & potassium dichromate & water (synonym: dichromate etch)

sulfuric acid & sodium dichromate, 3% & water

* 830 Lacquer Products

acrylate UV - lacquer in butyl acetate

melamine-formaldehyde resin in solution (butyl alcohol, 1% & water, 50%)

phenol & methyl ethyl ketone, 30-70% & methyl isobutyl ketone, <30%

phenol-formaldehyde resin in solution (methyl alcohol, 5% & water)

* 840 Pesticides, Mixtures and Formulations

2,4-D dimethylamine ammonium salt (see Class 550)

2-(2,4-dichlorophenoxy)propionic acid, 15% & 2,4-D, 8% & water (see Class 103)

2-(2,4-dichlorophenoxy)propionic acid, 40% & MCPA, 20% & water (see Class 103)

2-(2,4-dichlorophenoxy)propionic acid, formulation (see Class 103)

2-(4-chloro-2-methylphenoxy)propionic acid, formulation (see Class 103)

4-chloro-2-methylphenoxyacetic acid, formulation (see Class 103)

Ambush

Benlate (see Class 274)

chlordane, >70% (see Class 261)

Cymbush

cypermethrin

dinoseb, 48% & xylene, 52% (see Class 316)

diquat dibromide (see Class 274)

ethyl parathion, 30-70% (see Class 462)

fusilade 250EC (synonym: fluazifop-butyl)

lindane & chloroform

lindane & xylene

methomyl (see Class 233)

methyl parathion, 30-70% (see Class 462)

methyl parathion, <30% (see Class 462)

Orthocid 83

Pramitol
propyzamide, <30% (see Class 135)
sevin 50W (see Class 233)
treflan EC (see Class 146)

* 850 Cutting Fluids
emulsions with metals
pure cutting fluid (no emulsions) with metals

* 860 Coals, Charcoals, Oils
castor oil
coal tar & benzene, 1:1
coal tar extract
corn oil
hydraulic oil
liquified coal
lubricating oil
pentachlorophenol, 4.3% in diesel oil
petroleum, Shell speciality
shale oil
sodium pentachlorophenate, 4.2% in diesel oil
soybean oil
teak oil
transformer oil, Nytro 10X
transmission oil, opel
vegetable oil

* 870 Antineoplastic Drugs and Other Pharmaceuticals
2',3'-dideoxycytidine
2',3'-dideoxycytidine in 0.2% methyl cellulose & water
AZT
AZT, 200 mg/mL in 0.2% methyl cellulose & water
carmustine (synonym: 1,3-bis(2-chloroethyl)-1-nitrosocarbamide)
chlorpromazine hydrochloride
doxorubicin
methotrexate
vincristine

* 900 Miscellaneous Unclassified Chemicals
accumix
Aeroshell fluid 4
AFFF (synonym: Aqueous Fire Fighting Foam)
anionic detergent, <30%
biological detergent (saturated solution)
blood, human
cationic detergent (unspecified concentration)
corrosive fluid Dyrup 49685
Deep Woods Off
Degalan S309 (mixture)
Degalan S696
Dinol
electroless copper
electroless nickel
Glance (mixture)
Kovac's indol reagent
Monsanto santicizer 2037 & nonyltoluene & dinonyltoluene

Monsanto skydrol R
Monsanto therminol VP-1 heat transfer fluid
organo-tin paint
paint and varnish remover
Perma Fluid (synonym: ammonium thioglycolate, 8% in water)
petroleum, Shell speciality (see Class 860)
Photo Resist 1450 (synonym: 2-ethoxyethanol & xylene & butyl acetate)
polymer 14435W8
Posistrip LE (synonym: morpholine & butyrolactone & N-methyl-2-pyrrolidone)
Pro Strip (mixture)
Pyrotec HFD46
Sadofoss primer 17
Sadofoss primer 513
Solmaster (synonym: methylene chloride & ethyl alcohol)
solvent 60
Turco 5092 stripping agent
U-V resin 20074
water
witch hazel
Xylamon

There are a total of 758 chemicals and mixtures in the 108 classes that comprise the 1992 edition of the Chemical Protective Clothing Permeation and Degradation Compendium.

## CHAPTER 4

PERMEATION INDEX NUMBERS

A standard test method (ASTM Method F739) for measuring the permeation of chemicals through protective materials has been established under the American Society for Testing and Materials (ASTM). This method evaluates the chemical resistance of a chemical protective clothing material versus liquids or gases. Two key parameters are reported using this method, i.e., breakthrough time and permeation rate.

A table encompassing Permeation Index Numbers utilizing both breakthrough times and permeation rates was first proposed by Krister Forsberg at the First International ASTM Symposium on the Performance of Protective Clothing, 16-20 July, 1984 in Raleigh, North Carolina and published in, "Performance of Protective Clothing," (ASTM STP 900, p. 247, 1986) ASTM, Philadelphia, PA.

The following table is an extension of this philosophy which covers all possible combinations of breakthrough time and permeation rate.

Table 8. Permeation Index Numbers for Permeation Rates and Breakthrough Times

Permeation Rates*	Breakthrough Time in Minutes				
	<1m	1 to 14m	15 to 59m	60 to 239m	240+m
<1	4	3	2	1	0
1 to <10	4	3	2	1	1
10 to <100	5	4	3	2	2
100 to <1,000	5	4	3	3	3
1,000 to <10,000	5	5	4	4	4
10,000 and greater	5	5	5	5	5

*   Permeation Rate units are milligrams per square meter per minute

These Permeation Index Numbers are also used in the risk assessment equation of The National Toxicology Program's GlovES+ (an expert system also published by Lewis Publishers). They help GlovES+ to find the best selections. GlovES+ uses Index Number 5 to screen out totally unsatisfactory selections.

The Permeation Index Numbers can be translated in terms of general performance to the following descriptions in Table 9.

Division of chemicals versus protective materials into the above classifications is a logical and conservative progression. These index numbers provide a useful guide to help in selection of the best protective materials when there is more than one choice available or to provide some relative risk assessment if data is available for only one type of material. Final selection should also include degradation data when it is available.

Table 9.   Permeation Index Number Descriptions

Index	Level of Permeation Rate	Comments
0	None or Very Low	Best Selection
1	Very Low	Next Best Selection
2	Low	Sometimes Satisfactory - Change When Exposed
3	Moderate	Poor Choice - Change Quickly When Exposed
4	High	Very Poor - Splashes Only, Change Quickly When Exposed
5	Very High	Dangerous Choice

Index 0 and Index 1 apply to the most resistant materials towards a specific chemical or test mixture.

Index 1 indicates a highly resistant material and may often be accepted by an industrial hygienist for harmful chemicals (e.g., as toxic as toluene or as corrosive as concentrated sodium hydroxide or as irritating as acetaldehyde).

Index 2 requires a greater degree of judgement by an industrial hygienist before it will be accepted. Industrial hygienists will not feel comfortable in recommending Index 2 materials for chemicals that are harmful (where a relatively small dose has a medium to moderate toxicity).

Index 3 materials are not usually sufficiently protective to be recommended by industrial hygienists unless there is no other choice or unless the work involves protection only against occasional splashes or compounds that are not very harmful.

Materials in the Index 4 classification are usually not acceptable for protection in any situation except occasional splashes. When splashes occur with materials in either Index 3 or 4 the gloves, or other types of protective clothing, should be immediately removed and exchanged for new ones.

Index 5 materials may be considered dangerous by most industrial hygienists because of the lack of protection they offer towards the chemical(s) tested.

Clothing material test results with classification Index Numbers of 4 or 5 may exhibit one or more forms of degradation. It is important to note that a material which has an index number of 4 or 5 against one particular chemical may in fact be rated with an index number of 0 or 1 against some other chemical. Therefore, these index numbers are not to be construed as a rating of a protective material except in the case of its use for protection against the specific chemical referenced and under the conditions of the measurements.

A seventh code, Number 6, was added for the GlovES+ expert system so that data where permeation rates are missing could still be used, with due caution, by this program.

## CHAPTER 5

A COMPARISON OF CPC PERMEATION AND DEGRADATION DATABASE WITH GlovES+

As mentioned above, this book is derived from the electronic publication named the Chemical Protective Clothing Permeation and Degradation Database. This database, carefully compiled from as many sources as could be found, is also the information basis of the National Toxicology Program's GlovES+ expert system which is published separately by Lewis Publishers. Each of these programs handle the data in very different ways and each has its strengths and weaknesses. These two programs are synergistic rather than duplicative and in order for you to properly use them it is imperative that you have a good understanding of the contents and workings of each. They are much more powerful than this book because of their ability to be searched by more than chemical names in order to find the information you need for your work.

"Chemical Protective Clothing Permeation and Degradation Database" is a traditional computerized database presented in a report or text type format. The advantages of providing the data in this format are that it is compact, taking less disk space than a true database file, and it is less expensive. In addition, if you want to print all or portions of it in a report, this can automatically be performed by the Print Option of the Searching Program that is provided.

Searches can be made on any keyword or combination of keywords including glove materials, specific manufacturers and/or model numbers of their products, references, breakthrough times or other specific pieces of data, degradation notations, chemical mixtures, laminated materials or blended polymers, and special test conditions such as aqueous solutions, tests at elevated temperatures, preexposed materials, etc.

Advantages of the GlovES+ expert system are that it has a built-in knowledge base that uses both the task requirements that the protective material should meet and chemical attributes of the various chemicals against which you may require protection. Thus, it can take into consideration the real world needs of a protective material for a specific purpose and help locate specific materials and/or garments which may be candidates for selection. It also has the advantages of providing these searches very rapidly and consistently. It can perform multiple searches, remember which materials or models from each search conform to your requirements and provide you with complete reports within minutes.

Disadvantages of the expert system are that it can only be searched by chemical name or chemical class, it has over a thousand fewer records than the full database, and it does not have references. It contains fewer records because records with unknown manufacturers and mixtures cannot be handled by it. The expert system also lacks the comment field in the database and it excludes all records where a test exhibited degradation. It is designed to be a rapid screening tool for the CPC Permeation and Degradation Database and you MUST check candidate materials against their source in the database to avoid making dangerous assumptions.

Similarities between Forsberg's database and GlovES+ are that both can be used to search for protective materials by chemical name or chemical class, and both have breakthrough times, permeation rates and material thickness. Both also have Forsberg's Permeation Index Number but you can only see it in the database; in the GlovES+ expert system it is incorporated into the risk assessment equations. Also both programs automatically print summary reports of their search findings to a disk file or to paper using your printer.

Major differences in the two programs are that the expert system calculates an estimated protection time based on a risk assessment decision module. This risk assessment module is based upon the premise that some exposure to some chemicals may be acceptable! The expert system is faster to use and is a screening tool. It has incomplete information and must be checked to verify that the assumptions it made are acceptable. The database has only information but the GlovES+ expert system has knowledge and makes assumptions which you may agree or disagree with.

Both of these programs are designed to search for candidate garments that may meet your specific needs.

Both Forsberg's database and the GlovES+ expert system will be very valuable aids in providing you with the best information that can be found but neither will do your job — the final decision for selecting chemical protective clothing for a specific need still rests with you. The same caveat applies when you use the data in this book.

CHAPTER 6

WHERE TO FIND ADDITIONAL TECHNICAL INFORMATION

Technical Representatives

You may obtain information from ASTM Committee F-23 which is charged with evaluating and recommending test methods for chemical protective clothing. Mr. Arthur Schwope is the chairman of that committee. Another source of information is Mr. Steven P. Berardeinelli or Mr. Michael M. Roder at NIOSH. The addresses of these men are:

Mr. Arthur Schwope
Arthur D. Little, Inc.
Acorn Park
Cambridge MA 02140
(617) 864-5770

Mr. Steven P. Berardeinelli or Mr. Michael M. Roder
NIOSH
944 Chestnut Ridge Rd.
Morgantown, WV 26505-2888
(304) 291-4337

Courses in Chemical Protective Clothing

The American Industrial Hygiene Association (AIHA) arranges several courses in CPC every year. Contact Mr. Zack Mansdorf of Mansdorf & Associates or the American Industrial Hygiene Association for further details.

Manufacturers of Chemical Protective Clothing

You may also obtain information directly from the manufacturers of chemical protective clothing. The names, addresses and international telephone numbers below represent most of the manufacturers whose products are represented in this publication. You may be able to obtain additional information from these experts or from the Customer Service representatives at each company. We have found them to be very knowledgeable and helpful.

Ansell Edmont Industrial
Mr. Nelson Schlatter
National Sales Manager
Edmont Glove Company
1300 Walnut Street

Cochocton, OH 43812
Tel: (614) 622-4311
FAX: (614) 622-6222

Ansell Canada (formerly Edmont Canada)
Mr. Norm F. Babyak
Ansell Canada, Inc.
30 Boul. de l'Aeroport
Bromont, Quebec JOE 1LO
CANADA
Tel: (514) 538-1850
FAX: (514) 534-1848

Bata
Mr. Mark Rolses
Bata Shoe Company, Inc.
Bata Industrials Division
Belcamp, MD 21017
Tel: (410) 272-2000
FAX: (410) 272-3346

Best
Mr. Michael Garobo or Don Bruce
Best Manufacturing Co.
Edison Street
Menlo, GA 30731-0008
Tel: (800) 241-0323 or
Tel: (404) 862-2302
FAX: (404) 862-6000 or
FAX: (404) 862-2660 (Don Bruce)

Brunswick, Guardian Glove
Ms. Susan Leak
Brunswick Corporation
302 Conwell Ave.
Willard, OH 44890-9525
Tel: (419) 933-2711
FAX: (419) 935-8961

CHEMFAB
Mr. John J. Schramko
Chemical Fabrics Corp.
P. O. Box 1137
Merrimack, NH 03054
Tel: (603) 424-9000
FAX: (603) 424-9012

Chemron
Mr. Jim Bartasis

Chemron, Inc.
954 Corporate Woods Parkway
Vernon Hills, IL 60061
Tel: (708) 520-7300
FAX: (708) 520-9812

Comasec
Mr. Joseph Krocheski
Comasec Safety, Inc.
P.O. Drawer 1219
Enfield, CT 06082
Tel: (203) 741-2207
FAX: (203) 741-0881

Mssr. Claude Piat
Comasec S.A.
Route de Muisement
F-28103 Dreux
FRANCE
Tel: 33-374-668-40

Dr. G. T. Knight
Comasec Dipco Ltd.
Greatwestern Business Park
Yate Avon, U.K. B517 5RF
Tel: 44-045-432-3633
FAX: 44-045-432-4366

du Pont
Ms. Janet L. Thiem
E.I. du Pont de Nemours & Co., Inc.
Fibers Dept., Laurel Run Building
P.O. Box 80 705
705 Centerville Road
Wilmington, DE 19880-0705
Tel: 1-800-44 TYVEK or
   (302) 999-3907
FAX: (302) 999 4763

Drager
Mr. Joerg Steuer
Drager Aktiegesellschaft
Moislinger Allee 53/55
2400 Lubeck 1
Germany
Tel: 49-451-882-2658
Fax: 49-451-882-3955

Mr. Steve Grossman
National Draeger Inc.
101, Technology Drive
Pittsburgh, PA 15230
Tel: (412) 787-8383
Fax: (412) 787-2207

Fairprene
Mr. Andree Boyer
Fairprene Industrial Products Co.,
  Inc.
85 Mill Plain Road
Fairfield, CT 06430
Tel: (203) 259-3351
FAX: (203) 254-2481

Fyrepel
Ms. Joy Wagner
Fyrepel Products, Inc.
P.O. Box 518
Newark, OH 43005
Tel: (614) 344-0391
FAX: (614) 344-6025

Granet
Mr. Elwood R. Reynolds
Granet Glove Company
Highway 258 South
Snow Hill, NC 28580
Tel: 1-800-562-0600
FAX: (919) 747-3450

ILC Dover
Mr. Evan Hensley, Jr.
ILC Dover, Inc.
P.O. Box 266
Frederica, DE 19946
Tel: (302) 335-3911
FAX: (302) 335-0762

Kappler Life-Guard
Mr. John Langley
Kappler Safety Group
P.O. Box 27

Guntersville, AL 35976
Tel: (205) 582-2119
FAX: (205) 582-2263

Mr. Al Smith
Life-Guard Kappler, Ltd.
Mile End Rd.
Colwick, Nottingham NG4 2BN
England
Tel: 44-602-618-182
FAX: 44-602-615-676

Marigold
Mr. Patrick McLaughlin
LRC - Surety Products, Inc.
1819 Main St., P O Box 4703
Sarasota, FL 34230
Tel: (800) 733-0987
FAX: (813) 366-9592

Mr. Tom Forsyth
London International Group plc
Unit 205, Cambridge Science Park
Cambridge, CB4 4GZ
ENGLAND
Tel: 44-22-342-3232
FAX: 44-22-342-3310

MAPA
Ms. Marianne Rodot
B.P. 158
92305 Levallois Perret Cedex
FRANCE
Tel: 33-1-40895-331
FAX: 33-1-47481-386 or
  33-1-47590-168

MAPA-Pioneer
(See Pioneer)

Molnlycke
Ms. Karina Hasselblad
Molnlycke Tissue AB
S-405 03 Göteborg
SWEDEN

Tel: 46-316-795-00
FAX: 46-318-777-92
Telex: 21628 Molnab S

MSA
Mr. Chris Kairys
Mine Safety Appliances Co.
P.O. Box 426
Pittsburgh, PA 15230
Tel: (412) 967-3148
FAX: (412) 967-3460

North
Mr. W. Fred Seebode
North Hand Protection
4090 Azalea Drive
P.O. Box 70729
Charleston, SC 29415
Tel: (803) 554-0660
FAX: (803) 744-2857

Pioneer
Mr. Radhan Radhakrishnan or Mr.
    John A. Varos
Pioneer Industrial Products Co.
512 East Tiffin Street
Willard, OH 44890
Tel: (419) 933-2211
FAX: (419) 933-2710

Playtex
Mr. Roger Blose
Playtex Family Products, Inc.
Industrial Gloves Division
700 Fairfield Avenue
Stamford, CT 06904
Tel: (203) 356-8050
FAX: (203) 356-8448

Safety 4
Ms. Kirsten Folkersen
Safety 4 A/S
P. O. Box 238
DK-2800 Lyngby

DENMARK
Tel: 45-459-309-57
FAX: 45-459-315-18
Telex: 31260-31382 Broeste

Mrs. Tina Hoff
Safety 4 Inc.
9765 Widmer, Bldg. 5
Lenexa, KS 66215
Tel: (913) 492-0860
FAX: (913) 492-0584

Standard Safety
Mr. Doug Barnhart
Standard Safety Equipment Co.
P. O. Box 188
Palatine, IL 60078
Tel: (708) 359-1400
FAX: (708) 359-2885

Trelleborg
Mr. Mike Kulberch
Goodall Rubber Co.
Trellchem Division
Quakerbridge Executive Center
Suite 203
Grovers Mill Road
Lawrenceville, NJ 08648, USA
Tel: (609) 799-2000
FAX: (609) 799-4582

Mr. Bengt Wittsell or Mr. John
    Ekelund
Trelleborg AB
Protective Products Division
23181 Trelleborg, Sweden
Tel: 46-410-510-00

Wheeler
Mr. Don Schmidt
Wheeler Protective Apparel, Inc.
4330 West Belmont Avenue
Chicago, IL 60641
Tel: (312) 685-5551
FAX: (312) 685-8897

CHAPTER 7

REFERENCES AND SOURCES OF THE DATA

The references below appear in the last column of the chemical permeation data table. Some are in U.S. journals, some in other international publications and some are in various reports the author has obtained or from private communications with various scientists. Each of these has been duly noted below.

1. Henry, N.W. and C.N. Schlatter, "The Development of a Standard Method for Evaluating Chemical Protective Clothing to Permeation by Liquids," Am. Ind. Hyg. Assoc. J., 42:202, 1981.
2. Schwope, A.D., M.A. Randel and M.G. Broome, "Dimethyl Sulfoxide Permeation Through Glove Materials," Am. Ind. Hyg. Assoc. J., 42:722 1981.
3. Coletta, G.C., A.D. Schwope, I. Arons, J. King and A. Sivak, "Development of Performance Criteria for Protective Clothing Used Against Carcinogenic Liquids," Arthur D. Little, Inc., Report to NIOSH Under Contract 210-76-0130, (October, 1978).
4. Weeks, R.W. and M.J. McLeod, "Permeation of Protective Garmet Material by Liquid Halogenated Ethanes and a Polychlorinated Biphenyl," Los Alamos Scientific Laboratory Report LA-8572-MS to NIOSH, (1980).
5. Nelson, G.O., B. Lum, G. Carlson, C. Wong and J. Johnson, "Glove Permeation by Organic Solvents," Am. Ind. Hyg. Assoc. J., 42:217, 1981.
6. "Ansell Edmont Chemical Resistant Guide," Ansell Edmont Gloves and Protective Equipment, Coshocton, Ohio 43812 (1991 and earlier Edmont Publications in 1983, 1986, 1989, and 1991).
7. North Permeation Resistance Guide 7-300, North Hand Protection Inc., Charleston, S. C. 29405 (1983, 1986 and 1989).
8. duPont Co., Supnbonded Product Division, Wilmington, Delaware, 19898, (1983) and Permeation Guide for Du Pont Protective Apparel Fabrics, Du Pont Co., Protective Apparel Fabrics of TYVEK, P.O. Box 80.705, Wilmington, DE 19880-0705 (Revised October 1991).
9. Bush, D.G., L.E. Tersegno, J.E. Winter and D.H. Schoch, "A Method for Testing Permeability of Protective Clothing to Acids and Bases," Eastman Kodak Co., Rochester, N.Y. 14650, (1982).
10. Schoch D.H., L.K. Tersegno, J.E. Winter and D.G. Bush, "Testing of 'Impervious' Gloves for Permeation by Organic Solvents," Eastman Kodak Co., Rochester, N. Y. 14650, (1982).
11. Forsberg, K., A. Linnarson, K.G. Olsson and L. Sperling, "Utveckling av skyddshandskar. Arbete med losningsmedel i djup- och offsettryck", (Translated from Swedish: "Development of Protective Gloves. Gloves for Printers"), Ergolab Report S 81:10, ASF Contract 80-220, (November 1981).
12. Forsberg, K., K.G. Olsson and B. Carlmark, "Utveckling av skyddshandskar. Arbete med smorj och kylmedel vid metallbearbetning," (Translated from Swedish: "Development of Protective Gloves. Gloves for Work With Metal Cutting Fluids"), Ergolab Report S 82:17, ASF-Contract 81-1187, (October 1982).

13. Forsberg K., B. Carlmark and S. Hoglund, "Undersokning av skyddshandskar och skyddsoveralls (nonwoven) motstand mot permeation av fenoxisyror," Ergolab Report S SLO-kontrakt 200, (1983). [In Swedish. Translation: "Permeation of Gloves and Disposable Chemical Protective Clothing Materials by Chlorophenoxy Acids".]

14. Henriksen, H.R., "Udvaelgelse af materialer til beskyttelseshandsker. Polymermembraner til vaern mod kontakt med epoxyprodukter," Direktor- atet for arbejdstilsynet. Rapport 8/1982, (1982). [In Danish with English summary. Translation: "Selection of Glove Materials for Protection from Epoxy Products".]

15. Jensen, A., "Provning af beskyttelsehandskers resistens mod organiske oplosningsmidler," Dantest Rapport, Arbejdsmiljofondet kontrakt 80-11, Marts, (1982). [In Danish. Translation: "Permeation Testing of Glove Materials' Resistance to Organic Solvents".]

16. Linnarson, A. and K. Halvarson, "Studie av polymermaterials genomslapplighet for organiske foreningar," FOA Rapport C 20414-H2, ASF- kontrakt 75-65, (June 1981). [In Swedish. Translation: "The Study of Polymer Materials Permeation by Organic Solvents".]

17. Nelson, G.O., G.J. Carlson and A.L. Buerer, "Glove Permeation by Shale Oil and Coal Tar Extract," Lawrence Livermore Laboratory, UCRL, California, 52893, (1980). 18 Bennett, R.D., C.E. Feigley, E.O. Oswald and R.H. Hill, "The Permeation by Liquefied Coal of Gloves Used in Coal Liquefacation Pilot Plants," Am. Ind. Hyg. Assoc. J., 44:447, 1983.

19. Spence M.W., "An Evaluation of Several Critical Variables," presented at the Am. Ind. Hyg. Conf., Portland, Oregan, (May 1981).

20. Wakefield, M.E. and M. W. Hall, "Development of a Specification for an Improved Ensamble for Propellant Handlers," NASA Report, Contract NASA10-9714, (1980).

21. Abernathy, R.N., R.B. Cohen and J.J. Shirtz, "Measurements of Hypergolic Fuels and Oxidants Permeation Through Commercial Protective Materials - Part I, Inhibited Red Fuming Nitric Acid and Unsymmetrical Dimethylhydrazine," Am. Ind. Hyg. Assoc. J., 44:505, 1983.

22. Weeks, R.W., Jr. and B.J. Dean, "Permeation of Methabolic Aromatic Amine Solutions Through Commercially Available Glove Materials," Am. Ind. Hyg. Assoc. J., 38:721, 1977.

23. Weeks, R.W., Jr. and M.J. McLeod, "Permeation of Protective Garmet Material by Liquid Benzene and by Tritiated Water," Am. Ind. Hyg. Assoc. J., 43:201, 1982.

24. Stampfer, J.F., M.J. McLeod, A.M. Martinez and M.R. Betts," "Permeation of Aroclor 1254 and Solutions of This Substance Through Selected Protective Clothing Materials," Am. Ind. Hyg. Assoc. J., 45:634, 1984.

25. Stampfer, J.F., M.J. McLeod, M.R. Betts, A.M. Martinez and S.P. Berardinelli, "The Permeation of Eleven Protective Garmet Materials by Four Organic Solvents," Am. Ind. Hyg. Assoc. J., 45:642, 1984.

26. Williams, J.R., "Permeation of Glove Materials by Physiologically Harmful Chemicals," Am. Ind. Hyg. Assoc. J., 40:877, 1979.

27. Williams, J.R., "Chemical Permeation of Protective Clothing," Am. Ind. Hyg. Assoc. J., 41:884, 1980.

28. Williams, J.R., "Evaluation of Intact Gloves and Boots for Chemical Permeation," Am. Ind. Hyg. Assoc. J., 42:468, 1981.

29. Algera, R., "Development of a Hazardous Chemical Protective Ensamble: Phase I.," Interim Report, U.S. Coast Guard Contract DTCG 23-81-C-20003, (November 1982).

30. Scandlab, "Unpublished data," Information available from Scandlab, Box 7013, 191 07 Sollentuna, Sweden.

31. Schlatter, C.N., "Permeation Resistance of Gloves After Repeated Cleaning and Exposure to Liquid Chemicals," Presented at the Am. Ind. Hyg. Conference, Philadelphia, Pa., USA, (May 1983).

32. Waegemaekers, T., E. Seutter, J. den Arend and K. Malten, "Permeability of Surgeon's Gloves to Methyl Methacrylate," Acta Ortoped. Scand., 54:790, 1983.

33. Forsberg, K. and K.G. Olsson, "Faststallande av riktlinjer for val av kemikalieskyddshandskar," ASF-kontrakt 83-0750, (May, 1985). [In Swedish. Translation: "Establishment of Guidelines for the Selection of Chemical Protective Gloves".]

34. Keith, L.H., M. Conoley, R.L. Nolen, D.B. Walters and A.T. Prokopetz, "Chemical Permeation and Degradation Data From the National Toxicology Program Measured by Radian Corporation, Austin, Texas," data from a series of NTP/Radian progress reports and private communications from 1984 to 1990. [Unpublished data]

35. Spence, M. W., "Glove Materials for Chlorinated Solvents: Permeation Resistance Comparison for Four Solvents," presented at the Am. Ind. Hyg. Assoc. Conference, Detroit, Michigan, USA, (May 1984).

36. Pioneer Chemical Performance Guide, Pioneer Industrial Products Co., Willard, Ohio, 44890 (1984, 1987, and 1991).

37. Elingsen, F., "Unpublished data," Information available from FFI, Box 25, N-2007 Kjeller, Norway.

38. Zetterstrom, B., "Unpublished data," Information available from FOA, 4, 901 82, Umea, Sweden.

39. Mikatavage, M., S.S. Que Hee and H.E. Ayer, "Permeation of Chlorinated Aromatic Compounds Through Viton and Nitrile Glove Materials," Am. Ind. Hyg. Assoc. J., 45:617, 1984.

40. Comasec Chemical Resistance Guide, Comasec SA, F-75883, Paris, Cedex 18, France; and Comasec, Inc., Drawer 10, 8 Niblick Road Enfield, CN, 06082 (1986, 1989, 1990).

41. Mickelsen, R. L., M.M. Roder and S.P. Berardinelli, "Permeation of Chemical Protective Clothing by Three Binary Solvent Mixtures," Am. Ind. Hyg. Assoc. J., 47:236-240, 1986.

42. Edmont High Technology Products Chemical Resistance Guide, available from Edmont Division of Becton, Dickinson & Co., 1300 Walnut St., Coshocton, Ohio, 43812. (Oct. 6, 1986).

43. Bentz, A. and V. Mann, "Critical Variables Regarding Permeability of Materials for Total Encapsulated Suits," presented at the Scandinavian Symposium on Protective Clothing Against Chemicals, Lyngby, Denmark, (November 1984).

44. Kashi, K.P., M. Muther and S.K. Majumder, "Rapid Evaluation of Phosphine Permeability Through Various Flexible Films and Coated Fibers," Pestic. Sci., 8:492, 1977.

45. Forskov, T., "Beskyttelsehandskers gennemtraenglighed for organiske oplosningsmidler," Kemisk Lab. A, Denmark Tekniske Hojskole, DK-2800 Lyngby, Denmark (1980). [In Danish. Translation: "The Permeation of Inorganic Solvents Through Gloves".]

46. Stull, J.O., "Early Development of Hazardous Chemical Protective Ensemble," Technical Report CG-D-24-86, U. S. Coast Guard, Washington D. C., October 1986.

47. Henriksen, H.R., "Development of Protective Gloves Against Epoxides and Solvents," presented at the Scandinavian Symposium on Protective Clothing Against Chemicals, Lyngby, Denmark (November 1984).

48. MSA, Data Sheet 13-02-07, MSA, Pittsburgh, PA, 15235, USA, (1984).

49. Laursen, H., "Provning af beskyttelsehandskers resistens mod organiske oplosningsmidler II," Dantest Rapport, Arbejdsmiljofondet kontrakt 83-21, (March 1985). [In Danish. Translation: "Testing of Glove Materials' Resistance to Organic Solvents".]

50. Johnson, T.C. and W.D. Merciez, "Permeation of Halogenated Solvents Through Drybox Gloves," Dow Chemical Co. Report, U. S. Atomic Energy Commission Contract, (1971).

51. Union Carbide Technical Bulletin, "Protective Clothing (Gloves) for Glycol Ethers," Union Carbide Corporation, Danbury, CT, 06817 USA, (1984).

52. Dillon, I.G. and E. Obasuyi, "Permeation of Hexane Through Butyl Nomex," Am. Ind. Hyg. Assoc. J., 46:233, 1985.

53. Best Permeation Resistance Guide, Best Manufacturing Company, Menlo, GA., 30731, USA, (1985, 1989, and 1991).

54. Forsberg, K., K.G. Olsson, and M. Svensson, "Utveckling av nya fagangs-plagg mot kemikalier. Forsok med nya materialkombinationer baserade pa nonwoven", rapport TRITA-AAV-1012 (March, 1985). [In Swedish. Translation: "Development of Disposable Chemical Protective Clothing".]

55. Ansell Industrial Gloves Chemical Permeation Guide, Ansell Industrial Products, P. O. Box 1252 Dothan, AL., 36302, USA, (1986) and (1988).

56. Product Release - Permeation Resistance 1986, Durafab Division of Texel Industries, Inc. (1986 and 1989).

57. Akerblom, M, L. Jansson, and K. Forsberg, "Permeation of Pesticides Through Gloves and Disposable Materials", Presented at the Second International Symposium on the Performance of Protective Clothing, ASTM Committee F-23, Tampa, FL (January, 1987).

58. Mickelsen, R.L., and R.C. Hall, "A Breakthrough Time Comparison of Nitrile and Neoprene Glove Materials Produced by Different Glove Manufacturers", Am. Ind. Hyg. Assoc. J, 48:941, 1987.

59. Stjernstrom, G., "Permeation av cytostatica genom handskmaterial", Apoteksbolaget, (1986). [In Swedish. Translation: "The Permeation of Antineoplastic Drugs Through Glove Materials".]

60. 4H Chemical Protection Guide. Safety 4 A/S, Box 238, 2800 Lynby, Denmark (1986, 1989, 1990, and 1991).

61. Henry, Norman W., III, "How Protective is Protective Clothing?", ASTM STP 900, Performance of Protective Clothing, R. L. Barker and G. C. Coletta, Editors, ASTM, 1916 Race St., Philadelphia, PA, 19103, USA (1986).

62. Perkins, J.L., M.C. Ridge, A.B. Holcombe, M.K. Wang and W.E. Nonidez, "Skin Protection, Viton and Solubility Parameters," Am. Ind. Hyg.Assoc. J., 47:805, 1986

63. Perkins, J.L., "Predicting Permeation Properties of Butyl, Viton and Nitrile Polymers," Presented at the Second International Symposium on the Performance of Protective Clothing, ASTM Committee F-23, Tampa, FL (January, 1987).

64. Silkowski, J.B., S.W. Horstman and M.S. Morgan, "Permeation Through Five Commercially Available Glove Materials by Two Pentachlorophenol Formulations," Am. Ind. Hyg. Assoc. J., 45:501-504, 1984.

65. Stull, J.O., R.A. Jemke and M. G. Steckel, "Development of a U.S. Coast Guard Chemical Protective Response Suit," Technical Report No. CG-D-24-86, U. S. Coast Guard, Washington, D.C., (July, 1987).

66. Vahdat, N., "Permeation of Polymeric Materials by Toluene," Am. Ind. Hyg. Assoc. J.," 48:155-159, 1987.

67. Chemrel Protective Clothing Selection Guide, Chemron Inc., Buffalo Grove, IL 60015, (1987).

68. Coyne, L.B., M.W. Spence and S.K. Norwood, "Protective Equipment Effectiveness for Downol (TM) Glycol Ethers: Gloves and Respirators," Presented at the Am. Ind. Hyg. Conference, Philadelphia, PN, (May, 1983).

69. Berardinelli, S.P., and E.S. Moyer, "Methyl Isocyanate Liquid and Vapor Permeation Through Selected Respirator Diaphragms and Chemical Protective Clothing," Am. Ind. Hyg. Assoc. J., 48:324-329, 1987.

70. Vahdat, N., "Permeation of Protective Clothing Materials by Methylene Chloride and Perchloroethylene," Am. Ind. Hyg. Assoc. J., 48:646-651, 1987.

71. Trelleborg Trellchem Resistance Table with Instructions (1986), Technical Data Package for Trellchem HPS (December 1990), Trelleborg Industri AB, Protective Products Division, S-231 81 Trelleborg, Sweden.

72. Playtex Permeation and Chemical Resistance Guide, Playtex Gloves, International Playtex, Inc., Stamford, CT 06902 (1987).

73. Granet permeation test results. Granet Co, P.O. Box 337, Snow Hill, NC 28580, Personal communication from E. R. Reynolds, January, 1988.

74. Moody, R.P., J.M. Caroll and L. Ritter, "Comparison of ASTM F-739 Test Method With an Automated HPLC Analysis of Pesticide Penetration Through Protective Clothing," Am. Ind. Hyg. Assoc. J., 51:79-83 (1990).

75. StaSafe CPE Chemical Testing, Standard Safety Equipment Company, Palatine, IL 60078, January 1988 and December 1990.

76. Monsanto Glove Facts, Monsanto Company, 800 N. Lindbergh Blvd., St. Louis, MO 63167, (October, 1987).

77. Life-Guard Product Sheets, Life-Guard Company, Life-Guard, Inc., and Kappler Product Sheets. Kappler Safety Group, Guntersville, AL 35976 (1988, 1989, and 1991).

78. Menke R., and C.F. Chelton, "Evaluation of Glove Material Resistance to Ethylene Glycol Dimethyl Ether Permeation" Am. Ind. Hyg. Assoc. J., 49:386-389, 1988.

79. Marigold Industrial Gloves - Glove Resistance Chart, LRC Products Ltd., London E4 8QA, England (Second Edition, 1988 and Fourth edition, 1990).

80. Challenge 5200 Chemical Permeation Performance, Chemfab Technical Report No. 3-0788-13, (1988); Challenge 6000 NFPA 1991 Certified Vapor Protective Suit for Hazardous Chemical Emergencies (1991), Chemical Fabrics Corp., Merrimack, NH 03054.

81. Vahdat, N. and R. Delaney, "Decontamination of Chemical Protective Clothing," Am. Ind. Hyg. Assoc. J., 50:152-156, 1989.

82. Drolet, D. and J. Lara, "Permeation of Protective Gloves to Nitroglycerine (NG) and Ethylene Glycol Dinitrate (EGDN)," Third Scandinavian Symposium on Protective Clothing Against Chemicals and Other Health Risks. Gausdal, Norway, (September 27-30, 1989).

83. Stull, J.O. and B. Herring, "Selection and Testing of Glove Combination for Use With the U. S. Coast Guard's Chemical Response Suit," Am. Ind. Hyg. Assoc. J., 51:378-383 (1990).

84. Stull, J.O., S. Pinette and R.E. Green, "Permeation of Ethylene Oxide Through Protective Clothing Materials for Chemical Emergency Response," Appl. Ocup. Environ. Hyg., 5:448-452 (1990).

85. Wheeler Protective Apparel, Inc. Chemical/Physical Guideline Tables, 1990.

86. BATA Permeation Data Resistance (1990, 1991), Bata Shoe Company, Inc., Belcamp, MD 21017.

87. MAPA Laboratory Permeation Test Data (1990 and 1991), MAPA, B.P. 158, 92305 Levallois-Perret, FRANCE.

88. Berardinelli, S.P., E.S. Moyer and C.H. Rotha, "Ammonia and Ethylene Oxide Permeation Through Selected Protective Clothing," Am. Ind. Hyg. Assoc. J., 51:595-600, 1990.

89. "Guardian Gloves Permeation Test Results," Brunswick Corporation, 302 Conwell, Willard, OH 44890 (1991).

90. TRI/Environmental Inc., "Support for Development of Protective Clothing Standards," Interim Report, TRI, Inc., Austin, TX (1991). 91 Drager Chemical Resistance Data (1991). Dragerwerk Aktiegesellschaft. 2400 Lubeck 1, Germany.

91. Drager Chemical Resistance Data, Dragerwerk Aktiegesellschaft, 2400 Lubeck 1, Germany, 1991

92. Perkins, J.L. and You Ming-Jia, "Predicting Temperature Effects of Chemical Protective Clothing Permeation," Am. Ind. Hyg. Assoc. J., 53(2):77-83 (1992).

93. Zeller, E.T., H. Ke, D. Smigiel, R. Sulewski, S.J. Patrash, M. Han, and G.Z. Zhang, "Glove Pemreation by Semiconductor Processing Mixtures contiaing Glycol-Ether Derivatives," Am. Ind. Hyg. Assoc. J., 53(2): 105-116 (1992).

94. Renard, E.P., R. Goydan, and T. Stolki, "Permeation of Multifunctional Acrylates Through Selected Protective Glove Materials," Am. Ind. Hyg. Assoc. J., 53(2):117-123 (1992).

G				Perm-	I		
A				eation	N	Thick-	R
R			Break-	Rate in	D	ness	degrad.	e
M	Class & Number/	MANUFACTURER	through	mg/sq m	E	in	and	f
E	test chemical/	& PRODUCT	Time in	/min.	X	mm	comments	s
N	MATERIAL NAME	IDENTIFICATION	Minutes				

1,1,1,2-tetrachloroethane
CAS Number: 630-20-6
Primary Class: 261   Halogen Compounds, Aliphatic and Alicyclic

g	BUTYL	NORTH B-174	140m	1405	4	0.51	degrad.	34
g	NATURAL RUBBER	ACKWELL 5-109				0.15	degrad.	34
g	NEOPRENE	ANSELLEDMONT 29-870				0.48	degrad.	34
g	NITRILE	ANSELLEDMONT 37-155				0.40	degrad.	34
g	PVAL	EDMONT 25-950	>480m	0	0	0.56		34
g	PVC	ANSELLEDMONT 34-100	3m	3280	5	0.20	degrad.	34
g	VITON	NORTH F-091	>480m	0	0	0.25		34

1,1,1,3,3,3-hexamethyldisilazane (synonyms: bis(trimethylsilyl)amine; HMDS)
CAS Number: 999-97-3
Primary Class: 142   Amines, Aliphatic and Alicyclic, Secondary
Related Class: 480   Organosilicon Compounds

g	NATURAL RUBBER	ANSELLEDMONT 36-124	40m			0.46		6
g	NATURAL RUBBER	ANSELLEDMONT 392	15m			0.48		6
g	NEOPRENE	ANSELLEDMONT 29-840	50m			0.38		6
g	NEOPRENE	ANSELLEDMONT NEOX	60m			n.a.		6
g	NITRILE	ANSELLEDMONT 37-165	>360m	0	0	0.60		6
g	PE/EVAL/PE	SAFETY4 4H	>240m	0	0	0.07	35°C	60
g	PVAL	ANSELLEDMONT PVA	>360m		0	n.a.		6
g	PVAL	EDMONT 15-552	>360m	0	0	n.a.		6
g	PVC	ANSELLEDMONT SNORKE				n.a.	degrad.	6

1,1,1-TCE, 66% & propyleneglycol monoethylether acetate, 34%
Primary Class: 261   Halogen Compounds, Aliphatic and Alicyclic
Related Class: 222   Esters, Carboxylic, Acetates
              245   Ethers, Glycols

g	PE/EVAL/PE	SAFETY4 4H	>240m	0	0	0.07	35°C	60

1,1,1-TCE, 73% & dichloromethane, 17% & dodecylbenzene sulfonic acid, 10%
Primary Class: 800   Multicomponent Mixtures With > 2 Components

g	PE/EVAL/PE	SAFETY4 4H	>240m	0	0	0.07		60

1,1,1-trichloroethane (synonyms: 1,1,1-TCE; methylchloroform)
CAS Number: 71-55-6
Primary Class: 261   Halogen Compounds, Aliphatic and Alicyclic

g	BUTYL	BEST 878	72m	4590	4	0.75	degrad.	53
g	BUTYL	BRUNSWICK BUTYL-XTR	65m	>2500	4	0.63		89
g	BUTYL	COMASEC BUTYL	12m	7860	5	0.50		40
g	BUTYL	NORTH B-161	60m			0.56	degrad.	4

G A R M E N T	Class & Number/ test chemical/ MATERIAL NAME	MANUFACTURER & PRODUCT IDENTIFICATION	Break-through Time in Minutes	Perm-eation Rate in mg/sq m /min.	I N D E X	Thick-ness in mm	degrad. and comments	R e f s
g	BUTYL	NORTH B-161	29m	9180	4	0.40		35
g	BUTYL	NORTH B-174				0.65	degrad.	34
g	BUTYL/ECO	BRUNSWICK BUTYL-POL	65m	1310	4	0.63		89
g	BUTYL/NEOPRENE	COMASEC BUTYL PLUS	12m	7860	5	0.60		40
s	CPE	STD. SAFETY				n.a.	degrad.	75
g	HYPALON	COMASEC DIPCO	61m	8214	4	0.58		40
g	NAT+NEOPR+NITRILE	MAPA-PIONEER TRIONI	>11m			0.48	degrad.	36
g	NATURAL RUBBER	ACKWELL 5-109				0.15	degrad.	34
g	NATURAL RUBBER	ANSELL PVL 040				n.a.	degrad.	55
g	NATURAL RUBBER	ANSELLEDMONT 36-124				0.46	degrad.	6
g	NATURAL RUBBER	ANSELLEDMONT 392				0.48	degrad.	6
g	NATURAL RUBBER	ANSELLEDMONT 46-320	4m	26000	5	0.31		5
g	NATURAL RUBBER	BEST 65NFW				n.a.	degrad.	53
g	NATURAL RUBBER	COMASEC FLEXIGUM	24m	6600	4	0.95		40
g	NATURAL RUBBER	MAPA-PIONEER 258	10m	70	4	0.64	30°C	87
g	NATURAL RUBBER	MARIGOLD ORANGE SUP	16m	13200	5	0.75		12
g	NATURAL+NEOPRENE	ANSELL OMNI 276	8m	30000	5	0.45		5
g	NATURAL+NEOPRENE	ANSELL OMNI 276				0.56	degrad.	55
g	NEOPRENE	ANSELL NEOPRENE 520				0.46	degrad.	55
g	NEOPRENE	ANSELLEDMONT 29-840				0.38	degrad.	6
g	NEOPRENE	ANSELLEDMONT 29-870	19m	7440	4	0.50	degrad.	34
g	NEOPRENE	ANSELLEDMONT 29-870	24m	8940	4	0.48		35
g	NEOPRENE	BEST 32	19m	3320	4	n.a.	degrad.	53
g	NEOPRENE	BEST 6780	37m	1840	4	n.a.	degrad.	53
g	NEOPRENE	BRUNSWICK NEOPRENE	23m	>1000	4	0.88		89
g	NEOPRENE	COMASEC COMAPRENE	12m	7800	5	n.a.		40
g	NEOPRENE	MAPA-PIONEER 420	22m	12000	5	0.82		12
g	NEOPRENE	MAPA-PIONEER GF-N-	19m	10000	5	0.47		5
g	NEOPRENE	MAPA-PIONEER N-44	27m	11820	5	0.56	degrad.	36
g	NEOPRENE	MAPA-PIONEER N-73	25m	9000	4	0.46		5
u	NEOPRENE	Unknown	45m			0.58	degrad.	4
g	NEOPRENE/NATURAL	ANSELL CHEMI-PRO				0.72	degrad.	55
g	NITRILE	ANSELL CHALLENGER	18m	5	2	0.38		55
g	NITRILE	ANSELLEDMONT 37-155	11m	41000	5	0.35		5
g	NITRILE	ANSELLEDMONT 37-165	>60m			0.55		5
g	NITRILE	ANSELLEDMONT 37-165	90m	<90000	5	0.60		6
g	NITRILE	ANSELLEDMONT 37-175	56m	<500	3	0.37		5
g	NITRILE	ANSELLEDMONT 37-175	35m	9000	4	0.34		12
g	NITRILE	ANSELLEDMONT 37-175	17m	2820	4	0.34		35
g	NITRILE	BEST 727	49m	1740	4	0.38		53
g	NITRILE	COMASEC COMATRIL	80m	6720	4	0.55		40
g	NITRILE	COMASEC COMATRIL SU	120m	6000	4	0.60		40
g	NITRILE	COMASEC FLEXITRIL	10m	1860	5	n.a.		40
g	NITRILE	ERISTA SPECIAL	28m	7200	4	0.37		12
g	NITRILE	MAPA-PIONEER 370	140m	1	1	n.a.	30°C	87
g	NITRILE	MAPA-PIONEER 490	32m	220	3	0.37	30°C	87
g	NITRILE	MAPA-PIONEER 492	46m	45	3	0.48	30°C	87
g	NITRILE	MAPA-PIONEER 493	46m	9	2	0.56	30°C	87
g	NITRILE	MAPA-PIONEER A-14	132m	2628	4	0.56	degrad.	36
g	NITRILE	MAPA-PIONEER A-15	75m	243	3	0.36		36

G A R M E Class & Number/ N test chemical/ T MATERIAL NAME	MANUFACTURER & PRODUCT IDENTIFICATION	Break-through Time in Minutes	Perm-eation Rate in mg/sq m /min.	I N D E X	Thick-ness in mm	degrad. and comments	R e f s
g NITRILE	MARIGOLD NITROSOLVE	145m	179	3	0.75	degrad.	79
u NITRILE	NOLATO	22m	7000	4	0.38		12
g NITRILE	NORTH LA-142G	37m	4584	4	0.38	degrad.	34
u NITRILE	Unknown	30m			0.20		4
g NITRILE+PVC	COMASEC MULTIMAX	65m	2526	4	n.a.		40
g NITRILE+PVC	COMASEC MULTIPLUS	56m	960	3	n.a.		40
g PE	ANSELLEDMONT 35-125	2m	1300	5	0.03		5
g PE	ANSELLEDMONT 35-125	<1m	1542	5	0.07		35
g PE/EVAL/PE	SAFETY4 4H	>240m	0	0	0.07	35°C	60
g PE/EVAL/PE	SAFETY4 4H	>480m	0	0	0.07		60
g PVAL	ANSELLEDMONT PVA	>360m	0		n.a.		6
g PVAL	COMASEC SOLVATRIL	>480m	0	0	n.a.		40
g PVAL	EDMONT 15-552	60m	<9	1	n.a.		6
g PVAL	EDMONT 25-545	>360m	0	0	n.a.		6
g PVAL	EDMONT 25-545	>480m	0	0	0.45		35
g PVAL	EDMONT 25-950	>960m	0	0	0.33		34
g PVC	ANSELLEDMONT 34-500	1m	>10000	5	0.17		12
g PVC	ANSELLEDMONT SNORKE				n.a.	degrad.	6
g PVC	COMASEC MULTIPOST	25m	1200	4	n.a.		40
g PVC	COMASEC MULTITOP	40m	1200	4	n.a.		40
g PVC	COMASEC NORMAL	30m	1200	4	n.a.		40
g PVC	COMASEC OMNI	18m	1560	4	n.a.		40
g PVC	KID 490	37m	6000	4	n.a.		12
g PVC	KID VINYLPRODUKTER	7m	13000	5	0.55		12
g PVC	MAPA-PIONEER V-5	1m	15900	5	0.10	degrad.	35
g PVC	SEMPERIT	12m	12000	5	0.65		12
u PVDC/PE/PVDC	MOLNLYCKE	64m	390	3	0.07		54
s TEFLON	CHEMFAB CHALL. 5100	>180m	0	1	n.a.		65
s UNKNOWN MATERIAL	DU PONT BARRICADE	>480m	0	0	n.a.		8
s UNKNOWN MATERIAL	KAPPLER CPF III	>480m	0	0	n.a.		77
s UNKNOWN MATERIAL	LIFE-GUARD RESPONDE	>480m	0	0	n.a.		77
g UNKNOWN MATERIAL	NORTH SILVERSHIELD	>360m	0	0	0.08		7
g VITON	NORTH F-091	>3600m	0	0	0.24		4
g VITON	NORTH F-091	>120m			0.24		30
g VITON	NORTH F-091	>900m	0	0	0.25		34
g VITON	NORTH F-091	1450m	31	2	0.30		62
g VITON	NORTH F-121	>480m	0	0	0.30		35

1,1,1-trichloroethane & ethyl alcohol & turpentine, unknown %
Primary Class: 800   Multicomponent Mixtures With > 2 Components

g PE/EVAL/PE	SAFETY4 4H	>240m	0	0	0.07	35°C	60

1,1,1-trichloroethane, 97% & 1,4-dioxane, 3%
Primary Class: 800   Multicomponent Mixtures With > 2 Components

g PE/EVAL/PE	SAFETY4 4H	>240m	0	0	0.07	35°C	60

G A R M E Class & Number/ N test chemical/ T MATERIAL NAME	MANUFACTURER & PRODUCT IDENTIFICATION	Break- through Time in Minutes	Perm- eation Rate in mg/sq m /min.	I N D E X	Thick- ness degrad. in and mm comments	R e f s
1,1,2,2-tetrachloroethane						
CAS Number: 79-34-5						
Primary Class: 261	Halogen Compounds, Aliphatic and Alicyclic					
g BUTYL	NORTH B-174	276m	420	3	0.66 degrad.	34
u CPE	ILC DOVER	64m			n.a.	29
g NATURAL RUBBER	ACKWELL 5-109				0.15 degrad.	34
g NATURAL RUBBER	ANSELLEDMONT 46-320	6m	26000	5	0.31	5
g NATURAL+NEOPRENE	ANSELL OMNI 276	9m	32000	5	0.45	5
g NEOPRENE	ANSELLEDMONT 29-870				0.48 degrad.	34
g NEOPRENE	ANSELLEDMONT NEOX	6m	5000	5	n.a.	5
g NEOPRENE	MAPA-PIONEER N-73	18m	14000	5	0.46	5
g NITRILE	ANSELLEDMONT 37-165	74m	>3000	4	0.55	5
g NITRILE	ANSELLEDMONT 37-175	22m	32000	5	0.35	5
g PE	ANSELLEDMONT 35-125	4m	100	4	0.03	5
g PVAL	EDMONT 25-950	>480m	0	0	0.35	34
g PVC	ANSELLEDMONT 34-100	<1m	420	4	0.20 degrad.	34
g PVC	MAPA-PIONEER V-20	6m	25000	5	0.31	5
s SARANEX-23	DU PONT TYVEK SARAN	75m	120	3	n.a.	8
s TEFLON	CHEMFAB CHALL. 5100	>900m	0	0	n.a.	65
g VITON	NORTH F-091	>480m	0	0	0.30	34
u VITON/CHLOROBUTYL	Unknown	>180m			n.a.	29
1,1,2-trichloroethane						
CAS Number: 79-00-5						
Primary Class: 261	Halogen Compounds, Aliphatic and Alicyclic					
g BUTYL	NORTH B-131	45m			0.38	4
g BUTYL	NORTH B-161	50m			0.56 degrad.	4
g BUTYL	NORTH B-174	340m	407	3	0.68	34
g NATURAL RUBBER	ACKWELL 5-109				0.15 degrad.	34
u NATURAL RUBBER	Unknown	1m			0.23 degrad.	4
g NEOPRENE	ANSELLEDMONT 29-870				0.48 degrad.	34
g NEOPRENE	MAPA-PIONEER N-44	11m	39660	5	0.74	36
u NEOPRENE	Unknown	7m			0.58 degrad.	4
g NITRILE	ANSELLEDMONT 37-155				0.40 degrad.	34
g NITRILE	MAPA-PIONEER A-14	9m	3720	5	0.54 degrad.	36
u NITRILE	Unknown	2m			0.20 degrad.	4
u PE	Unknown	3m			0.05	4
u POLYURETHANE	Unknown	<1m			0.10 degrad.	4
g PVC	ANSELLEDMONT 34-100	2m	52325	5	0.20 degrad.	34
g TEFLON	CLEAN ROOM PRODUCTS	>1440m	0	0	0.05	4
g TEFLON	CLEAN ROOM PRODUCTS	137m			0.05 crumpled	4
g VITON	NORTH F-091	>1440m	0	0	0.25	4
g VITON	NORTH F-091	>480m	0	0	0.26	34
1,1-dichloroethane (synonym: ethylidene chloride)						
CAS Number: 75-34-3						
Primary Class: 261	Halogen Compounds, Aliphatic and Alicyclic					
g BUTYL	NORTH B-174	91m	1860	4	0.66 degrad.	34

G A R M E Class & Number/ N test chemical/ T MATERIAL NAME	MANUFACTURER & PRODUCT IDENTIFICATION	Break- through Time in Minutes	Perm- eation Rate in mg/sq m /min.	I N D E X	Thick- ness in mm	degrad. and comments	R e f s
g NATURAL RUBBER	ACKWELL 5-109				0.15	degrad.	34
g NEOPRENE	ANSELLEDMONT 29-870				0.48	degrad.	34
g NITRILE	ANSELLEDMONT 37-155				0.40	degrad.	34
g PVAL	EDMONT 25-950	160m			0.58		34
g PVC	ANSELLEDMONT 34-100	1m	22320	5	0.20	degrad.	34
g VITON	NORTH F-091	145m	348	3	0.30		34

1,1-dimethylhydrazine (synonym: unsymmetrical hydrazine)
CAS Number: 57-14-7
Primary Class: 280   Hydrazines

g BUTYL	NORTH B-174	>90m			0.43		21
u BUTYL	Unknown	1380m	29	2	0.84		3
m CHLOROBUTYL	ARROWHEAD PRODUCTS	>90m			0.51		21
s CPE	ILC DOVER CLOROPEL	30m			n.a.		21
g NEOPRENE	ANSELLEDMONT 29-870	38m			0.46		21
u NEOPRENE	Unknown	25m	4500	4	0.76		3
g NITRILE	ANSELLEDMONT 37-155	9m			0.38		21
g PVAL	EDMONT 15-552	12m			n.a.		21
g PVC	ANSELLEDMONT SNORKE	35m			n.a.		21
m PVC	RICH INDUSTRIES	5m			0.38		21
u PVC	Unknown	5m	1900	5	0.43		3
s SARANEX-23	DU PONT TYVEK SARAN	12m	60	4	n.a.		8
s UNKNOWN MATERIAL	DU PONT BARRICADE	>480m	0	0	n.a.		8
s UNKNOWN MATERIAL	LIFE-GUARD RESPONDE	>480m	0	0	n.a.		77
g VITON	NORTH F-121	12m			0.30		21

1,2,3-trichloropropane
CAS Number: 96-18-4
Primary Class: 261   Halogen Compounds, Aliphatic and Alicyclic

g BUTYL	NORTH B-174	>480m	0	0	0.61		34
g NITRILE	ANSELLEDMONT 37-155	20m	120	3	0.38	degrad.	34
g PVAL	EDMONT 25-950	>480m	0	0	0.33		34
g VITON	NORTH F-091	>480m	0	0	0.25		34

1,2,4,5-tetrachlorobenzene
CAS Number: 95-94-3
Primary Class: 264   Halogen Compounds, Aromatic

g NITRILE	ANSELLEDMONT 37-155	>480m	0	0	0.38		34
g PVAL	EDMONT 25-545	>480m	0	0	0.40		34

1,2,4-trichlorobenzene
CAS Number: 120-82-1
Primary Class: 264   Halogen Compounds, Aromatic

g BUTYL	BEST 878				0.75	degrad.	53
g BUTYL	NORTH B-161	5m			0.38		24

G A R M E N T	Class & Number/ test chemical/ MATERIAL NAME	MANUFACTURER & PRODUCT IDENTIFICATION	Break- through Time in Minutes	Perm- eation Rate in mg/sq m /min.	I N D E X	Thick- ness in mm	degrad. and comments	R e f s
g	NATURAL RUBBER	BEST 65NFW				n.a.	degrad.	53
u	NATURAL RUBBER	Unknown	5m			0.23		24
g	NEOPRENE	ANSELLEDMONT 29-840	60m			0.43		24
g	NEOPRENE	BEST 32				n.a.	degrad.	53
g	NEOPRENE	BEST 6780				n.a.	degrad.	53
g	NITRILE	BEST 22R				n.a.	degrad.	53
g	NITRILE	SURETY	240m	0		0.30		24
g	PE	ANSELLEDMONT 35-125	10m			0.03		24
c	PE	DU PONT TYVEK QC	<1m	80	4	n.a.		8
g	PVAL	EDMONT 15-552	60m			n.a.		24
s	SARANEX-23	DU PONT TYVEK SARAN	115m	9	1	n.a.		8
s	SARANEX-23	DU PONT TYVEK SARAN	60m			n.a.		24
g	TEFLON	CLEAN ROOM PRODUCTS	>1440m	0	0	0.05		24
c	UNKNOWN MATERIAL	CHEMRON CHEMTUFF	48m	35	3	n.a.		67
s	UNKNOWN MATERIAL	DU PONT BARRICADE	>480m	0	0	n.a.		8
g	VITON	NORTH F-091	10m			0.23		24
g	VITON/NITRILE	NORTH VITRILE	240m		0	0.20		24

1,2,4-trimethylbenzene   (synonym: pseudocumene)
CAS Number: 526-73-8
Primary Class: 292   Hydrocarbons, Aromatic

s	UNKNOWN MATERIAL	LIFE-GUARD RESPONDE	>40m	0	0	n.a.		77

1,2-dichlorobenzene   (synonym: o-dichlorobenzene)
CAS Number: 95-50-1
Primary Class: 264   Halogen Compounds, Aromatic

g	BUTYL	BEST 878				0.75	degrad.	53
s	BUTYL	TRELLEBORG TRELLCHE				n.a.	degrad.	71
u	CPE	ILC DOVER	39m			0.05		29
g	NATURAL RUBBER	BEST 65NFW				n.a.	degrad.	53
g	NEOPRENE	BEST 6780				n.a.	degrad.	53
g	NITRILE	BEST 727				0.38	degrad.	53
g	NITRILE	GRANET N11F	20m	10133	5	n.a.		39
g	NITRILE	MAPA-PIONEER A-15	37m	11383	5	0.35		39
g	PE/EVAL/PE	SAFETY4 4H	>240m	0	0	0.07	35°C	60
g	PVC	BEST 725R				n.a.	degrad.	53
g	VITON	NORTH F-121	>240m	0	0	0.31		39
u	VITON/CHLOROBUTYL	Unknown	>180m			0.36		29

1,2-dichloroethyl ether
CAS Number: 623-46-1
Primary Class: 241   Ethers, Aliphatic and Alicyclic

s	TEFLON	CHEMFAB CHALL. 5100	>180m	0	1	n.a.		65

G A R M E Class & Number/ N test chemical/ T MATERIAL NAME	MANUFACTURER & PRODUCT IDENTIFICATION	Break- through Time in Minutes	Perm- eation Rate in mg/sq m /min.	I N D E X	Thick- ness in mm	degrad. and comments	R e f s

1,2-dichloropropane (synonym: propylene dichloride)
CAS Number: 78-87-5
Primary Class: 261   Halogen Compounds, Aliphatic and Alicyclic

g BUTYL	NORTH B-174	185m	1134	4	0.66	degrad.	34
u CPE	ILC DOVER	36m			n.a.		29
g NATURAL RUBBER	ACKWELL 5-109				0.15	degrad.	34
g NEOPRENE	ANSELLEDMONT 29-870				0.48	degrad.	34
g NITRILE	ANSELLEDMONT 37-155				0.40	degrad.	34
g PVAL	EDMONT 25-950	>480m	0	0	0.66		34
g PVC	ANSELLEDMONT 34-100	2m	68580	5	0.20	degrad.	34
s TEFLON	CHEMFAB CHALL. 5100	>186m	0	1	n.a.		65
g VITON	NORTH F-091	>480m	0	0	0.25		34
u VITON/CHLOROBUTYL	Unknown	>180m			n.a.		29

1,2-epoxybutane (synonym: 1,2-butylene oxide)
CAS Number: 106-88-7
Primary Class: 275   Heterocyclic Compounds, Oxygen, Epoxides

g BUTYL	NORTH B-174	200m	422	3	0.58	degrad.	34
g NATURAL RUBBER	ACKWELL 5-109	<1m	19680	5	0.15	degrad.	34
g NEOPRENE	ANSELLEDMONT 29-870				0.48	degrad.	34
g NITRILE	ANSELLEDMONT 37-155				0.40	degrad.	34
g PVAL	EDMONT 25-950	4m	9	3	0.33		34
g PVC	ANSELLEDMONT 34-100				0.20	degrad.	34
s UNKNOWN MATERIAL	LIFE-GUARD RESPONDE	>240m		0	n.a.		77
g VITON	NORTH F-091	<1m	107500	5	0.30	degrad.	34
s VITON/CHLOROBUTYL	LIFE-GUARD VC-100	14m			n.a.		77

1,2-propylene oxide (synonym: 1,2-epoxypropane)
CAS Number: 75-56-9
Primary Class: 275   Heterocyclic Compounds, Oxygen, Epoxides

g BUTYL	NORTH B-174	68m	3384	4	0.61		34
g NATURAL RUBBER	ACKWELL 5-109	<1m	19700	5	0.20	degrad.	34
g NATURAL RUBBER	ANSELLEDMONT 36-124				0.46	degrad.	6
g NATURAL RUBBER	ANSELLEDMONT 392				0.48	degrad.	6
g NEOPRENE	ANSELLEDMONT 29-840				0.38	degrad.	6
g NEOPRENE	ANSELLEDMONT 29-870				0.48	degrad.	34
g NEOPRENE/NATURAL	ANSELL CHEMI-PRO				n.a.	degrad.	55
g NITRILE	ANSELLEDMONT 37-155				0.40	degrad.	34
g NITRILE	ANSELLEDMONT 37-165				0.60	degrad.	6
c PE	DU PONT TYVEK QC	10m	15360	5	n.a.		8
g PE/EVAL/PE	SAFETY4 4H	>240m	0	0	0.07		60
g PE/EVAL/PE	SAFETY4 4H	26m	30	3	0.07	H₂O/35°C	60
g PVAL	ANSELLEDMONT PVA	35m	<900	3	n.a.		6

G A R M E N T Class & Number/ test chemical/ MATERIAL NAME	MANUFACTURER & PRODUCT IDENTIFICATION	Break- through Time in Minutes	Perm- eation Rate in mg/sq m /min.	I N D E X	Thick- ness in mm	degrad. and comments	R e f s
g PVAL	EDMONT 15-552	35m	<900	3	n.a.		6
g PVAL	EDMONT 25-545	>360m	0	0	n.a.		6
g PVAL	EDMONT 25-950	10m	18	4	0.33		34
g PVC	ANSELLEDMONT 34-100				0.20	degrad.	34
g PVC	ANSELLEDMONT SNORKE				n.a.	degrad.	6
s SARANEX-23	DU PONT TYVEK SARAN	10m	3000	5	n.a.		8
s TEFLON	CHEMFAB CHALL. 5100	138m	0.24	1	n.a.		65
s UNKNOWN MATERIAL	DU PONT BARRICADE	>480m	0	0	n.a.		8
s UNKNOWN MATERIAL	LIFE-GUARD RESPONDE	>180m	0	1	n.a.		77
g VITON	NORTH F-091	2m	180	4	0.30	degrad.	34

1,3-butadiene
CAS Number: 106-99-0
Primary Class: 294   Hydrocarbons, Aliphatic and Alicyclic, Unsaturated

g BUTYL	BRUNSWICK BUTYL STD	75m	1.1	1	0.73		89
g BUTYL	BRUNSWICK BUTYL-XTR	>480m	0	0	0.63		89
g BUTYL	NORTH B-174	>480m	0	0	0.67		34
s BUTYL	WHEELER ACID KING	7m	<90	4	n.a.		85
g BUTYL/ECO	BRUNSWICK BUTYL-POL	107m	>5000	4	0.63		89
g NATURAL RUBBER	ACKWELL 5-109	1m	6360	5	0.15		34
g NEOPRENE	ANSELLEDMONT 29-870	47m	18	3	0.51		34
g NEOPRENE	BRUNSWICK NEOPRENE	71m	>500	3	0.90		89
g PVC	ANSELLEDMONT 34-100	1m	1260	5	0.23		34
s PVC	WHEELER ACID KING	13m	<900	4	n.a.		85
s SARANEX-23	DU PONT TYVEK SARAN	>480m	0	0	n.a.		8
s TEFLON	LIFE-GUARD TEFGUARD	49m	<0.033	2	n.a.		77
s TEFLON	WHEELER ACID KING	180m	<9	1	n.a.		85
s UNKNOWN MATERIAL	DU PONT BARRICADE	>480m	0	0	n.a.		8
s UNKNOWN MATERIAL	LIFE-GUARD RESPONDE	>480m	0	0	n.a.		77
g VITON	NORTH F-091	>480m	0	0	0.31		34
s VITON/BUTYL	DRAGER 500 OR 710	>240m	0		n.a.		91
s VITON/CHLOROBUTYL	LIFE-GUARD VC-100	402m	0		n.a.		77

1,3-dichloro-2-butene
CAS Number: 926-57-8
Primary Class: 262   Halogen Compounds, Allylic and Benzylic

g PE/EVAL/PE	SAFETY4 4H	>240m	0	0	0.07	35°C	60

1,3-dichlorobenzene (synonym: m-dichlorobenzene)
CAS Number: 541-73-1
Primary Class: 264   Halogen Compounds, Aromatic

s BUTYL	TRELLEBORG TRELLCHE				n.a.	degrad.	71
g NITRILE	GRANET N11F	17m	11283	5	n.a.		39
g NITRILE	MAPA-PIONEER A-14	73m	1740	4	0.56	degrad.	36
g NITRILE	MAPA-PIONEER A-15	30m	11550	5	0.35		39

G A R M E Class & Number/ N test chemical/ T MATERIAL NAME	MANUFACTURER & PRODUCT IDENTIFICATION	Break- through Time in Minutes	Perm- eation Rate in mg/sq m /min.	I N D E X	Thick- ness in mm	degrad. and comments	R e f s
g VITON	NORTH F-121	>240m	0	0	0.31		39

1,3-dichloropropene
CAS Number: 142-28-9
Primary Class: 262   Halogen Compounds, Allylic and Benzylic

g BUTYL	NORTH B-174	77m	1938	4	0.67	degrad.	34
g NATURAL RUBBER	ACKWELL 5-109				0.15	degrad.	34
g NEOPRENE	ANSELLEDMONT 29-870				0.48	degrad.	34
g NITRILE	ANSELLEDMONT 37-155				0.40	degrad.	34
g PVAL	EDMONT 25-950	>480m	0	0	0.56		34
g PVC	ANSELLEDMONT 34-100	1m	39000	5	0.20	degrad.	34
s TEFLON	CHEMFAB CHALL. 5100	>180m	0	1	n.a.		65
s UNKNOWN MATERIAL	KAPPLER CPF III	4m	1030	5	n.a.		77
g VITON	NORTH F-091	>480m	0	0	0.25		34

1,3-dimethylbutylamine
CAS Number: 108-09-8
Primary Class: 141   Amines, Aliphatic and Alicyclic, Primary

g BUTYL	NORTH B-174	101m	1920	4	0.63	degrad.	34
g NATURAL RUBBER	ACKWELL 5-109				0.15	degrad.	34
g NEOPRENE	ANSELLEDMONT 29-870				0.48	degrad.	34
g NITRILE	ANSELLEDMONT 37-155	80m	4260	4	0.40	degrad.	34
g PVAL	EDMONT 25-950	20m	840	3	0.76		34
g PVC	ANSELLEDMONT 34-100	2m	15420	5	0.20		34
g VITON	NORTH F-091				0.26	degrad.	34

1,3-dioxane
CAS Number: 505-22-6
Primary Class: 278   Heterocyclic Compounds, Oxygen, Others

g BUTYL	NORTH B-174	>480m	0	0	0.61		34
g NATURAL RUBBER	ACKWELL 5-109	<1m	1800	5	0.15	degrad.	34
g NEOPRENE	ANSELLEDMONT 29-870				0.48	degrad.	34
g NITRILE	ANSELLEDMONT 37-155				0.40	degrad.	34
g PVAL	EDMONT 25-950	22m	<1	2	0.30	degrad.	34
g PVC	ANSELLEDMONT 34-100				0.20	degrad.	34
g UNKNOWN MATERIAL	NORTH SILVERSHIELD	>480m	0	0	0.08		7
g VITON	NORTH F-091	32m	2160	4	0.33	degrad.	34

1,3-diphenylguanidine
CAS Number: 102-06-7
Primary Class: 137   Amides, Carbamides, and Guanidines

g NATURAL RUBBER	ANSELL CANNERS 334	>480m	0	0	n.a.		34
g NEOPRENE	ANSELLEDMONT 29-870	>480m	0	0	0.46		34

G A R M E Class & Number/ N test chemical/ T MATERIAL NAME	MANUFACTURER & PRODUCT IDENTIFICATION	Break- through Time in Minutes	Perm- eation Rate in mg/sq m /min.	I N D E X	Thick- ness in mm	degrad. and comments	R e f s

1,3-propanediamine (synonym: 1,3-diaminopropane)
CAS Number: 109-76-2
Primary Class: 148   Amines, Poly, Aliphatic and Alicyclic
Related Class: 141   Amines, Aliphatic and Alicyclic, Primary

g BUTYL	NORTH B-174	>480m	0	0	0.63		34
g NATURAL RUBBER	ACKWELL 5-109	3m	2640	5	0.15		34
g NEOPRENE	ANSELLEDMONT 29-870	>270m	200	3	0.46		34
g NITRILE	ANSELLEDMONT 37-155				0.40	degrad.	34
g PVAL	EDMONT 25-950				0.40	degrad.	34
g PVC	ANSELLEDMONT 34-100	6m	606	4	0.20		34
g VITON	NORTH F-091				0.26	degrad.	34

1,4-butanedioldiglycidyl ether
CAS Number: 2425-79-8
Primary Class: 275   Heterocyclic Compounds, Oxygen, Epoxides

g PE/EVAL/PE	SAFETY4 4H	>240m	0	0	0.07	35°C	60

1,4-butanediolglycidyl ether, 50% & methyl ethyl ketone, 50%
Primary Class: 275   Heterocyclic Compounds, Oxygen, Epoxides
Related Class: 810   Epoxy Products

1,4-butyleneglycol (synonym: 1,4-butanediol)
CAS Number: 110-63-4
Primary Class: 314   Hydroxy Compounds, Aliphatic and Alicyclic, Polyols

g BUTYL	NORTH B-161	>480m	0	0	0.43		34
g NITRILE	ANSELLEDMONT 37-155	>480m	0	0	0.38		34

1,4-dichloro-2-butene (synonym: 2-butylene dichloride)
CAS Number: 764-41-0
Primary Class: 262   Halogen Compounds, Allylic and Benzylic

u BUTYL	Unknown	>1440m	0	0	0.68		27
u CPE	ILC DOVER	35m	4000	4	0.50		27
u CPE	ILC DOVER	45m			n.a.		29
g NEOPRENE	ANSELLEDMONT 29-865	29m	1260	4	0.51		27
u NEOPRENE	Unknown	34m	1180	4	n.a.		26
u NEOPRENE	Unknown	74m	1210	4	1.30		27
u NEOPRENE	Unknown	34m	438	3	n.a.		27
u NEOPRENE	Unknown	54m	400	3	n.a.	pre- exposed	27
u NEOPRENE	Unknown	96m	1210	4	1.40		28

G A R M E N T Class & Number/ test chemical/ MATERIAL NAME	MANUFACTURER & PRODUCT IDENTIFICATION	Break- through Time in Minutes	Perm- eation Rate in mg/sq m /min.	I N D E X	Thick- ness in mm	degrad. and comments	R e f s
u NEOPRENE	Unknown	210m	400	3	1.80		28
u NEOPRENE	Unknown	18m	380	3	0.44		28
g NITRILE	ANSELLEDMONT 37-175	27m	1560	4	0.41		27
u NITRILE	Unknown	2m	3300	5	n.a.		27
u NITRILE	Unknown	26m	1560	4	0.35		28
c PE	DU PONT TYVEK QC	75m	2500	4	n.a.		8
c PE	DU PONT TYVEK QC	>1440m	0	0	0.10		27
c PE	DU PONT TYVEK QC	2m	330	4	n.a.		27
g PVAL	EDMONT 15-554	>4980m	0	0	0.87		27
u PVC	Unknown	31m	1080	4	n.a.		26
u PVC	Unknown	6m	3800	5	0.50		27
u PVC	Unknown	142m	1440	4	2.00		27
u PVC	Unknown	5m	2500	5	n.a.		27
u PVC	Unknown	1m	4300	5	0.24		27
u PVC	Unknown	29m	840	3	n.a.		27
u PVC	Unknown	2m	638	4	n.a.pre- exposed		27
u PVC	Unknown	162m	1220	4	2.00		28
u PVC	Unknown	35m	3100	4	n.a.		28
s SARANEX-23	DU PONT TYVEK SARAN	>1440m	0	0	n.a.		27
s UNKNOWN MATERIAL	LIFE-GUARD RESPONDE	>480m	0	0	n.a.		77
g VITON	NORTH F-091	>498m	0	0	0.30		28
u VITON	Unknown	>4320m	0	0	0.26		27
u VITON/CHLOROBUTYL	Unknown	>180m			n.a.		29

1,4-dichlorobenzene (synonym: p-dichlorobenzene)
CAS Number: 106-46-7
Primary Class: 264   Halogen Compounds, Aromatic

g NEOPRENE	ANSELLEDMONT NEOX	160m	30	2	n.a.		76
g NITRILE	MAPA-PIONEER A-14	195m	180	3	0.50		76
g PVC	BEST 7712R	45m	360	3	n.a.		76

1,4-dioxane
CAS Number: 123-91-1
Primary Class: 278   Heterocyclic Compounds, Oxygen, Others

g BUTYL	BEST 878	>480m	0	0	0.75		53
g BUTYL	BRUNSWICK BUTYL STD	>480m	0	0	0.63		89
g BUTYL	BRUNSWICK BUTYL-XTR	>480m	0	0	0.63		89
g BUTYL	COMASEC BUTYL	>480m	0	0	0.50		40
g BUTYL	NORTH B-161	457m	390	3	0.42		63
g BUTYL	NORTH B-174	>1200m	0	0	0.71		34
g BUTYL/ECO	BRUNSWICK BUTYL-POL	>480m	0	0	0.63		89
g NATURAL RUBBER	ACKWELL 5-109				0.15	degrad.	34
g NATURAL RUBBER	ANSELL FL 200 254	8m	3	3	0.51		55
g NATURAL RUBBER	ANSELL ORANGE 208	16m	2	2	0.76		55
g NATURAL RUBBER	ANSELL PVL 040	13m	2	3	0.46		55

G A R M E N T			Perm-	I			R
		Break-	eation	N	Thick-		R
E Class & Number/	MANUFACTURER	through	Rate in	D	ness	degrad.	e
N test chemical/	& PRODUCT	Time in	mg/sq m	E	in	and	f
T MATERIAL NAME	IDENTIFICATION	Minutes	/min.	X	mm	comments	s
g NATURAL RUBBER	ANSELLEDMONT 36-124	15m	<900	3	0.46		6
g NATURAL RUBBER	ANSELLEDMONT 392	5m	<9000	5	0.48		6
g NATURAL RUBBER	ANSELLEDMONT 46-320	9m	4200	5	0.31		5
g NATURAL RUBBER	MARIGOLD BLACK HEVY	8m	3648	5	0.65	degrad.	79
g NATURAL RUBBER	MARIGOLD FEATHERLIT	1m	6305	5	0.15	degrad.	79
g NATURAL RUBBER	MARIGOLD MED WT EXT	5m	4133	5	0.45	degrad.	79
g NATURAL RUBBER	MARIGOLD MEDICAL	1m	5400	5	0.28	degrad.	79
g NATURAL RUBBER	MARIGOLD ORANGE SUP	9m	3600	5	0.73	degrad.	79
g NATURAL RUBBER	MARIGOLD R SURGEONS	1m	5400	5	0.28	degrad.	79
g NATURAL RUBBER	MARIGOLD RED LT WT	4m	4314	5	0.43	degrad.	79
g NATURAL RUBBER	MARIGOLD SENSOTECH	1m	5400	5	0.28	degrad.	79
g NATURAL RUBBER	MARIGOLD SUREGRIP	7m	3794	5	0.58	degrad.	79
g NATURAL RUBBER	MARIGOLD YELLOW	3m	4676	5	0.38	degrad.	79
g NATURAL+NEOPRENE	ANSELL OMNI 276	17m	3400	4	0.45		5
g NATURAL+NEOPRENE	ANSELL OMNI 276	11m	1	3	0.56		55
g NATURAL+NEOPRENE	ANSELL TECHNICIANS	6m	6	3	0.43		55
g NATURAL+NEOPRENE	MARIGOLD FEATHERWT	3m	4857	5	0.35	degrad.	79
g NEOPRENE	ANSELLEDMONT 29-840				0.38	degrad.	6
g NEOPRENE	ANSELLEDMONT 29-870	16m	5600	4	0.46	degrad.	34
g NEOPRENE	BEST 32	62m	1120	4	n.a.		53
g NEOPRENE	BEST 6780	63m	1120	4	n.a.		53
g NEOPRENE	MAPA-PIONEER N-73	28m	3700	4	0.46		5
g NITRILE	ANSELLEDMONT 37-165	63m	30	2	0.60		5
g NITRILE	ANSELLEDMONT 37-165				0.60	degrad.	6
g NITRILE	ANSELLEDMONT 37-175	27m	8200	4	0.37		5
g NITRILE	BEST 22R		20640	5	n.a.	degrad.	53
g NITRILE	BEST 727	4m	1670	5	0.38	degrad.	53
g NITRILE	MARIGOLD NITROSOLVE				0.75	degrad.	79
g NITRILE	NORTH LA-142G	24m	4626	4	0.36	degrad.	34
g NITRILE+PVC	COMASEC MULTIMAX	83m	120	3	n.a.		40
g PE	ANSELLEDMONT 35-125	1m	3000	5	0.03		5
g PE/EVAL/PE	SAFETY4 4H	>240m	0	0	0.07	35°C	60
g PVAL	ANSELLEDMONT PVA				n.a.	degrad.	6
g PVAL	EDMONT 15-552				n.a.	degrad.	6
g PVAL	EDMONT 25-950	>960m	0	0	0.30		34
g PVC	ANSELLEDMONT 34-100				0.20	degrad.	34
g PVC	ANSELLEDMONT SNORKE				n.a.	degrad.	6
g PVC	BEST 812		4020	5	n.a.		53
g PVC	MAPA-PIONEER V-20	6m	15000	5	0.31		5
s SARANEX-23	DU PONT TYVEK SARAN	50m	174	3	n.a.		8
s TEFLON	CHEMFAB CHALL. 5100	>180m	0	1	n.a.		65
s UNKNOWN MATERIAL	CHEMRON CHEMREL	>1440m	0	0	n.a.		67
s UNKNOWN MATERIAL	CHEMRON CHEMREL MAX	>1440m	0	0	n.a.		67
c UNKNOWN MATERIAL	CHEMRON CHEMTUFF	250m	1	1	n.a.		67
g UNKNOWN MATERIAL	NORTH SILVERSHIELD	>8m			0.10		7
g VITON	NORTH F-091	23m	1600	4	0.25		34
g VITON	NORTH F-091	31m	3070	4	0.26		38
g VITON	NORTH F-091	105m			0.30		62
g VITON/NEOPRENE	ERISTA VITRIC	53m	1110	4	0.60		38

G A R M E Class & Number/ N test chemical/ T MATERIAL NAME	MANUFACTURER & PRODUCT IDENTIFICATION	Break-through Time in Minutes	Perm-eation Rate in mg/sq m /min.	I N D E X	Thick-ness in mm	degrad. and comments	R e f s

1,6-hexanediamine (synonym: 1,6-diaminohexane)
CAS Number: 124-09-4
Primary Class: 148   Amines, Poly, Aliphatic and Alicyclic

g BUTYL	NORTH B-161	>480m	0	0	0.38		34
s BUTYL/NEOPRENE	MSA BETEX	>240m	0	0	n.a.		48
g NATURAL RUBBER	ACKWELL 5-109	5m	2580	5	0.12		34
g NEOPRENE	ANSELLEDMONT 29-870	>480m	0	0	0.46		34
g NITRILE	ANSELLEDMONT 37-155	453m	180	3	0.46		34
s VITON/NEOPRENE	MSA VAUTEX	>240m	0	0	n.a.		48

1,6-hexanediamine, 30-70%
Primary Class: 148   Amines, Poly, Aliphatic and Alicyclic

g NATURAL RUBBER	ACKWELL 5-109	168m	42	2	0.15	in H_2O	34
g NITRILE	ANSELLEDMONT 37-155	>480m	0	0	0.40	in H_2O	34
g UNKNOWN MATERIAL	NORTH SILVERSHIELD	>480m	0	0	0.10	in H_2O	34

1,6-hexanediamine, <30%
Primary Class: 148   Amines, Poly, Aliphatic and Alicyclic

g NATURAL RUBBER	ACKWELL 5-109	>480m	0	0	0.15	1% in H_2O	34
g NITRILE	ANSELLEDMONT 37-155	>480m	0	0	0.42	1% in H_2O	34
g UNKNOWN MATERIAL	NORTH SILVERSHIELD	>480m	0	0	0.10	1% in H_2O	34

1,6-hexanediol diacrylate (synonym: HDDA)
CAS Number: 13048-33-4
Primary Class: 223   Esters, Carboxylic, Acrylates and Methacrylates

g BUTYL	NORTH B-161	>480m	0	0	0.46		94
g NATURAL RUBBER	MAPA-PIONEER L-118	30-60m	9	4	0.45		94
g NITRILE	ANSELLEDMONT 37-155	>480m	0	0	0.37		94

1,6-hexanediol diacrylate, 50% & 2-ethylhexyl acrylate, 50%
Primary Class: 223   Esters, Carboxylic, Acrylates and Methacrylates

g BUTYL	NORTH B-161	>480m	0	0	0.46		94
g NATURAL RUBBER	MAPA-PIONEER L-118	30-60m	117	5	0.45		94
g NITRILE	ANSELLEDMONT 37-155	>480m	0	0	0.37		94

1-(2-aminoethyl)piperazine (synonym: N-(beta-aminoethyl)piperazine)
CAS Number: 140-31-8
Primary Class: 148   Amines, Poly, Aliphatic and Alicyclic
Related Class: 141   Amines, Aliphatic and Alicyclic, Primary
            274   Heterocyclic Compounds, Nitrogen, Others

G A R M E Class & Number/ N test chemical/ T MATERIAL NAME	MANUFACTURER & PRODUCT IDENTIFICATION	Break- through Time in Minutes	Perm- eation Rate in mg/sq m /min.	I N D E X	Thick- ness in mm	degrad. and comments	R e f s
g BUTYL	NORTH B-161	>240m	0	0	0.45		33
g NITRILE	ERISTA SPECIAL				0.42	degrad.	33

1-bromo-2-propanol
CAS Number: 19686-73-8
Primary Class: 315   Hydroxy Compounds, Aliphatic and Alicyclic, Substituted
Related Class: 261   Halogen Compounds, Aliphatic and Alicyclic

g BUTYL	NORTH B-174	>480m	0	0	0.63		34
g NATURAL RUBBER	ACKWELL 5-109	1m	456	4	0.20		34
g NITRILE	ANSELLEDMONT 37-155				0.40	degrad.	34
g PVAL	EDMONT 15-552	>480m	0	0	0.23		34
g VITON	NORTH F-091	>480m	0	0	0.23		34

1-bromoethyl ethyl carbonate
CAS Number: 89766-09-6
Primary Class: 232   Esters, Non-Carboxylic, Carbonates

g PE/EVAL/PE	SAFETY4 4H	>240m		0	0.07	35°C	60

1-chloro-2-propanol
CAS Number: 127-00-4
Primary Class: 315   Hydroxy Compounds, Aliphatic and Alicyclic, Substituted
Related Class: 261   Halogen Compounds, Aliphatic and Alicyclic

g BUTYL	NORTH B-174	>480m	0	0	0.63		34
g NATURAL RUBBER	ACKWELL 5-109	<1m	258	4	0.13		34
g NITRILE	ANSELLEDMONT 37-155				0.40	degrad.	34
g NITRILE	ANSELLEDMONT 37-175	201m	770	3	0.64		34
g PVAL	EDMONT 25-950	>480m	0	0	0.91		34
g PVC	ANSELLEDMONT 34-100	1m	2304	5	0.15		34
g VITON	NORTH F-091	>480m	0	0	0.25		34

1-chloronaphthalene (synonym: naphthyl chloride)
CAS Number: 90-13-1
Primary Class: 264   Halogen Compounds, Aromatic

g BUTYL	NORTH B-161				0.43	degrad.	7
g NATURAL RUBBER	ANSELLEDMONT 36-124				0.46	degrad.	6
g NATURAL RUBBER	ANSELLEDMONT 392				0.48	degrad.	6
g NEOPRENE	ANSELLEDMONT 29-840				0.46	degrad.	6
g NITRILE	ANSELLEDMONT 37-165				0.63	degrad.	6
g NITRILE	NORTH LA-142G	173m	79	2	0.36		7
g PVAL	ANSELLEDMONT PVA	>360m		0	n.a.		6
g PVAL	EDMONT 15-552	>360m	0	0	n.a.		6
g PVAL	EDMONT 25-545	90m	<9	1	n.a.		6

G A R M E Class & Number/ N test chemical/ T MATERIAL NAME	MANUFACTURER & PRODUCT IDENTIFICATION	Break-through Time in Minutes	Perm-eation Rate in mg/sq m /min.	I N D E X	Thick-ness in mm	degrad. and comments	R e f s
g PVC	ANSELLEDMONT SNORKE				n.a.	degrad.	6
g UNKNOWN MATERIAL	NORTH SILVERSHIELD	>480m	0	0	0.10		7
g VITON	NORTH F-091	>960m	0	0	0.25		7

1-ethoxy-2-propanol (synonym: ethyl 1-isopropanol ether)
CAS Number: 1569-02-4
Primary Class: 245   Ethers, Glycols
Related Class: 241   Ethers, Aliphatic and Alicyclic
              311   Hydroxy Compounds, Aliphatic and Alicyclic, Primary

g NITRILE	ANSELLEDMONT 37-155				0.40	degrad.	34

1-methoxy-2-propanol (synonym: propylene glycol monomethyl ether)
CAS Number: 107-98-2
Primary Class: 245   Ethers, Glycols

g BUTYL	NORTH B-161	>480m	0	0	0.40		68
g NEOPRENE	ANSELLEDMONT NEOX	>480m	0	0	n.a.		68
g NITRILE	MAPA-PIONEER AF-18	108m	354	3	0.56		68
g PE/EVAL/PE	SAFETY4	4H	>240m	0	0.07	35°C	60
g PVC	ANSELLEDMONT SNORKE	94m	14	2	n.a.		68

1-methoxy-2-propyl acetate
CAS Number: 41448-83-3
Primary Class: 245   Ethers, Glycols
Related Class: 222   Esters, Carboxylic, Acetates
              241   Ethers, Aliphatic and Alicyclic

g PE/EVAL/PE	SAFETY4 4H	>240m	0	0	0.07	35°C	60

1-nitropropane
CAS Number: 108-03-2
Primary Class: 441   Nitro Compounds, Unsubstituted

g BUTYL	NORTH B-161	>480m	0	0	0.43		34
g NATURAL RUBBER	ANSELLEDMONT 36-124	5m	<900	4	0.46		6
g NATURAL RUBBER	ANSELLEDMONT 392	5m	<900	4	0.48		6
g NEOPRENE	ANSELLEDMONT 29-840	5m	<9000	5	0.38		6
g NEOPRENE	ANSELLEDMONT 29-870				0.48	degrad.	34
g NEOPRENE	ANSELLEDMONT NEOX	60m	<900	3	n.a.		6
g NITRILE	ANSELLEDMONT 37-165				0.60	degrad.	6
g NITRILE	NORTH LA-142G	12m	1770	5	0.38	degrad.	34
g PVAL	ANSELLEDMONT PVA	>360m		0	n.a.		6
g PVAL	EDMONT 15-552	>360m	0	0	n.a.		6
g PVAL	EDMONT 15-552	>900m	0	0	0.28		34
g PVAL	EDMONT 25-545	>360m	0	0	n.a.		6

G A R M E Class & Number/ N test chemical/ T MATERIAL NAME	MANUFACTURER & PRODUCT IDENTIFICATION	Break- through Time in Minutes	Perm- eation Rate in mg/sq m /min.	I N D E X	Thick- ness in mm	degrad. and comments	R e f s
g PVC	ANSELLEDMONT 34-100				0.20	degrad.	34
g PVC	ANSELLEDMONT SNORKE				n.a.	degrad.	6
g UNKNOWN MATERIAL	NORTH SILVERSHIELD	>480m	0	0	0.08		7
g VITON	NORTH F-091	17m	1556	4	0.25		34

2',3'-dideoxycytidine
CAS Number: 7481-89-2
Primary Class: 870   Antineoplastic Drugs and Other Pharmaceuticals

g NATURAL RUBBER	ANSELL CANNERS 334	>480m	0	0	n.a.		34
g PVC	ANSELLEDMONT 34-100	>480m	0	0	0.20		34

2',3'-dideoxycytidine in 0.2% methyl cellulose & water
Primary Class: 870   Antineoplastic Drugs and Other Pharmaceuticals

g NATURAL RUBBER	ANSELL CANNERS 334	>480m	0	0	n.a.		34
g PVC	ANSELLEDMONT 34-100	>480m	0	0	0.20		34

2,2'-dichloroethyl ether
CAS Number: 111-44-4
Primary Class: 241   Ethers, Aliphatic and Alicyclic
Related Class: 261   Halogen Compounds, Aliphatic and Alicyclic

u CPE	ILC DOVER	80m			0.05		29
u VITON/CHLOROBUTYL	Unknown	>180m			n.a.		29

2,2,2-trichloroethanol
CAS Number: 115-20-8
Primary Class: 315   Hydroxy Compounds, Aliphatic and Alicyclic, Substituted
Related Class: 261   Halogen Compounds, Aliphatic and Alicyclic

s SARANEX-23	DU PONT TYVEK SARAN	19m	132	3	n.a.		8
s UNKNOWN MATERIAL	DU PONT BARRICADE	>480m	0	0	n.a.		8

2,2,2-trifluoroethanol
CAS Number: 75-89-8
Primary Class: 315   Hydroxy Compounds, Aliphatic and Alicyclic, Substituted
Related Class: 261   Halogen Compounds, Aliphatic and Alicyclic

g NATURAL RUBBER	ANSELLEDMONT 46-320	> 60m			0.31		5
g NATURAL+NEOPRENE	ANSELL OMNI 276	>99m			0.45		5
g NEOPRENE	ANSELLEDMONT NEOX	>60m			n.a.		5
g NEOPRENE	MAPA-PIONEER N-73	>60m			0.46		5
g NITRILE	ANSELLEDMONT 37-175	17m	23000	5	0.37		5
g PE	ANSELLEDMONT 35-125	>60m			0.03		5
c PE	DU PONT TYVEK QC	6m			n.a.		8

G A R M E Class & Number/ N test chemical/ T MATERIAL NAME	MANUFACTURER & PRODUCT IDENTIFICATION	Break- through Time in Minutes	Perm- eation Rate in mg/sq m /min.	I N D E X	Thick- ness in mm	degrad. and comments	R e f s
g PVC	MAPA-PIONEER V-20	15m	13000	5	0.31		5
s UNKNOWN MATERIAL	DU PONT BARRICADE	>480m	0	0	n.a.		8

2,3-dichloro-1-propene
CAS Number: 78-88-6
Primary Class: 262   Halogen Compounds, Allylic and Benzylic

g BUTYL	NORTH B-174	113m	834	3	0.67	degrad.	34
g NATURAL RUBBER	ACKWELL 5-109				0.15	degrad.	34
g NEOPRENE	ANSELLEDMONT 29-870				0.48	degrad.	34
g NITRILE	ANSELLEDMONT 37-155				0.40	degrad.	34
g PVAL	EDMONT 25-950	>480m	0	0	0.55		34
g PVC	ANSELLEDMONT 34-100	1m	31920	5	0.20	degrad.	34
s UNKNOWN MATERIAL	DU PONT BARRICADE	>480m	0	0	n.a.		8
g VITON	NORTH F-091	>480m	0	0	0.25		34

2,4-D dimethylamine ammonium salt (synonym: 2,4-D amine 96)
CAS Number: 2008-39-1
Primary Class: 550   Organic Salts (Solutions)
Related Class: 840   Pesticides, Mixtures and Formulations

g NATURAL RUBBER	ANSELLEDMONT 36-124	>480m	0	0	0.51		74
g NEOPRENE	ANSELLEDMONT 29-865	>480m	0	0	0.51		74
g NITRILE	ANSELLEDMONT 37-175	>480m	0	0	0.46		74
g PVC	EDMONT CANADA 14112	>480m	0	0	n.a.		74

2,4-dinitrotoluene (synonym: 2,4-DNT)
CAS Number: 121-14-2
Primary Class: 441   Nitro Compounds, Unsubstituted

s SARANEX-23	DU PONT TYVEK SARAN	>1440m	0	0	n.a.		8

2,6-dimethylmorpholine
CAS Number: 141-91-3
Primary Class: 142   Amines, Aliphatic and Alicyclic, Secondary
Related Class: 274   Heterocyclic Compounds, Nitrogen, Others

g NATURAL RUBBER	ACKWELL 5-109				0.15	degrad.	34
g NEOPRENE	ANSELLEDMONT 29-870				0.48	degrad.	34
g NITRILE	ANSELLEDMONT 37-155				0.40	degrad.	34

2-(2,4-dichlorophenoxy)propionic acid, 15% & 2,4-D, 8% & water
Primary Class: 103   Acids, Carboxylic, Aliphatic and Alicyclic, Substituted
Related Class: 840   Pesticides, Mixtures and Formulations

g NITRILE	ANSELLEDMONT 37-175	>180m			0.37		13

G A R M E Class & Number/ N test chemical/ T MATERIAL NAME	MANUFACTURER & PRODUCT IDENTIFICATION	Break- through Time in Minutes	Perm- eation Rate in mg/sq m /min.	I N D E X	Thick- ness in mm	degrad. and comments	R e f s
c PE	MOLNLYCKE H D	97m			n.a.		13

2-(2,4-dichlorophenoxy)propionic acid, 40% & MCPA, 20% & water
Primary Class: 103   Acids, Carboxylic, Aliphatic and Alicyclic, Substituted
Related Class: 840   Pesticides, Mixtures and Formulations

g NEOPRENE	ANSELLEDMONT 29-840	>420m	0	0	0.43	35°C	57
g NITRILE	ANSELLEDMONT 37-155	240m	0	0	0.35	degrad. 35°C	57
c PE	MOLNLYCKE H D	<60m	0	1	n.a.	35°C	57
g PE/EVAL/PE	SAFETY44H	>240m	0	0	0.07	35°C	57

2-(2,4-dichlorophenoxy)propionic acid, formulation
CAS Number: 120-36-5
Primary Class: 103   Acids, Carboxylic, Aliphatic and Alicyclic, Substituted
Related Class: 243   Ethers, Alkyl-Aryl
            550   Organic Salts (Solutions)
            840   Pesticides, Mixtures and Formulations

g NITRILE	ANSELLEDMONT 37-145	>120m			0.32		13
g NITRILE	ERISTA SPECIAL	>120m			0.35		13
c PE	MOLNLYCKE H D	>120m			0.15		13
g PVC	KID 490	>120m			n.a.		13

2-(2-aminoethoxy)ethanol (synonym: diglycolamine)
CAS Number: 929-06-0
Primary Class: 141   Amines, Aliphatic and Alicyclic, Primary
Related Class: 241   Ethers, Aliphatic and Alicyclic
            311   Hydroxy Compounds, Aliphatic and Alicyclic, Primary

g PE/EVAL/PE	SAFETY4 4H	>240m	0	0	0.07	35°C	60

2-(2-ethoxyethoxy)ethanol (synonym: diethyleneglycol mono ethyl ether)
CAS Number: 111-90-0
Primary Class: 245   Ethers, Glycols

s SARANEX-23	DU PONT TYVEK SARAN	>480m	0	0	n.a.		8

2-(2-methoxyethoxy)ethanol (synonym: diethylene glycol monomethyl ether)
CAS Number: 111-77-3
Primary Class: 245   Ethers, Glycols

g BUTYL	NORTH B-161	>480m	0	0	0.40		68
g NEOPRENE	ANSELLEDMONT NEOX	>480m	0	0	n.a.		68
g NITRILE	MAPA-PIONEER AF-18	175m	10	2	0.56		68
g PVC	ANSELLEDMONT SNORKE	170m	8	1	n.a.		68

G A R M E N T Class & Number/ test chemical/ MATERIAL NAME	MANUFACTURER & PRODUCT IDENTIFICATION	Break- through Time in Minutes	Perm- eation Rate in mg/sq m /min.	I N D E X	Thick- ness in mm	degrad. and comments	R e f s

2-(4-chloro-2-methylphenoxy)propionic acid, formulation
CAS Number: 93-65-2
Primary Class: 103   Acids, Carboxylic, Aliphatic and Alicyclic, Substituted
Related Class: 243   Ethers, Alkyl-Aryl
            550   Organic Salts (Solutions)
            840   Pesticides, Mixtures and Formulations

g NITRILE	ANSELLEDMONT 37-145	>120m			0.32		13
g NITRILE	ERISTA SPECIAL	>120m			0.35		13
c PE	MOLNLYCKE H D	>120m			0.15		13
g PVC	KID 490	>120m			n.a.		13

2-(diethylamino)ethanol (synonym: N,N-diethylaminoethanol)
CAS Number: 100-37-8
Primary Class: 143   Amines, Aliphatic and Alicyclic, Tertiary
Related Class: 311   Hydroxy Compounds, Aliphatic and Alicyclic, Primary

g BUTYL	NORTH B-174	>480m	0	0	0.69		34
g NATURAL RUBBER	ACKWELL 5-109				0.15	degrad.	34
g NEOPRENE	ANSELLEDMONT 29-870				0.48	degrad.	34
g NITRILE	ANSELLEDMONT 37-155	>480m	0	0	0.40		34
g PVAL	EDMONT 25-950	>480m	0	0	0.71		34
g PVC	ANSELLEDMONT 34-100				0.20	degrad.	34
g VITON	NORTH F-091	>480m	0	0	0.26		34

2-(dimethylamino)ethanol (synonym: N,N-dimethylethanolamine)
CAS Number: 108-01-0
Primary Class: 143   Amines, Aliphatic and Alicyclic, Tertiary
Related Class: 311   Hydroxy Compounds, Aliphatic and Alicyclic, Primary

g BUTYL	NORTH B-174	>480m	0	0	0.63		34
g NATURAL RUBBER	ACKWELL 5-109	5m	608	4	0.15		34
g NEOPRENE	ANSELLEDMONT 29-870	234m	164	3	0.33	degrad.	34
g NITRILE	ANSELLEDMONT 37-155	>480m	0	0	0.32	degrad.	34
g PE/EVAL/PE	SAFETY4 4H	>240m	0	0	0.07	35°C	60
g PVC	ANSELLEDMONT 34-100				0.20	degrad.	34
g VITON	NORTH F-091				0.26	degrad.	34

2-bromoethanol
CAS Number: 540-51-2
Primary Class: 315   Hydroxy Compounds, Aliphatic and Alicyclic, Substituted
Related Class: 261   Halogen Compounds, Aliphatic and Alicyclic

g BUTYL	NORTH B-174	>480m	0	0	0.68		34
g NATURAL RUBBER	ACKWELL 5-109	1m	660	4	0.15		34
g NITRILE	ANSELLEDMONT 37-155				0.40	degrad.	34

G A R M E Class & Number/ N test chemical/ T MATERIAL NAME	MANUFACTURER & PRODUCT IDENTIFICATION	Break- through Time in Minutes	Perm- eation Rate in mg/sq m /min.	I N D E X	Thick- ness in mm	degrad. and comments	R e f s
g PVAL	EDMONT 25-950				0.50	degrad.	34
g PVC	ANSELLEDMONT 34-100	2m	4560	5	0.20		34
g VITON	NORTH F-091	>480m	0	0	0.26		34

2-butanone peroxide
CAS Number: 1338-23-4
Primary Class: 300   Peroxides

g BUTYL	NORTH B-174	>240m	0	0	0.62		34
g NATURAL RUBBER	ACKWELL 5-109	45m	60	3	0.16		34
g NEOPRENE	ANSELLEDMONT 29-870	>240m	0	0	0.48		34
g VITON	NORTH F-091	>240m	0	0	0.40		34

2-butoxyethanol (synonyms: butylcellosolve; butylglycol)
CAS Number: 111-76-2
Primary Class: 245   Ethers, Glycols
Related Class: 241   Ethers, Aliphatic and Alicyclic
              311   Hydroxy Compounds, Aliphatic and Alicyclic, Primary

g BUTYL	NORTH B-131	>1608m	0	0	0.30		51
g BUTYL	NORTH B-161	>480m	0	0	0.40		68
g NAT+NEOPR+NITRILE	MAPA-PIONEER TRIONI	53m	228	3	0.46		36
g NATURAL RUBBER	ANSELLEDMONT 36-124	45m	<900	3	0.46		6
g NATURAL RUBBER	ANSELLEDMONT 392	45m	<900	3	0.48		6
g NATURAL RUBBER	MAPA-PIONEER L-118	12m	1620	5	0.46		36
g NEOPRENE	ANSELLEDMONT 29-840	90m	<90	2	0.38		6
g NEOPRENE	ANSELLEDMONT NEOX	>360m	0	0	n.a.		6
g NEOPRENE	ANSELLEDMONT NEOX	>480m	0	0	n.a.		68
g NEOPRENE	BEST 6780	45m	20020	5	0.80		51
g NEOPRENE	MAPA-PIONEER N-44	150m	300	3	0.54		36
g NITRILE	ANSELLEDMONT 37-155	420m	6820	4	0.50		51
g NITRILE	ANSELLEDMONT 37-165	90m	<90	2	0.60		6
g NITRILE	MAPA-PIONEER A-14	>480m	0	0	0.56		36
g NITRILE	MAPA-PIONEER AF-18	>480m	0	0	0.56		68
g NITRILE	NORTH LA-102G	>240m	0	0	0.28		11
g PE/EVAL/PE	SAFETY4 4H	>240m	0	0	0.07	35°C	60
g PVAL	ANSELLEDMONT PVA	120m	<900	3	n.a.		6
g PVAL	EDMONT 15-552	10m	<900	4	n.a.		6
g PVAL	EDMONT 15-552	>1080m	0	0	n.a.		11
g PVAL	EDMONT 15-552	3m	1173	5	n.a.		51
g PVC	ANSELLEDMONT SNORKE				n.a.	degrad.	6
g PVC	ANSELLEDMONT SNORKE	126m	16	2	n.a.		68
s SARANEX-23	DU PONT TYVEK SARAN	>480m	0	0	n.a.		8

2-butoxyethanol, 50% & naphtha, 50% (<3% aromatics, b.p. 150-200 degrees C)
Primary Class: 245   Ethers, Glycols

G A R M E Class & Number/ N test chemical/ T MATERIAL NAME	MANUFACTURER & PRODUCT IDENTIFICATION	Break- through Time in Minutes	Perm- eation Rate in mg/sq m /min.	I N D E X	Thick- ness in mm	degrad. and comments	R e f s

2-butoxyethyl acetate (synonyms: butylcellosolve acetate; butylglycol acetate)
CAS Number: 112-07-2
Primary Class: 245   Ethers, Glycols
Related Class: 222   Esters, Carboxylic, Acetates
              241   Ethers, Aliphatic and Alicyclic

g NAT+NEOPR+NITRILE	MAPA-PIONEER TRIONI	35m	5382	4	0.46		36

2-chloro-2-nitropropane
CAS Number: 594-71-8
Primary Class: 442   Nitro Compounds, Substituted
Related Class: 261   Halogen Compounds, Aliphatic and Alicyclic

g BUTYL	NORTH B-174	>480m	0	0	0.69		34
g NATURAL RUBBER	ACKWELL 5-109	1m	2820	5	0.15	degrad.	34
g NEOPRENE	ANSELLEDMONT 29-870				0.48	degrad.	34
g NITRILE	ANSELLEDMONT 37-155				0.40	degrad.	34
g PVAL	EDMONT 25-950	>480m	0	0	0.66		34
g PVC	ANSELLEDMONT 34-100				0.20	degrad.	34
g VITON	NORTH F-091	120m	1220	4	0.30	degrad.	34

2-chloroethanol (synonym: ethylene chlorohydrin)
CAS Number: 107-07-3
Primary Class: 315   Hydroxy Compounds, Aliphatic and Alicyclic, Substituted
Related Class: 261   Halogen Compounds, Aliphatic and Alicyclic

g BUTYL	NORTH B-174	>480m	0	0	0.61		34
g NEOPRENE	ANSELLEDMONT 29-870	300m	4	1	0.48		34
g NITRILE	ANSELLEDMONT 37-155				0.40	degrad.	34
c PE	DU PONT TYVEK QC	3m	31	4	n.a.		8
g PE/EVAL/PE	SAFETY4 4H	>240m	0	0	0.07	35°C	60
g PVAL	EDMONT 25-950	>480m	0	0	0.50		34
s UNKNOWN MATERIAL	DU PONT BARRICADE	>480m	0	0	n.a.		8
g VITON	NORTH F-091	>480m	0	0	0.30		34

2-ethoxy-1-propanol (synonym: alpha-propylene glycol monoethyl ether)
CAS Number: 19089-47-5
Primary Class: 245   Ethers, Glycols
Related Class: 311   Hydroxy Compounds, Aliphatic and Alicyclic, Primary

g PE/EVAL/PE	SAFETY4 4H	>240m	0	0	0.07	35°C	60

2-ethoxyethanol (synonyms: cellosolve; ethylglycol)
CAS Number: 110-80-5
Primary Class: 245   Ethers, Glycols

g BUTYL	BEST 878	>480m	0	0	0.75		53

G A R M E N T	Class & Number/ test chemical/ MATERIAL NAME	MANUFACTURER & PRODUCT IDENTIFICATION	Break- through Time in Minutes	Perm- eation Rate in mg/sq m /min.	I N D E X	Thick- ness in mm	degrad. and comments	R e f s
g	BUTYL	NORTH B-131	>1260m	0	0	0.30		51
g	BUTYL	NORTH B-161	>480m	0	0	0.40		68
g	BUTYL	NORTH B-174	>480m	0	0	0.61		34
g	NAT+NEOPR+NITRILE	MAPA-PIONEER TRIONI	4m	768	4	0.46		36
g	NATURAL RUBBER	ACKWELL 5-109	<1m	492	4	0.13		34
g	NATURAL RUBBER	ANSELLEDMONT 36-124	45m	<900	3	0.46		6
g	NATURAL RUBBER	ANSELLEDMONT 392	25m	<90	3	0.48		6
g	NATURAL RUBBER	BEST 65NFW		12	5	n.a.		53
g	NATURAL RUBBER	BEST 67NFW		6	4	n.a.		53
g	NATURAL RUBBER	MAPA-PIONEER L-118	10m	120	4	0.46		36
g	NATURAL RUBBER	MAPA-PIONEER L-118	37m	20	3	0.51		92
g	NATURAL RUBBER	MAPA-PIONEER L-118	12m	140	4	0.51	37°C	92
g	NATURAL RUBBER	MAPA-PIONEER L-118	5m	410	4	0.51	50°C	92
g	NATURAL RUBBER	MARIGOLD BLACK HEVY	28m	151	3	0.65		79
g	NATURAL RUBBER	MARIGOLD FEATHERLIT	1m	19	4	0.15	degrad.	79
g	NATURAL RUBBER	MARIGOLD MED WT EXT	18m	430	3	0.45	degrad.	79
g	NATURAL RUBBER	MARIGOLD MEDICAL	5m	190	4	0.28	degrad.	79
g	NATURAL RUBBER	MARIGOLD ORANGE SUP	29m	123	3	0.73		79
g	NATURAL RUBBER	MARIGOLD R SURGEONS	5m	190	4	0.28	degrad.	79
g	NATURAL RUBBER	MARIGOLD RED LT WT	16m	396	3	0.43	degrad.	79
g	NATURAL RUBBER	MARIGOLD SENSOTECH	5m	190	4	0.28	degrad.	79
g	NATURAL RUBBER	MARIGOLD SUREGRIP	25m	235	3	0.58	degrad.	79
g	NATURAL RUBBER	MARIGOLD YELLOW	12m	327	4	0.38	degrad.	79
g	NATURAL+NEOPRENE	MARIGOLD FEATHERWT	10m	293	4	0.35	degrad.	79
g	NEOPRENE	ANSELLEDMONT 29-840	45m	<9	2	0.38		6
g	NEOPRENE	ANSELLEDMONT 29-870	245m	186	3	0.64		34
g	NEOPRENE	ANSELLEDMONT NEOX	240m	<9	1	n.a.		6
g	NEOPRENE	ANSELLEDMONT NEOX	>480m	0	0	n.a.		68
g	NEOPRENE	BEST 6780	11m	18800	5	0.80		51
g	NEOPRENE	BEST 6780	60		5	n.a.		53
g	NEOPRENE	MAPA-PIONEER N-44	352m	180	3	0.74		36
g	NITRILE	ANSELLEDMONT 37-155	92m	564	3	0.38	degrad.	34
g	NITRILE	ANSELLEDMONT 37-155	137m	1600	4	0.50		51
g	NITRILE	ANSELLEDMONT 37-165	210m	<900	3	0.60		6
g	NITRILE	BEST 22R	420		5	n.a.		53
g	NITRILE	BEST 727	540		5	n.a.		53
g	NITRILE	MAPA-PIONEER A-14	416m	240	3	0.54		36
g	NITRILE	MAPA-PIONEER AF-18	94m	300	3	0.56		68
g	NITRILE	MARIGOLD BLUE	281m	270	3	0.45	degrad.	79
g	NITRILE	MARIGOLD GREEN SUPA	163m	430	3	0.30	degrad.	79
g	NITRILE	MARIGOLD NITROSOLVE	305m	237	3	0.75	degrad.	79
g	PE/EVAL/PE	SAFETY4 4H	>240m	0	0	0.07	35°C	60
g	PVAL	ANSELLEDMONT PVA	80m	<900	3	n.a.		6
g	PVAL	EDMONT 15-552	75m	<900	3	n.a.		6
g	PVAL	EDMONT 15-552	6m	9083	5	n.a.		51
g	PVAL	EDMONT 25-950	30m	660	3	0.75	degrad.	34
g	PVC	ANSELLEDMONT 34-100	4m	1650	5	0.20	degrad.	34
g	PVC	ANSELLEDMONT SNORKE				n.a.	degrad.	6

GARMENT Class & Number/ test chemical/ MATERIAL NAME	MANUFACTURER & PRODUCT IDENTIFICATION	Break-through Time in Minutes	Perm-eation Rate in mg/sq m /min.	INDEX	Thick-ness in mm	degrad. and comments	Refs
g PVC	ANSELLEDMONT SNORKE	105m	6	1	n.a.		68
g PVC	BEST 812	60		5	n.a.		53
s SARANEX-23	DU PONT TYVEK SARAN	>480m	0	0	n.a.		8
s UNKNOWN MATERIAL	DU PONT BARRICADE	>480m	0	0	n.a.		8
g VITON	NORTH F-091	385m	83	2	0.30		62

2-ethoxyethyl acetate (synonyms: cellosolve acetate; ethylglycol acetate)
CAS Number: 111-15-9
Primary Class: 245   Ethers, Glycols
Related Class: 222   Esters, Carboxylic, Acetates
              241   Ethers, Aliphatic and Alicyclic

g BUTYL	BEST 878	>480m	0	0	0.75		53
g BUTYL	NORTH B-131	>1278m	0	0	0.30		51
g BUTYL	NORTH B-161	>240m		0	0.38		93
g NAT+NEOPR+NITRILE	MAPA-PIONEER TRIONI	4m	768	4	0.48		36
g NAT+NEOPR+NITRILE	MAPA-PIONEER TRIONI	13m	880	4	0.46		93
g NATURAL RUBBER	ANSELLEDMONT 30-139	14m	850	4	0.51		93
g NATURAL RUBBER	ANSELLEDMONT 36-124	11m	<900	4	0.46		6
g NATURAL RUBBER	ANSELLEDMONT 392	10m	<900	4	0.48		6
g NATURAL RUBBER	MAPA-PIONEER L-118	13m			0.46		36
g NATURAL RUBBER	MARIGOLD BLACK HEVY	27m	1228	4	0.65	degrad.	79
g NATURAL RUBBER	MARIGOLD FEATHERLIT	1m	1906	5	0.15		79
g NATURAL RUBBER	MARIGOLD MED WT EXT	12m	507	4	0.45	degrad.	79
g NATURAL RUBBER	MARIGOLD MEDICAL	4m	1323	5	0.28	degrad.	79
g NATURAL RUBBER	MARIGOLD ORANGE SUP	29m	1300	4	0.73	degrad.	79
g NATURAL RUBBER	MARIGOLD R SURGEONS	4m	1323	5	0.28	degrad.	79
g NATURAL RUBBER	MARIGOLD RED LT WT	11m	624	4	0.43	degrad.	79
g NATURAL RUBBER	MARIGOLD SENSOTECH	4m	1323	5	0.28	degrad.	79
g NATURAL RUBBER	MARIGOLD SUREGRIP	23m	1012	4	0.58	degrad.	79
g NATURAL RUBBER	MARIGOLD YELLOW	9m	857	4	0.38	degrad.	79
g NATURAL+NEOPRENE	MARIGOLD FEATHERWT	8m	973	4	0.35	degrad.	79
g NATURAL+NEOPRENE	PLAYTEX 827	14m	1280	5	0.38		93
g NEOPRENE	ANSELLEDMONT 29-840	25m	<900	3	0.38		6
g NEOPRENE	ANSELLEDMONT NEOX	90m	<90	2	n.a.		6
g NEOPRENE	BEST 32	46m	340	3	n.a.		53
g NEOPRENE	BEST 6780	41m	21520	5	0.80		51
g NEOPRENE	BEST 6780	228m	150	3	n.a.		53
g NEOPRENE	MAPA-PIONEER N-44	76m	2520	4	0.74		36
g NITRILE	ANSELLEDMONT 37-155	24m	11880	5	0.50		51
g NITRILE	ANSELLEDMONT 37-165	90m	<900	3	0.60		6
g NITRILE	ANSELLEDMONT 49-155	112m	1180	4	0.43		93
g NITRILE	BEST 727	47m	860	3	0.38		53
g NITRILE	MAPA-PIONEER A-14	162m	720	3	0.54	degrad.	36
g NITRILE	MARIGOLD BLUE	110m	1312	4	0.45	degrad.	79
g NITRILE	MARIGOLD GREEN SUPA	50m	1300	4	0.30	degrad.	79
g NITRILE	MARIGOLD NITROSOLVE	123m	1315	4	0.75	degrad.	79
g PVAL	ANSELLEDMONT PVA	>360m		0	n.a.		6

G A R M E Class & Number/ N test chemical/ T MATERIAL NAME	MANUFACTURER & PRODUCT IDENTIFICATION	Break- through Time in Minutes	Perm- eation Rate in mg/sq m /min.	I N D E X	Thick- ness in mm	degrad. and comments	R e f s
g PVAL	EDMONT 15-552	40m	<90	3	n.a.		6
g PVAL	EDMONT 15-552	6m	380	4	n.a.		51
g PVC	ANSELLEDMONT SNORKE				n.a.	degrad.	6
s SARANEX-23	DU PONT TYVEK SARAN	39m	18	3	n.a.		8

2-ethoxyethyl acetate, 68% & butyl acetate, 7.5% & xylene, 7.5%
Primary Class: 800   Multicomponent Mixtures With >2 Components

g BUTYL	NORTH B-161	>240m		0	0.38		93
g NAT+NEOPR+NITRILE	MAPA-PIONEER TRIONI	14m	900	4	0.46		93
g NATURAL RUBBER	ANSELLEDMONT 30-139	14m	780	4	0.51		93
g NATURAL+NEOPRENE	PLAYTEX 827	12m	1010	5	0.36		93
g NITRILE	ANSELLEDMONT 49-155	89m	680	3	0.36		93
g NITRILE	ANSELLEDMONT 49-155	52m	1200	4	0.36	37°C	93

2-ethoxyethyl acetate, 82% & butyl acetate, 9% & xylene, 9%
Primary Class: 800   Multicomponent Mixtures With >2 Components

g BUTYL	NORTH B-161	>300m		0	0.38		93
g BUTYL	NORTH B-161	238m			0.38	37°C	93
g NAT+NEOPR+NITRILE	MAPA-PIONEER TRIONI	8m	1540	5	0.46		93
g NAT+NEOPR+NITRILE	MAPA-PIONEER TRIONI	3m	2800	5	0.46	37°C	93
g NATURAL RUBBER	ANSELLEDMONT 30-139	11m	1880	5	0.51		93
g NATURAL RUBBER	ANSELLEDMONT 30-139	3m	2700	5	0.51	37°C	93
g NATURAL+NEOPRENE	PLAYTEX 827	14m	1350	5	0.38		93
g NITRILE	ANSELLEDMONT 49-155	64m	1180	4	0.36		93
g NITRILE	ANSELLEDMONT 49-155	33m	2500	4	0.36	37°C	93

2-ethyl-1-hexanol (synonym: 2-ethylhexyl alcohol)
CAS Number: 104-76-7
Primary Class: 311   Hydroxy Compounds, Aliphatic and Alicyclic, Primary

g BUTYL	NORTH B-174	>480m	0	0	0.51		34
g NATURAL RUBBER	ACKWELL 5-109				0.15	degrad.	34
g NEOPRENE	ANSELLEDMONT 29-870	>480m	0	0	0.34		34
g PVAL	EDMONT 25-950	>480m	0	0	0.66		34
g VITON	NORTH F-091	>480m	0	0	0.30		34

2-ethylhexanoic acid (synonym: butylethylacetic acid)
CAS Number: 149-57-5
Primary Class: 102   Acids, Carboxylic, Aliphatic and Alicyclic, Unsubstituted

g NEOPRENE	ANSELLEDMONT 29-840	>240m	0	0	0.50		33
g NITRILE	ANSELLEDMONT 37-175	>240m	0	0	0.40		33
g PVC	KID VINYLPRODUKTER	>240m	0	0	0.53		33

G A R M E Class & Number/ N test chemical/ T MATERIAL NAME	MANUFACTURER & PRODUCT IDENTIFICATION	Break- through Time in Minutes	Perm- eation Rate in mg/sq m /min.	I N D E X	Thick- ness in mm	R e degrad. f and s comments
2-ethylhexyl acrylate, 75% & 1,6-hexanediol diacrylate, 25%						
Primary Class: 223	Esters, Carboxylic, Acrylates and Methacrylates					
g BUTYL	NORTH B-161	>480m	0	0	0.46	94
g NATURAL RUBBER	MAPA-PIONEER L-118	15-30m	200	5	0.45	94
g NITRILE	ANSELLEDMONT 37-155	>480m	0	0	0.37	94
2-hydroxyethyl acrylate (synonym: ethylene glycol acrylate)						
CAS Number: 818-61-1						
Primary Class: 223	Esters, Carboxylic, Acrylates and Methacrylates					
g PE/EVAL/PE	SAFETY4 4H	>240m	0	0	0.07	35°C 60
2-hydroxyethyl methacrylate (synonym: HEMA)						
CAS Number: 868-77-9						
Primary Class: 223	Esters, Carboxylic, Acrylates and Methacrylates					
g PE/EVAL/PE	SAFETY4 4H	>240m		0	0.07	60
2-hydroxyethyl-N,N,N-trimethylammonium hydroxide (synonym: choline)						
Primary Class: 550	Organic Salts (Solutions)					
g PE/EVAL/PE	SAFETY4 4H	>240m	0	0	0.07	35°C 60
2-mercaptoethanol (synonyms: 2-hydroxyethyl mercaptan; thioglycol)						
CAS Number: 60-24-2						
Primary Class: 501	Sulfur Compounds, Thiols					
Related Class: 311	Hydroxy Compounds, Aliphatic and Alicyclic, Primary					
g PE/EVAL/PE	SAFETY4 4H	>240m	0	0	0.07	35°C 60
2-methoxyethanol (synonyms: methylcellosolve; methylglycol)						
CAS Number: 109-86-4						
Primary Class: 245	Ethers, Glycols					
Related Class: 241	Ethers, Aliphatic and Alicyclic					
311	Hydroxy Compounds, Aliphatic and Alicyclic, Primary					
g BUTYL	NORTH B-131	>1386m	0	0	0.30	51
g BUTYL	NORTH B-161	>1200m	0	0	0.45	33
g BUTYL	NORTH B-161	>480m	0	0	0.40	68
g BUTYL	NORTH B-161	>240m		0	0.38	93
g NAT+NEOPR+NITRILE	MAPA-PIONEER TRIONI	40m	96	3	0.48	36
g NAT+NEOPR+NITRILE	MAPA-PIONEER TRIONI	27m	60	3	0.46	93
g NATURAL RUBBER	ANSELLEDMONT 30-139	43m	40	3	0.51	93

G A R M E N T	Class & Number/ test chemical/ MATERIAL NAME	MANUFACTURER & PRODUCT IDENTIFICATION	Break-through Time in Minutes	Perm-eation Rate in mg/sq m /min.	I N D E X	Thick-ness in mm	degrad. and comments	R e f s
g	NATURAL RUBBER	ANSELLEDMONT 36-124	4m	<90	4	0.46		6
g	NATURAL RUBBER	ANSELLEDMONT 392	20m	<90	3	0.48		6
g	NATURAL+NEOPRENE	PLAYTEX 827	48m	120	3	0.38		93
g	NATURAL+NEOPRENE	PLAYTEX ARGUS 123	36m	60	3	n.a.		72
g	NEOPRENE	ANSELLEDMONT 29-840	25m	<900	3	0.38		6
g	NEOPRENE	ANSELLEDMONT NEOX	70m	<90	2	n.a.		6
g	NEOPRENE	ANSELLEDMONT NEOX	>480m	0	0	n.a.		68
g	NEOPRENE	BEST 6780	14m	14030	5	0.80		51
g	NITRILE	ANSELLEDMONT 37-155				0.40	degrad.	34
g	NITRILE	ANSELLEDMONT 37-155	105m	6450	4	0.50		51
g	NITRILE	ANSELLEDMONT 37-165	11m	<900	4	0.60		6
g	NITRILE	ANSELLEDMONT 37-175	40m	600	3	0.40		33
g	NITRILE	ANSELLEDMONT 49-155	126m	880	3	0.43		93
g	NITRILE	MAPA-PIONEER AF-18	66m	720	3	0.56		68
g	PE/EVAL/PE	SAFETY4 4H	>240m	0	0	0.07	35°C	60
g	PVAL	ANSELLEDMONT PVA	30m	<900	3	n.a.		6
g	PVAL	EDMONT 15-552	6m	<900	4	n.a.		6
g	PVAL	EDMONT 15-552	60m	540	3	n.a.		33
g	PVAL	EDMONT 15-552	3m	41020	5	n.a.		51
g	PVC	ANSELLEDMONT SNORKE				n.a.	degrad.	6
g	PVC	ANSELLEDMONT SNORKE	85m	4	1	n.a.		68
s	SARANEX-23	DU PONT TYVEK SARAN	80m	1092	4	n.a.		8
g	VITON	NORTH F-091				0.25	degrad.	34

2-methoxyethanol, 96% & ninhydrin, 4%
Primary Class: 245   Ethers, Glycols

G	Class	MANUFACTURER	Break	Perm	I	Thick	degrad.	R
g	PE/EVAL/PE	SAFETY4 4H	>240m	0	0	0.07	35°C	60

2-methoxyethyl acetate (synonym: methylcellosolve acetate)
CAS Number: 110-49-6
Primary Class: 245   Ethers, Glycols
Related Class: 222   Esters, Carboxylic, Acetates
              241   Ethers, Aliphatic and Alicyclic

G	Class	MANUFACTURER	Break	Perm	I	Thick	degrad.	R
g	BUTYL	NORTH B-131	1301m	1	1	0.30		51
g	BUTYL	NORTH B-161	>240m		0	0.38	37°C	93
g	NAT+NEOPR+NITRILE	MAPA-PIONEER TRIONI	27m	543	3	0.46		36
g	NEOPRENE	BEST 6780	6m	76070	5	0.80		51
g	NITRILE	ANSELLEDMONT 37-155	197m	3600	4	0.50		51
g	NITRILE	ANSELLEDMONT 49-155	42m	2500	4	0.36		93
g	NITRILE	ANSELLEDMONT 49-155	4m	2560	5	0.36	pre-exp 20h	93
g	NITRILE	ANSELLEDMONT 49-155	27m	4600	4	0.36	37°C	93
g	PE/EVAL/PE	SAFETY4 4H	>240m	0	0	0.07	35°C	60
g	PVAL	EDMONT 15-552	5m	540	4	n.a.		51
s	SARANEX-23	DU PONT TYVEK SARAN	257m	11	2	n.a.		8

G							
A				Perm-	I		
R			Break-	eation	N	Thick-	R
M			through	Rate in	D	ness degrad.	e
E Class & Number/	MANUFACTURER	Time in	mg/sq m	E	in and	f
N test chemical/	& PRODUCT	Minutes	/min.	X	mm comments	s
T MATERIAL NAME	IDENTIFICATION					

2-nitropropane
CAS Number: 79-46-9
Primary Class: 441   Nitro Compounds, Unsubstituted

g BUTYL	BRUNSWICK BUTYL STD	>480m	0	0	0.63		89
g BUTYL	BRUNSWICK BUTYL-XTR	>480m	0	0	0.63		89
g BUTYL	NORTH B-161	1100m	<1	1	0.42		63
g BUTYL	NORTH B-174	>480m	0	0	0.71		34
u BUTYL	Unknown	>101m			0.80		3
g BUTYL/ECO	BRUNSWICK BUTYL-POL	>480m	0	0	0.63		89
g NATURAL RUBBER	ACKWELL 5-109	2m	1920	5	0.15		34
g NEOPRENE	ANSELLEDMONT 29-870				0.48	degrad.	34
g NITRILE	ANSELLEDMONT 37-155				0.40	degrad.	34
u NITRILE	Unknown	25m	2000	4	0.90		3
g PE/EVAL/PE	SAFETY4 4H	>240m	0	0	0.07	35°C	60
g PVAL	EDMONT 25-950	>480m	0	0	0.60		34
u PVAL	Unknown	<5m	440	4	0.20		3
g PVC	ANSELLEDMONT 34-100				0.20	degrad.	34
s TEFLON	CHEMFAB CHALL. 5100	>180m	0	1	n.a.		65
g VITON	NORTH F-091				0.26	degrad.	34

2-nitrotoluene (synonym: o-nitrotoluene)
CAS Number: 88-72-2
Primary Class: 441   Nitro Compounds, Unsubstituted

g BUTYL	NORTH B-174	>480m	0	0	0.43		34
g NITRILE	ANSELLEDMONT 37-175	30m	433	3	0.38		34

3,3',4,4'-benzophenonetetracarboxylic dianhydride
CAS Number: 2421-28-5
Primary Class: 162   Anhydrides, Aromatic

g PE/EVAL/PE	SAFETY4 4H	>240m	0	0	0.07		60

3,3'-iminobis(propylamine) (synonym: azaheptamethylenediamine)
CAS Number: 56-18-8
Primary Class: 148   Amines, Poly, Aliphatic and Alicyclic
Related Class: 141   Amines, Aliphatic and Alicyclic, Primary
                142   Amines, Aliphatic and Alicyclic, Secondary

g BUTYL	NORTH B-174	>480m	0	0	0.63		34
g NATURAL RUBBER	ACKWELL 5-109	5m	847	4	0.15		34
g NEOPRENE	ANSELLEDMONT 29-870	>480m	0	0	0.37		34
g NITRILE	ANSELLEDMONT 37-155				0.40	degrad.	34
g PVAL	EDMONT 25-950				0.40	degrad.	34
g PVC	ANSELLEDMONT 34-100				0.20	degrad.	34
g VITON	NORTH F-091	>480m	0	0	0.30		34

G A R M E N T MATERIAL NAME	MANUFACTURER & PRODUCT IDENTIFICATION	Break-through Time in Minutes	Perm-eation Rate in mg/sq m /min.	I N D E X	Thick-ness in mm	degrad. and comments	R e f s

3-(dimethylamino)propylamine
CAS Number: 109-55-7
Primary Class: 148   Amines, Poly, Aliphatic and Alicyclic
Related Class: 141   Amines, Aliphatic and Alicyclic, Primary
            143   Amines, Aliphatic and Alicyclic, Tertiary

g BUTYL	NORTH B-174	>480m	0	0	0.68		34
g NATURAL RUBBER	ACKWELL 5-109	1m	12680	5	0.15	degrad.	34
g NEOPRENE	ANSELLEDMONT 29-870	29m	2797	4	0.40	degrad.	34
g NITRILE	ANSELLEDMONT 37-155				0.40	degrad.	34
g PVAL	EDMONT 25-950				0.50	degrad.	34
g PVC	ANSELLEDMONT 34-100	2m	6600	5	0.20	degrad.	34
g VITON	NORTH F-091				0.26	degrad.	34

3-bromo-1-propanol
CAS Number: 627-18-9
Primary Class: 315   Hydroxy Compounds, Aliphatic and Alicyclic, Substituted
Related Class: 261   Halogen Compounds, Aliphatic and Alicyclic

g BUTYL	NORTH B-174	>480m	0	0	0.63		34
g NEOPRENE	ANSELLEDMONT 29-870	>480m	0	0	0.50		34
g NITRILE	ANSELLEDMONT 37-155				0.40	degrad.	34
g PVAL	EDMONT 25-950	>480m	0	0	0.33		34
g VITON	NORTH F-091	>480m	0	0	0.23		34

3-bromopropionic acid
CAS Number: 590-92-1
Primary Class: 103   Acids, Carboxylic, Aliphatic and Alicyclic, Substituted
Related Class: 261   Halogen Compounds, Aliphatic and Alicyclic

g NATURAL RUBBER	ANSELLEDMONT 36-124	210m			0.46		6
g NATURAL RUBBER	ANSELLEDMONT 392	190m			0.48		6
g NEOPRENE	ANSELLEDMONT 29-840	180m			0.38		6
g NEOPRENE	ANSELLEDMONT NEOX	240m		0	n.a.		6
g NITRILE	ANSELLEDMONT 37-165	120m			0.60		6
g PVAL	ANSELLEDMONT PVA				n.a.		6
g PVAL	EDMONT 15-552				n.a.	degrad.	6
g PVC	ANSELLEDMONT M GRIP	180m			n.a.		6

3-chloro-1-propanol
CAS Number: 627-30-5
Primary Class: 315   Hydroxy Compounds, Aliphatic and Alicyclic, Substituted
Related Class: 261   Halogen Compounds, Aliphatic and Alicyclic

g BUTYL	NORTH B-174	>480m	0	0	0.63		34

Class & Number/ test chemical/ MATERIAL NAME	MANUFACTURER & PRODUCT IDENTIFICATION	Break- through Time in Minutes	Perm- eation Rate in mg/sq m /min.	I N D E X	Thick- ness in mm	degrad. and comments	R e f s
g NITRILE	ANSELLEDMONT 37-155				0.40	degrad.	34
g PVAL	EDMONT 25-950	48m	924	3	0.35		34
g PVC	ANSELLEDMONT 34-100	11m	4080	5	0.15		34
g VITON	NORTH F-091	>480m	0	0	0.27		34

3-chloro-2-methylpropene (synonym: 2-methylallyl chloride)
CAS Number: 563-47-3
Primary Class: 262   Halogen Compounds, Allylic and Benzylic

g BUTYL	NORTH B-174	30m	720	3	0.63	degrad.	34
g NATURAL RUBBER	ACKWELL 5-109				0.15	degrad.	34
g NEOPRENE	ANSELLEDMONT 29-870				0.48	degrad.	34
g PVAL	EDMONT 25-950	1m	460	4	0.36		34
g PVC	ANSELLEDMONT 34-100	<1m	720	4	0.20	degrad.	34
g VITON	NORTH F-091	>480m	0	0	0.33		34

3-methylaminopropylamine
CAS Number: 6291-84-5
Primary Class: 148   Amines, Poly, Aliphatic and Alicyclic
Related Class: 141   Amines, Aliphatic and Alicyclic, Primary
               142   Amines, Aliphatic and Alicyclic, Secondary

g BUTYL	NORTH B-174	>480m	0	0	0.51		34
g NATURAL RUBBER	ACKWELL 5-109	3m	4380	5	0.15	degrad.	34
g NEOPRENE	ANSELLEDMONT 29-870	63m	960	3	0.33	degrad.	34
g NITRILE	ANSELLEDMONT 37-155				0.40	degrad.	34
g PVAL	EDMONT 25-950				0.40	degrad.	34
g PVC	ANSELLEDMONT 34-100	2m	4037	5	0.20	degrad.	34
g VITON	NORTH F-091				0.26	degrad.	34

4,4'-diaminodiphenylsulfone (synonym: 4,4'-sulfonyldianiline)
CAS Number: 80-08-0
Primary Class: 145   Amines, Aromatic, Primary

g NAT+NEOPR+NITRILE	MAPA-PIONEER TRIONI	>480m	0	0	0.48		36

4,4'-methylene dianiline, 15% & methyl ethyl ketone, 85%
Primary Class: 145   Amines, Aromatic, Primary

g NAT+NEOPR+NITRILE	MAPA-PIONEER TRIONI	23m	45	3	0.51		36

4,4'-methylenebis(2-chloroaniline), 50% & acetone, 50%
Primary Class: 145   Amines, Aromatic, Primary

g PE/EVAL/PE	SAFETY4 4H	>240m	0	0	0.07		60

G A R M E Class & Number/ N test chemical/ T MATERIAL NAME	MANUFACTURER & PRODUCT IDENTIFICATION	Break- through Time in Minutes	Perm- eation Rate in mg/sq m /min.	I N D E X	Thick- ness in mm	R degrad. e and f comments s

4,4'-methylenedianiline (synonyms: MDA; p,p'-diaminodiphenylmethane)
CAS Number: 101-77-9
Primary Class: 145   Amines, Aromatic, Primary

| c PE | DU PONT TYVEK QC | < 60m | | | n.a. | 90-95°C 8 |
| g PE/EVAL/PE | SAFETY4 4H | >480m | 0 | 0 | 0.07 | 60 |

4,4'-methylenedianiline, 10% & isopropyl alcohol
Primary Class: 145   Amines, Aromatic, Primary

| g PE/EVAL/PE | SAFETY4 4H | >240m | 0 | 0 | 0.07 | 60 |

4,4'-methylenedianiline, 15% & toluene, 85%
Primary Class: 145   Amines, Aromatic, Primary

| g NAT+NEOPR+NITRILE | MAPA-PIONEER TRIONI | <15m | 10 | 3 | 0.50 | 36 |

4,4'-methylenedianiline, 50% & methyl ethyl ketone, 50%
Primary Class: 145   Amines, Aromatic, Primary
Related Class: 391   Ketones, Aliphatic and Alicyclic

| g PE/EVAL/PE | SAFETY4 4H | >240m | 0 | 0 | 0.07 | 35°C 60 |
| g PE/EVAL/PE | SAFETY4 4H | >1440m | 0 | 0 | 0.07 | 60 |

4,4'-oxidianiline
CAS Number: 101-80-4
Primary Class: 145   Amines, Aromatic, Primary

| c PE | DU PONT TYVEK QC | 270m | | 0 | n.a. | 8 |

4-chloro-2-methylphenoxyacetic acid, formulation (synonym: MCPA)
CAS Number: 3653-48-3
Primary Class: 103   Acids, Carboxylic, Aliphatic and Alicyclic, Substituted
Related Class: 243   Ethers, Alkyl-Aryl
           550   Organic Salts (Solutions)
           840   Pesticides, Mixtures and Formulations

g NEOPRENE	MAPA-PIONEER 420	>180m			0.85	13
g NITRILE	ANSELLEDMONT 37-145	>120m			0.32	13
g NITRILE	ERISTA SPECIAL	>180m			0.35	13
c PE	DU PONT TYVEK QC	167m			0.15	13
c PE	MOLNLYCKE H D	>120m			0.15	13
g PVC	KID 490	>180m			n.a.	13

			Perm-	I			R
G							
A							
R							
M			Break-	eation	N	Thick-	R
E Class & Number/	MANUFACTURER	through	Rate in	D	ness	degrad.	e
N test chemical/	& PRODUCT	Time in	mg/sq m	E	in	and	f
T MATERIAL NAME	IDENTIFICATION	Minutes	/min.	X	mm	comments	s

4-chlorobenzotrifluoride (synonym: 1-chloro-4-(trifluoromethyl)benzene)
CAS Number: 98-56-6
Primary Class: 264   Halogen Compounds, Aromatic

g BUTYL	NORTH B-161	46m	28	3	0.38		34
g NEOPRENE	ANSELLEDMONT 29-870	36m	2	2	0.49		34
g NITRILE	ANSELLEDMONT 37-165	278m	2	1	0.66		34
g PVAL	EDMONT 25-250	>480m	0	0	0.49		34
g VITON	NORTH F-091	91m	5	1	0.41		34

4-hydroxy-4-methyl-2-pentanone (synonym: diacetone alcohol)
CAS Number: 123-42-2
Primary Class: 391   Ketones, Aliphatic and Alicyclic

g NATURAL RUBBER	ANSELLEDMONT 36-124	20m	<90	3	0.46		6
g NATURAL RUBBER	ANSELLEDMONT 392	15m	<90	3	0.48		6
g NEOPRENE	ANSELLEDMONT 29-840	300m	<9	1	0.38		6
g NEOPRENE	ANSELLEDMONT NEOX	>360m		0	n.a.		6
g NITRILE	ANSELLEDMONT 37-165	240m	<9	1	0.60		6
g PE/EVAL/PE	SAFETY4 4H	>240m	0	0	0.07	35°C	60
g PVAL	ANSELLEDMONT PVA	150m	<900	3	n.a.		6
g PVAL	EDMONT 15-552	120m	<90	2	n.a.		6
g PVC	ANSELLEDMONT SNORKE				n.a.	degrad.	6

4-methoxy-4-methyl-2-pentanone
CAS Number: 107-70-0
Primary Class: 391   Ketones, Aliphatic and Alicyclic

g BUTYL	NORTH B-174	>780m	0	0	0.66		34
g NATURAL RUBBER	ACKWELL 5-109				0.15	degrad.	34
g NEOPRENE	ANSELLEDMONT 29-870	99m	330	3	0.46	degrad.	34
g NITRILE	ANSELLEDMONT 37-155				0.40	degrad.	34
g PVAL	EDMONT 25-950	>840m	0	0	0.30		34
g PVC	ANSELLEDMONT 34-100				0.20	degrad.	34
g VITON	NORTH F-091	24m	1158	4	0.27	degrad.	34

4-nitrodiphenylamine (synonym: p-nitrodiphenylamine)
CAS Number: 836-30-6
Primary Class: 146   Amines, Aromatic, Secondary and Tertiary
Related Class: 442   Nitro Compounds, Substituted

g PE/EVAL/PE	SAFETY4 4H	>240m	0	0	0.07	35°C	60

G A R M E Class & Number/ N test chemical/ T MATERIAL NAME	MANUFACTURER & PRODUCT IDENTIFICATION	Break- through Time in Minutes	Perm- eation Rate in mg/sq m /min.	I N D E X	Thick- ness in mm	degrad. and comments	R e f s
4-vinyl-1-cyclohexene							
CAS Number: 100-40-3							
Primary Class: 294	Hydrocarbons, Aliphatic and Alicyclic, Unsaturated						
g BUTYL	NORTH B-174	31m	3540	4	0.66	degrad.	34
g NITRILE	ANSELLEDMONT 37-155	390m	12	2	0.38		34
g PVAL	EDMONT 25-950	54m	<0.01	1	0.86		34
g VITON	NORTH F-091	>480m	0	0	0.41		34
5-methyl-2-hexanone (synonym: isoamyl methyl ketone)							
CAS Number: 110-12-3							
Primary Class: 391	Ketones, Aliphatic and Alicyclic						
g PE/EVAL/PE	SAFETY4 4H	>240m	0	0	0.07	35°C	60
g PE/EVAL/PE	SAFETY4 4H	>480m	0	0	0.07		60
9-aminoacridine hydrochloride							
CAS Number: 134-50-9							
Primary Class: 550	Organic Salts (Solutions)						
Related Class: 274	Heterocyclic Compounds, Nitrogen, Others						
g BUTYL	NORTH B-174	>480m	0	0	0.63		34
g NATURAL RUBBER	ACKWELL 5-109	>480m	0	0	0.15		34
g NEOPRENE	ANSELLEDMONT 29-870	>480m	0	0	0.51		34
g PVC	ANSELLEDMONT 34-100	>480m	0	0	0.20		34
accumix							
Primary Class: 900	Miscellaneous Unclassified Chemicals						
g PE/EVAL/PE	SAFETY4 4H	>240m		0	0.07		60
acetaldehyde (synonym: ethanal)							
CAS Number: 75-07-0							
Primary Class: 121	Aldehydes, Aliphatic and Alicyclic						
g BUTYL	BEST 878	>480m	0	0	0.75		53
g BUTYL	BRUNSWICK BUTYL STD	4m	21	4	0.63		89
g BUTYL	BRUNSWICK BUTYL-XTR	193m	110	3	0.63		89
g BUTYL	NORTH B-174	>575m	4	1	0.41		34
g BUTYL/ECO	BRUNSWICK BUTYL-POL	85m	28	2	0.63		89
s CPE	ILC DOVER CLOROPEL	20m			n.a.		46
g NATURAL RUBBER	ANSELLEDMONT 36-124	7m	<9000	5	0.46		6
g NATURAL RUBBER	ANSELLEDMONT 392	7m	<9000	5	0.48		6
g NATURAL RUBBER	BEST 65NFW		60	5	n.a.		53
g NATURAL RUBBER	BEST 65NFW	55m	500	3	n.a.		53
g NATURAL RUBBER	BEST 67NFW		120	5	n.a.	degrad.	53
g NATURAL RUBBER	MAPA-PIONEER L-118	2m	780	4	0.46		36

G A R M E N T	Class & Number/ test chemical/ MATERIAL NAME	MANUFACTURER & PRODUCT IDENTIFICATION	Break-through Time in Minutes	Perm-eation Rate in mg/sq m /min.	I N D E X	Thick-ness in mm	degrad. and comments	R e f s
g	NATURAL+NEOPRENE	PLAYTEX ARGUS 123	4m	600	4	n.a.		72
g	NEOPRENE	ANSELLEDMONT 29-840	10m	<90000	5	0.38		6
g	NEOPRENE	ANSELLEDMONT 29-870	12m	1986	5	0.48		34
g	NEOPRENE	ANSELLEDMONT NEOX	17m	<90000	5	n.a.		6
g	NEOPRENE	BEST 32	<1m	170	4	n.a.		53
g	NEOPRENE	BEST 6780	1m	820	4	n.a.		53
g	NEOPRENE	MAPA-PIONEER N-44	21m	1080	4	0.74		36
g	NITRILE	ANSELLEDMONT 37-165				0.60	degrad.	6
g	NITRILE	BEST 22R	9m	1620	5	n.a.	degrad.	53
g	NITRILE	BEST 727				0.38	degrad.	53
g	NITRILE	NORTH LA-142G	<1m	9660	5	0.33	degrad.	34
g	PE/EVAL/PE	SAFETY4 4H	>240m	0	0	0.07		60
g	PVAL	ANSELLEDMONT PVA				n.a.	degrad.	6
g	PVAL	COMASEC SOLVATRIL	>480m	0	0	n.a.		40
g	PVAL	EDMONT 15-552				n.a.	degrad.	6
g	PVAL	EDMONT 25-950	16m	2820	4	0.25		34
g	PVC	ANSELLEDMONT SNORKE				n.a.	degrad.	6
g	PVC	BEST 725R	13m	2120	5	n.a.		53
g	PVC	BEST 812		2640	5	n.a.	degrad.	53
s	TEFLON	CHEMFAB CHALL. 5100	>180m			n.a.		65
s	UNKNOWN MATERIAL	LIFE-GUARD RESPONDE	>480m	0	0	n.a.		77
g	UNKNOWN MATERIAL	NORTH SILVERSHIELD	>360m	0	0	0.08		7
g	VITON	NORTH F-091	<1m	17160	5	0.28	degrad.	34
u	VITON/CHLOROBUTYL	Unknown	35m			0.36		46

acetic acid, 30-70%
Primary Class: 102   Acids, Carboxylic, Aliphatic and Alicyclic, Unsubstituted

g	NATURAL RUBBER	MAPA-PIONEER L-118	31m	162	3	0.46		36
g	NEOPRENE	MAPA-PIONEER N-44	>480m	0	0	0.56		36
g	NITRILE	MAPA-PIONEER A-14	>480m	0	0	0.56		36
g	PVC	MAPA-PIONEER V-20	47m	4	2	0.51		36

acetic acid, glacial (synonym: ethanoic acid)
CAS Number: 64-19-7
Primary Class: 102   Acids, Carboxylic, Aliphatic and Alicyclic, Unsubstituted

g	BUTYL	ANSELL LAMPRECHT	>60m			0.70		38
g	BUTYL	BEST 878	>480m	0	0	0.75	84%	53
g	BUTYL	BRUNSWICK BUTYL STD	>480m	0	0	0.63		89
g	BUTYL	BRUNSWICK BUTYL-XTR	>480m	0	0	0.63		89
s	BUTYL	MSA CHEMPRUF	>480m	0	0	n.a.		48
m	BUTYL	PLYMOUTH RUBBER	>180m			n.a.		46
g	BUTYL/ECO	BRUNSWICK BUTYL-POL	>480m	0	0	0.63		89
g	BUTYL/NEOPRENE	COMASEC BUTYL PLUS	60m	60	2	0.50		40
s	CPE	ILC DOVER CLOROPEL	>180m			n.a.		46
g	NAT+NEOPR+NITRILE	MAPA-PIONEER TRIONI	32m	1278	4	0.48		36
g	NATURAL RUBBER	ANSELL CONFORM 4205	30m			0.13		55
g	NATURAL RUBBER	ANSELL CONT ENVIRON	270m		0	0.53		55

G
A
R

M			Break-	Perm-eation	I N D	Thick-		R e
E	Class & Number/	MANUFACTURER	through	Rate in		ness	degrad.	
N	test chemical/	& PRODUCT	Time in	mg/sq m	E	in	and	f
T	MATERIAL NAME	IDENTIFICATION	Minutes	/min.	X	mm	comments	s
g	NATURAL RUBBER	ANSELL FL 200	254	90m		0.51		55
g	NATURAL RUBBER	ANSELL ORANGE 208	300m		0	0.76		55
g	NATURAL RUBBER	ANSELL PVL 040	90m			0.46		55
g	NATURAL RUBBER	ANSELL STERILE 832	240m		0	0.23		55
g	NATURAL RUBBER	ANSELLEDMONT 30-139	135m			0.25		42
g	NATURAL RUBBER	ANSELLEDMONT 36-124	150m			0.46		6
g	NATURAL RUBBER	ANSELLEDMONT 392	140m			0.48		6
g	NATURAL RUBBER	BEST 65NFW	>480m	0	0	n.a.		53
g	NATURAL RUBBER	COMASEC FLEXIGUM	120m	150	3	0.95		40
g	NATURAL RUBBER	MAPA-PIONEER L-118	21m	120	3	0.46		36
g	NATURAL RUBBER	MARIGOLD BLACK HEVY	43m	687	3	0.65	degrad.	79
g	NATURAL RUBBER	MARIGOLD FEATHERLIT	4m	5103	5	0.15	degrad.	79
g	NATURAL RUBBER	MARIGOLD MED WT EXT	14m	2380	5	0.45	degrad.	79
g	NATURAL RUBBER	MARIGOLD MEDICAL	8m	3968	5	0.28	degrad.	79
g	NATURAL RUBBER	MARIGOLD ORANGE SUP	46m	518	3	0.73	degrad.	79
g	NATURAL RUBBER	MARIGOLD R SURGEONS	8m	3968	5	0.28	degrad.	79
g	NATURAL RUBBER	MARIGOLD RED LT WT	13m	2607	5	0.43	degrad.	79
g	NATURAL RUBBER	MARIGOLD SENSOTECH	8m	3968	5	0.28	degrad.	79
g	NATURAL RUBBER	MARIGOLD SUREGRIP	34m	1195	4	0.58	degrad.	79
g	NATURAL RUBBER	MARIGOLD YELLOW	11m	3061	5	0.38	degrad.	79
u	NATURAL RUBBER	Unknown	41m			n.a.		9
u	NATURAL RUBBER	Unknown	51m			0.43		9
g	NATURAL+NEOPRENE	ANSELL OMNI 276	90m			0.56		55
g	NATURAL+NEOPRENE	ANSELL TECHNICIANS	90m			0.43		55
g	NATURAL+NEOPRENE	MARIGOLD FEATHERWT	11m	3288	5	0.35	degrad.	79
g	NATURAL+NEOPRENE	PLAYTEX ARGUS 123	76m	60	2	0.91		72
u	NATURAL+NEOPRENE	Unknown	>60m			n.a.		9
g	NEOPRENE	ANSELL NEOPRENE 530	180m			0.46		55
g	NEOPRENE	ANSELLEDMONT 29-840	420m	0	0	0.38		6
g	NEOPRENE	ANSELLEDMONT NEOX	>360m	0	0	n.a.		6
g	NEOPRENE	BEST 32	>480m	0	0	n.a.	84%	53
g	NEOPRENE	BEST 6780	>480m	0	0	n.a.	84%	53
g	NEOPRENE	BRUNSWICK NEOPRENE	>480m	0	0	0.88		89
g	NEOPRENE	COMASEC COMAPRENE	>360m	0	0	n.a.		40
u	NEOPRENE	Unknown	>60m			0.61		9
g	NEOPRENE/NATURAL	ANSELL CHEMI-PRO	180m			0.72		55
g	NITRILE	ANSELL CHALLENGER	>360m		0	0.38		55
g	NITRILE	ANSELLEDMONT 37-165	270m	0	0	0.60		6
g	NITRILE	ANSELLEDMONT 37-175	60m	2340	4	0.42		38
g	NITRILE	ANSELLEDMONT 49-125	45m			0.23		42
g	NITRILE	ANSELLEDMONT 49-155	480m		0	0.38		42
g	NITRILE	BEST 22R	300m		0	n.a.	84%	53
g	NITRILE	BEST 727	240m	17790	5	0.38	84%	53
g	NITRILE	COMASEC COMATRIL	250m	1080	4	0.55		40
g	NITRILE	COMASEC COMATRIL SU	>480m	0	0	0.60		40
g	NITRILE	COMASEC FLEXITRIL	>480m	0	0	n.a.		40
g	NITRILE	MAPA-PIONEER A-14	118m	13260	5	0.56		36
g	NITRILE	MARIGOLD BLUE	118m	1315	4	0.45	degrad.	79
g	NITRILE	MARIGOLD GREEN SUPA	55m	903	3	0.30	degrad.	79
g	NITRILE	MARIGOLD NITROSOLVE	131m	1398	4	0.75	degrad.	79

G A R M E Class & Number/ N test chemical/ T MATERIAL NAME	MANUFACTURER & PRODUCT IDENTIFICATION	Break- through Time in Minutes	Perm- eation Rate in mg/sq m /min.	I N D E X	Thick- ness in mm	degrad. and comments	R e f s
u NITRILE	Unknown	>60m			0.45	degrad.	9
g NITRILE+PVC	COMASEC MULTIPLUS	>480m	0	0	n.a.		40
c PE	DU PONT TYVEK QC	7m	30	4	n.a.		8
g PE/EVAL/PE	SAFETY4 4H	>240m	0	0	0.07		60
g PE/EVAL/PE	SAFETY4 4H	53m	24	3	0.07	35°C	60
g PVAL	ANSELLEDMONT PVA				n.a.	degrad.	6
g PVAL	EDMONT 15-552				n.a.	degrad.	6
g PVC	ANSELLEDMONT M GRIP	180m			n.a.		6
g PVC	BEST 725R	300m	31	2	n.a.		53
g PVC	COMASEC MULTIPOST	290m	60	2	n.a.		40
g PVC	COMASEC MULTITOP	>480m	0	0	n.a.		40
g PVC	COMASEC NORMAL	270m	0	0	n.a.		40
g PVC	COMASEC OMNI	216m	18	2	n.a.		40
g PVC	MAPA-PIONEER V-20	85m	18	2	0.51		36
s PVC	MSA UPC	>480m	0	0	0.20		48
u PVC	Unknown	4m			0.22	degrad.	9
u PVC	Unknown	>60m			n.a.		9
s SARANEX-23	DU PONT TYVEK SARAN	>480m	0	0	n.a.		8
s TEFLON	CHEMFAB CHALL. 5100	>480m	0	0	n.a.		56
s TEFLON	CHEMFAB CHALL. 5100	>240m	0	0	n.a.		65
s UNKNOWN MATERIAL	DU PONT BARRICADE	145m	39	2	n.a.		8
s UNKNOWN MATERIAL	LIFE-GUARD RESPONDE	>480m	0	0	n.a.		77
s UNKNOWN MATERIAL	LIFE-GUARD RESPONDE	>480m	0	0	n.a.		77
g VITON	NORTH F-091	2m	5300	5	0.26		38
g VITON	NORTH F-121	>60m			0.31		9
s VITON/BUTYL	DRAGER 500 OR 710	>360m		0	n.a.		91
s VITON/CHLOROBUTYL	LIFE-GUARD VC-100	>240m		0	n.a.		77
u VITON/CHLOROBUTYL	Unknown	>180m			n.a.		46
g VITON/NEOPRENE	ERISTA VITRIC	>60m			0.60		38

acetic anhydride (synonym: acetyl oxide)
CAS Number: 108-24-7
Primary Class: 161   Anhydrides, Aliphatic and Alicyclic

g BUTYL	NORTH B-174	>480m	0	0	0.65		34
g NATURAL RUBBER	ACKWELL 5-109	3m	74	4	0.15		34
g NEOPRENE	ANSELLEDMONT 29-870	210m	36	2	0.46		34
g NITRILE	ANSELLEDMONT 37-155				0.40	degrad.	34
g PE/EVAL/PE	SAFETY4 4H	>240m	0	0	0.07	35°C	60
g PVC	ANSELLEDMONT 34-100	4m	744	4	0.20		34
s TEFLON	CHEMFAB CHALL. 5100	>180m	0	1	n.a.		65
g VITON	NORTH F-091				0.26	degrad.	34

acetone
CAS Number: 67-64-1
Primary Class: 391   Ketones, Aliphatic and Alicyclic

g BUTYL	BEST 878	>480m	0	0	0.75		53
g BUTYL	BRUNSWICK BUTYL STD	>480m	0	0	0.80		89

G A R M E N T	Class & Number/ test chemical/ MATERIAL NAME	MANUFACTURER & PRODUCT IDENTIFICATION	Break-through Time in Minutes	Perm-eation Rate in mg/sq m /min.	I N D E X	Thick-ness in mm	degrad. and comments	R e f s
g	BUTYL	BRUNSWICK BUTYL-XTR	>480m	0	0	0.80		89
g	BUTYL	COMASEC BUTYL	>480m	0	0	0.65		40
s	BUTYL	LIFE-GUARD BUTYL	125m	0.1		n.a.		77
g	BUTYL	NORTH B-161	575m	4	1	0.45		7
g	BUTYL	NORTH B-161	367m	15	2	0.42		63
g	BUTYL	NORTH B-174	>1218m	0	0	0.76		34
m	BUTYL	PLYMOUTH RUBBER	>180m			n.a.		46
s	BUTYL	WHEELER ACID KING	182m	<9	1	n.a.		85
g	BUTYL/ECO	BRUNSWICK BUTYL-POL	324m	0.3	0	0.63		89
g	BUTYL/NEOPRENE	COMASEC BUTYL PLUS	>480m	0	0	0.50		40
s	BUTYL/NEOPRENE	MSA BETEX	135m	1080	4	n.a.		48
s	CPE	ILC DOVER CLOROPEL	23m			n.a.		43
s	CPE	STD. SAFETY	35m	13	3	n.a.	degrad.	75
g	HYPALON	COMASEC DIPCO	38m	462	3	0.59		40
g	NAT+NEOPR+NITRILE	MAPA-PIONEER TRIONI	12m	1200	5	0.48		36
g	NATURAL RUBBER	ACKWELL 5-109	<1m	4620	5	0.15		34
g	NATURAL RUBBER	ANSELL CONFORM 4205	1m	29	4	0.13		55
g	NATURAL RUBBER	ANSELL CONT ENVIRON	8m			0.53		55
g	NATURAL RUBBER	ANSELL FL 200 254	6m	0.5	3	0.51		55
g	NATURAL RUBBER	ANSELL ORANGE 208	14m	0.7	3	0.76		55
g	NATURAL RUBBER	ANSELL PVL 040	8m	53	4	0.46		55
g	NATURAL RUBBER	ANSELL STERILE 832	8m	0.5	3	0.23		55
g	NATURAL RUBBER	ANSELLEDMONT 30-139	10m	<900	4	0.25		42
g	NATURAL RUBBER	ANSELLEDMONT 36-124	10m	<9000	5	0.46		6
g	NATURAL RUBBER	ANSELLEDMONT 392	10m	<9000	5	0.48		6
g	NATURAL RUBBER	ANSELLEDMONT 46-320	5m	1100	5	0.31		5
g	NATURAL RUBBER	BEST 65NFW		1320	5	n.a.		53
g	NATURAL RUBBER	BEST 67NFW		2880	5	n.a.		53
g	NATURAL RUBBER	COMASEC FLEXIGUM	99m	300	3	0.95		40
g	NATURAL RUBBER	MAPA-PIONEER L-118	7m	300	4	0.40		36
g	NATURAL RUBBER	MARIGOLD BLACK HEVY	5m	641	4	0.65		79
g	NATURAL RUBBER	MARIGOLD FEATHERLIT	1m	2405	5	0.15		79
g	NATURAL RUBBER	MARIGOLD MED WT EXT	1m	1103	5	0.45		79
g	NATURAL RUBBER	MARIGOLD MEDICAL	1m	1862	5	0.28		79
g	NATURAL RUBBER	MARIGOLD ORANGE SUP	5m	595	4	0.73		79
g	NATURAL RUBBER	MARIGOLD R SURGEONS	1m	1862	5	0.28		79
g	NATURAL RUBBER	MARIGOLD RED LT WT	1m	1211	5	0.43		79
g	NATURAL RUBBER	MARIGOLD SENSOTECH	1m	1862	5	0.28		79
g	NATURAL RUBBER	MARIGOLD SUREGRIP	4m	780	4	0.58		79
g	NATURAL RUBBER	MARIGOLD YELLOW	1m	1428	5	0.38		79
g	NATURAL+NEOPRENE	ANSELL OMNI 276	5m	1000	5	0.45		5
g	NATURAL+NEOPRENE	ANSELL OMNI 276	7m	84	4	0.56		55
g	NATURAL+NEOPRENE	ANSELL TECHNICIANS	5m	1.2	3	0.43		55
g	NATURAL+NEOPRENE	MARIGOLD FEATHERWT	1m	1537	5	0.35		79
g	NATURAL+NEOPRENE	PLAYTEX ARGUS 123	3m	1260	5	n.a.		72
g	NEOPRENE	ANSELL NEOPRENE 520	8m	0.5	3	0.46		55
g	NEOPRENE	ANSELLEDMONT 29-840	5m	<9000	5	0.38		6
g	NEOPRENE	ANSELLEDMONT 29-870	14m	3336	5	n.a.		34

G A R M E Class & Number/ N test chemical/ T MATERIAL NAME	MANUFACTURER & PRODUCT IDENTIFICATION	Break- through Time in Minutes	Perm- eation Rate in mg/sq m /min.	I N D E X	Thick- ness in mm	degrad. and comments	R e f s
g NEOPRENE	ANSELLEDMONT NEOX	10m	<9000	5	n.a.		6
g NEOPRENE	BEST 32	5m	1020	5	n.a.		53
g NEOPRENE	BEST 6780	25m	830	3	n.a.		53
g NEOPRENE	BRUNSWICK NEOPRENE	49m	>5000	4	0.84		89
g NEOPRENE	COMASEC COMAPRENE	20m	4800	4	n.a.		40
s NEOPRENE	LIFE-GUARD NEOPRENE	18m	4	2	n.a.		77
g NEOPRENE	MAPA-PIONEER GF-N-	19m	1400	4	0.47		5
g NEOPRENE	MAPA-PIONEER N-44	12m	2100	5	0.56		36
g NEOPRENE	MAPA-PIONEER N-54	32m	1700	4	0.70		5
g NEOPRENE	MAPA-PIONEER N-73	33m	900	3	0.46		5
b NEOPRENE	RAINFAIR	48m	190	3	n.a.		90
b NEOPRENE	RANGER	57m	180	3	n.a.		90
b NEOPRENE	SERVUS NO 22204	85m	910	3	n.a.		90
u NEOPRENE	Unknown	>60m			0.93		10
u NEOPRENE	Unknown	21m	2205	4	0.51		30
u NEOPRENE	Unknown	15m	3600	4	0.52		49
g NITRILE	ANSELLEDMONT 37-155	5m	20000	5	0.36		5
g NITRILE	ANSELLEDMONT 37-155	7m	720	4	0.38	degrad.	34
g NITRILE	ANSELLEDMONT 37-165				0.60	degrad.	6
g NITRILE	ANSELLEDMONT 49-155				0.38	degrad.	42
g NITRILE	BEST 727	3m	2650	5	0.38		53
g NITRILE	COMASEC COMATRIL	11m	5400	5	0.55		40
g NITRILE	COMASEC COMATRIL SU	21m	4800	4	0.60		40
g NITRILE	COMASEC FLEXITRIL	6m	4800	5	n.a.		40
g NITRILE	MAPA-PIONEER A-14	6m	5220	5	0.56		36
g NITRILE	MARIGOLD BLUE	9m	15428	5	0.45	degrad.	79
g NITRILE	MARIGOLD GREEN SUPA	3m	19764	5	0.30	degrad.	79
g NITRILE	MARIGOLD NITROSOLVE	10m	14561	5	0.75	degrad.	79
g NITRILE	NORTH LA-102G	2m	18000	5	0.32		30
g NITRILE	NORTH LA-142G				0.38	degrad.	7
b NITRILE+PUR+PVC	BATA HAZMAX	124m	131	3	n.a.		86
g NITRILE+PVC	COMASEC MULTIMAX	28m	8	2	n.a.		40
g NITRILE+PVC	COMASEC MULTIPLUS	19m	3120	4	n.a.		40
b NITRILE+PVC	SERVUS NO 73101	53m	>220	3	n.a.		90
b NITRILE+PVC	TINGLEY	119m	420	3	n.a.		90
g PE	ANSELLEDMONT 35-125	>60m			0.03		5
c PE	DU PONT TYVEK QC	<1m	78	4	n.a.		8
g PE/EVAL/PE	SAFETY4 4H	>240m	0	0	0.07	35°C	60
g PE/EVAL/PE	SAFETY4 4H	>1440m	0	0	0.07		60
g PVAL	ANSELLEDMONT PVA				n.a.	degrad.	6
g PVAL	COMASEC SOLVATRIL	>480m	0	0	n.a.		40
g PVAL	EDMONT 15-552				n.a.	degrad.	6
g PVAL	EDMONT 15-552	1m	>1500	5	n.a.		10
g PVAL	EDMONT 15-552	30m	700	3	n.a.		10
g PVAL	EDMONT 15-552	>240m	0	0	n.a.	degrad.	30
g PVAL	EDMONT 25-950	4m	138	4	0.25	degrad.	34
g PVC	ANSELLEDMONT 34-100				0.20	degrad.	34
g PVC	ANSELLEDMONT SNORKE				n.a.	degrad.	6

G A R M E Class & Number/ N test chemical/ T MATERIAL NAME	MANUFACTURER & PRODUCT IDENTIFICATION	Break- through Time in Minutes	Perm- eation Rate in mg/sq m /min.	I N D E X	Thick- ness in mm	degrad. and comments	R e f s
b PVC	BATA STANDARD	5m	745	4	n.a.		86
g PVC	COMASEC MULTIPOST	17m	4560	4	n.a.		40
g PVC	COMASEC MULTITOP	18m	3660	4	n.a.		40
g PVC	COMASEC NORMAL	18m	5400	4	n.a.		40
g PVC	COMASEC OMNI	13m	6300	5	n.a.		40
b PVC	JORDAN DAVID	>180m			n.a.		90
g PVC	MAPA-PIONEER V-20	<1m	>90000	5	0.31		5
g PVC	MAPA-PIONEER V-20				0.51	degrad.	36
b PVC	STD. SAFETY GL-20	4m	>4000	5	n.a.		40
s PVC	STD. SAFETY WG-20	4m	>4000	5	n.a.		40
u PVC	Unknown	9m	>1500	5	n.a.		10
u PVC	Unknown	6m	20100	5	0.48		49
s PVC	WHEELER ACID KING	5m	<90000	5	n.a.	degrad.	85
b PVC+POLYURETHANE	BATA POLYBLEND	24m	808	3	n.a.		86
b PVC+POLYURETHANE	BATA POLYBLEND	88m	520	3	n.a.		90
b PVC+POLYURETHANE	BATA POLYMAX	106m	68	2	n.a.		86
b PVC+POLYURETHANE	BATA SUPER POLY	33m	723	3	n.a.		86
u PVDC/PE/PVDC	MOLNLYCKE	>240m	0	0	0.07		54
s SARANEX-23	DU PONT TYVEK SARAN	29m	120	3	n.a.		8
s SARANEX-23 2-PLY	DU PONT TYVEK SARAN	180m	18	2	n.a.		8
u TEFLON	CHEMFAB	2m	23	4	0.10		65
s TEFLON	CHEMFAB CHALL. 5100	>210m	0	1	n.a.		65
s TEFLON	CHEMFAB CHALL. 5200	>380m	0	0	n.a.		80
s TEFLON	CHEMFAB CHALL. 6000	>180m			n.a.		80
s TEFLON	LIFE-GUARD TEFGUARD	>480m	0	0	n.a.		72
s TEFLON	WHEELER ACID KING	>480m	<9	1	n.a.		85
v TEFLON-FEP	CHEMFAB CHALL.	>180m			0.25		65
s UNKNOWN MATERIAL	CHEMRON CHEMREL	>1440m	0	0	n.a.		67
s UNKNOWN MATERIAL	CHEMRON CHEMREL MAX	>1440m	0	0	n.a.		67
c UNKNOWN MATERIAL	CHEMRON CHEMTUFF	20m	0.8	2	n.a.		67
s UNKNOWN MATERIAL	DU PONT BARRICADE	>480m	0	0	n.a.		8
s UNKNOWN MATERIAL	KAPPLER CPF III	>480m		0	n.a.		77
s UNKNOWN MATERIAL	LIFE-GUARD RESPONDE	>480m	0	0	n.a.		77
g UNKNOWN MATERIAL	NORTH SILVERSHIELD	>360m	0	0	0.08		7
g VITON	NORTH F-091				0.25	degrad.	7
g VITON	NORTH F-091	4m	18000	5	0.25		30
g VITON	NORTH F-091	<1m	48340	5	0.23	degrad.	34
s VITON/BUTYL/UNKN.	TRELLEBORG HPS	>180m			n.a.		71
s VITON/CHLOROBUTYL	LIFE-GUARD VC-100	28m			n.a.		77
s VITON/CHLOROBUTYL	LIFE-GUARD VNC 200	90m	0.1		n.a.		77
u VITON/CHLOROBUTYL	Unknown	57m			n.a.		43
g VITON/NEOPRENE	ERISTA VITRIC	15m	2787	4	0.60		38
s VITON/NEOPRENE	MSA VAUTEX	15m	334	3	n.a.		48

acetone & toluene & methylated spirits & conc. ammonia, 2:1:1:1
Primary Class: 800   Multicomponent Mixtures With >2 Components

| g PE/EVAL/PE | SAFETY4 4H | 40m | | | 0.07 | 35°C | 60 |

G A R M E Class & Number/ N test chemical/ T MATERIAL NAME	MANUFACTURER & PRODUCT IDENTIFICATION	Break- through Time in Minutes	Perm- eation Rate in mg/sq m /min.	I N D E X	Thick- ness in mm	degrad. and comments	R e f s
acetone, 50% & gasoline, 50%							
Primary Class: 391	Ketones, Aliphatic and Alicyclic						
g PE/EVAL/PE	SAFETY4 4H	>240m	0	0	0.07		60
g PE/EVAL/PE	SAFETY4 4H	3m			0.07	35°C	60
acetone, 50% & hexane, 50%							
Primary Class: 391	Ketones, Aliphatic and Alicyclic						
u VITON/CHLOROBUTYL	Unknown	3m			0.36		43
acetone, 88% & propylene glycol, 10% & 2,3-diphenylcyclopropene-1-one, 2%							
Primary Class: 800	Multicomponent Mixtures With >2 Components						
g PE/EVAL/PE	SAFETY4 4H	>240m	0	0	0.07	35°C	60
acetonitrile (synonym: methyl cyanide)							
CAS Number: 75-05-8							
Primary Class: 431	Nitriles, Aliphatic and Alicyclic						
g BUTYL	BEST 878	>480m	0	0	0.75		53
g BUTYL	BRUNSWICK BUTYL STD	>480m	0	0	0.66		89
g BUTYL	BRUNSWICK BUTYL-XTR	>480m	0	0	0.66		89
g BUTYL	COMASEC BUTYL	>480m	0	0	0.70		40
s BUTYL	DRAGER 500 OR 710	>60m			n.a.		91
s BUTYL	LIFE-GUARD BUTYL	120m	0.1		n.a.		77
s BUTYL	MSA CHEMPRUF	>480m	0	0	n.a.		48
g BUTYL	NORTH B-174	>480m	0	0	0.71		34
m BUTYL	PLYMOUTH RUBBER	>180m			n.a.		46
s BUTYL	WHEELER ACID KING	>480m	0	0	n.a.		85
g BUTYL/ECO	BRUNSWICK BUTYL-POL	450m	0.2	0	0.63		89
s BUTYL/NEOPRENE	MSA BETEX	165m	0.5	1	n.a.		48
s CPE	ILC DOVER CLOROPEL	82m			n.a.		46
s CPE	STD. SAFETY	111m	9	1	n.a.		75
g NATURAL RUBBER	ACKWELL 5-109	<1m	1170	5	0.13		34
g NATURAL RUBBER	ANSELLEDMONT 36-124	4m	<90	4	0.46		6
g NATURAL RUBBER	ANSELLEDMONT 392	4m	<90	4	0.48		6
g NATURAL RUBBER	BEST 65NFW	16m	70	3	n.a.		53
g NATURAL RUBBER	BEST 67NFW		1500	5	n.a.	degrad.	53
g NATURAL RUBBER	MARIGOLD BLACK HEVY	6m	158	4	0.65		79
g NATURAL RUBBER	MARIGOLD FEATHERLIT	1m	38	4	0.15		79
g NATURAL RUBBER	MARIGOLD MED WT EXT	5m	267	4	0.45		79
g NATURAL RUBBER	MARIGOLD MEDICAL	2m	133	4	0.28		79
g NATURAL RUBBER	MARIGOLD ORANGE SUP	6m	147	4	0.73		79
g NATURAL RUBBER	MARIGOLD R SURGEONS	2m	133	4	0.28		79

G A R M E Class & Number/ N test chemical/ T MATERIAL NAME	MANUFACTURER & PRODUCT IDENTIFICATION	Break- through Time in Minutes	Perm- eation Rate in mg/sq m /min.	I N D E X	Thick- ness in mm	R degrad. and e comments	R e f s
g NATURAL RUBBER	MARIGOLD RED LT WT	5m	248	4	0.43		79
g NATURAL RUBBER	MARIGOLD SENSOTECH	2m	133	4	0.28		79
g NATURAL RUBBER	MARIGOLD SUREGRIP	6m	190	4	0.58		79
g NATURAL RUBBER	MARIGOLD YELLOW	4m	210	4	0.38		79
u NATURAL RUBBER	Unknown	2m	840	4	0.64		38
g NATURAL+NEOPRENE	MARIGOLD FEATHERWT	3m	190	4	0.35		79
g NATURAL+NEOPRENE	PLAYTEX ARGUS 123	10m	120	4	n.a.		72
g NEOPRENE	ANSELLEDMONT 29-840	30m	<90	3	0.38		6
g NEOPRENE	ANSELLEDMONT 29-870	76m	108	3	0.63		34
g NEOPRENE	ANSELLEDMONT NEOX	90m	<9	1	n.a.		6
g NEOPRENE	BEST 32	27m	160	3	n.a.		53
g NEOPRENE	BEST 6780	53m	50	3	n.a.		53
g NEOPRENE	BRUNSWICK NEOPRENE	120m	95	2	0.84		89
s NEOPRENE	LIFE-GUARD NEOPRENE	42m	0.9	2	n.a.		77
g NEOPRENE	MAPA-PIONEER N-44	40m	420	3	0.56		36
b NEOPRENE	RAINFAIR	147m	4	1	n.a.		90
b NEOPRENE	RANGER	127m	8	1	n.a.		90
b NEOPRENE	SERVUS NO 22204	168m	5	1	n.a.		90
g NITRILE	ANSELLEDMONT 37-155				0.40	degrad.	34
g NITRILE	ANSELLEDMONT 37-165	30m	<9000	4	0.60		6
g NITRILE	ANSELLEDMONT 37-175	9m	7580	5	0.42		38
g NITRILE	BEST 22R	14m	540	4	n.a.		53
g NITRILE	BEST 727	6m	1820	5	0.38		53
g NITRILE	MARIGOLD BLUE	15m	9030	4	0.45	degrad.	79
g NITRILE	MARIGOLD GREEN SUPA	7m	11767	5	0.30	degrad.	79
g NITRILE	MARIGOLD NITROSOLVE	17m	8483	4	0.75	degrad.	79
b NITRILE+PUR+PVC	BATA HAZMAX	>180m			n.a.		86
b NITRILE+PVC	SERVUS NO 73101	>180m			n.a.		40
b NITRILE+PVC	TINGLEY	>180m			n.a.		90
c PE	DU PONT TYVEK QC	1m	130	4	n.a.		8
g PE/EVAL/PE	SAFETY4 4H	>240m	0	0	0.07	35°C	60
g PE/EVAL/PE	SAFETY4 4H	>1440m	0	0	0.07		60
g PVAL	ANSELLEDMONT PVA	150m	<900	3	n.a.		6
g PVAL	EDMONT 15-552	60m <	9	1	n.a.		6
g PVAL	EDMONT 25-950	>480m	0	0	0.36		34
g PVC	ANSELLEDMONT 34-100				0.20	degrad.	34
g PVC	ANSELLEDMONT SNORKE				n.a.	degrad.	6
b PVC	BATA STANDARD	61m	0.3	1	n.a.		86
g PVC	BEST 725R	24m	1890	4	n.a.		53
g PVC	BEST 812		660	5	n.a.	degrad.	53
b PVC	JORDAN DAVID	95m	>400	3	n.a.		90
s PVC	MSA UPC	<180m			0.20		48
b PVC	STD. SAFETY GL-20	30m	1290	4	n.a.		75
s PVC	STD. SAFETY WG-20	30m	1400	4	n.a.		75
s PVC	WHEELER ACID KING	16m	<90000	5	n.a.		85
b PVC+POLYURETHANE	BATA POLYBLEND	131m	0.2	1	n.a.		86
b PVC+POLYURETHANE	BATA POLYBLEND	169m	9	1	n.a.		90
b PVC+POLYURETHANE	BATA POLYMAX	18m	3	2	n.a.		86

G A R M E N T Class & Number/ test chemical/ MATERIAL NAME	MANUFACTURER & PRODUCT IDENTIFICATION	Break- through Time in Minutes	Perm- eation Rate in mg/sq m /min.	I N D E X	Thick- ness in mm	degrad. and comments	R e f s
b PVC+POLYURETHANE	BATA SUPER POLY	74m	0.06	1	n.a.		86
s SARANEX-23	DU PONT TYVEK SARAN	97m	5	1	n.a.		8
s SARANEX-23 2-PLY	DU PONT TYVEK SARAN	>480m		0	n.a.		8
u TEFLON	CHEMFAB	5m	10	4	0.10		65
s TEFLON	CHEMFAB CHALL. 5100	>480m	0	0	n.a.		56
s TEFLON	CHEMFAB CHALL. 5100	>270m	0	0	n.a.		65
s TEFLON	CHEMFAB CHALL. 5200	>300m	0	0	n.a.		80
s TEFLON	CHEMFAB CHALL. 6000	>180m			n.a.		80
s TEFLON	LIFE-GUARD TEFGUARD	>480m	0	0	n.a.		77
s TEFLON	WHEELER ACID KING	>480m	0	0	n.a.		85
v TEFLON-FEP	CHEMFAB CHALL.	>180m			0.25		65
s UNKNOWN MATERIAL	CHEMRON CHEMREL	>1440m	0	0	n.a.		67
s UNKNOWN MATERIAL	CHEMRON CHEMREL MAX	>1440m	0	0	n.a.		67
s UNKNOWN MATERIAL	DU PONT BARRICADE	>480m	0	0	n.a.		8
s UNKNOWN MATERIAL	KAPPLER CPF III	7m	8	3	n.a.		77
s UNKNOWN MATERIAL	LIFE-GUARD RESPONDE	>480m	0	0	n.a.		77
g UNKNOWN MATERIAL	NORTH SILVERSHIELD	>480m	0	0	0.10		7
g VITON	NORTH F-091				0.25	degrad.	34
g VITON	NORTH F-091	6m	7580	5	0.26		38
s VITON/BUTYL	DRAGER 500 OR 710	>150m			n.a.		91
s VITON/BUTYL/UNKN.	TRELLEBORG HPS	>180m			n.a.		71
s VITON/CHLOROBUTYL	LIFE-GUARD VC-100	100m			n.a.		77
s VITON/CHLOROBUTYL	LIFE-GUARD VNC 200	120m	0.1		n.a.		77
u VITON/CHLOROBUTYL	Unknown	97m			n.a.		46
g VITON/NEOPRENE	ERISTA VITRIC	>60m			0.60		38
s VITON/NEOPRENE	MSA VAUTEX	20m	40	3	n.a.		48

acetophenone
CAS Number: 98-86-2
Primary Class: 392   Ketones, Aromatic

g PE/EVAL/PE	SAFETY4 4H	>240m	0	0	0.07	35°C	60
s TEFLON	CHEMFAB CHALL. 5100	>180m	0	1	n.a.		65

acetyl chloride
CAS Number: 75-36-5
Primary Class: 111   Acid Halides, Carboxylic, Aliphatic and Alicyclic

g BUTYL	NORTH B-174	180m	360	3	0.64	degrad.	34
g NATURAL RUBBER	ACKWELL 5-109	<1m			0.15	degrad.	34
g NEOPRENE	ANSELLEDMONT 29-870				0.48	degrad.	34
g NITRILE	ANSELLEDMONT 37-155				0.40	degrad.	34
g PVAL	EDMONT 25-950				n.a.	degrad.	34
g PVC	ANSELLEDMONT 34-100	<1m			0.20	degrad.	34
s SARANEX-23	DU PONT TYVEK SARAN	37m	11	3	n.a.		8
s TEFLON	CHEMFAB CHALL. 5100	>190m			n.a.		65
s UNKNOWN MATERIAL	CHEMRON CHEMREL	58m	120	3	n.a.		67

G A R M E Class & Number/ N test chemical/ T MATERIAL NAME	MANUFACTURER & PRODUCT IDENTIFICATION	Break-through Time in Minutes	Perm-eation Rate in mg/sq m /min.	I N D E X	Thick-ness in mm	degrad. and comments	R e f s
s UNKNOWN MATERIAL	DU PONT BARRICADE	164m	9	1	n.a.		8
s UNKNOWN MATERIAL	LIFE-GUARD RESPONDE	>240m	0	0	n.a.		77
g VITON	NORTH F-091	8m	40980	5	0.26	degrad.	34
s VITON/CHLOROBUTYL	LIFE-GUARD VC-100	76m			n.a.		77

acrolein (synonym: acrylaldehyde)
CAS Number: 107-02-8
Primary Class: 121   Aldehydes, Aliphatic and Alicyclic

g BUTYL	NORTH B-174	>900m	0	0	0.61		34
g NEOPRENE	ANSELLEDMONT 29-870				0.48	degrad.	34
g NITRILE	ANSELLEDMONT 37-155	4m	9600	5	0.38	degrad.	34
g PE/EVAL/PE	SAFETY4 4H	>240m	0	0	0.07	35°C	60
g PVAL	EDMONT 25-950	15m	30	3	0.25	degrad.	34
g PVC	ANSELLEDMONT 34-100				0.20	degrad.	34
s TEFLON	CHEMFAB CHALL. 5100	38m	<1	2	n.a.		65
s UNKNOWN MATERIAL	LIFE-GUARD RESPONDE	>180m	0	1	n.a.		77
g VITON	NORTH F-091	<1m	4320	5	0.23	degrad.	34
s VITON/BUTYL	DRAGER 500 OR 710	>30m			n.a.		91

acrylamide (synonym: 2-propenamide)
CAS Number: 79-06-1
Primary Class: 133   Acrylamides

g PE/EVAL/PE	SAFETY4 4H	>240m	0	0	0.07	35°C	60

acrylamide, 15% & methyl ethyl ketone, 85%
Primary Class: 133   Acrylamides

acrylamide, 88% & N,N-methylenebisacrylamide & water
Primary Class: 133   Acrylamides

g PE/EVAL/PE	SAFETY4 4H	>240m	0	0	0.07	35°C	60

acrylate UV - lacquer in butyl acetate
Primary Class: 830   Lacquer Products

g PE/EVAL/PE	SAFETY4 4H	>480m	0	0	0.07	35°C	60

acrylic acid
CAS Number: 79-10-7
Primary Class: 102   Acids, Carboxylic, Aliphatic and Alicyclic, Unsubstituted

g BUTYL	NORTH B-174	>480m	0	0	0.63		34

G A R M E Class & Number/ N test chemical/ T MATERIAL NAME	MANUFACTURER & PRODUCT IDENTIFICATION	Break- through Time in Minutes	Perm- eation Rate in mg/sq m /min.	I N D E X	Thick- ness in mm	degrad. and comments	R e f s
g NATURAL RUBBER	ACKWELL 5-109	<1m	2820	5	0.15	degrad.	34
g NATURAL RUBBER	ANSELLEDMONT 36-124	35m			0.46		6
g NATURAL RUBBER	ANSELLEDMONT 392	80m			0.48		6
g NEOPRENE	ANSELLEDMONT 29-840	70m			0.38		6
g NEOPRENE	ANSELLEDMONT 29-870				0.50	degrad.	34
g NEOPRENE	ANSELLEDMONT NEOX	>360m		0	n.a.		6
g NITRILE	ANSELLEDMONT 37-155				0.40	degrad.	34
g NITRILE	ANSELLEDMONT 37-165	120m			0.60		6
c PE	DU PONT TYVEK QC	7m	54	4	n.a.		8
g PE/EVAL/PE	SAFETY4 4H	210m			0.07	35°C	60
g PVAL	ANSELLEDMONT PVA				n.a.	degrad.	6
g PVAL	EDMONT 15-552				n.a.	degrad.	6
g PVC	ANSELLEDMONT 34-100	9m	1500	5	0.20		34
g PVC	ANSELLEDMONT SNORKE				n.a.	degrad.	6
s SARANEX-23	DU PONT TYVEK SARAN	>480m	0	0	n.a.		8
s TEFLON	CHEMFAB CHALL. 5100	>180m			n.a.		65
s UNKNOWN MATERIAL	DU PONT BARRICADE	79m	60	2	n.a.		8
s UNKNOWN MATERIAL	LIFE-GUARD RESPONDE	>480m	0	0	n.a.		77
g VITON	NORTH F-091	354m	12	2	0.28		34

acrylonitrile (synonym: propenenitrile)
CAS Number: 107-13-1
Primary Class: 431   Nitriles, Aliphatic and Alicyclic

g BUTYL	BEST 878	>480m	0	0	0.75		53
g BUTYL	NORTH B-161	180m	<0.1	1	0.31		7
g BUTYL	NORTH B-174	>480m	0	0	0.70		34
s BUTYL/NEOPRENE	MSA BETEX	125m	2	1	n.a.		48
u CPE	ILC DOVER	17m			n.a.		29
g HYPALON	NORTH Y-1532	2m	2380	5	0.40		7
g NATURAL RUBBER	ACKWELL 5-109	1m	7800	5	0.15		34
g NEOPRENE	ANSELLEDMONT 29-870	20m	1860	4	0.48		34
g NEOPRENE	BEST 32	27m	130	3	n.a.		53
g NEOPRENE	BEST 6780	23m	590	3	n.a.		53
g NITRILE	BEST 727				0.38	degrad.	53
g NITRILE	NORTH LA-142G	3m	10560	5	0.36		7
c PE	DU PONT TYVEK QC	5m	0.1	3	n.a.		8
g PE/EVAL/PE	SAFETY4 4H	>240m	0	0	0.07	35°C	60
g PE/EVAL/PE	SAFETY4 4H	>480m	0	0	0.07		60
g PVAL	EDMONT 25-950	42m	60	3	0.76		34
s SARANEX-23	DU PONT TYVEK SARAN	23m	<1	2	n.a.		8
s TEFLON	CHEMFAB CHALL. 5100	54m	0.9	2	n.a.		65
c UNKNOWN MATERIAL	CHEMRON CHEMTUFF	8m	1.5	3	n.a.		67
s UNKNOWN MATERIAL	DU PONT BARRICADE	>480m	0	0	n.a.		8
s UNKNOWN MATERIAL	LIFE-GUARD RESPONDE	>240m	0	0	n.a.		77
g VITON	NORTH F-091	1m	10560	5	0.40		7
s VITON/CHLOROBUTYL	LIFE-GUARD VC-100	26m			n.a.		77
u VITON/CHLOROBUTYL	Unknown	70m			0.36		29

G A R M E Class & Number/ N test chemical/ T MATERIAL NAME	MANUFACTURER & PRODUCT IDENTIFICATION	Break- through Time in Minutes	Perm- eation Rate in mg/sq m /min.	I N D E X	Thick- ness in mm	degrad. and comments	R e f s
s VITON/NEOPRENE	MSA VAUTEX	<5m	230	4	n.a.		48

adiponitrile
CAS Number: 111-69-3
Primary Class: 431   Nitriles, Aliphatic and Alicyclic

s BUTYL/NEOPRENE	MSA BETEX	>240m	0	0	n.a.		48
s TEFLON	CHEMFAB CHALL. 5100	>180m	0	1	n.a.		65
s VITON/NEOPRENE	MSA VAUTEX	>240m	0	0	n.a.		48

Aeroshell fluid 4
Primary Class: 900   Miscellaneous Unclassified Chemicals

g PE/EVAL/PE	SAFETY4 4H	>240m	0	0	0.07	35°C	60

AFFF (synonym: Aqueous Fire Fighting Foam)
Primary Class: 900   Miscellaneous Unclassified Chemicals

s UNKNOWN MATERIAL	LIFE-GUARD RESPONDE	>240m		0	n.a.		77

allyl acrylate (synonym: allyl 2-propenoate)
CAS Number: 999-55-3
Primary Class: 223   Esters, Carboxylic, Acrylates and Methacrylates

g NATURAL RUBBER	ACKWELL 5-109				0.15	degrad.	34
g NEOPRENE	ANSELLEDMONT 29-870				0.48	degrad.	34
g NITRILE	ANSELLEDMONT 37-155				0.40	degrad.	34
g PVC	ANSELLEDMONT 34-100				0.20	degrad.	34
g VITON	NORTH F-091				0.25	degrad.	34

allyl alcohol (synonym: 2-propenol)
CAS Number: 107-18-6
Primary Class: 311   Hydroxy Compounds, Aliphatic and Alicyclic, Primary

g BUTYL	NORTH B-161	>491m	0	0	0.41		19
s BUTYL/NEOPRENE	MSA BETEX	>480m	0	0	n.a.		48
u CPE	ILC DOVER	120m			n.a.		29
g NEOPRENE	ANSELLEDMONT 29-870	4m			0.50		19
g PVAL	EDMONT 25-545	14m			n.a.		19
s TEFLON	CHEMFAB CHALL. 5100	>840m	0	0	n.a.		65
s UNKNOWN MATERIAL	DU PONT BARRICADE	>480m	0	0	n.a.		8
s UNKNOWN MATERIAL	LIFE-GUARD RESPONDE	>480m	0	0	n.a.		77
u VITON/CHLOROBUTYL	Unknown	>180m			n.a.		29
s VITON/NEOPRENE	MSA VAUTEX	>480m	0	0	n.a.		48

G			Perm-	I			R
A				eation	N	Thick-		
R			Break-	Rate in	D	ness	degrad.	e
M	Class & Number/	MANUFACTURER	through					
E	Class & Number/	MANUFACTURER	through	Rate in	D	ness	degrad.	e
N	test chemical/	& PRODUCT	Time in	mg/sq m	E	in	and	f
T	MATERIAL NAME	IDENTIFICATION	Minutes	/min.	X	mm	comments	s

allyl bromide (synonym: 3-bromopropene)
CAS Number: 106-95-6
Primary Class: 262   Halogen Compounds, Allylic and Benzylic

s	BUTYL	TRELLEBORG TRELLCHE				n.a.	degrad.	71

allyl chloride (synonym: 3-chloropropene)
CAS Number: 107-05-1
Primary Class: 262   Halogen Compounds, Allylic and Benzylic

g	BUTYL	NORTH B-131	40m			0.35		83
g	BUTYL	NORTH B-174	50m	16860	5	0.71	degrad.	34
u	CPE	ILC DOVER	75m			n.a.		29
g	PE/EVAL/PE	SAFETY4 4H	>240m	0	0	0.07	35°C	60
g	PVAL	EDMONT 25-950	81m	0.3	1	0.71		34
g	PVC	ANSELLEDMONT 34-100	1m	14100	5	0.20	degrad.	34
s	TEFLON	CHEMFAB CHALL. 5100	>102m	<1	1	n.a.		65
v	TEFLON-FEP	CHEMFAB CHALL.	>180m			0.25		65
s	UNKNOWN MATERIAL	KAPPLER CPF III	12m	12	4	n.a.		77
s	UNKNOWN MATERIAL	LIFE-GUARD RESPONDE	>180m	0	1	n.a.		77
g	UNKNOWN MATERIAL	NORTH SILVERSHIELD	5m			0.08		83
g	VITON	NORTH F-091	31m	960	3	0.36		34
g	VITON	NORTH F-091	8m			0.25		83
u	VITON/CHLOROBUTYL	Unknown	3m			n.a.		29

allylamine (synonym: monoallylamine)
CAS Number: 107-11-9
Primary Class: 141   Amines, Aliphatic and Alicyclic, Primary

g	BUTYL	NORTH B-174	234m	416	3	0.63		34
g	NATURAL RUBBER	ACKWELL 5-109	1m	39700	5	0.15	degrad.	34
g	NEOPRENE	ANSELLEDMONT 29-870				0.48	degrad.	34
g	NITRILE	ANSELLEDMONT 37-155				0.40	degrad.	34
g	PE/EVAL/PE	SAFETY4 4H	15m			0.07		60
g	PVAL	EDMONT 25-950	12m	72540	5	0.64	degrad.	34
g	PVC	ANSELLEDMONT 34-100	1m	58800	5	0.20	degrad.	34
g	VITON	NORTH F-091				0.26	degrad.	34

alpha-methylstyrene (synonyms: 2-phenylpropene; isopropenylbenzene)
CAS Number: 98-83-9
Primary Class: 292   Hydrocarbons, Aromatic
Related Class: 294   Hydrocarbons, Aliphatic and Alicyclic, Unsaturated

g	BUTYL	NORTH B-161	15m	45	3	0.37		34
g	NEOPRENE	ANSELLEDMONT 29-870	<15m	137	3	0.28		34
g	NITRILE	ANSELLEDMONT 37-175	20m	22	3	0.32		34
g	PVAL	EDMONT 25-250	<15m	0.02	2	0.50		34
g	VITON	NORTH F-091	>480m	0	0	0.33		34

G A R M E Class & Number/ N test chemical/ T MATERIAL NAME	MANUFACTURER & PRODUCT IDENTIFICATION	Break- through Time in Minutes	Perm- eation Rate in mg/sq m /min.	I N D E X	Thick- ness in mm	R degrad. and e comments s	R e f s

Ambush
Primary Class: 840   Pesticides, Mixtures and Formulations

g NEOPRENE	ANSELLEDMONT 29-840	120m	4	1	0.40		57
g NITRILE	ANSELLEDMONT 37-145	240m	<1	1	0.30		57
c PE	MOLNLYCKE H D	<5m	4	3	n.a.		57
g PE/EVAL/PE	SAFETY4 4H	>400m	0	0	0.07		57
g VITON	NORTH F-091	<30m			0.27		57

ammonia (synonym: ammonia gas)
CAS Number: 7664-41-7
Primary Class: 350   Inorganic Gases and Vapors

g BUTYL	BRUNSWICK BUTYL STD	>480m	0	0	0.63		89
g BUTYL	BRUNSWICK BUTYL-XTR	>480m	0	0	0.71		89
s BUTYL	DRAGER 500 OR 710	>60m			n.a.		91
s BUTYL	WHEELER ACID KING	>480m	0	0	n.a.		85
s BUTYL	WHEELER ACID KING	>566m	0	0	n.a.	100%	88
g BUTYL/ECO	BRUNSWICK BUTYL-POL	>480m	0	0	0.63		89
s BUTYL/NEOPRENE	MSA BETEX	>480m	0	0	n.a.		48
s BUTYL/NEOPRENE	MSA BETEX	>850m	0	0	n.a.	100%	88
s CPE	ILC DOVER CLOROPEL	154m	3	1	n.a.	100%	88
s CPE	ILC DOVER CLOROPEL	273m	2	1	n.a.	0.2%	88
s CPE	STD. SAFETY	120m	15	2	n.a.		75
g NATURAL RUBBER	DAYTON SURGICAL	2m	265	4	0.19	100%	88
g NATURAL RUBBER	DAYTON SURGICAL	5m	78	4	0.19	0.2%	88
g NEOPRENE	BRUNSWICK NEOPRENE	>480m	0	0	0.88		89
s NEOPRENE	FAIRPRENE	163m	6	1	n.a.	100%	88
s NEOPRENE	FAIRPRENE	303m	2	1	n.a.	0.2%	88
b NEOPRENE	RAINFAIR	>180m			n.a.		90
b NEOPRENE	RANGER	>180m			n.a.		90
b NEOPRENE	SERVUS NO 22204	>180m			n.a.		90
g NITRILE	MAPA-PIONEER A-15	288m	70	2	0.35	100%	88
g NITRILE	MAPA-PIONEER A-15	621m	4	1	0.35	0.2%	88
b NITRILE+PUR+PVC	BATA HAZMAX	>180m			n.a.		86
b NITRILE+PVC	SERVUS NO 73101	>180m			2.40		90
b NITRILE+PVC	TINGLEY	157m	3	1	3.30		90
c PE	DU PONT TYVEK QC	11m	1.2	3	n.a.		8
m POLYURETHANE	ILC DOVER	53m	10	3	0.43	100%	88
m POLYURETHANE	ILC DOVER	58m	4	2	0.43	0.2%	88
b PVC	JORDAN DAVID	175m	1.4	1	n.a.		90
b PVC	STD. SAFETY GL-20	25m	14	3	n.a.		75
s PVC	STD. SAFETY WG-20	13m	23	4	n.a.		75
s PVC	WHEELER ACID KING	25m	<9000	4	n.a.		85
s PVC	WHEELER ACID KING	19m	16	3	n.a.	100%	88
s PVC	WHEELER ACID KING	21m	11	3	n.a.	0.2%	88
b PVC+POLYURETHANE	BATA POLYBLEND	151m	3	1	n.a.		90
s SARANEX-23	DU PONT TYVEK SARAN	19m	2	2	n.a.	2°C	8

G A R M E Class & Number/ N test chemical/ T MATERIAL NAME	MANUFACTURER & PRODUCT IDENTIFICATION	Break- through Time in Minutes	Perm- eation Rate in mg/sq m /min.	I N D E X	Thick- ness in mm	degrad. and comments	R e f s
s SARANEX-23	DU PONT TYVEK SARAN	21m	0.9	2	n.a.	100%	88
s SARANEX-23	DU PONT TYVEK SARAN	>300m	0	0	n.a.	0.2%	88
s TEFLON	CHEMFAB CHALL. 5100	254m	0.3	0	n.a.	100%	88
s TEFLON	CHEMFAB CHALL. 5200	>300m	0	0	n.a.		80
s TEFLON	CHEMFAB CHALL. 6000	>180m			n.a.		80
s TEFLON	LIFE-GUARD TEFGUARD	>480m	0	0	n.a.		77
s TEFLON	WHEELER ACID KING	>480m	0	0	n.a.		85
s UNKNOWN MATERIAL	CHEMRON CHEMREL	33m	2	2	n.a.	100%	88
s UNKNOWN MATERIAL	CHEMRON CHEMREL	141m	0.2	1	n.a.	0.2%	88
s UNKNOWN MATERIAL	DU PONT BARRICADE	68m	17	2	n.a.	2°C	8
s UNKNOWN MATERIAL	KAPPLER CPF III	12m	15	4	n.a.		77
s UNKNOWN MATERIAL	LIFE-GUARD RESPONDE	>480m	0	0	n.a.		77
s VITON/BUTYL	DRAGER 500 OR 710	>60m			n.a.		91
s VITON/BUTYL/UNKN.	TRELLEBORG HPS	>180m			n.a.		71
s VITON/CHLOROBUTYL	LIFE-GUARD VC-100	452m	0	0	n.a.		77
s VITON/NEOPRENE	MSA VAUTEX	>480m	0	0	n.a.		48

ammonia, liquid
Primary Class: 380   Inorganic Bases

s UNKNOWN MATERIAL	LIFE-GUARD RESPONDE	>480m	0	0	n.a.		77

ammonium acetate (saturated solution) (synonym: acetic acid ammonium salt)
CAS Number: 631-61-8
Primary Class: 340   Inorganic Salts

g NATURAL RUBBER	MARIGOLD BLACK HEVY	>480m	0	0	0.65		79
g NATURAL RUBBER	MARIGOLD FEATHERLIT	>480m	0	0	0.15		79
g NATURAL RUBBER	MARIGOLD MED WT EXT	>480m	0	0	0.45		79
g NATURAL RUBBER	MARIGOLD MEDICAL	>480m	0	0	0.28		79
g NATURAL RUBBER	MARIGOLD ORANGE SUP	>480m	0	0	0.73		79
g NATURAL RUBBER	MARIGOLD R SURGEONS	>480m	0	0	0.28		79
g NATURAL RUBBER	MARIGOLD RED LT WT	>480m	0	0	0.43		79
g NATURAL RUBBER	MARIGOLD SENSOTECH	>480m	0	0	0.28		79
g NATURAL RUBBER	MARIGOLD SUREGRIP	>480m	0	0	0.58		79
g NATURAL RUBBER	MARIGOLD YELLOW	>480m	0	0	0.43		79
g NATURAL+NEOPRENE	MARIGOLD FEATHERWT	>480m	0	0	0.35		79
g NITRILE	MARIGOLD BLUE	>480m	0	0	0.45		79
g NITRILE	MARIGOLD GREEN SUPA	>480m	0	0	0.30		79
g NITRILE	MARIGOLD NITROSOLVE	>480m	0	0	0.75		79
g UNKNOWN MATERIAL	MARIGOLD R MEDIGLOV	>480m	0	0	0.03		79

ammonium carbonate (saturated solution)
CAS Number: 10361-29-2
Primary Class: 340   Inorganic Salts

g NATURAL RUBBER	MARIGOLD BLACK HEVY	>480m	0	0	0.65		79
g NATURAL RUBBER	MARIGOLD FEATHERLIT	>480m	0	0	0.15	degrad.	79

G A R M E N T	Class & Number/ test chemical/ MATERIAL NAME	MANUFACTURER & PRODUCT IDENTIFICATION	Break- through Time in Minutes	Perm- eation Rate in mg/sq m /min.	I N D E X	Thick- ness in mm	degrad. and comments	R e f s
g	NATURAL RUBBER	MARIGOLD MED WT EXT	>480m	0	0	0.45		79
g	NATURAL RUBBER	MARIGOLD MEDICAL	>480m	0	0	0.28	degrad.	79
g	NATURAL RUBBER	MARIGOLD ORANGE SUP	>480m	0	0	0.73		79
g	NATURAL RUBBER	MARIGOLD R SURGEONS	>480m	0	0	0.28	degrad.	79
g	NATURAL RUBBER	MARIGOLD RED LT WT	>480m	0	0	0.43		79
g	NATURAL RUBBER	MARIGOLD SENSOTECH	>480m	0	0	0.28	degrad.	79
g	NATURAL RUBBER	MARIGOLD SUREGRIP	>480m	0	0	0.58		79
g	NATURAL RUBBER	MARIGOLD YELLOW	>480m	0	0	0.38		79
g	NATURAL+NEOPRENE	MARIGOLD FEATHERWT	>480m	0	0	0.35		79
g	NITRILE	MARIGOLD BLUE	>480m	0	0	0.45		79
g	NITRILE	MARIGOLD GREEN SUPA	>480m	0	0	0.30		79
g	NITRILE	MARIGOLD NITROSOLVE	>480m	0	0	0.75		79
g	UNKNOWN MATERIAL	MARIGOLD R MEDIGLOV	>480m	0	0	0.03		79

ammonium fluoride <30%
Primary Class: 340   Inorganic Salts

g	NATURAL+NEOPRENE	PLAYTEX ARGUS 123	>480m	0	0	n.a.		72

ammonium fluoride, 30-70%
CAS Number: 12125-01-8
Primary Class: 340   Inorganic Salts

g	NATURAL RUBBER	ANSELL CONFORM 4205	>360m	0	0	0.13		55
g	NATURAL RUBBER	ANSELL FL 200 254	>360m	0	0	0.51		55
g	NATURAL RUBBER	ANSELL ORANGE 208	>360m	0	0	0.76		55
g	NATURAL RUBBER	ANSELL PVL 040	>360m	0	0	0.46		55
g	NATURAL RUBBER	ANSELL STERILE 832	>360m	0	0	0.23		55
g	NATURAL RUBBER	ANSELLEDMONT 30-139	>360m	0	0	0.25		42
g	NATURAL RUBBER	ANSELLEDMONT 392	>360m	0	0	0.48		6
g	NATURAL+NEOPRENE	ANSELL OMNI 276	>360m	0	0	0.56		55
g	NATURAL+NEOPRENE	ANSELL TECHNICIANS	>360m	0	0	0.43		55
g	NEOPRENE	ANSELL NEOPRENE 530	>360m	0	0	0.46		55
g	NEOPRENE/NATURAL	ANSELL CHEMI-PRO	>360m	0	0	0.72		55
g	NITRILE	ANSELL CHALLENGER	>360m	0	0	0.38		55
g	NITRILE	ANSELLEDMONT 49-125	>360m	0	0	0.23		42
g	NITRILE	ANSELLEDMONT 49-155	>360m	0	0	0.38		42
g	PE/EVAL/PE	SAFETY4 4H	>240m	0	0	0.07	35°C	60
g	PVAL	ANSELLEDMONT PVA	>360m		0	n.a.		6

ammonium hydroxide, 30-70%
CAS Number: 1336-21-6
Primary Class: 380   Inorganic Bases

g	BUTYL	BRUNSWICK BUTYL STD	>480m	0	0	0.63		89
g	BUTYL	BRUNSWICK BUTYL-XTR	>480m	0	0	0.63		89
s	BUTYL	MSA CHEMPRUF	>480m	0	0	n.a.		48
g	BUTYL/ECO	BRUNSWICK BUTYL-POL	>480m	0	0	0.63		89
g	HYPALON	COMASEC DIPCO	>480m		0	0.60		40

G A R M E Class & Number/ N test chemical/ T MATERIAL NAME	MANUFACTURER & PRODUCT IDENTIFICATION	Break- through Time in Minutes	Perm- eation Rate in mg/sq m /min.	I N D E X	Thick- ness in mm	degrad. and comments	R e f s
g NATURAL RUBBER	ANSELL CONFORM 4205	>360m	0	0	0.13		55
g NATURAL RUBBER	ANSELL FL 200 254	>360m	0	0	0.51		55
g NATURAL RUBBER	ANSELL ORANGE 208	>360m	0	0	0.76		55
g NATURAL RUBBER	ANSELL PVL 040	>360m	0	0	0.46		55
g NATURAL RUBBER	ANSELL STERILE 832	>360m	0	0	0.23		55
g NATURAL RUBBER	ANSELLEDMONT 36-124	102m			0.46		6
g NATURAL RUBBER	ANSELLEDMONT 392	90m	<9	1	0.48		6
g NATURAL RUBBER	COMASEC FLEXIGUM	330m	3000	4	0.95		40
g NATURAL RUBBER	MARIGOLD BLACK HEVY	12m	158	4	0.65		79
g NATURAL RUBBER	MARIGOLD FEATHERLIT	1m	70	4	0.15		79
g NATURAL RUBBER	MARIGOLD MED WT EXT	10m	525	4	0.45		79
g NATURAL RUBBER	MARIGOLD MEDICAL	4m	260	4	0.28		79
g NATURAL RUBBER	MARIGOLD ORANGE SUP	12m	122	4	0.73		79
g NATURAL RUBBER	MARIGOLD R SURGEONS	4m	260	4	0.28		79
g NATURAL RUBBER	MARIGOLD RED LT WT	9m	487	4	0.43		79
g NATURAL RUBBER	MARIGOLD SENSOTECH	4m	260	4	0.28		79
g NATURAL RUBBER	MARIGOLD SUREGRIP	11m	268	4	0.58		79
g NATURAL RUBBER	MARIGOLD YELLOW	8m	411	4	0.38		79
g NATURAL+NEOPRENE	ANSELL OMNI 276	>360m	0	0	0.46		55
g NATURAL+NEOPRENE	ANSELL TECHNICIANS	>360m	0	0	0.43		55
g NATURAL+NEOPRENE	MARIGOLD FEATHERWT	7m	373	4	0.35		79
g NEOPRENE	ANSELL NEOPRENE 530	>360m	0	0	0.46		55
g NEOPRENE	ANSELLEDMONT 29-840	>360m	0	0	0.46		6
g NEOPRENE	ANSELLEDMONT NEOX	>360m	0	0	n.a.		6
g NEOPRENE	BRUNSWICK NEOPRENE	>480m	0	0	0.63		89
g NEOPRENE	COMASEC COMAPRENE	>360m	0	0	n.a.		40
g NEOPRENE/NATURAL	ANSELL CHEMI-PRO	>360m	0	0	0.72		55
g NITRILE	ANSELL CHALLENGER	>360m	0	0	0.38		55
g NITRILE	ANSELLEDMONT 37-165	>360m	0	0	0.60		6
g NITRILE	COMASEC COMATRIL	>480m	0	0	0.55		40
g NITRILE	COMASEC COMATRIL SU	>480m	0	0	0.60		40
g NITRILE	COMASEC FLEXITRIL	45m	7860	4	n.a.		40
g NITRILE	MARIGOLD BLUE	276m	58	2	0.45		79
g NITRILE	MARIGOLD GREEN SUPA	154m	92	2	0.30		79
g NITRILE+PVC	COMASEC MULTIPLUS	180m	3000	4	n.a.		40
g PVAL	ANSELLEDMONT PVA				n.a.	degrad.	6
g PVAL	EDMONT 15-552				n.a.	degrad.	6
g PVC	ANSELLEDMONT M GRIP	240m	0	0	n.a.		6
g PVC	COMASEC MULTIPOST	130m	3840	4	n.a.		40
g PVC	COMASEC MULTITOP	160m	3300	4	n.a.		40
g PVC	COMASEC NORMAL	45m	3840	4	n.a.		40
g PVC	COMASEC OMNI	90m	15300	5	n.a.		40
s PVC	MSA UPC	>480m	0	0	0.20		48
g UNKNOWN MATERIAL	MARIGOLD R MEDIGLOV	1m	106	4	0.03		79

ammonium hydroxide, <30%
Primary Class: 380   Inorganic Bases

g BUTYL	BEST 878	>480m	0	0	0.75		53
g NATURAL RUBBER	BEST 65NFW	120m	800	3	n.a.		53

G A R M E Class & Number/ N test chemical/ T MATERIAL NAME	MANUFACTURER & PRODUCT IDENTIFICATION	Break- through Time in Minutes	Perm- eation Rate in mg/sq m /min.	I N D E X	Thick- ness in mm	degrad. and comments	R e f s
g NATURAL RUBBER	MAPA-PIONEER L-118	58m	1080	4	0.46		36
u NATURAL RUBBER	Unknown	> 60m			n.a.		9
u NATURAL RUBBER	Unknown	<1m			0.43	degrad.	9
g NATURAL+NEOPRENE	PLAYTEX ARGUS 123	27m	180	3	n.a.		72
g NEOPRENE	BEST 32	180m	270	3	n.a.		53
g NEOPRENE	BEST 6780	>480m	0	0	n.a.		53
g NEOPRENE	MAPA-PIONEER N-44	>480m	0	0	0.56		36
u NEOPRENE	Unknown	>60m			0.61		9
g NITRILE	BEST 727	240m	274	3	0.38		53
g NITRILE	MAPA-PIONEER A-14	>480m	0	0	0.56		36
g NITRILE	MARIGOLD NITROSOLVE	300m	51	2	0.75		79
u NITRILE	Unknown	>60m			0.45	degrad.	9
c PE	DU PONT TYVEK QC	<1m	603	4	n.a.		8
g PE/EVAL/PE	SAFETY4 4H	40m			0.07	35°C	60
g PE/EVAL/PE	SAFETY4 4H	110m			0.07		60
g PVC	MAPA-PIONEER V-20	>480m	0	0	0.51		36
u PVC	Unknown	1m			0.22	degrad.	9
u PVC	Unknown	<1m			n.a.		9
s UNKNOWN MATERIAL	DU PONT BARRICADE	100m	11	2	n.a.		8
g VITON	NORTH F-121	>60m			0.31		9

ammonium nitrate
CAS Number: 6884-52-2
Primary Class: 340   Inorganic Salts

g BUTYL	BRUNSWICK BUTYL STD	>480m	0	0	0.63		89
g BUTYL	BRUNSWICK BUTYL-XTR	>480m	0	0	0.63		89
g BUTYL/ECO	BRUNSWICK BUTYL-POL	>480m	0	0	0.63		89
g NEOPRENE	BRUNSWICK NEOPRENE	>480m	0	0	0.88		89

amyl acetate (synonyms: n-amyl acetate; n-pentyl acetate)
CAS Number: 628-63-7
Primary Class: 222   Esters, Carboxylic, Acetates

g BUTYL	BEST 878	158m	510	3	0.75		53
g BUTYL	COMASEC BUTYL	190m	426	3	0.50		40
g BUTYL/NEOPRENE	COMASEC BUTYL PLUS	190m	426	3	0.50		40
g NATURAL RUBBER	ANSELL ORANGE 208	30m	0.2		0.76		55
g NATURAL RUBBER	ANSELLEDMONT 36-124				0.46	degrad.	6
g NATURAL RUBBER	ANSELLEDMONT 392				0.48	degrad.	6
g NATURAL RUBBER	BEST 65NFW				n.a.	degrad.	53
g NATURAL RUBBER	COMASEC FLEXIGUM	55m	360	3	0.95		40
g NATURAL RUBBER	MARIGOLD BLACK HEVY	8m	2733	5	0.65	degrad.	79
g NATURAL RUBBER	MARIGOLD FEATHERLIT	1m	3324	5	0.15	degrad.	79
g NATURAL RUBBER	MARIGOLD MED WT EXT	3m	2067	5	0.45	degrad.	79
g NATURAL RUBBER	MARIGOLD MEDICAL	1m	2800	5	0.28	degrad.	79
g NATURAL RUBBER	MARIGOLD ORANGE SUP	8m	2800	5	0.73	degrad.	79

G A R M E Class & Number/ N test chemical/ T MATERIAL NAME	MANUFACTURER & PRODUCT IDENTIFICATION	Break-through Time in Minutes	Perm-eation Rate in mg/sq m /min.	I N D E X	Thick-ness in mm	degrad. and comments	R e f s
g NATURAL RUBBER	MARIGOLD R SURGEONS	1m	2800	5	0.28	degrad.	79
g NATURAL RUBBER	MARIGOLD RED LT WT	3m	2172	5	0.43	degrad.	79
g NATURAL RUBBER	MARIGOLD SENSOTECH	1m	2800	5	0.28	degrad.	79
g NATURAL RUBBER	MARIGOLD SUREGRIP	6m	2533	5	0.65	degrad.	79
g NATURAL RUBBER	MARIGOLD YELLOW	2m	2381	5	0.38	degrad.	79
g NATURAL+NEOPRENE	MARIGOLD FEATHERWT	2m	2486	5	0.35	degrad.	79
g NATURAL+NEOPRENE	PLAYTEX ARGUS 123	9m	3720	5	n.a.		72
g NEOPRENE	ANSELLEDMONT 29-840				0.38	degrad.	6
g NEOPRENE	ANSELLEDMONT NEOX				n.a.	degrad.	6
g NEOPRENE	BEST 32	63m	1140	4	n.a.		53
g NEOPRENE	BEST 6780	71m	1040	4	n.a.		53
g NEOPRENE	COMASEC COMAPRENE	15m	660	3	n.a.		40
g NITRILE	ANSELL CHALLENGER	>360m	0	0	0.38		55
g NITRILE	ANSELLEDMONT 37-165	60m	<900	3	0.60		6
g NITRILE	BEST 22R				n.a.	degrad.	53
g NITRILE	BEST 727	77m	1190	4	0.38		53
g NITRILE	COMASEC COMATRIL	25m	900	3	0.55		40
g NITRILE	COMASEC COMATRIL SU	40m	300	3	0.60		40
g NITRILE	COMASEC FLEXITRIL	15m	1200	4	n.a.		40
g NITRILE	MARIGOLD BLUE	108m	1449	4	0.43	degrad.	79
g NITRILE	MARIGOLD GREEN SUPA	50m	1200	4	0.30	degrad.	79
g NITRILE	MARIGOLD NITROSOLVE	120m	1499	4	0.75	degrad.	79
g NITRILE+PVC	COMASEC MULTIPLUS	50m	420	3	n.a.		40
g PVAL	ANSELLEDMONT PVA	>360m	0		n.a.		6
g PVAL	EDMONT 15-552	>360m	0	0	n.a.		6
g PVAL	EDMONT 25-545	>360m	0	0	n.a.		6
g PVC	ANSELLEDMONT SNORKE				n.a.	degrad.	6
g PVC	COMASEC MULTIPOST	50m	420	3	n.a.		40
g PVC	COMASEC MULTITOP	35m	420	3	n.a.		40
g PVC	COMASEC NORMAL	30m	480	3	n.a.		40
g PVC	COMASEC OMNI	21m	600	3	n.a.		40
s UNKNOWN MATERIAL	DU PONT BARRICADE	>480m	0	0	n.a.		8
s UNKNOWN MATERIAL	LIFE-GUARD RESPONDE	>480m	0	0	n.a.		77

amyl alcohol (synonyms: 1-pentanol; n-pentanol; pentyl alcohol)
CAS Number: 71-41-0
Primary Class: 311   Hydroxy Compounds, Aliphatic and Alicyclic, Primary

g BUTYL	BEST 878	>480m	0	0	0.75		53
g BUTYL	NORTH B-174	>480m	0	0	0.60		34
g NATURAL RUBBER	ANSELL CONFORM 4205	1m	22	4	0.13		55
g NATURAL RUBBER	ANSELL FL 200 254	22m	0.9	2	0.51		55
g NATURAL RUBBER	ANSELL ORANGE 208	60m	0.2		0.76		55
g NATURAL RUBBER	ANSELL PVL 040	77m	0.5	1	0.46		55
g NATURAL RUBBER	ANSELL STERILE 832	3m	7	3	0.23		55
g NATURAL RUBBER	ANSELLEDMONT 36-124	7m	<90	4	0.46		6

G				Perm-	I			R
A				eation	N	Thick-		
R			Break-	Rate in	D	ness		e
M	Class & Number/	MANUFACTURER	through	mg/sq m	E	in	degrad.	f
E	test chemical/	& PRODUCT	Time in	/min.	X	mm	and	s
N	MATERIAL NAME	IDENTIFICATION	Minutes				comments	
g	NATURAL RUBBER	ANSELLEDMONT 392	25m	<90	3	0.48		6
g	NATURAL RUBBER	BEST 65NFW		5160	5	n.a.		53
g	NATURAL RUBBER	BEST 67NFW		24	5	n.a.		53
g	NATURAL RUBBER	MARIGOLD BLACK HEVY	56m	57	3	0.65		79
g	NATURAL RUBBER	MARIGOLD FEATHERLIT	1m	2036	5	0.15		79
g	NATURAL RUBBER	MARIGOLD MED WT EXT	23m	510	3	0.45	degrad.	79
g	NATURAL RUBBER	MARIGOLD MEDICAL	7m	1400	5	0.28		79
g	NATURAL RUBBER	MARIGOLD ORANGE SUP	59m	12	3	0.73		79g
	NATURAL RUBBER	MARIGOLD R SURGEONS	7m	1400	5	0.28		79
g	NATURAL RUBBER	MARIGOLD RED LT WT	21m	637	3	0.43		79
g	NATURAL RUBBER	MARIGOLD SENSOTECH	7m	1400	5	0.28		79
g	NATURAL RUBBER	MARIGOLD SUREGRIP	46m	193	3	0.58		79
g	NATURAL RUBBER	MARIGOLD YELLOW	16m	891	3	0.38		79
g	NATURAL+NEOPRENE	ANSELL OMNI 276	31m	0.7	2	0.56		55
g	NATURAL+NEOPRENE	ANSELL TECHNICIANS	31m	0.5	2	0.43		55
g	NATURAL+NEOPRENE	MARIGOLD FEATHERWT	14m	1019	5	0.35		79
g	NEOPRENE	ANSELL NEOPRENE 530	240m	0.1		0.46		55
g	NEOPRENE	ANSELLEDMONT 29-840	>360m	0	0	0.31		6
g	NEOPRENE	ANSELLEDMONT 29-870	320m	5	1	0.38		34
g	NEOPRENE	ANSELLEDMONT NEOX	>360m	0	0	n.a.		6
g	NEOPRENE	BEST 32	>480m	0	0	n.a.		53
g	NEOPRENE	BEST 6780	>480m	0	0	n.a.		53
g	NEOPRENE/NATURAL	ANSELL CHEMI-PRO	60m	0.7	1	0.72		55
g	NITRILE	ANSELLEDMONT 37-155	>480m	0	0	0.33		34
g	NITRILE	ANSELLEDMONT 37-165	30m	<9	2	0.60		6
g	NITRILE	BEST 22R	238m	7	1	n.a.		53
g	NITRILE	BEST 727	>480m	0	0	0.38		53
g	NITRILE	MARIGOLD BLUE	>480m	0	0	0.45		79
g	NITRILE	MARIGOLD GREEN SUPA	>480m	0	0	0.30		79
g	NITRILE	MARIGOLD NITROSOLVE	>480m	0	0	0.75		79
g	PVAL	ANSELLEDMONT PVA	>360m		0	n.a.		6
g	PVAL	EDMONT 15-552	>360m	0	0	n.a.		6
g	PVAL	EDMONT 25-545	>360m	0	0	n.a.		6
g	PVC	ANSELLEDMONT M GRIP	12m	9	3	n.a.		6
g	PVC	BEST 725R	116m	30	2	n.a.		53
g	PVC	BEST 812	48		5	n.a.		53
g	UNKNOWN MATERIAL	MARIGOLD R MEDIGLOV	1m	87	4	0.03		79
g	VITON	NORTH F-091	>480m	0	0	0.30		34

			Perm-	I		R
G						
A						
R			Break-	eation	N	Thick-	R
M			through	Rate in	D	ness degrad.	e
E Class & Number/	MANUFACTURER		through	Rate in	D	ness degrad.	e
N test chemical/	& PRODUCT		Time in	mg/sq m	E	in and	f
T MATERIAL NAME	IDENTIFICATION		Minutes	/min.	X	mm comments	s

aniline (synonyms: benzamine; phenylamine)
CAS Number: 62-53-3
Primary Class: 145   Amines, Aromatic, Primary

g	BUTYL	BEST 878	>480m	0		0	0.75		53
g	BUTYL	BRUNSWICK BUTYL STD	>480m	0		0	0.63		89
g	BUTYL	BRUNSWICK BUTYL-XTR	>480m	0		0	0.63		89
g	BUTYL	COMASEC BUTYL	>480m	0		0	0.55		40
s	BUTYL	MSA CHEMPRUF	>480m	0		0	n.a.		48
g	BUTYL	NORTH B-161	>480m	0		0	0.40		34
g	BUTYL/ECO	BRUNSWICK BUTYL-POL	>480m	0		0	0.63		89
g	BUTYL/NEOPRENE	COMASEC BUTYL PLUS	>480m	0		0	0.50		40
s	BUTYL/NEOPRENE	MSA BETEX	>480m	0		0	n.a.		48
g	NATURAL RUBBER	ANSELL ORANGE 208	>360m	0		0	0.76		55
g	NATURAL RUBBER	ANSELLEDMONT 30-139	30m	<90		3	0.25		42
g	NATURAL RUBBER	ANSELLEDMONT 36-124	30m	<90		3	0.46		6
g	NATURAL RUBBER	ANSELLEDMONT 392	25m	<90		3	0.48		6
g	NATURAL RUBBER	COMASEC FLEXIGUM	190m	24		2	0.95		40
g	NATURAL RUBBER	MAPA-PIONEER L-118	>480m	0		0	0.46		36
g	NATURAL RUBBER	MARIGOLD BLACK HEVY	24m	115		3	0.65		79
g	NATURAL RUBBER	MARIGOLD FEATHERLIT	15m	434		3	0.15		79
g	NATURAL RUBBER	MARIGOLD MED WT EXT	15m	184		3	0.45	degrad.	79
g	NATURAL RUBBER	MARIGOLD MEDICAL	15m	330		3	0.28		79
g	NATURAL RUBBER	MARIGOLD ORANGE SUP	25m	108		3	0.73		79
g	NATURAL RUBBER	MARIGOLD R SURGEONS	15m	330		3	0.28		79
g	NATURAL RUBBER	MARIGOLD RED LT WT	15m	205		3	0.43	degrad.	79
g	NATURAL RUBBER	MARIGOLD SENSOTECH	15m	330		3	0.28		79
g	NATURAL RUBBER	MARIGOLD SUREGRIP	21m	136		3	0.58		79
g	NATURAL RUBBER	MARIGOLD YELLOW	15m	247		3	0.38		79
u	NATURAL RUBBER	Unknown	32m				n.a.		10
g	NATURAL+NEOPRENE	MARIGOLD FEATHERWT	15m	267		3	0.35		79
g	NATURAL+NEOPRENE	PLAYTEX ARGUS 123	60m	2520		4	n.a.		72
g	NEOPRENE	ANSELL NEOPRENE 530	>360m	0		0	0.46		55
g	NEOPRENE	ANSELLEDMONT 29-840	35m	<90		3	0.38		6
g	NEOPRENE	ANSELLEDMONT NEOX	180m	<90		2	n.a.		6
g	NEOPRENE	BRUNSWICK NEOPRENE	53m	>140		3	0.88		89
g	NEOPRENE	COMASEC COMAPRENE	30m	120		3	n.a.		40
g	NEOPRENE	MAPA-PIONEER N-44	>480m	0		0	0.56		36
g	NEOPRENE/NATURAL	ANSELL CHEMI-PRO	>360m	0		0	0.72		55
g	NITRILE	ANSELLEDMONT 37-165					0.60	degrad.	6
g	NITRILE	ANSELLEDMONT 37-175	96m	1200		4	0.37		5
g	NITRILE	COMASEC COMATRIL	130m	30		2	0.55		40
g	NITRILE	COMASEC COMATRIL SU	160m	12		2	0.60		40
g	NITRILE	COMASEC FLEXITRIL	80m	18		2	n.a.		40
g	NITRILE	MAPA-PIONEER A-14	72m	180		3	0.56	degrad.	36
g	NITRILE	MARIGOLD BLUE	92m	6864		4	0.45	degrad.	79
g	NITRILE	MARIGOLD GREEN SUPA	30m	9355		4	0.30	degrad.	79
g	NITRILE	MARIGOLD NITROSOLVE	104m	6365		4	0.75	degrad.	79

G A R M E N T	Class & Number/ test chemical/ MATERIAL NAME	MANUFACTURER & PRODUCT IDENTIFICATION	Break- through Time in Minutes	Perm- eation Rate in mg/sq m /min.	I N D E X	Thick- ness in mm	degrad. and comments	R e f s
g	NITRILE	NORTH LA-142G	63m	2700	4	0.36	degrad.	34
g	NITRILE+PVC	COMASEC MULTIPLUS	>240m		0	n.a.		40
g	PE	ANSELLEDMONT 35-125	>60m			0.03		5
c	PE	DU PONT TYVEK QC	<1m	21	4	n.a.		8
g	PE/EVAL/PE	SAFETY4 4H	>240m	0	0	0.07	35°C	60
g	PE/EVAL/PE	SAFETY4 4H	>1440m	0	0	0.07		60
g	PVAL	ANSELLEDMONT PVA	>360m		0	n.a.		6
g	PVAL	COMASEC SOLVATRIL	>480m	0	0	n.a.		40
g	PVAL	EDMONT 15-552	90m	<90	2	n.a.		6
g	PVAL	EDMONT 25-545	>360m	0	0	n.a.		6
g	PVAL	EDMONT 25-950	>960m	0	0	0.33		34
g	PVC	ANSELLEDMONT 34-100				0.20	degrad.	34
g	PVC	ANSELLEDMONT SNORKE	180m	<90	2	n.a.		6
g	PVC	COMASEC MULTIPOST	235m	6	1	n.a.		40
g	PVC	COMASEC MULTITOP	>480m	0	0	n.a.		40
g	PVC	COMASEC NORMAL	290m	6	1	n.a.		40
g	PVC	COMASEC OMNI	220m	48	2	n.a.		40
g	PVC	MAPA-PIONEER V-20	18m	1600	4	0.31		5
g	PVC	MAPA-PIONEER V-20	>480m	0	0	0.51		36
s	PVC	MSA UPC	<60m			0.20		48
s	SARANEX-23	DU PONT TYVEK SARAN	330m	12	2	n.a.		8
s	TEFLON	CHEMFAB CHALL. 5100	>200m	0	1	n.a.		65
s	UNKNOWN MATERIAL	DU PONT BARRICADE	>480m	0	0	n.a.		8
s	UNKNOWN MATERIAL	KAPPLER CPF III	36m	11	3	n.a.		77
s	UNKNOWN MATERIAL	LIFE-GUARD RESPONDE	>480m	0	0	n.a.		77
g	UNKNOWN MATERIAL	NORTH SILVERSHIELD	>480m	0	0	0.10		7
g	VITON	NORTH F-091	6m	1122	5	0.25		34
g	VITON	NORTH F-121	>60m			0.31		10
s	VITON/NEOPRENE	MSA VAUTEX	>480m	0	0	n.a.		48

anionic detergent, <30%
Primary Class: 900   Miscellaneous Unclassified Chemicals

g	NATURAL RUBBER	MARIGOLD BLACK HEVY	>480m	0	0	0.65		79
g	NATURAL RUBBER	MARIGOLD FEATHERLIT	>480m	0	0	0.15		79
g	NATURAL RUBBER	MARIGOLD MED WT EXT	>480m	0	0	0.45		79
g	NATURAL RUBBER	MARIGOLD MEDICAL	>480m	0	0	0.28		79
g	NATURAL RUBBER	MARIGOLD ORANGE SUP	>480m	0	0	0.73		79
g	NATURAL RUBBER	MARIGOLD R SURGEONS	>480m	0	0	0.28		79
g	NATURAL RUBBER	MARIGOLD RED LT WT	>480m	0	0	0.43		79
g	NATURAL RUBBER	MARIGOLD SENSOTECH	>480m	0	0	0.28		79
g	NATURAL RUBBER	MARIGOLD SUREGRIP	>480m	0	0	0.58		79
g	NATURAL RUBBER	MARIGOLD YELLOW	>480m	0	0	0.38		79
g	NATURAL+NEOPRENE	MARIGOLD FEATHERWT	>480m	0	0	0.35		79
g	NITRILE	MARIGOLD BLUE	>480m	0	0	0.45		79
g	NITRILE	MARIGOLD GREEN SUPA	>480m	0	0	0.30		79
g	NITRILE	MARIGOLD NITROSOLVE	>480m	0	0	0.75		79
g	UNKNOWN MATERIAL	MARIGOLD R MEDIGLOV	>480m	0	0	0.03		79

G A R M E Class & Number/ N test chemical/ T MATERIAL NAME	MANUFACTURER & PRODUCT IDENTIFICATION	Break- through Time in Minutes	Perm- eation Rate in mg/sq m /min.	I N D E X	Thick- ness in mm	degrad. and comments	R e f s

antimony pentachloride
CAS Number: 7647-18-9
Primary Class: 360   Inorganic Acid Halides

| s SARANEX-23 | DU PONT TYVEK SARAN | >480m | | 0 | 0 | n.a. | 8 |

aqua regia (hydrochloric acid, 25-37% & nitric acid, 63-75%)
CAS Number: 8007-56-5
Primary Class: 370   Inorganic Acids

g NATURAL RUBBER	ANSELL ORANGE 208	>360m	0	0	0.76		55
g NATURAL RUBBER	ANSELLEDMONT 36-124	>360m	0	0	0.46		6
g NATURAL RUBBER	ANSELLEDMONT 392				0.48	degrad.	6
g NATURAL RUBBER	MARIGOLD BLACK HEVY	469m	2	1	0.65	degrad.	79
g NATURAL RUBBER	MARIGOLD FEATHERLIT	169m	65	2	0.15	degrad.	79
g NATURAL RUBBER	MARIGOLD MED WT EXT	357m	19	2	0.45	degrad.	79
g NATURAL RUBBER	MARIGOLD MEDICAL	247m	46	2	0.28	degrad.	79
g NATURAL RUBBER	MARIGOLD ORANGE SUP	>480m	0	0	0.73	degrad.	79
g NATURAL RUBBER	MARIGOLD R SURGEONS	247m	46	2	0.28	degrad.	79
g NATURAL RUBBER	MARIGOLD RED LT WT	341m	23	2	0.43		79
g NATURAL RUBBER	MARIGOLD SENSOTECH	247m	46	2	0.28	degrad.	79
g NATURAL RUBBER	MARIGOLD SUREGRIP	435m	7	1	0.58	degrad.	79
g NATURAL RUBBER	MARIGOLD YELLOW	310m	31	2	0.38	degrad.	79
g NATURAL+NEOPRENE	MARIGOLD FEATHERWT	294m	34	2	0.35	degrad.	79
g NEOPRENE	ANSELL NEOPRENE 530	>360m	0	0	0.46		55
g NEOPRENE	ANSELLEDMONT 29-840	45m			0.31		6
g NEOPRENE	ANSELLEDMONT NEOX	>360m	0	0	n.a.		6
g NEOPRENE/NATURAL	ANSELL CHEMI-PRO	>360m	0	0	0.72		55
g NITRILE	ANSELLEDMONT 37-165	>360m	0	0	0.60		6
g NITRILE	MARIGOLD BLUE	>480m	0	0	0.45	degrad.	79
g NITRILE	MARIGOLD GREEN SUPA	>480m	0	0	0.30	degrad.	79
g NITRILE	MARIGOLD NITROSOLVE	>480m	0	0	0.75	degrad.	79
g PE/EVAL/PE	SAFETY4 4H	>240m	0	0	0.07	37°C	60
g PVAL	ANSELLEDMONT PVA				n.a.	degrad.	6
g PVAL	EDMONT 15-552				n.a.	degrad.	6
g PVC	ANSELLEDMONT M GRIP	120m			n.a.		6

arsine (synonym: arsenic trihydride)
CAS Number: 7784-42-1
Primary Class: 350   Inorganic Gases and Vapors

| s UNKNOWN MATERIAL | LIFE-GUARD RESPONDE | >180m | | 0 | 1 | n.a. | 77 |

AZT
CAS Number: 8066-44-2
Primary Class: 870   Antineoplastic Drugs and Other Pharmaceuticals

| g NATURAL RUBBER | ANSELL CANNERS 334 | >480m | | 0 | 0 | n.a. | 34 |

G A R M E Class & Number/ N test chemical/ T MATERIAL NAME	MANUFACTURER & PRODUCT IDENTIFICATION	Break-through Time in Minutes	Perm-eation Rate in mg/sq m /min.	I N D E X	Thick-ness in mm	degrad. and comments	R e f s

AZT, 200 mg/mL in 0.2% methyl cellulose & water
Primary Class: 870   Antineoplastic Drugs and Other Pharmaceuticals

g NATURAL RUBBER	ANSELL CANNERS 334	>480m		0	0	n.a.	34

Benlate (synonym: Benomyl)
CAS Number: 17804-35-2
Primary Class: 274   Heterocyclic Compounds, Nitrogen, Others
Related Class: 840   Pesticides, Mixtures and Formulations

g NITRILE	ANSELLEDMONT 37-145	<15m			0.30		57
c PE	MOLNLYCKE H D	<15m	0.03	2	0.24		57
g PE/EVAL/PE	SAFETY4 4H	>240m	0	0	0.07		60

benzaldehyde
CAS Number: 100-52-7
Primary Class: 122   Aldehydes, Aromatic

g BUTYL	BEST 878	>480m	0	0	0.75		53
g BUTYL	NORTH B-174	>540m	0	0	0.66		34
g NATURAL RUBBER	ACKWELL 5-109				0.15	degrad.	34
g NATURAL RUBBER	ANSELLEDMONT 36-124	14m	< 900	4	0.46		6
g NATURAL RUBBER	ANSELLEDMONT 392	10m	<90	4	0.48		6
g NEOPRENE	ANSELLEDMONT 29-840				0.38	degrad.	6
g NEOPRENE	ANSELLEDMONT 29-870	39m	240	3	0.48	degrad.	34
g NEOPRENE	ANSELLEDMONT NEOX				n.a.	degrad.	6
g NEOPRENE	BEST 32				n.a.	degrad.	53
g NEOPRENE	BEST 6780	75m	150	3	n.a.		53
g NITRILE	ANSELLEDMONT 37-155	24m	258	3	0.33		34
g NITRILE	ANSELLEDMONT 37-165				0.60	degrad.	6
g NITRILE	BEST 22R				n.a.	degrad.	53
g NITRILE	BEST 727				0.38	degrad.	53
g PE/EVAL/PE	SAFETY4 4H	>240m	0	0	0.07	35°C	60
g PE/EVAL/PE	SAFETY4 4H	>480m	0	0	0.07		60
g PVAL	ANSELLEDMONT PVA	>360m		0	n.a.		6
g PVAL	EDMONT 15-552	>360m	0	0	n.a.		6
g PVAL	EDMONT 25-545	>360m	0	0	n.a.		6
g PVAL	EDMONT 25-950	>960m	0	0	0.33		34
g PVC	ANSELLEDMONT 34-100				0.20	degrad.	34
g PVC	ANSELLEDMONT SNORKE				n.a.	degrad.	6
g PVC	BEST 725R				n.a.	degrad.	53
g VITON	NORTH F-091	594m	240	3	0.25		34

benzene
CAS Number: 71-43-2
Primary Class: 292   Hydrocarbons, Aromatic

g BUTYL	ANSELL LAMPRECHT	52m	3760	4	0.70		38

	Class & Number/ test chemical/ MATERIAL NAME	MANUFACTURER & PRODUCT IDENTIFICATION	Break- through Time in Minutes	Perm- eation Rate in mg/sq m /min.	I N D E X	Thick- ness in mm	degrad. and comments	R e f s
g	BUTYL	BEST 878	34m	1810	4	0.75		53
s	BUTYL	MSA CHEMPRUF	<30m			n.a.		48
g	BUTYL	NORTH B-161	20m			0.43		23
g	BUTYL	NORTH B-174	31m	1920	4	0.58		34
u	BUTYL	Unknown	88m	1300	4	0.84		3
u	BUTYL	Unknown	54m	2000	4	0.84		3
u	BUTYL	Unknown	51m	2000	4	0.84	benzene exp	3
u	BUTYL	Unknown	60m	900	3	0.56		3
g	BUTYL/ECO	BRUNSWICK BUTYL-POL	33m	>5000	4	0.63		89
g	BUTYL/NEOPRENE	COMASEC BUTYL PLUS	35m	1800	4	0.50		40
s	BUTYL/NEOPRENE	MSA BETEX	<5m	146	4	n.a.		48
s	CPE	ILC DOVER CLOROPEL	26m			n.a.		29
s	CPE	ILC DOVER CLOROPEL	73m			n.a.		46
s	CPE	STD. SAFETY				n.a.	degrad.	75
u	EVAC	Unknown	1m			0.23	degrad.	23
g	HYPALON	COMASEC DIPCO	23m	18000	5	0.55		40
g	NATURAL RUBBER	ANSELLEDMONT 36-124				0.46	degrad.	6
g	NATURAL RUBBER	ANSELLEDMONT 392				0.48	degrad.	6
g	NATURAL RUBBER	ANSELLEDMONT 46-320	2m	32000	5	0.31		5
g	NATURAL RUBBER	BEST 65NFW				n.a.	degrad.	53
g	NATURAL RUBBER	COMASEC FLEXIGUM	33m	1260	4	0.95		40
g	NATURAL RUBBER	MARIGOLD BLACK HEVY	3m	13303	5	0.82	degrad.	79
g	NATURAL RUBBER	MARIGOLD ORANGE SUP	3m	4633	5	0.73	degrad.	79
g	NATURAL RUBBER	MARIGOLD SUREGRIP	2m	39312	5	0.89	degrad.	79
g	NATURAL+NEOPRENE	ANSELL OMNI 276	3m	28000	5	0.45		5
g	NATURAL+NEOPRENE	ANSELL OMNI 276				0.45	degrad.	55
g	NATURAL+NEOPRENE	PLAYTEX ARGUS 123	3m	22500	5	n.a.		72
g	NEOPRENE	ANSELLEDMONT 29-840				0.38	degrad.	6
g	NEOPRENE	ANSELLEDMONT 29-865	11m	18900	5	0.51		26
g	NEOPRENE	ANSELLEDMONT 29-870	17m	1656	4	0.53		34
g	NEOPRENE	ANSELLEDMONT NEOX				n.a.	degrad.	6
g	NEOPRENE	BEST 32	9m	2640	5	n.a.	degrad.	53
g	NEOPRENE	BEST 6780	8m	2580	5	n.a.	degrad.	53
g	NEOPRENE	COMASEC COMAPRENE	16m	5580	4	n.a.		40
g	NEOPRENE	DAVIDS GLOVES	20m			0.64		23
g	NEOPRENE	MAPA-PIONEER N-73	16m	8000	4	0.46		5
u	NEOPRENE	Unknown	40m	1900	4	0.80	7°C	3
u	NEOPRENE	Unknown	24m	2300	4	0.80	22°C	3
u	NEOPRENE	Unknown	16m	3300	4	0.80	37°C	3
u	NEOPRENE	Unknown	24m	2300	4	0.76		3
u	NEOPRENE	Unknown	26m	3100	4	0.76	benzene exp	3
g	NITRILE	ANSELL CHALLENGER	9m	8	3	0.38		55
g	NITRILE	ANSELLEDMONT 37-155	19m	<1	2	0.38		34
g	NITRILE	ANSELLEDMONT 37-165	46m	5100	4	0.55		5
g	NITRILE	ANSELLEDMONT 37-165				0.60	degrad.	6
g	NITRILE	ANSELLEDMONT 37-175	19m	8500	4	0.37		5
g	NITRILE	ANSELLEDMONT 37-175	10m			0.36		23
g	NITRILE	ANSELLEDMONT 37-175	14m	8690	5	0.40		26
g	NITRILE	BEST 22R				n.a.	degrad.	53

G A R M E Class & Number/ N test chemical/ T MATERIAL NAME	MANUFACTURER & PRODUCT IDENTIFICATION	Break- through Time in Minutes	Perm- eation Rate in mg/sq m /min.	I N D E X	Thick- ness in mm	R degrad. and comments	R e f s
g NITRILE	BEST 727	16m	1640	4	0.38	degrad.	53
g NITRILE	COMASEC COMATRIL	14m	8640	5	0.55		40
g NITRILE	COMASEC COMATRIL SU	24m	9000	4	0.60		40
g NITRILE	COMASEC FLEXITRIL	6m	7200	5	n.a.		60
g NITRILE	MAPA-PIONEER A-14	27m	5820	4	0.54		36
g NITRILE	MARIGOLD BLUE	25m	6055	4	0.45	degrad.	79
g NITRILE	MARIGOLD GREEN SUPA	12m	5826	5	0.30	degrad.	79
g NITRILE	MARIGOLD NITROSOLVE	28m	6101	4	0.75	degrad.	79
g NITRILE	NORTH LA-142G				0.38	degrad.	7
g NITRILE+PVC	COMASEC MULTIMAX	16m	1416	4	n.a.		40
g NITRILE+PVC	COMASEC MULTIPLUS	36m	2640	4	n.a.		40
g PE	ANSELLEDMONT 35-125	<1m	2500	5	0.03		5
c PE	DU PONT TYVEK QC	1m	2200	5	n.a.		3
c PE	DU PONT TYVEK QC	2m			n.a.		23
g PE/EVAL/PE	SAFETY4 4H	>240m	0	0	0.07	35°C	60
g PE/EVAL/PE	SAFETY4 4H	>1440m	0	0	0.07		60
u POLYURETHANE	Unknown	2m	1100	5	0.20		3
g PVAL	ANSELLEDMONT PVA	>360m		0	n.a.		6
g PVAL	COMASEC SOLVATRIL	>480m	0	0	n.a.		40
g PVAL	EDMONT 15-552	7m	<9	3	n.a.		6
g PVAL	EDMONT 15-552	20m			n.a.		23
g PVAL	EDMONT 15-552	>1980m	0	0	n.a.		26
g PVAL	EDMONT 25-545	>360m	0	0	n.a.		6
g PVAL	EDMONT 25-950	49m	0.06	1	0.25		34
g PVC	ANSELLEDMONT 34-100	<1m	11800	5	0.15		34
g PVC	ANSELLEDMONT SNORKE				n.a.	degrad.	6
g PVC	COMASEC MULTITOP	40m	3120	4	n.a.		40
g PVC	COMASEC MULTITOP	22m	3420	4	n.a.		40
g PVC	COMASEC NORMAL	48m	1920	4	n.a.		40
g PVC	COMASEC OMNI	30m	2100	4	n.a.		40
g PVC	MAPA-PIONEER V-20	2m	15000	5	0.31		5
g PVC	MAPA-PIONEER V-5	1m			0.13		23
s PVC	MSA UPC	<3m			0.20		48
u PVC	Unknown	6m	1500	5	n.a.		3
u PVC	Unknown	19m	4210	4	n.a.		26
s SARANEX-23	DU PONT TYVEK SARAN	10m			n.a.		23
s TEFLON	CHEMFAB CHALL. 5100	>200m	0	1	n.a.		65
g TEFLON	CLEAN ROOM PRODUCTS	10m			0.05		23
s UNKNOWN MATERIAL	CHEMRON CHEMREL	17m	<1	2	n.a.		67
s UNKNOWN MATERIAL	CHEMRON CHEMREL MAX	>480m	0	0	n.a.		67
s UNKNOWN MATERIAL	DU PONT BARRICADE	>480m	0	0	n.a.		8
s UNKNOWN MATERIAL	KAPPLER CPF III	>480m	0	0	n.a.		77
s UNKNOWN MATERIAL	LIFE-GUARD RESPONDE	>480m	0	0	n.a.		77
g UNKNOWN MATERIAL	NORTH SILVERSHIELD	>480m	0	0	0.08		7
g VITON	NORTH F-091	30m			0.25		23
g VITON	NORTH F-091	354m	1	1	0.23		34
u VITON	Unknown	900m	5	1	0.95		3
s VITON/CHLOROBUTYL	LIFE-GUARD VC-100	>240m		0	n.a.		77
u VITON/CHLOROBUTYL	Unknown	>180m			n.a.		29
g VITON/NEOPRENE	ERISTA VITRIC	>60m			0.60		38
s VITON/NEOPRENE	MSA VAUTEX	98m	0.8	1	n.a.		48

G A R M E N T Class & Number/ test chemical/ MATERIAL NAME	MANUFACTURER & PRODUCT IDENTIFICATION	Break- through Time in Minutes	Perm- eation Rate in mg/sq m /min.	I N D E X	Thick- ness in mm	degrad. and comments	R e f s

benzethonium chloride (synonym: hyamine)
CAS Number: 121-54-0
Primary Class: 550   Organic Salts (Solutions)

g BUTYL	NORTH B-174	>480m	0	0	0.63		34
g NATURAL RUBBER	ACKWELL 5-109	>480m	0	0	0.18		34
g NEOPRENE	ANSELLEDMONT 29-870	>480m	0	0	0.49		34
g NEOPRENE	ANSELLEDMONT 29-870	>480m	0	0	0.46	.5% in EtOH	34
g PVC	ANSELLEDMONT 34-100	>480m	0	0	0.20		34
g PVC	ANSELLEDMONT 34-100	>480m	0	0	0.16	.5% in EtOH	34
g UNKNOWN MATERIAL	NORTH SILVERSHIELD	>480m	0	0	0.10	.5% in EtOH	34

benzonitrile (synonym: phenyl cyanide)
CAS Number: 100-47-0
Primary Class: 432   Nitriles, Aromatic

g BUTYL	NORTH B-174	>480m	0	0	0.61		34
g NATURAL RUBBER	ACKWELL 5-109	<1m	240	4	0.10	degrad.	34
g NEOPRENE	ANSELLEDMONT 29-870				0.48	degrad.	34
g NITRILE	ANSELLEDMONT 37-155				0.40	degrad.	34
g PVAL	EDMONT 25-950	>480m	0	0	0.25		34
g PVC	ANSELLEDMONT 34-100				0.20	degrad.	34
s UNKNOWN MATERIAL	DU PONT BARRICADE	>480m	0	0	n.a.		8
s UNKNOWN MATERIAL	LIFE-GUARD RESPONDE	>480m	0	0	n.a.		77
g VITON	NORTH F-091	56m	240	3	0.27		34

benzoyl chloride (synonym: benzoic acid chloride)
CAS Number: 98-88-4
Primary Class: 112   Acid Halides, Carboxylic, Aromatic

g BUTYL	NORTH B-174	377m	996	3	0.61	degrad.	34
g HYPALON	COMASEC DIPCO	20m			0.60	degrad.	33
g NATURAL RUBBER	ACKWELL 5-109				0.15	degrad.	34
g NEOPRENE	ANSELLEDMONT 29-840	15m			0.45	degrad.	33
g NEOPRENE	ANSELLEDMONT 29-870				0.48	degrad.	34
g NITRILE	ANSELLEDMONT 37-155				0.40	degrad.	34
g PVAL	EDMONT 25-950	>480m	0	0	0.48		34
g PVC	ANSELLEDMONT 34-100	<1m	595	4	0.15	degrad.	34
s UNKNOWN MATERIAL	DU PONT BARRICADE	>480m	0	0	n.a.		8
g VITON	NORTH F-091	45m			0.25	degrad.	33
g VITON	NORTH F-091	>480m	0	0	0.23		34

benzyl acetate
CAS Number: 140-11-4
Primary Class: 222   Esters, Carboxylic, Acetates

g NATURAL RUBBER	ACKWELL 5-109	15m	2040	4	0.15		34
g NITRILE	ANSELLEDMONT 37-155	60m	2460	4	0.40		34
g UNKNOWN MATERIAL	NORTH SILVERSHIELD	>480m	0	0	0.10		34

G A R M E Class & Number/ N test chemical/ T MATERIAL NAME	MANUFACTURER & PRODUCT IDENTIFICATION	Break- through Time in Minutes	Perm- eation Rate in mg/sq m /min.	I N D E X	Thick- ness in mm	degrad. and comments	R e f s
benzyl acetate, 30-70% Primary Class: 222 Esters, Carboxylic, Acetates							
g NATURAL RUBBER	ACKWELL 5-109	15m	480	3	0.15	in H_2O	34
g NITRILE	ANSELLEDMONT 37-155	15m	1560	4	0.42	in H_2O	34
g UNKNOWN MATERIAL	NORTH SILVERSHIELD	>480m	0	0	0.10	in H_2O	34
benzyl acetate, <30% Primary Class: 222 Esters, Carboxylic, Acetates							
g NATURAL RUBBER	ACKWELL 5-109	30m	6	2	0.13	1% in H_2O	34
g NATURAL RUBBER	ACKWELL 5-109	15m	2040	4	0.15	5% in EtOH	34
g NATURAL RUBBER	ACKWELL 5-109	15m	1560	4	0.15	3% in EtOH	34
g NATURAL RUBBER	ACKWELL 5-109	15m	1500	4	0.15	1% in EtOH	34
g NATURAL RUBBER	ACKWELL 5-109	15m	30	3	0.15	.5% in EtOH	34
g NITRILE	ANSELLEDMONT 37-155	15m	30	3	0.36	1% in H_2O	34
g NITRILE	ANSELLEDMONT 37-155	60m	2460	4	0.38	5% in EtOH	34
g NITRILE	ANSELLEDMONT 37-155	15m	480	3	0.38	3% in EtOH	34
g NITRILE	ANSELLEDMONT 37-155	15m	72	3	0.38	1% in EtOH	34
g NITRILE	ANSELLEDMONT 37-155	15m	6	2	0.38	.5% in EtOH	34
g UNKNOWN MATERIAL	NORTH SILVERSHIELD	>480m	0	0	0.10	1% in H_2O	34
g UNKNOWN MATERIAL	NORTH SILVERSHIELD	>480m	0	0	0.10	5% in EtOH	34
g UNKNOWN MATERIAL	NORTH SILVERSHIELD	>480m	0	0	0.10	3% in EtOH	34
g UNKNOWN MATERIAL	NORTH SILVERSHIELD	>480m	0	0	0.10	1% in EtOH	34
g UNKNOWN MATERIAL	NORTH SILVERSHIELD	>480m	0	0	0.10	.5% in EtOH	34
benzyl alcohol CAS Number: 100-51-6 Primary Class: 311 Hydroxy Compounds, Aliphatic and Alicyclic, Primary							
g BUTYL	ERISTA BX	>240m	0	0	0.68		33
g NITRILE	ANSELLEDMONT 37-175				0.40	degrad.	33
g PE/EVAL/PE	SAFETY4 4H	>240m		0	0.07		60
g VITON	NORTH F-091	>1200m	0	0	0.25		33
benzyl chloride (synonyms: alpha-chloromethylbenzene; alpha-chlorotoluene) CAS Number: 100-44-7 Primary Class: 262 Halogen Compounds, Allylic and Benzylic							
u CPE	ILC DOVER	47m			n.a.		29
g NATURAL RUBBER	ACKWELL 5-109				0.15	degrad.	34
g NEOPRENE	ANSELLEDMONT 29-870				0.48	degrad.	34
g NITRILE	ANSELLEDMONT 37-155				0.40	degrad.	34
g PE/EVAL/PE	SAFETY4 4H	>240m	0	0	0.07	35°C	60
g PE/EVAL/PE	SAFETY4 4H	>480m	0	0	0.07		60
g PVC	ANSELLEDMONT 34-100				0.20	degrad.	34
s TEFLON	CHEMFAB CHALL. 5100	>192m			n.a.		65

G A R M E Class & Number/ N test chemical/ T MATERIAL NAME	MANUFACTURER & PRODUCT IDENTIFICATION	Break- through Time in Minutes	Perm- eation Rate in mg/sq m /min.	I N D E X	Thick- ness in mm	degrad. and comments	R e f s
s UNKNOWN MATERIAL	LIFE-GUARD RESPONDE	>480m	0	0	n.a.		77
u VITON/CHLOROBUTYL	Unknown	>180m			n.a.		29

benzyl ether
CAS Number: 103-50-4
Primary Class: 241   Ethers, Aliphatic and Alicyclic

| s VITON/BUTYL | DRAGER 500 OR 710 | >30m | | | n.a. | | 91 |

benzyl neocaprate
CAS Number: 66794-75-0
Primary Class: 224   Esters, Carboxylic, Aliphatic, Others

| g NITRILE | ANSELLEDMONT 37-145 | >240m | 0 | 0 | 0.32 | | 30 |
| g NITRILE | ANSELLEDMONT 37-145 | >240m | 0 | 0 | 0.32 | preexp. 24h | 30 |

benzylnitrile (synonyms: benzyl cyanide; phenylacetonitrile)
CAS Number: 140-29-4
Primary Class: 432   Nitriles, Aromatic

| g PE/EVAL/PE | SAFETY4 4H | >240m | 0 | 0 | 0.07 | 35°C | 60 |

beta-butyrolactone (synonym: 3-hydroxybutyric acid lactone)
CAS Number: 3068-88-0
Primary Class: 225   Esters, Carboxylic, Lactones

g NATURAL RUBBER	ANSELLEDMONT 36-124	60m	<900	3	0.46		6
g NATURAL RUBBER	ANSELLEDMONT 392	60m	<900	3	0.48		6
g NEOPRENE	ANSELLEDMONT 29-840	10m	<90	4	0.38		6
g NITRILE	ANSELLEDMONT 37-165				0.60	degrad.	6
g PVAL	ANSELLEDMONT PVA	120m	<90	2	n.a.		6
g PVAL	EDMONT 15-552	10m	<90	4	n.a.		6
g PVC	ANSELLEDMONT SNORKE				n.a.	degrad.	6

beta-ionone
CAS Number: 14901-07-6
Primary Class: 391   Ketones, Aliphatic and Alicyclic

g BUTYL	NORTH B-174	>540m	0	0	0.64		34
g NATURAL RUBBER	ACKWELL 5-109				0.15	degrad.	34
g NEOPRENE	ANSELLEDMONT 29-870				0.48	degrad.	34
g NITRILE	ANSELLEDMONT 37-155	>840m	0	0	0.38		34
g PVAL	EDMONT 25-950	>480m	0	0	0.27		34
g PVC	ANSELLEDMONT 34-100				0.20	degrad.	34
g VITON	NORTH F-091	>480m	0	0	0.25		34

G								
A								
R				Perm-	I			
M			Break-	eation	N	Thick-		R
E Class & Number/	MANUFACTURER		through	Rate in	D	ness	degrad.	e
N test chemical/	& PRODUCT		Time in	mg/sq m	E	in	and	f
T MATERIAL NAME	IDENTIFICATION		Minutes	/min.	X	mm	comments	s

beta-picoline (synonym: 3-methylpyridine)
CAS Number: 108-99-6
Primary Class: 271   Heterocyclic Compounds, Nitrogen, Pyridines

g NATURAL RUBBER	ACKWELL 5-109					0.15	degrad.	34
g NEOPRENE	ANSELLEDMONT 29-870					0.48	degrad.	34
g NITRILE	ANSELLEDMONT 37-155					0.40	degrad.	34
g PVC	ANSELLEDMONT 34-100					0.20	degrad.	34
g VITON	NORTH F-091					0.25	degrad.	34

beta-propiolactone
CAS Number: 57-57-8
Primary Class: 225   Esters, Carboxylic, Lactones

u NATURAL RUBBER	Unknown		15m	43	3	0.30		3
u PE	Unknown		10m	12	4	0.06		3
u POLYURETHANE	Unknown		<5m	8300	5	n.a.		3

beta-tetralone (synonym: tetralone, beta-)
CAS Number: 530-93-8
Primary Class: 391   Ketones, Aliphatic and Alicyclic

s UNKNOWN MATERIAL	LIFE-GUARD RESPONDE		>480m	0	0	n.a.		77

biological detergent (saturated solution)
Primary Class: 900   Miscellaneous Unclassified Chemicals

g NATURAL RUBBER	MARIGOLD BLACK HEVY		>480m	0	0	0.65		79
g NATURAL RUBBER	MARIGOLD FEATHERLIT		>480m	0	0	0.15	degrad.	79
g NATURAL RUBBER	MARIGOLD MED WT EXT		>480m	0	0	0.45		79
g NATURAL RUBBER	MARIGOLD MEDICAL		>480m	0	0	0.28		79
g NATURAL RUBBER	MARIGOLD ORANGE SUP		>480m	0	0	0.73		79
g NATURAL RUBBER	MARIGOLD R SURGEONS		>480m	0	0	0.28		79
g NATURAL RUBBER	MARIGOLD RED LT WT		>480m	0	0	0.43		79
g NATURAL RUBBER	MARIGOLD SENSOTECH		>480m	0	0	0.28		79
g NATURAL RUBBER	MARIGOLD SUREGRIP		>480m	0	0	0.58		79
g NATURAL RUBBER	MARIGOLD YELLOW		>480m	0	0	0.38		79
g NATURAL+NEOPRENE	MARIGOLD FEATHERWT		>480m	0	0	0.35		79
g NITRILE	MARIGOLD BLUE		>480m	0	0	0.45		79
g NITRILE	MARIGOLD GREEN SUPA		>480m	0	0	0.30	degrad.	79
g NITRILE	MARIGOLD NITROSOLVE		>480m	0	0	0.75		79
g UNKNOWN MATERIAL	MARIGOLD R MEDIGLOV		>480m	0	0	0.03		79

bisphenol A diglycidyl ether (synonyms: DGBA; diglycidyl ether of bisphenol A)
CAS Number: 2238-07-5
Primary Class: 275   Heterocyclic Compounds, Oxygen, Epoxides

u BUTYL	ARSIMA		46m			0.48	40°C TL	14

G A R M E Class & Number/ N test chemical/ T MATERIAL NAME	MANUFACTURER & PRODUCT IDENTIFICATION	Break- through Time in Minutes	Perm- eation Rate in mg/sq m /min.	I N D E X	Thick- ness in mm	degrad. and comments	R e f s
g NAT+NEOPR+NITRILE	MAPA-PIONEER TRIONI	>480m	0	0	0.50		36
u NATURAL RUBBER	Unknown	24m			1.10	40°C TL	14
u NEOPRENE	Unknown	38m			0.54	40°C TL	14
u PE	ALLHABO	4m			0.08	40°C TL	14
g PE/EVAL/PE	SAFETY4 4H	>240m	0	0	0.05	40°C TL	14
g PVAL	KURASHIKI	>240m	0	0	0.07	40°C TL	14

blood, human
Primary Class: 900   Miscellaneous Unclassified Chemicals

c PE	DU PONT TYVEK QC	>480m	0	0	n.a.		8

boric acid
CAS Number: 10043-35-3
Primary Class: 370   Inorganic Acids

g BUTYL	NORTH B-174	>480m	0	0	0.74		34
g NEOPRENE	ANSELLEDMONT 29-870	>480m	0	0	0.50		34
g NITRILE	ANSELLEDMONT 37-155	>480m	0	0	0.40		34
g VITON	NORTH F-091	>480m	0	0	0.30		34

bromine
CAS Number: 7726-95-6
Primary Class: 330   Elements

g NEOPRENE	MAPA-PIONEER GF-N-	240m	0	0	0.38		38
g NEOPRENE	PEDI	210m			0.50		38
g NEOPRENE	REX	150m			0.50		38
c PE	DU PONT TYVEK QC	<1m	>10000	5	n.a.		8
s TEFLON	CHEMFAB CHALL. 5100	>200m	0	1	n.a.		65
s UNKNOWN MATERIAL	CHEMRON CHEMREL	3m	16020	5	n.a.		67
s UNKNOWN MATERIAL	DU PONT BARRICADE	9m	5168	5	n.a.		8
s UNKNOWN MATERIAL	LIFE-GUARD RESPONDE	18m	5333	4	n.a.		77

bromoacetonitrile
CAS Number: 590-17-0
Primary Class: 431   Nitriles, Aliphatic and Alicyclic
Related Class: 261   Halogen Compounds, Aliphatic and Alicyclic

g BUTYL	NORTH B-174	>480m	0	0	0.63		34
g NATURAL RUBBER	ACKWELL 5-109	<1m	570	4	0.10		34
g PVAL	EDMONT 25-950	>480m	0	0	0.25		34
g VITON	NORTH F-091	>480m	0	0	0.23		34

bromobenzene
CAS Number: 108-86-1
Primary Class: 264   Halogen Compounds, Aromatic

g BUTYL	NORTH B-174	32m	2388	4	0.63	degrad.	34

G A R M E Class & Number/ N test chemical/ T MATERIAL NAME	MANUFACTURER & PRODUCT IDENTIFICATION	Break-through Time in Minutes	Perm-eation Rate in mg/sq m /min.	I N D E X	Thick-ness in mm	degrad. and comments	R e f s
g NATURAL RUBBER	ACKWELL 5-109				0.15	degrad.	34
g NEOPRENE	ANSELLEDMONT 29-870				0.48	degrad.	34
g NITRILE	NORTH LA-142G	13m	546	4	0.38	degrad.	34
g PVAL	EDMONT 25-950	>480m	0	0	0.23		34
g PVC	ANSELLEDMONT 34-100				0.20	degrad.	34
g VITON	NORTH F-091	>480m	0	0	0.25		34

bromochloromethane (synonym: chlorobromomethane)
CAS Number: 74-97-5
Primary Class: 261   Halogen Compounds, Aliphatic and Alicyclic

s UNKNOWN MATERIAL	LIFE-GUARD RESPONDE	>180m		0	1	n.a.	77

bromodichloromethane (synonym: dichlorobromomethane)
CAS Number: 75-27-4
Primary Class: 261   Halogen Compounds, Aliphatic and Alicyclic

g BUTYL	NORTH B-174	41m	18960	5	0.58	degrad.	34
g NATURAL RUBBER	ACKWELL 5-109				0.15	degrad.	34
g NEOPRENE	ANSELLEDMONT 29-870				0.48	degrad.	34
g PVAL	EDMONT 25-950	106m	0.02	1	0.55		34
g PVC	ANSELLEDMONT 34-100	1m	69300	5	0.20	degrad.	34
g VITON	NORTH F-091	470m	4	1	0.26		34

bromotrifluoride
CAS Number: 7787-71-5
Primary Class: 360   Inorganic Acid Halides

s BUTYL	TRELLEBORG TRELLCHE				n.a.	degrad.	71
s PVC	TRELLEBORG TRELLCHE				n.a.	degrad.	71
s VITON/BUTYL	TRELLEBORG TRELLCHE				n.a.	degrad.	71

butyl acetate (synonym: n-butyl acetate)
CAS Number: 123-86-4
Primary Class: 222   Esters, Carboxylic, Acetates

g BUTYL	BEST 878	125m	830	3	0.75		53
g BUTYL	BRUNSWICK BUTYL STD	94m	100	3	0.63		89
g BUTYL	BRUNSWICK BUTYL-XTR	169m	1000	4	0.63		89
g BUTYL	COMASEC BUTYL	84m			0.50		40
g BUTYL	NORTH B-161	113m	457	3	0.45		7
g BUTYL	NORTH B-161	82m	620	3	0.38		93
g BUTYL/ECO	BRUNSWICK BUTYL-POL	84m	>150	3	0.63		89
g BUTYL/NEOPRENE	COMASEC BUTYL PLUS	140m	480	3	0.50		40
g HYPALON	COMASEC DIPCO	55m	5274	4	0.60		40
g NAT+NEOPR+NITRILE	MAPA-PIONEER TRIONI	7m	1212	5	0.48	degrad.	36
g NAT+NEOPR+NITRILE	MAPA-PIONEER TRIONI	5m	7890	5	0.46		93
g NATURAL RUBBER	ANSELL CONT ENVIRON	4m	720	4	0.53		55
g NATURAL RUBBER	ANSELL FL 200 254				0.51	degrad.	55

G A R M E N T	Class & Number/ test chemical/ MATERIAL NAME	MANUFACTURER & PRODUCT IDENTIFICATION	Break- through Time in Minutes	Perm- eation Rate in mg/sq m /min.	I N D E X	Thick- ness in mm	degrad. and comments	R e f s
g	NATURAL RUBBER	ANSELL PVL 040				0.46	degrad.	55
g	NATURAL RUBBER	ANSELLEDMONT 30-139	7m	8440	5	0.51		93
g	NATURAL RUBBER	ANSELLEDMONT 36-124				0.46	degrad.	6
g	NATURAL RUBBER	ANSELLEDMONT 392				0.48	degrad.	6
g	NATURAL RUBBER	COMASEC FLEXIGUM	45m	1800	4	0.95		40
g	NATURAL RUBBER	MAPA-PIONEER 258	10m	80	4	0.60	30°C	87
g	NATURAL RUBBER	MAPA-PIONEER 300	6m	40	4	n.a.	30°C	87
g	NATURAL RUBBER	MAPA-PIONEER 730	34m	20	3	1.13	30°C	87
g	NATURAL RUBBER	MARIGOLD BLACK HEVY	7m	6594	5	0.65	degrad.	79
g	NATURAL RUBBER	MARIGOLD FEATHERLIT	3m	17149	5	0.15	degrad.	79
g	NATURAL RUBBER	MARIGOLD MED WT EXT	3m	6594	5	0.45	degrad.	79
g	NATURAL RUBBER	MARIGOLD MEDICAL	3m	12751	5	0.28	degrad.	79
g	NATURAL RUBBER	MARIGOLD ORANGE SUP	7m	6594	5	0.73	degrad.	79
g	NATURAL RUBBER	MARIGOLD R SURGEONS	3m	12751	5	0.28	degrad.	79
g	NATURAL RUBBER	MARIGOLD RED LT WT	3m	7474	5	0.43	degrad.	79
g	NATURAL RUBBER	MARIGOLD SENSOTECH	3m	12751	5	0.28	degrad.	79
g	NATURAL RUBBER	MARIGOLD SUREGRIP	6m	6594	5	0.58	degrad.	79
g	NATURAL RUBBER	MARIGOLD YELLOW	3m	9233	5	0.38	degrad.	79
g	NATURAL+NEOPRENE	ANSELL OMNI 276	6m	6400	5	0.45		5
g	NATURAL+NEOPRENE	ANSELL OMNI 276				0.56	degrad.	55
g	NATURAL+NEOPRENE	MAPA-PIONEER 424	10m	85	4	0.63	30°C	87
g	NATURAL+NEOPRENE	MARIGOLD FEATHERWT	3m	10112	5	0.35	degrad.	79
g	NATURAL+NEOPRENE	PLAYTEX 827	7m	6790	5	0.38		93
g	NEOPRENE	ANSELL NEOPRENE 520	14m			0.38		58
g	NEOPRENE	ANSELLEDMONT 29-840				0.38	degrad.	6
g	NEOPRENE	ANSELLEDMONT 29-870	20m			0.48		58
g	NEOPRENE	ANSELLEDMONT NEOX				n.a.	degrad.	6
g	NEOPRENE	BEST 32				n.a.	degrad.	53
g	NEOPRENE	BEST 723	13m			0.64		58
g	NEOPRENE	COMASEC COMAPRENE	15m	720	3	n.a.		40
g	NEOPRENE	GRANET 2001	13m			0.55		58
g	NEOPRENE	MAPA-PIONEER 360	36m	15	3	n.a.	30°C	87
g	NEOPRENE	MAPA-PIONEER 420	22m	35	3	0.79	30°C	87
g	NEOPRENE	MAPA-PIONEER GF-N-	29m	3200	4	0.47		5
g	NEOPRENE	MAPA-PIONEER N-36	30m			0.54		58
g	NEOPRENE	MAPA-PIONEER N-44	52m	3180	4	0.74		36
g	NEOPRENE	MAPA-PIONEER N-54	52m	3200	4	0.70		5
g	NEOPRENE	MAPA-PIONEER N-73	>60m			0.46		5
g	NEOPRENE/NATURAL	ANSELL CHEMI-PRO	4m	720	4	0.72		55
g	NITRILE	ANSELL 632	17m			0.36		58
g	NITRILE	ANSELL CHALLENGER	112m	5	1	0.38		55
g	NITRILE	ANSELLEDMONT 37-155	33m	4800	4	0.35		5
g	NITRILE	ANSELLEDMONT 37-155	59m	5460	4	0.35		41
t	NITRILE	ANSELLEDMONT 37-155	44m			0.32		58
g	NITRILE	ANSELLEDMONT 37-165	58m	2500	4	0.55		5
g	NITRILE	ANSELLEDMONT 37-165	90m	<9000	4	0.60		6
g	NITRILE	ANSELLEDMONT 37-175	40m	4500	4	0.37		5
g	NITRILE	ANSELLEDMONT 49-125	4m	<90	4	0.23		42
g	NITRILE	ANSELLEDMONT 49-155	40m	<900	3	0.38		42
g	NITRILE	ANSELLEDMONT 49-155	69m	2390	4	0.43		93
g	NITRILE	BEST 22R				n.a.	degrad.	53

G A R M E N T							
Class & Number/ test chemical/ MATERIAL NAME	MANUFACTURER & PRODUCT IDENTIFICATION	Break-through Time in Minutes	Perm-eation Rate in mg/sq m /min.	I N D E X	Thick-ness in mm	degrad. and comments	R e f s
g NITRILE	BEST 727	48m	1350	4	0.38		53
g NITRILE	BEST 727	28m			0.28		58
g NITRILE	COMASEC COMATRIL	40m	1320	4	0.55		40
g NITRILE	COMASEC COMATRIL SU	80m	900	3	0.60		40
g NITRILE	COMASEC FLEXITRIL	18m	1860	4	n.a.		40
g NITRILE	GRANET 490	32m			0.32		58
g NITRILE	MAPA-PIONEER 490	28m	90	3	0.36	30°C	87
g NITRILE	MAPA-PIONEER A-14	101m	1440	4	0.54		36
g NITRILE	MAPA-PIONEER A-15	50m			0.33		58
g NITRILE	MARIGOLD BLUE	77m	6457	4	0.45	degrad.	79
g NITRILE	MARIGOLD GREEN SUPA	35m	6376	4	0.30	degrad.	79
g NITRILE	MARIGOLD NITROSOLVE	85m	6474	4	0.75	degrad.	79
g NITRILE	NORTH LA-142G	29m	3264	4	0.36		7
g NITRILE	NORTH LA-142G	38m			0.33		58
g NITRILE	SURETY 315R	38m			0.32		58
g NITRILE	UNIROYAL 4-15	39m			0.32		58
g NITRILE+PVC	COMASEC MULTIMAX	36m	960	3	n.a.		40
g NITRILE+PVC	COMASEC MULTIPLUS	40m	600	3	n.a.		40
g PE	ANSELLEDMONT 35-125	2m	200	4	0.03		5
g PE/EVAL/PE	SAFETY4 4H	>240m	0	0	0.07	35°C	60
g PE/EVAL/PE	SAFETY4 4H	>480m	0	0	0.07		60
g PVAL	ANSELLEDMONT PVA	>360m		0	n.a.		6
g PVAL	EDMONT 15-552	>360m	0	0	n.a.		6
g PVAL	EDMONT 25-545	>360m	0	0	n.a.		6
g PVC	ANSELLEDMONT SNORKE				n.a.	degrad.	6
g PVC	COMASEC MULTIPOST	19m	600	3	n.a.		40
g PVC	COMASEC MULTITOP	30m	660	3	n.a.		40
g PVC	COMASEC NORMAL	20m	720	3	n.a.		40
g PVC	COMASEC OMNI	14m	780	4	n.a.		40
g PVC	MAPA-PIONEER V-20	2m	33000	5	0.31		5
s TEFLON	CHEMFAB CHALL. 5100	>180m	0	1	n.a.		65
g UNKNOWN MATERIAL	NORTH SILVERSHIELD	>360m	0	0	0.08		7
g VITON	NORTH F-091				0.24	degrad.	7
g VITON/BUTYL	NORTH	73m	6419	4	0.46		30

butyl acetate, 50% & methyl alcohol, 50%
Primary Class: 222   Esters, Carboxylic, Acetates

g NITRILE	ANSELLEDMONT 37-155	42/35m	/3780		0.35		41

butyl acetate, 90% & methyl alcohol, 10%
Primary Class: 222   Esters, Carboxylic, Acetates

g NITRILE	ANSELLEDMONT 37-155	39/39m	5460/	4	0.35		41

butyl acrylate (synonyms: n-butyl 2-propenoate; n-butyl acrylate)
CAS Number: 141-32-2
Primary Class: 223   Esters, Carboxylic, Acrylates and Methacrylates

g BUTYL	NORTH B-174				0.66	degrad.	34

G A R M E Class & Number/ N test chemical/ T MATERIAL NAME	MANUFACTURER & PRODUCT IDENTIFICATION	Break- through Time in Minutes	Perm- eation Rate in mg/sq m /min.	I N D E X	Thick- ness in mm	degrad. and comments	R e f s
g NATURAL RUBBER	ACKWELL 5-109				0.15	degrad.	34
g NEOPRENE	ANSELLEDMONT 29-870				0.48	degrad.	34
g NITRILE	ANSELLEDMONT 37-155				0.40	degrad.	34
g PE/EVAL/PE	SAFETY4 4H	>480m	0	0	0.07		60
g PE/EVAL/PE	SAFETY4 4H	>240m	0		0.07	35°C	60
g PVC	ANSELLEDMONT 34-100				0.20	degrad.	34
s TEFLON	CHEMFAB CHALL. 5100	>180m	0	1	n.a.		65
g VITON	NORTH F-091				0.25	degrad.	34

butyl alcohol (synonyms: 1-butanol; n-butanol)
CAS Number: 71-36-3
Primary Class: 311   Hydroxy Compounds, Aliphatic and Alicyclic, Primary

g BUTYL	BEST 878	>480m	0	0	0.75		53
g HYPALON	COMASEC DIPCO	>480m		0	0.60		40
g NATURAL RUBBER	ANSELL CONFORM 4205	1m	5	3	0.13		55
g NATURAL RUBBER	ANSELL FL 200 254	19m	0.2		0.51		55
g NATURAL RUBBER	ANSELL ORANGE 208	46m	0.1		0.76		55
g NATURAL RUBBER	ANSELL PVL 040	74m	0.06		0.46		55
g NATURAL RUBBER	ANSELL STERILE 832	8m	0.5	3	0.23		55
g NATURAL RUBBER	ANSELLEDMONT 36-124	15m	<900	3	0.46		6
g NATURAL RUBBER	ANSELLEDMONT 392	20m	<90	3	0.48		6
g NATURAL RUBBER	COMASEC FLEXIGUM	>480m	0	0	0.95		40
g NATURAL RUBBER	MARIGOLD BLACK HEVY	20m	66	3	0.65		79
g NATURAL RUBBER	MARIGOLD FEATHERLIT	1m	160	4	0.15		79
g NATURAL RUBBER	MARIGOLD MED WT EXT	21m	79	3	0.45		79
g NATURAL RUBBER	MARIGOLD MEDICAL	9m	127	4	0.28		79
g NATURAL RUBBER	MARIGOLD ORANGE SUP	19m	64	3	0.73		79
g NATURAL RUBBER	MARIGOLD R SURGEONS	9m	127	4	0.28		79
g NATURAL RUBBER	MARIGOLD RED LT WT	19m	86	3	0.43		79
g NATURAL RUBBER	MARIGOLD SENSOTECH	9m	127	4	0.28		79
g NATURAL RUBBER	MARIGOLD SUREGRIP	20m	70	3	0.58		79
g NATURAL RUBBER	MARIGOLD YELLOW	16m	100	3	0.38		79
g NATURAL+NEOPRENE	ANSELL OMNI 276	42m	0.1		0.56		55
g NATURAL+NEOPRENE	ANSELL TECHNICIANS	15m	0.2		0.43		55
g NATURAL+NEOPRENE	MARIGOLD FEATHERWT	14m	106	4	0.35		79
g NATURAL+NEOPRENE	PLAYTEX ARGUS 123	35m	<60	3	n.a.		72
g NEOPRENE	ANSELL NEOPRENE 530	>360m	0	0	0.46		55
g NEOPRENE	ANSELLEDMONT 29-840	240m	<90	2	0.31		6
g NEOPRENE	ANSELLEDMONT NEOX	>480m	0	0	n.a.		6
g NEOPRENE	BEST 32	>480m	0	0	n.a.		53
g NEOPRENE	BEST 6780	>480m	0	0	n.a.		53
g NEOPRENE	COMASEC COMAPRENE	>360m	0	0	n.a.		40
g NEOPRENE/NATURAL	ANSELL CHEMI-PRO	70m	0.06		0.72		55
g NITRILE	ANSELL CHALLENGER	>360m	0	0	0.38		55
g NITRILE	ANSELLEDMONT 37-165	>360m	0	0	0.60		6
g NITRILE	BEST 22R	192m	4	1	n.a.		53
g NITRILE	BEST 727	>480m	0	0	0.38		53
g NITRILE	COMASEC COMATRIL SU	>360m	0	0	0.55		40
g NITRILE	COMASEC FLEXITRIL	>480m	0	0	n.a.		40
g NITRILE	MARIGOLD BLUE	>480m	0	0	0.45	degrad.	79

G A R M E Class & Number/ N test chemical/ T MATERIAL NAME	MANUFACTURER & PRODUCT IDENTIFICATION	Break- through Time in Minutes	Perm- eation Rate in mg/sq m /min.	I N D E X	Thick- ness in mm	degrad. and comments	R e f s
g NITRILE	MARIGOLD GREEN SUPA	>480m	0	0	0.30	degrad.	79
g NITRILE	MARIGOLD NITROSOLVE	>480m	0	0	0.75	degrad.	79
g NITRILE	NORTH LA-142G	302m	83	2	0.38		63
g NITRILE+PVC	COMASEC MULTIPLUS	>360m	0	0	n.a.		40
c PE	DU PONT TYVEK QC	3m	16	4	n.a.		8
g PE/EVAL/PE	SAFETY4 4H	>240m	0	0	0.07	35°C	40
g PE/EVAL/PE	SAFETY4 4H	>480m	0	0	0.07		60
g PVAL	ANSELLEDMONT PVA	75m	<9	1	n.a.		6
g PVAL	COMASEC SOLVATRIL	300m	54	2	n.a.		40
g PVAL	EDMONT 15-552	30m	<900	3	n.a.		6
g PVAL	EDMONT 15-552	>480m	0	0	n.a.		33
g PVAL	EDMONT 25-545	>360m	0	0	n.a.		6
g PVC	ANSELLEDMONT SNORKE	180m	<90	2	n.a.		6
g PVC	BEST 725R	155m	30	2	n.a.		53
g PVC	COMASEC MULTIPOST	>480m	0	0	n.a.		40
g PVC	COMASEC MULTITOP	>480m	0	0	n.a.		40
g PVC	COMASEC NORMAL	>480m	0	0	n.a.		40
g PVC	COMASEC OMNI	95m	150	3	n.a.		40
s TEFLON	CHEMFAB CHALL. 5100	>940m	0	0	n.a.		65
s UNKNOWN MATERIAL	DU PONT BARRICADE	>480m	0	0	n.a.		8
s UNKNOWN MATERIAL	LIFE-GUARD RESPONDE	>480m	0	0	n.a.		77
g UNKNOWN MATERIAL	MARIGOLD R MEDIGLOV	2m	207	4	0.03	degrad.	79

butyl benzyl phthalate (synonym: benzyl butyl phthalate)
CAS Number: 85-68-7
Primary Class: 226   Esters, Carboxylic, Benzoates and Phthalates

g NITRILE	ANSELLEDMONT 37-155	>480m	0	0	0.38		34

butyl ether (synonym: di-n-butyl ether)
CAS Number: 142-96-1
Primary Class: 241   Ethers, Aliphatic and Alicyclic

s UNKNOWN MATERIAL	DU PONT BARRICADE	>480m	0	0	n.a.		8
s UNKNOWN MATERIAL	KAPPLER CPF III	>480m	0	0	n.a.		77

butyraldehyde
CAS Number: 123-72-8
Primary Class: 121   Aldehydes, Aliphatic and Alicyclic

g BUTYL	NORTH B-131	>900m	0	0	0.66		34
g NEOPRENE	ANSELLEDMONT 29-870	44m	756	3	0.46		34
c PE	DU PONT TYVEK QC	1m	220	4	n.a.		8
g PE/EVAL/PE	SAFETY4 4H	>240m	0	0	0.07	35°C	60
g PVAL	EDMONT 25-950	16m	8	2	0.26		34
s SARANEX-23	DU PONT TYVEK SARAN	47m	61	3	n.a.		8
s TEFLON	CHEMFAB CHALL. 5100	>450m	0	0	n.a.		65

G A R M E Class & Number/ N test chemical/ T MATERIAL NAME	MANUFACTURER & PRODUCT IDENTIFICATION	Break- through Time in Minutes	Perm- eation Rate in mg/sq m /min.	I N D E X	Thick- ness in mm	degrad. and comments	R e f s
s UNKNOWN MATERIAL	CHEMRON CHEMREL	256m	<1	1	n.a.		67
s UNKNOWN MATERIAL	DU PONT BARRICADE	>480m	0	0	n.a.		8
s UNKNOWN MATERIAL	LIFE-GUARD RESPONDE	>480m	0	0	n.a.		77
g VITON	NORTH F-091	54m	540	3	0.29		34

butyric acid
CAS Number: 107-92-6
Primary Class: 102   Acids, Carboxylic, Aliphatic and Alicyclic, Unsubstituted

g BUTYL	NORTH B-174	>480m	0	0	0.64		34
g NATURAL RUBBER	ACKWELL 5-109				0.15	degrad.	34
g NEOPRENE	ANSELLEDMONT 29-870	150m	240	3	0.50		34
g PVC	ANSELLEDMONT 34-100	4m	2040	5	0.20		34
g VITON	NORTH F-091	>480m	0	0	0.25		34

cadmium oxide (solid)
CAS Number: 1306-19-0
Primary Class: 340   Inorganic Salts

g NEOPRENE	ANSELLEDMONT 29-870	>480m	0	0	0.46		34
g NITRILE	ANSELLEDMONT 37-155	>480m	0	0	0.38		34

calcium chloride
CAS Number: 10043-52-4
Primary Class: 340   Inorganic Salts

s UNKNOWN MATERIAL	LIFE-GUARD RESPONDE	>240m	0	0	n.a.		77

calcium hydrogen phosphate (synonym: dicalcium phosphate)
CAS Number: 7757-93-9
Primary Class: 340   Inorganic Salts

calcium hydroxide (saturated solution)
CAS Number: 1305-62-0
Primary Class: 380   Inorganic Bases

g NATURAL RUBBER	MARIGOLD BLACK HEVY	>480m	0	0	0.65		79
g NATURAL RUBBER	MARIGOLD FEATHERLIT	>480m	0	0	0.15		79
g NATURAL RUBBER	MARIGOLD MED WT EXT	>480m	0	0	0.45		79
g NATURAL RUBBER	MARIGOLD MEDICAL	>480m	0	0	0.28		79
g NATURAL RUBBER	MARIGOLD ORANGE SUP	>480m	0	0	0.73		79
g NATURAL RUBBER	MARIGOLD R SURGEONS	>480m	0	0	0.28		79
g NATURAL RUBBER	MARIGOLD RED LT WT	>480m	0	0	0.43		79
g NATURAL RUBBER	MARIGOLD SENSOTECH	>480m	0	0	0.28		79
g NATURAL RUBBER	MARIGOLD SUREGRIP	>480m	0	0	0.58		79
g NATURAL RUBBER	MARIGOLD YELLOW	>480m	0	0	0.38		79

G A R M E Class & Number/ N test chemical/ T MATERIAL NAME	MANUFACTURER & PRODUCT IDENTIFICATION	Break- through Time in Minutes	Perm- eation Rate in mg/sq m /min.	I N D E X	Thick- ness in mm	degrad. and comments	R e f s
g NATURAL+NEOPRENE	MARIGOLD FEATHERWT	>480m	0	0	0.35		79
g NITRILE	MARIGOLD BLUE	>480m	0	0	0.45	degrad.	79
g NITRILE	MARIGOLD GREEN SUPA	>480m	0	0	0.30	degrad.	79
g NITRILE	MARIGOLD NITROSOLVE	>480m	0	0	0.75	degrad.	79
g UNKNOWN MATERIAL	MARIGOLD R MEDIGLOV	>480m	0	0	0.03		79

caprylic acid (synonym: octanoic acid)
CAS Number: 124-07-2
Primary Class: 102   Acids, Carboxylic, Aliphatic and Alicyclic, Unsubstituted

g BUTYL	NORTH B-174				0.63	degrad.	34
g NATURAL RUBBER	ACKWELL 5-109				0.15	degrad.	34
g NEOPRENE	ANSELLEDMONT 29-870	>480m	0	0	0.50		34
g NITRILE	ANSELLEDMONT 37-155	>480m	0	0	0.38		34
g PVC	ANSELLEDMONT 34-100	18m	18	3	0.15		34
g VITON	NORTH F-091	>480m	0	0	0.27		34

carbon disulfide (synonym: carbon bisulfide)
CAS Number: 75-15-0
Primary Class: 508   Sulfur Compounds, Thiones
Related Class: 502   Sulfur Compounds, Sulfides and Disulfides

g BUTYL	ANSELL LAMPRECHT	3m	61300	5	0.70		38
g BUTYL	BRUNSWICK BUTYL STD	<4m	>5000	5	0.73		89
g BUTYL	BRUNSWICK BUTYL-XTR	10m	>5000	5	0.88		89
s BUTYL	LIFE-GUARD BUTYL	2m	38	4	n.a.		77
g BUTYL	NORTH B-131				n.a.	degrad.	65
g BUTYL	NORTH B-161	7m			0.43		34
g BUTYL	NORTH B-174	3m	5904	5	0.62	degrad.	34
s BUTYL	WHEELER ACID KING	1m	<90	4	n.a.		85
g BUTYL/ECO	BRUNSWICK BUTYL-POL	4m	>5000	5	0.63		89
s BUTYL/NEOPRENE	MSA BETEX	<5m	58	4	n.a.		48
u CPE	ILC DOVER	8m			n.a.		29
s CPE	STD. SAFETY	12m	117	4	n.a.		75
g NATURAL RUBBER	ACKWELL 5-109				0.15	degrad.	34
g NATURAL RUBBER	ANSELLEDMONT 36-124				0.46	degrad.	6
g NATURAL RUBBER	ANSELLEDMONT 392				0.48	degrad.	6
u NATURAL RUBBER	Unknown	1m	>80000	5	0.64		38
g NATURAL+NEOPRENE	PLAYTEX ARGUS 123	1m	8880	5	n.a.		72
g NEOPRENE	ANSELLEDMONT 29-840				0.38	degrad.	6
g NEOPRENE	ANSELLEDMONT 29-870				0.48	degrad.	34
g NEOPRENE	ANSELLEDMONT NEOX				n.a.	degrad.	6
g NEOPRENE	BRUNSWICK NEOPRENE	4m	>5000	5	0.93		89
s NEOPRENE	LIFE-GUARD NEOPRENE	5m	38	4	n.a.		77
b NEOPRENE	RAINFAIR	12m	>120	4	n.a.		90
b NEOPRENE	RANGER	9m	>320	4	n.a.		90
b NEOPRENE	SERVUS NO 22204	13m	>330	4	n.a.		90
g NITRILE	ANSELL CHALLENGER	8m	4	3	0.38		55
g NITRILE	ANSELLEDMONT 37-165	30m	<9000	4	0.60		6

G A R M E N T	Class & Number/ test chemical/ MATERIAL NAME	MANUFACTURER & PRODUCT IDENTIFICATION	Break- through Time in Minutes	Perm- eation Rate in mg/sq m /min.	I N D E X	Thick- ness in mm	degrad. and comments	R e f s
g	NITRILE	ANSELLEDMONT 37-175	9m	6130	5	0.42		38
g	NITRILE	MAPA-PIONEER A-14	20m	5160	4	0.56		36
g	NITRILE	MARIGOLD NITROSOLVE	25m	1485	4	0.75	degrad.	79
g	NITRILE	NORTH LA-142G	9m	3060	5	0.36	degrad.	34
b	NITRILE+PUR+PVC	BATA HAZMAX	85m	250	3	n.a.		86
b	NITRILE+PVC	SERVUS NO 73101	40m	>380	3	n.a.		90
b	NITRILE+PVC	TINGLEY	41m	170	3	n.a.		90
c	PE	DU PONT TYVEK QC	<1m			n.a.		8
g	PE/EVAL/PE	SAFETY4 4H	>1440m	0	0	0.07		60
g	PE/EVAL/PE	SAFETY4 4H	>240m		0	0.07	35°C	60
g	PVAL	ANSELLEDMONT PVA	>360m		0	n.a.		6
g	PVAL	EDMONT 15-552	>360m	0	0	n.a.		6
g	PVAL	EDMONT 25-545	>360m	0	0	n.a.		6
g	PVAL	EDMONT 25-545	>480m	0	0	n.a.		34
g	PVAL	EDMONT 25-950	>960m	0	0	0.30		34
g	PVC	ANSELLEDMONT 34-100				n.a.		34
g	PVC	ANSELLEDMONT SNORKE				n.a.	degrad.	6
b	PVC	BATA STANDARD	38m	0.2	2	n.a.		86
b	PVC	JORDAN DAVID	35m	220	3	n.a.		90
b	PVC	STD. SAFETY GL-20	4m	>4000	5	n.a.		75
s	PVC	STD. SAFETY WG-20	4m	>4000	5	n.a.		75
s	PVC	WHEELER ACID KING	1m	<90	4	n.a.		85
b	PVC+POLYURETHANE	BATA POLYBLEND	49m	0.1	2	n.a.		86
b	PVC+POLYURETHANE	BATA POLYBLEND	43m	>320	3	n.a.		90
b	PVC+POLYURETHANE	BATA POLYMAX	37m	0.9	2	n.a.		86
b	PVC+POLYURETHANE	BATA SUPER POLY	61m	0.09	1	n.a.		86
s	SARANEX-23	DU PONT TYVEK SARAN	<1m			n.a.		8
s	SARANEX-23 2-PLY	DU PONT TYVEK SARAN	5m	2820	5	n.a.		8
s	TEFLON	CHEMFAB CHALL. 5100	18m	0.6	2	n.a.	degrad.	65
s	TEFLON	CHEMFAB CHALL. 5200	143m	0.5	1	n.a.		80
s	TEFLON	CHEMFAB CHALL. 6000	>180m			n.a.		80
s	TEFLON	LIFE-GUARD TEFGUARD	>480m	0	0	n.a.		77
s	TEFLON	WHEELER ACID KING	>480m	0	0	n.a.		85
v	TEFLON-FEP	CHEMFAB CHALL.	94m	1.8	1	0.25		65
s	UNKNOWN MATERIAL	CHEMRON CHEMREL	5m	36	4	n.a.		67
s	UNKNOWN MATERIAL	CHEMRON CHEMREL MAX	>480m	0	0	n.a.		67
c	UNKNOWN MATERIAL	CHEMRON CHEMTUFF	4m	112	4	n.a.		67
s	UNKNOWN MATERIAL	DU PONT BARRICADE	>480m	0	0	n.a.		8
s	UNKNOWN MATERIAL	KAPPLER CPF III	16m	5	2	n.a.		77
s	UNKNOWN MATERIAL	LIFE-GUARD RESPONDE	>480m	0	0	n.a.		77
g	VITON	NORTH F-091	>960m	0	0	0.26		34
g	VITON	NORTH F-091	>480m	0	0	0.26		34
g	VITON	NORTH F-091	336m	2	1	0.30		62
s	VITON/BUTYL/UNKN.	TRELLEBORG HPS	>180m			n.a.		71
s	VITON/CHLOROBUTYL	LIFE-GUARD VC-100	132m			n.a.		77
s	VITON/CHLOROBUTYL	LIFE-GUARD VNC 200	>480m	0	0	n.a.		77
u	VITON/CHLOROBUTYL	Unknown	>180m			n.a.		29
u	VITON/CHLOROBUTYL	Unknown	13m			n.a.		46
g	VITON/NEOPRENE	ERISTA VITRIC	>60m			0.60		38
s	VITON/NEOPRENE	MSA VAUTEX	>240m	0	0	n.a.		48

```
G
A
R                                                Perm-  I
M                                        Break-  eation N Thick-              R
E  Class & Number/      MANUFACTURER     through Rate in D ness  degrad.     e
N  test chemical/       & PRODUCT        Time in mg/sq m E in     and        f
T  MATERIAL NAME        IDENTIFICATION   Minutes /min.  X mm  comments       s
```

carbon tetrachloride (synonym: tetrachloromethane)
CAS Number: 56-23-5
Primary Class: 261   Halogen Compounds, Aliphatic and Alicyclic

	Class & Number / test chemical / MATERIAL NAME	MANUFACTURER & PRODUCT IDENTIFICATION	Breakthrough Time in Minutes	Permeation Rate in mg/sq m /min.	INDEX	Thickness in mm	degrad. and comments	Refs
g	BUTYL	ANSELL LAMPRECHT	50m	9050	4	0.70		38
g	BUTYL	BEST 878	53m	7650	4	0.75	degrad.	53
s	BUTYL	MSA CHEMPRUF	<60m			n.a.		48
g	BUTYL	NORTH B-161				0.43	degrad.	34
s	CPE	ILC DOVER CLOROPEL	>180m			n.a.		46
g	NATURAL RUBBER	ANSELLEDMONT 36-124				0.46	degrad.	6
g	NATURAL RUBBER	ANSELLEDMONT 392				0.48	degrad.	6
g	NATURAL RUBBER	ANSELLEDMONT 46-320	4m	160000	5	0.31		5
g	NATURAL RUBBER	BEST 65NFW				n.a.	degrad.	53
g	NATURAL RUBBER	MARIGOLD BLACK HEVY	6m	13242	5	0.82	degrad.	79
g	NATURAL RUBBER	MARIGOLD ORANGE SUP	6m	4567	5	0.73	degrad.	79
g	NATURAL RUBBER	MARIGOLD SUREGRIP	4m	39269	5	0.89	degrad.	79
g	NATURAL+NEOPRENE	ANSELL OMNI 276	4m	46000	5	0.45		5
g	NATURAL+NEOPRENE	ANSELL OMNI 276				0.45	degrad.	34
g	NEOPRENE	ANSELLEDMONT 29-840				0.38	degrad.	6
g	NEOPRENE	ANSELLEDMONT 29-865	19m	4960	4	0.51		26
g	NEOPRENE	ANSELLEDMONT NEOX	30m	1000	4	0.77		5
g	NEOPRENE	ANSELLEDMONT NEOX				n.a.	degrad.	6
g	NEOPRENE	BEST 32	28m	2920	4	n.a.	degrad.	53
g	NEOPRENE	BEST 6780	66m	1980	4	n.a.	degrad.	53
g	NEOPRENE	MAPA-PIONEER N-44	31m	15120	5	0.56	degrad.	36
g	NITRILE	ANSELL CHALLENGER	270m	0.1		0.38		55
g	NITRILE	ANSELLEDMONT 37-165	150m	<900	3	0.60		6
g	NITRILE	ANSELLEDMONT 37-175	>60m			0.37		5
g	NITRILE	ANSELLEDMONT 37-175	>197m			n.a.		26
g	NITRILE	BEST 727	>480m	0	0	0.38		53
g	NITRILE	MAPA-PIONEER A-14	342m	450	3	0.56		36
g	NITRILE	MARIGOLD BLUE	385m	27	2	0.45		79
g	NITRILE	MARIGOLD GREEN SUPA	328m	44	2	0.30		79
g	NITRILE	MARIGOLD NITROSOLVE	396m	24	2	0.75		79
g	NITRILE	NORTH LA-132G	64m	16200	5	0.36		30
g	NITRILE	NORTH LA-142G	203m	300	3	0.36		7
g	PE	ANSELLEDMONT 35-125	2m	5000	5	0.03		5
g	PE/EVAL/PE	SAFETY4 4H	>240m	0	0	0.07	35°C	60
g	PE/EVAL/PE	SAFETY4 4H	>1440m	0	0	0.07		60
g	PVAL	ANSELLEDMONT PVA	>360m		0	n.a.		6
g	PVAL	EDMONT 15-552	>360m	0	0	n.a.		6
g	PVAL	EDMONT 15-552	>197m	30	2	n.a.		26
g	PVAL	EDMONT 25-545	>360m	0	0	n.a.		6
g	PVAL	EDMONT 25-950	>480m	0	0	0.36		34
g	PVC	ANSELLEDMONT M GRIP	25m	<9000	4	n.a.		6
u	PVC	Unknown	40m	2030	4	n.a.		26
s	TEFLON	CHEMFAB CHALL. 5100	>180m	0	1	n.a.		65
s	UNKNOWN MATERIAL	DU PONT BARRICADE	>480m	0	0	n.a.		8
s	UNKNOWN MATERIAL	LIFE-GUARD RESPONDE	>480m	0	0	n.a.		77

G A R M E Class & Number/ N test chemical/ T MATERIAL NAME	MANUFACTURER & PRODUCT IDENTIFICATION	Break- through Time in Minutes	Perm- eation Rate in mg/sq m /min.	I N D E X	Thick- ness in mm	R e degrad. and f comments s
g UNKNOWN MATERIAL	NORTH SILVERSHIELD	>360m	0	0	0.08	7
g VITON	NORTH F-091	>780m	0	0	0.24	7
g VITON	NORTH F-091	>480m	0	0	0.25	34
g VITON	NORTH F-091	5355m	0.21	1	0.30	62
g VITON/NEOPRENE	ERISTA VITRIC	>60m			0.60	38

---

carmustine (synonym: 1,3-bis(2-chloroethyl)-1-nitrosocarbamide)
CAS Number: 154-93-8
Primary Class: 870   Antineoplastic Drugs and Other Pharmaceuticals

g EMA	Unknown	3m			0.07	59
u NATURAL RUBBER	Unknown	10m			0.18	59
u NATURAL RUBBER	Unknown	3m			0.18 preexp EtOH	59
g NEOPRENE	ANSELL NEOPRENE 530	120m			0.28	59
u PVC	Unknown	15m			0.19	59

---

castor oil
CAS Number: 8001-79-4
Primary Class: 860   Coals, Charcoals, Oils

g NEOPRENE	ANSELLEDMONT 29-870	>480m	0	0	0.46	34

---

cationic detergent (unspecified concentration)
CAS Number: 6004-24-6
Primary Class: 900   Miscellaneous Unclassified Chemicals

g NATURAL RUBBER	MARIGOLD BLACK HEVY	>480m	0	0	0.65	79
g NATURAL RUBBER	MARIGOLD FEATHERLIT	>480m	0	0	0.15	79
g NATURAL RUBBER	MARIGOLD MED WT EXT	>480m	0	0	0.45	79
g NATURAL RUBBER	MARIGOLD MEDICAL	>480m	0	0	0.28	79
g NATURAL RUBBER	MARIGOLD ORANGE SUP	>480m	0	0	0.73	79
g NATURAL RUBBER	MARIGOLD R SURGEONS	>480m	0	0	0.28	79
g NATURAL RUBBER	MARIGOLD ED LT WT	>480m	0	0	0.43	79
g NATURAL RUBBER	MARIGOLD SENSOTECH	>480m	0	0	0.28	79
g NATURAL RUBBER	MARIGOLD SUREGRIP	>480m	0	0	0.58	79
g NATURAL RUBBER	MARIGOLD YELLOW	>480m	0	0	0.38	79
g NATURAL+NEOPRENE	MARIGOLD FEATHERWT	>480m	0	0	0.35	79
g NITRILE	MARIGOLD BLUE	>480m	0	0	0.45	79
g NITRILE	MARIGOLD GREEN SUPA	>480m	0	0	0.30 degrad.	79
g NITRILE	MARIGOLD NITROSOLVE	>480m	0	0	0.75	79
g UNKNOWN MATERIAL	MARIGOLD R MEDIGLOV	>480m	0	0	0.03	79

---

cationic detergent (unspecified concentration)
CAS Number: 75-87-6
Primary Class: 900   Miscellaneous Unclassified Chemicals
Related Class: 261   Halogen Compounds, Aliphatic and Alicyclic

g BUTYL	NORTH B-174	200m	294	3	0.71 degrad.	34

G A R M E Class & Number/ N test chemical/ T MATERIAL NAME	MANUFACTURER & PRODUCT IDENTIFICATION	Break- through Time in Minutes	Perm- eation Rate in mg/sq m /min.	I N D E X	Thick- ness in mm	degrad. and comments	R e f s
g NATURAL RUBBER	ACKWELL 5-109				0.15	degrad.	34
g NEOPRENE	ANSELLEDMONT 29-870				0.48	degrad.	34
g NITRILE	ANSELLEDMONT 37-155				0.40	degrad.	34
g PVAL	EDMONT 25-950	>480m	0	0	0.76		34
g PVC	ANSELLEDMONT 34-100	4m	16890	5	0.20	degrad.	34
g VITON	NORTH F-091	295m	0	0	0.25		34

chlordane, >70%
Primary Class: 261   Halogen Compounds, Aliphatic and Alicyclic
Related Class: 840   Pesticides, Mixtures and Formulations

s TEFLON	CHEMFAB CHALL. 5100	>200m			n.a.		65

chlorine (synonym: chlorine gas)
CAS Number: 7782-50-5
Primary Class: 350   Inorganic Gases and Vapors
Related Class: 330   Elements

g BUTYL	BRUNSWICK BUTYL STD	60m	>5000	4	0.63		89
g BUTYL	BRUNSWICK BUTYL-XTR	50m	7500	4	0.73		89
s BUTYL	DRAGER 500 OR 710	>60m			n.a.		91
s BUTYL	WHEELER ACID KING	>480m	0	0	n.a.		85
g BUTYL/ECO	BRUNSWICK BUTYL-POL	>180m			0.63		89
s BUTYL/NEOPRENE	MSA BETEX	>480m	0	0	n.a.		48
s CPE	STD. SAFETY	>180m			n.a.		75
u NATURAL RUBBER	Unknown	>480m	0	0	0.46	0.1%	61
g NEOPRENE	BEST 6780	>480m	0	0	n.a.		53
g NEOPRENE	BRUNSWICK NEOPRENE	120m	>500	3	0.88		89
b NEOPRENE	RAINFAIR	>180m			n.a.		90
b NEOPRENE	RANGER	>180m			n.a.		90
b NEOPRENE	SERVUS NO 22204	>180m			n.a.		90
u NEOPRENE	Unknown	>480m	0	0	0.41	0.1%	61
g NITRILE	BEST 22R	>480m	0	0	n.a.		53
b NITRILE+PUR+PVC	BATA HAZMAX	>180m			n.a.		86
b NITRILE+PVC	SERVUS NO 73101	>180m			n.a.		90
b NITRILE+PVC	TINGLEY	>180m			n.a.		90
c PE	DU PONT TYVEK QC	>480m	0	0	n.a.	20 ppm	8
c PE	DU PONT TYVEK QC	1m	180	4	n.a.	99%	8
g PE/EVAL/PE	SAFETY4 4H	>480m	0	0	0.07		60
g PVC	BEST 812		60	5	n.a.		53
b PVC	JORDAN DAVID	>180m			n.a.		90
b PVC	STD. SAFETY GL-20	90m	101	3	n.a.		75
s PVC	STD. SAFETY WG-20	60m	171	3	n.a.		75
s PVC	WHEELER ACID KING	45m	<0.9	2	n.a.		85
b PVC+POLYURETHANE	BATA POLYBLEND	>180m			n.a.		90
s SARANEX-23	DU PONT TYVEK SARAN	>480m	0	0	n.a.	20 ppm	8
s SARANEX-23	DU PONT TYVEK SARAN	>480m	0	0	n.a.	99%	8

G A R M E Class & Number/ N test chemical/ T MATERIAL NAME	MANUFACTURER & PRODUCT IDENTIFICATION	Break-through Time in Minutes	Perm-eation Rate in mg/sq m /min.	I N D E X	Thick-ness in mm	degrad. and comments	R e f s
s TEFLON	CHEMFAB CHALL. 5200	>300m	0	0	n.a.		80
s TEFLON	CHEMFAB CHALL. 6000	>180m			n.a.		80
s TEFLON	LIFE-GUARD TEFGUARD	>480m	0	0	n.a.		77
s TEFLON	WHEELER ACID KING	>480m	0	0	n.a.		85
s UNKNOWN MATERIAL	CHEMRON CHEMREL	>1440m	0	0	n.a.		67
s UNKNOWN MATERIAL	CHEMRON CHEMREL MAX	>1440m	0	0	n.a.		67
s UNKNOWN MATERIAL	DU PONT BARRICADE	>480m	0	0	n.a.	99%	8
s UNKNOWN MATERIAL	KAPPLER CPF III	>480m		0	n.a.		77
s UNKNOWN MATERIAL	LIFE-GUARD RESPONDE	>480m	0	0	n.a.		77
s VITON/BUTYL	DRAGER 500 OR 710	>60m			n.a.		91
s VITON/BUTYL/UNKN.	TRELLEBORG HPS	>180m			n.a.		71
s VITON/CHLOROBUTYL	LIFE-GUARD VC-100	>480m	0	0	n.a.		77
s VITON/NEOPRENE	MSA VAUTEX	>480m	0	0	n.a.		48

chlorine dioxide, <30%
CAS Number: 10049-04-4
Primary Class: 350   Inorganic Gases and Vapors

s UNKNOWN MATERIAL	LIFE-GUARD RESPONDE	>480m	0	0	n.a.		77

chlorine, liquid
Primary Class: 330   Elements
Related Class: 350   Inorganic Gases and Vapors

s UNKNOWN MATERIAL	LIFE-GUARD RESPONDE	>480m	0	0	n.a.		77

chloroacetic acid (synonym: monochloroacetic acid)
CAS Number: 79-11-8
Primary Class: 103   Acids, Carboxylic, Aliphatic and Alicyclic, Substituted

g BUTYL	NORTH B-174	>480m	0	0	0.61		34
g NEOPRENE	ANSELLEDMONT 29-840	>480m	0	0	0.50		34
g NITRILE	ANSELLEDMONT 37-155				0.40	degrad.	34
c PE	DU PONT TYVEK QC	>480m	0	0	n.a.		8
c PE	DU PONT TYVEK QC	5m			n.a.	65°C	8
g PVAL	EDMONT 25-950				n.a.	degrad.	34
g PVC	ANSELLEDMONT 34-100	26m	36	3	0.20		34
s SARANEX-23	DU PONT TYVEK SARAN	60m			n.a.	65°C	8
g VITON	NORTH F-091	>480m	0	0	0.27		34

chloroacetone (synonyms: 1-chloro-2-propanone; monochloroacetone)
CAS Number: 78-95-5
Primary Class: 391   Ketones, Aliphatic and Alicyclic
Related Class: 261   Halogen Compounds, Aliphatic and Alicyclic

g PE/EVAL/PE	SAFETY4 4H	>240m	0	0	0.07	35°C	60

G A R M E T	Class & Number/ test chemical/ MATERIAL NAME	MANUFACTURER & PRODUCT IDENTIFICATION	Break- through Time in Minutes	Perm- eation Rate in mg/sq m /min.	I N D E X	Thick- ness in mm	degrad. and comments	R e f s
s	SARANEX-23	DU PONT TYVEK SARAN	360m	0.8	0	n.a.		8

chloroacetonitrile
CAS Number: 107-14-2
Primary Class: 431   Nitriles, Aliphatic and Alicyclic
Related Class: 261   Halogen Compounds, Aliphatic and Alicyclic

G	Class & Number/ test chemical/ MATERIAL NAME	MANUFACTURER & PRODUCT IDENTIFICATION	Break- through Time	Perm- eation Rate	I N D E X	Thick- ness	degrad. and comments	R e f s
g	BUTYL	NORTH B-174	>480m	0	0	0.63		34
g	NATURAL RUBBER	ACKWELL 5-109	<1m	756	4	0.10		34
g	NITRILE	ANSELLEDMONT 37-155				0.40	degrad.	34
g	PVAL	EDMONT 25-950	>480m	0	0	0.27		34
g	VITON	NORTH F-091	>480m	0	0	0.26		34

chloroacetophenone (synonym: tear gas)
CAS Number: 532-27-4
Primary Class: 392   Ketones, Aromatic

s	UNKNOWN MATERIAL	LIFE-GUARD RESPONDE	>480m	0	0	n.a.		77

chlorobenzene
CAS Number: 108-90-7
Primary Class: 264   Halogen Compounds, Aromatic

G								
g	BUTYL	BEST 878				0.75	degrad.	53
g	BUTYL	NORTH B-174	34m	18480	5	0.66	degrad.	34
g	NATURAL RUBBER	ACKWELL 5-109				0.15	degrad.	34
g	NATURAL RUBBER	ANSELLEDMONT 36-124				0.46	degrad.	6
g	NATURAL RUBBER	ANSELLEDMONT 392				0.48	degrad.	6
g	NATURAL+NEOPRENE	ANSELL OMNI 276				0.45	degrad.	55
g	NEOPRENE	ANSELLEDMONT 29-840				0.38	degrad.	6
g	NEOPRENE	ANSELLEDMONT 29-870				0.48	degrad.	34
g	NEOPRENE	BEST 6780				n.a.	degrad.	53
g	NEOPRENE/NATURAL	ANSELL CHEMI-PRO				n.a.	degrad.	55
g	NITRILE	ANSELLEDMONT 37-155				0.40	degrad.	34
g	NITRILE	ANSELLEDMONT 37-165				0.60	degrad.	6
g	NITRILE	BEST 727				0.38	degrad.	53
g	NITRILE	GRANET N11F	12m	9383	5	n.a.		39
g	NITRILE	MAPA-PIONEER A-15	15m	9583	4	0.35		39
g	PVAL	ANSELLEDMONT PVA	>360m	0	0	n.a.		6
g	PVAL	EDMONT 15-552	15m	<900	3	n.a.		6
g	PVAL	EDMONT 25-545	>360m	0	0	n.a.		6
g	PVC	ANSELLEDMONT 34-100	1m	22500	5	0.20	degrad.	34
g	PVC	ANSELLEDMONT SNORKE				n.a.	degrad.	6
s	TEFLON	CHEMFAB CHALL. 5100	>180m	0	1	n.a.		65
s	UNKNOWN MATERIAL	DU PONT BARRICADE	>480m	0	0	n.a.		8
s	UNKNOWN MATERIAL	KAPPLER CPF III	63m	7	1	n.a.		77
s	UNKNOWN MATERIAL	LIFE-GUARD RESPONDE	>480m	0	0	n.a.		77

G A R M E Class & Number/ N test chemical/ T MATERIAL NAME	MANUFACTURER & PRODUCT IDENTIFICATION	Break- through Time in Minutes	Perm- eation Rate in mg/sq m /min.	I N D E X	Thick- ness in mm	degrad. and comments	R e f s
g VITON	NORTH F-091	>480m	0	0	0.26		34
g VITON	NORTH F-121	>240m	0	0	0.31		39

chlorodibromomethane (synonym: dibromochloromethane)
CAS Number: 124-48-1
Primary Class: 261 Halogen Compounds, Aliphatic and Alicyclic

g BUTYL	NORTH B-174	200m	1470	4	0.68	degrad.	34
g NATURAL RUBBER	ACKWELL 5-109				0.15	degrad.	34
g NEOPRENE	ANSELLEDMONT 29-870				0.48	degrad.	34
g NITRILE	ANSELLEDMONT 37-155				0.40	degrad.	34
g PVAL	EDMONT 25-950	36m	0.4	1	0.61		34
g PVC	ANSELLEDMONT 34-100	2m	11040	5	0.20	degrad.	34
g VITON	NORTH F-091	>480m	0	0	0.25		34

chloroform (synonym: trichloromethane)
CAS Number: 67-66-3
Primary Class: 261 Halogen Compounds, Aliphatic and Alicyclic

g BUTYL	BEST 878	21m	3000	4	0.75	degrad.	53
g BUTYL	NORTH B-161				0.43	degrad.	7
m BUTYL	PLYMOUTH RUBBER	13m			n.a.		46
u CPE	ILC DOVER	12m			n.a.		29
s CPE	ILC DOVER CLOROPEL	32m			n.a.		46
g NATURAL RUBBER	ANSELLEDMONT 30-139				0.25	degrad.	42
g NATURAL RUBBER	ANSELLEDMONT 36-124				0.46	degrad.	6
g NATURAL RUBBER	ANSELLEDMONT 392				0.48	degrad.	6
g NATURAL RUBBER	ANSELLEDMONT 46-320	2m	40000	5	0.31		5
g NATURAL RUBBER	BEST 65NFW				n.a.	degrad.	53
g NATURAL+NEOPRENE	ANSELL OMNI 276	3m	70000	5	0.45		5
g NATURAL+NEOPRENE	ANSELL OMNI 276				0.45	degrad.	55
g NEOPRENE	ANSELLEDMONT 29-840				0.38	degrad.	6
g NEOPRENE	ANSELLEDMONT NEOX	2m	27000	5	0.77		5
g NEOPRENE	BEST 32	4m	3030	5	n.a.	degrad.	53
g NEOPRENE	BEST 6780	11m	8600	5	n.a.	degrad.	53
g NEOPRENE	MAPA-PIONEER N-44	12m	13680	5	0.56	degrad.	36
u NEOPRENE	Unknown	11m			0.40	37°C TL	16
g NEOPRENE/NATURAL	ANSELL CHEMI-PRO				n.a.	degrad.	55
g NITRILE	ANSELLEDMONT 37-155	5m	94000	5	0.35		5
g NITRILE	ANSELLEDMONT 37-165				0.60	degrad.	6
g NITRILE	ANSELLEDMONT 49-155				0.38	degrad.	42
g NITRILE	BEST 727	6m	4780	5	0.38	degrad.	53
g NITRILE	MARIGOLD BLUE	9m	50083	5	0.45	degrad.	79
g NITRILE	MARIGOLD GREEN SUPA	4m	57434	5	0.30	degrad.	79
g NITRILE	MARIGOLD NITROSOLVE	10m	48613	5	0.75	degrad.	79
g NITRILE	NORTH LA-142G	4m	33120	5	0.36		7
u NITRILE	Unknown	9m			0.50	37°C TL	16

G A R M E N T	Class & Number/ test chemical/ MATERIAL NAME	MANUFACTURER & PRODUCT IDENTIFICATION	Break- through Time in Minutes	Perm- eation Rate in mg/sq m /min.	I N D E X	Thick- ness in mm	degrad. and comments	R e f s
g	PE	ANSELLEDMONT 35-125	1m	16000	5	0.03		5
c	PE	DU PONT TYVEK QC	<1m	3480	5	n.a.		8
u	PE	Unknown	3m			0.12	37°C TL	16
g	PE/EVAL/PE	SAFETY4 4H	>240m	0	0	0.07	35°C	60
g	PE/EVAL/PE	SAFETY4 4H	>1440m	0	0	0.07		60
g	PVAL	ANSELLEDMONT PVA	>360m		0	n.a.		6
g	PVAL	EDMONT 15-552	>360m	0	0	n.a.		6
g	PVAL	EDMONT 25-545	>360m	0	0	n.a.		6
g	PVAL	EDMONT 25-950	>480m	0	0	0.31		34
g	PVC	ANSELLEDMONT SNORKE				n.a.	degrad.	6
u	PVC	Unknown	8m			0.30	37°C TL	16
s	SARANEX-23	DU PONT TYVEK SARAN	<1m	2010	5	n.a.		8
s	TEFLON	CHEMFAB CHALL. 5100	>220m	0	1	n.a.		65
s	UNKNOWN MATERIAL	CHEMRON CHEMREL	4m	18	4	n.a.		67
s	UNKNOWN MATERIAL	DU PONT BARRICADE	>480m	0	0	n.a.		8
s	UNKNOWN MATERIAL	LIFE-GUARD RESPONDE	>480m	0	0	n.a.		77
g	UNKNOWN MATERIAL	NORTH SILVERSHIELD	10m	1	3	0.08		7
g	UNKNOWN MATERIAL	NORTH SILVERSHIELD	1906m	0.1	0	0.09		92
g	UNKNOWN MATERIAL	NORTH SILVERSHIELD	275m	2.4	1	0.09	37°C	92
g	UNKNOWN MATERIAL	NORTH SILVERSHIELD	42m	30	3	0.09	50°C	92
g	VITON	NORTH F-091	570m	28	2	0.24		7
u	VITON/CHLOROBUTYL	Unknown	162m			n.a.		29
u	VITON/CHLOROBUTYL	Unknown	>180m			n.a.		46

chloroform, 50% & toluene, 50%
Primary Class: 261   Halogen Compounds, Aliphatic and Alicyclic

chloroform, 96% & isopentyl alcohol, 4%
Primary Class: 261   Halogen Compounds, Aliphatic and Alicyclic
Related Class: 311   Hydroxy Compounds, Aliphatic and Alicyclic, Primary

g	PE/EVAL/PE	SAFETY4 4H	40m			0.07	35°C	60
g	PE/EVAL/PE	SAFETY4 4H	176m			0.07		60

chloroform, 97% & methyl alcohol, 3%
Primary Class: 261   Halogen Compounds, Aliphatic and Alicyclic

g	PE/EVAL/PE	SAFETY4 4H	>240m	0	0	0.07	35°C	60

chloropicrin
CAS Number: 76-06-2
Primary Class: 442   Nitro Compounds, Substituted

s	TEFLON	CHEMFAB CHALL. 5100	>186m	0	1	n.a.		65
s	UNKNOWN MATERIAL	LIFE-GUARD RESPONDE	>480m	0	0	n.a.		77

G A R M E Class & Number/ N test chemical/ T MATERIAL NAME	MANUFACTURER & PRODUCT IDENTIFICATION	Break- through Time in Minutes	Perm- eation Rate in mg/sq m /min.	I N D E X	Thick- ness in mm	degrad. and comments	R e f s

chloroprene (synonym: 2-chloro-1,3-butadiene)
CAS Number: 126-99-8
Primary Class: 263   Halogen Compounds, Vinylic
Related Class: 261   Halogen Compounds, Aliphatic and Alicyclic

g BUTYL	NORTH B-174	28m	1104	4	0.66	degrad.	34
g NATURAL RUBBER	ACKWELL 5-109				0.15	degrad.	34
g NEOPRENE	ANSELLEDMONT 29-865	6m	31580	5	0.51		26
g NEOPRENE	ANSELLEDMONT 29-870				0.48	degrad.	34
g NITRILE	ANSELLEDMONT 37-155	23m	912	3	0.38	degrad.	34
g NITRILE	ANSELLEDMONT 37-175	7m	20730	5	0.41		26
g PVAL	EDMONT 15-552	>1001m	0	0	n.a.		26
g PVAL	EDMONT 25-950	>480m	0	0	0.33		34
g PVC	ANSELLEDMONT 34-100				0.20	degrad.	34
u PVC	Unknown	5m	6680	5	n.a.		26
g VITON	NORTH F-091	>480m	0	0	0.30		34

chlorosulfonic acid
CAS Number: 7790-94-5
Primary Class: 370   Inorganic Acids
Related Class: 504   Sulfur Compounds, Sulfonic Acids

s BUTYL	TRELLEBORG TRELLCHE				n.a.	degrad.	71
c PE	DU PONT TYVEK QC	63m			n.a.		8
s PVC	TRELLEBORG TRELLCHE				n.a.	degrad.	71
s SARANEX-23	DU PONT TYVEK SARAN	347m	0	0	n.a.		8
s TEFLON	CHEMFAB CHALL. 5100	>180m	0	1	n.a.		8

chlorotrimethylsilane (synonym: trimethylchlorosilane)
CAS Number: 75-77-4
Primary Class: 480   Organosilicon Compounds

g BUTYL	NORTH B-161				0.65	degrad.	34
g NATURAL RUBBER	ACKWELL 5-109				0.15	degrad.	34
g NEOPRENE	ANSELLEDMONT 29-870				0.48	degrad.	34
g NITRILE	ANSELLEDMONT 37-155	>312m	42	2	0.38		34
g PVC	ANSELLEDMONT 34-100	<1m	25560	5	0.18		34
g VITON	NORTH F-091	>498m	60	2	0.30		34

chlorpromazine hydrochloride
CAS Number: 69-09-0
Primary Class: 870   Antineoplastic Drugs and Other Pharmaceuticals
Related Class: 550   Organic Salts (Solutions)

g NATURAL RUBBER	ANSELL 9070-5	>480m	0	0	0.13	0.4% in H_2O	34
g PVC	ANSELLEDMONT 34-100	>480m	0	0	0.18	0.4% in H_2O	34

G A R M E Class & Number/ N test chemical/ T MATERIAL NAME	MANUFACTURER & PRODUCT IDENTIFICATION	Break- through Time in Minutes	Perm- eation Rate in mg/sq m /min.	I N D E X	Thick- ness in mm	degrad. and comments	R e f s

chromic acid (synonym: chromium trioxide)
CAS Number: 1333-82-0
Primary Class: 370   Inorganic Acids

s BUTYL	MSA CHEMPRUF	>480m	0	0	n.a.		48
g NATURAL RUBBER	COMASEC FLEXIGUM	>480m		0	0.95		40
g NATURAL RUBBER	MARIGOLD BLACK HEVY	108m	476	3	0.65	degrad.	79
g NATURAL RUBBER	MARIGOLD FEATHERLIT	2m	4150	5	0.15	degrad.	79
g NATURAL RUBBER	MARIGOLD MED WT EXT	87m	3256	4	0.45	degrad.	79
g NATURAL RUBBER	MARIGOLD MEDICAL	37m	3777	4	0.28	degrad.	79
g NATURAL RUBBER	MARIGOLD ORANGE SUP	110m	198	3	0.73	degrad.	79
g NATURAL RUBBER	MARIGOLD R SURGEONS	37m	3777	4	0.28	degrad.	79
g NATURAL RUBBER	MARIGOLD RED LT WT	80m	3330	4	0.43	degrad.	79
g NATURAL RUBBER	MARIGOLD SENSOTECH	37m	3777	4	0.28	degrad.	79
g NATURAL RUBBER	MARIGOLD SUREGRIP	102m	1310	4	0.58	degrad.	79
g NATURAL RUBBER	MARIGOLD YELLOW	66m	3479	4	0.38	degrad.	79
g NATURAL+NEOPRENE	MARIGOLD FEATHERWT	59m	3554	4	0.35	degrad.	79
g NEOPRENE	COMASEC COMAPRENE	75m			n.a.		40
g NITRILE	COMASEC COMATRIL	300m	60	2	0.55		40
g NITRILE	COMASEC COMATRIL SU	>480m	0	0	0.60		40
g NITRILE	COMASEC FLEXITRIL	90m	252	3	n.a.		40
g NITRILE	MARIGOLD NITROSOLVE				0.75	degrad.	79
g NITRILE+PVC	COMASEC MULTIPLUS	>480m	0	0	n.a.		40
g PVC	COMASEC MULTIPOST	>480m	0	0	n.a.		40
g PVC	COMASEC MULTITOP	>480m	0	0	n.a.		40
g PVC	COMASEC NORMAL	>480m	0	0	n.a.		40
g PVC	COMASEC OMNI	330m	66	2	n.a.		40
s PVC	MSA UPC	<120m			0.20		48

chromic acid & sulfuric acid
Primary Class: 370   Inorganic Acids

g PE/EVAL/PE	SAFETY4 4H	>240m	0	0	0.07	35°C	60

chromic acid, 30-70%
CAS Number: 7738-94-5
Primary Class: 370   Inorganic Acids

g NATURAL RUBBER	ANSELL PVL 040				0.46	degrad.	55
g NATURAL RUBBER	ANSELLEDMONT 36-124				0.46	degrad.	6
g NATURAL RUBBER	ANSELLEDMONT 392	>360m		0	0.48	degrad.	6
g NATURAL+NEOPRENE	ANSELL OMNI 276				0.56	degrad.	55
g NEOPRENE	ANSELL NEOPRENE 520				0.46	degrad.	55
g NEOPRENE	ANSELLEDMONT 29-840				0.31	degrad.	6
g NEOPRENE	ANSELLEDMONT NEOX				n.a.	degrad.	6
g NEOPRENE/NATURAL	ANSELL CHEMI-PRO				0.72	degrad.	55
g NITRILE	ANSELL CHALLENGER	180m			0.38		55

G A R M E Class & Number/ N test chemical/ T MATERIAL NAME	MANUFACTURER & PRODUCT IDENTIFICATION	Break- through Time in Minutes	Perm- eation Rate in mg/sq m /min.	I N D E X	Thick- ness in mm	degrad. and comments	R e f s
g NITRILE	ANSELLEDMONT 37-165	240m	0	0	0.60		6
g NITRILE	MAPA-PIONEER A-14	180m	>10000	5	0.56		36
g PE/EVAL/PE	SAFETY4 4H	>240m	0	0	0.07	35°C	60
g PVAL	ANSELLEDMONT PVA				n.a.	degrad.	6
g PVAL	EDMONT 15-552				n.a.	degrad.	6
g PVC	ANSELLEDMONT M GRIP	>360m	0	0	n.a.		6
g PVC	MAPA-PIONEER V-20	>480m	0	0	0.51	degrad.	36

chromic acid, 30-70% & (sodium fluoride and potassium ferricyanide), 30-70%
Primary Class: 800   Multicomponent Mixtures With >2 Components

g BUTYL	ERISTA BX	>240m	0	0	0.69	preex EtOAC	33
g PVC	KID 490	>240m	0	0	n.a.		33
g PVC	KID VINYLPRODUKTER	>240m	0	0	0.50		33
g PVC	KID VINYLPRODUKTER	>240m	0	0	0.50	preexp. 20h	33

chromic acid, <30%
Primary Class: 370   Inorganic Acids

g NATURAL+NEOPRENE	PLAYTEX ARGUS 123	>480m	0	0	n.a.		72

cis,trans-1,2-dichlorethylene (synonym: 1,2-dichloroethylene)
CAS Number: 540-59-0
Primary Class: 263   Halogen Compounds, Vinylic

g BUTYL	NORTH B-174				0.63	degrad.	34
g NATURAL RUBBER	ACKWELL 5-109				0.15	degrad.	34
g NEOPRENE	ANSELLEDMONT 29-870				0.48	degrad.	34
g NITRILE	ANSELLEDMONT 37-155	7m	4680	5	0.40	degrad.	34
g PVAL	EDMONT 25-950	14m	3	3	0.41	degrad.	34
g PVC	ANSELLEDMONT 34-100	<1m	2134	5	0.20		34
s UNKNOWN MATERIAL	LIFE-GUARD RESPONDE	>180m	0	1	n.a.		77
g VITON	NORTH F-091	57m	318	3	0.25		34

cis-1,2-dichloroethylene
CAS Number: 156-59-2
Primary Class: 263   Halogen Compounds, Vinylic

g BUTYL	NORTH B-174	19m	1752	4	0.63	degrad.	34
g NATURAL RUBBER	ACKWELL 5-109				0.15	degrad.	34
g NEOPRENE	ANSELLEDMONT 29-870	5m	21240	5	0.46	degrad.	34
g NITRILE	ANSELLEDMONT 37-155				0.40	degrad.	34
g PVAL	EDMONT 25-950				0.60	degrad.	34
g PVC	ANSELLEDMONT 34-100	1m	19920	5	0.20	degrad.	34
g VITON	NORTH F-091	101m	180	3	0.28		34

G A R M E Class & Number/ N test chemical/ T MATERIAL NAME	MANUFACTURER & PRODUCT IDENTIFICATION	Break- through Time in Minutes	Perm- eation Rate in mg/sq m /min.	I N D E X	Thick- ness in mm	R degrad. e and f comments s

citric acid, 30-70%
Primary Class: 104   Acids, Carboxylic, Aliphatic and Alicyclic, Polybasic

g NATURAL RUBBER	ANSELL CONFORM 4205	>360m	0	0	0.13	55
g NATURAL RUBBER	ANSELL PVL 040	>360m	0	0	0.46	55
g NATURAL RUBBER	ANSELL STERILE 832	>360m	0	0	0.23	55
g NATURAL RUBBER	MARIGOLD BLACK HEVY	>480m	0	0	0.65	79
g NATURAL RUBBER	MARIGOLD FEATHERLIT	>480m	0	0	0.15	79
g NATURAL RUBBER	MARIGOLD MED WT EXT	>480m	0	0	0.45	79
g NATURAL RUBBER	MARIGOLD MEDICAL	>480m	0	0	0.28	79
g NATURAL RUBBER	MARIGOLD ORANGE SUP	>480m	0	0	0.73	79
g NATURAL RUBBER	MARIGOLD R SURGEONS	>480m	0	0	0.28	79
g NATURAL RUBBER	MARIGOLD RED LT WT	>480m	0	0	0.43	79
g NATURAL RUBBER	MARIGOLD SENSOTECH	>480m	0	0	0.28	79
g NATURAL RUBBER	MARIGOLD SUREGRIP	>480m	0	0	0.58	79
g NATURAL RUBBER	MARIGOLD YELLOW	>480m	0	0	0.38	79
g NATURAL+NEOPRENE	MARIGOLD FEATHERWT	>480m	0	0	0.35	79
g NATURAL+NEOPRENE	PLAYTEX ARGUS 123	>480m	0	0	n.a.	72
g NITRILE	MARIGOLD BLUE	>480m	0	0	0.45	79
g NITRILE	MARIGOLD GREEN SUPA	>480m	0	0	0.30	79
g NITRILE	MARIGOLD NITROSOLVE	>480m	0	0	0.75	79
g UNKNOWN MATERIAL	MARIGOLD R MEDIGLOV	>480m	0	0	0.03	79

citric acid, <30%
Primary Class: 104   Acids, Carboxylic, Aliphatic and Alicyclic, Polybasic
Related Class: 313   Hydroxy Compounds, Aliphatic and Alicyclic, Tertiary

s BUTYL	MSA CHEMPRUF	>480m	0	0	n.a.	48
g NATURAL RUBBER	ANSELL FL 200 254	>360m	0	0	0.51	55
g NATURAL RUBBER	ANSELL ORANGE 208	>360m	0	0	0.76	55
g NATURAL RUBBER	ANSELLEDMONT 36-124	>360m	0	0	0.46	10%	6
g NATURAL RUBBER	ANSELLEDMONT 392	>360m	0	0	0.48	10%	6
g NATURAL+NEOPRENE	ANSELL OMNI 276	>360m	0	0	0.56	55
g NATURAL+NEOPRENE	ANSELL TECHNICIANS	>360m	0	0	0.43	55
g NEOPRENE	ANSELL NEOPRENE 530	>360m	0	0	0.46	55
g NEOPRENE	ANSELLEDMONT 29-840	>360m	0	0	0.31	10%	6
g NEOPRENE	ANSELLEDMONT NEOX	>360m	0	0	n.a.	10%	6
g NEOPRENE/NATURAL	ANSELL CHEMI-PRO	>360m	0	0	0.72	55
g NITRILE	ANSELL CHALLENGER	>360m	0	0	0.38	55
g NITRILE	ANSELLEDMONT 37-165	>360m	0	0	0.60	10%	6
g PVAL	ANSELLEDMONT PVA	50m			n.a.	10%	6
g PVAL	EDMONT 15-552	50m			n.a.	10%	6
g PVC	ANSELLEDMONT M GRIP	>360m	0	0	n.a.	10%	6
s PVC	MSA UPC	>480m	0	0	n.a.	48

coal tar & benzene, 1:1
Primary Class: 860   Coals, Charcoals, Oils
Related Class: 292   Hydrocarbons, Aromatic

g PE/EVAL/PE	SAFETY4 4H	>240m	0	0	0.07	35°C	60

G A R M E Class & Number/ N test chemical/ T MATERIAL NAME	MANUFACTURER & PRODUCT IDENTIFICATION	Break- through Time in Minutes	Perm- eation Rate in mg/sq m /min.	I N D E X	Thick- ness in mm	degrad. and comments	R e f s

coal tar extract
CAS Number: 8007-45-2
Primary Class: 860   Coals, Charcoals, Oils

u NATURAL RUBBER	Unknown	62m			0.34		17
u NEOPRENE	Unknown	>60m			0.31		17
u NITRILE	Unknown	>60m			0.43		17
u PE	Unknown	24m			0.09		17
u PVC	Unknown	47m			0.31		17

cobalt sulfate heptahydrate
CAS Number: 10026-24-1
Primary Class: 340   Inorganic Salts

g NATURAL RUBBER	ANSELL CONFORM 4107	>480m	0	0	0.15		34
g PVC	ANSELLEDMONT 34-100	>480m	0	0	0.18		34

corn oil
CAS Number: 8001-30-7
Primary Class: 860   Coals, Charcoals, Oils

g NATURAL RUBBER	MARIGOLD BLACK HEVY	>480m	0	0	0.65		79
g NATURAL RUBBER	MARIGOLD FEATHERLIT	146m	314	3	0.15	degrad.	79
g NATURAL RUBBER	MARIGOLD MED WT EXT	>480m	0	0	0.45	degrad.	79
g NATURAL RUBBER	MARIGOLD MEDICAL	285m	183	3	0.28	degrad.	79
g NATURAL RUBBER	MARIGOLD ORANGE SUP	>480m	0	0	0.73		79
g NATURAL RUBBER	MARIGOLD R SURGEONS	285m	183	3	0.28	degrad.	79
g NATURAL RUBBER	MARIGOLD RED LT WT	452m	26	2	0.43	degrad.	79
g NATURAL RUBBER	MARIGOLD SENSOTECH	285m	183	3	0.28	degrad.	79
g NATURAL RUBBER	MARIGOLD SUREGRIP	>480m	0	0	0.58		79
g NATURAL RUBBER	MARIGOLD YELLOW	396m	78	2	0.38	degrad.	79
g NATURAL+NEOPRENE	MARIGOLD FEATHERWT	369m	105	3	0.35	degrad.	79
g NITRILE	MARIGOLD BLUE	>480m	0	0	0.45		79
g NITRILE	MARIGOLD GREEN SUPA	>480m	0	0	0.30		79
g NITRILE	MARIGOLD NITROSOLVE	>480m	0	0	0.75		79

corrosive fluid Dyrup 49685
Primary Class: 900   Miscellaneous Unclassified Chemicals

g PE/EVAL/PE	SAFETY4 4H	102m			0.07		60
g PE/EVAL/PE	SAFETY4 4H	20m			0.07	35°C	60

creosote
CAS Number: 8001-58-9
Primary Class: 316   Hydroxy Compounds, Aromatic (Phenols)

s BUTYL	MSA CHEMPRUF	<120m			n.a.		48
u BUTYL	Unknown	>5400m	0	0	0.81		3

G A R M E Class & Number/ N test chemical/ T MATERIAL NAME	MANUFACTURER & PRODUCT IDENTIFICATION	Break- through Time in Minutes	Perm- eation Rate in mg/sq m /min.	I N D E X	Thick- ness in mm	degrad. and comments	R e f s
u NEOPRENE	Unknown	270m	0	0	0.76		3
s PVC	MSA UPC	<120m			0.20		48
u VITON	Unknown	>5760m	0	0	0.43		3

cresols (isomeric mixture) (synonym: methylphenols (isomeric mixture))
CAS Number: 1319-77-3
Primary Class: 316   Hydroxy Compounds, Aromatic (Phenols)

g BUTYL	NORTH B-174	>480m	0	0	0.43		34
g NEOPRENE	ANSELLEDMONT 29-870	245m	96	2	0.46		34
g NITRILE	ANSELLEDMONT 37-155	75m	726	3	0.38		34
g NITRILE	BEST 727				0.38	degrad.	53
c PE	DU PONT TYVEK QC	37m	4	2	n.a.		8
s SARANEX-23	DU PONT TYVEK SARAN	>480m	0	0	n.a.		8
g VITON	NORTH F-091	>480m	0	0	0.23		34

crotonaldehyde (synonym: 2-butenal)
CAS Number: 4170-30-3
Primary Class: 121   Aldehydes, Aliphatic and Alicyclic

g BUTYL	NORTH B-174	>480m	0	0	0.66		34
u CPE	ILC DOVER	38m			0.05		29
g NATURAL RUBBER	ACKWELL 5-109	1m	2900	5	0.15		34
g NEOPRENE	ANSELLEDMONT 29-870	21m	209	3	0.50	degrad.	34
g NITRILE	ANSELLEDMONT 37-155	7m	>10000	5	0.36	degrad.	34
g PVAL	EDMONT 25-950	<1m	576	4	0.30		34
g PVC	ANSELLEDMONT 34-100				0.20	degrad.	34
s TEFLON	CHEMFAB CHALL. 5100	>190m	0	1	n.a.		65
g UNKNOWN MATERIAL	NORTH SILVERSHIELD	440m	12	2	0.10	distoration	34
g VITON	NORTH F-091	7m	3132	5	0.25	degrad.	34
u VITON/CHLOROBUTYL	Unknown	105m			0.36		29

crotonaldehyde, <30%
Primary Class: 121   Aldehydes, Aliphatic and Alicyclic

g NATURAL RUBBER	ACKWELL 5-109	1m	1620	5	0.14	15% in H_2O	34
g NATURAL RUBBER	ACKWELL 5-109	1m	480	4	0.13	1% in H_2O	34
g NATURAL RUBBER	ACKWELL 5-109	2m	6	3	0.13	.1% in H_2O	34
g NITRILE	ANSELLEDMONT 37-155	23m	600	3	0.37	15% in H_2O	34
g NITRILE	ANSELLEDMONT 37-155	187m	30	2	0.38	1% in H_2O	34
g NITRILE	ANSELLEDMONT 37-155	>480m	0	0	0.37	.1% in H_2O	34
g UNKNOWN MATERIAL	NORTH SILVERSHIELD	265m	24	2	0.10	15% in H_2O	34
g UNKNOWN MATERIAL	NORTH SILVERSHIELD	>480m	0	0	0.10	1% in H_2O	34

cumene (synonym: isopropylbenzene)
CAS Number: 98-82-8
Primary Class: 292   Hydrocarbons, Aromatic

s BUTYL	TRELLEBORG TRELLCHE				n.a.	degrad.	71

G A R M E Class & Number/ N test chemical/ T MATERIAL NAME	MANUFACTURER & PRODUCT IDENTIFICATION	Break- through Time in Minutes	Perm- eation Rate in mg/sq m /min.	I N D E X	Thick- ness in mm	degrad. and comments	R e f s
u CPE	ILC DOVER	78m			0.05		29
g NEOPRENE	MAPA-PIONEER N-44	41m	2160	4	0.56	degrad.	36
g NITRILE	MAPA-PIONEER A-14	271m	480	3	0.56		36
u VITON/CHLOROBUTYL Unknown		>180m			0.36		29

cumene hydroperoxide
CAS Number: 80-15-9
Primary Class: 300   Peroxides

g NATURAL RUBBER	ACKWELL 5-109				0.15	degrad.	34
g NEOPRENE	ANSELLEDMONT 29-870				0.48	degrad.	34
g NITRILE	ANSELLEDMONT 37-155				0.40	degrad.	34
g PVC	ANSELLEDMONT 34-100				0.20	degrad.	34
s TEFLON	CHEMFAB CHALL. 5100	>210m	0	1	n.a.		65

cyanide salt, 45% solution in water (synonym: sodium or potassium cyanide)
Primary Class: 345   Inorganic Cyano Compounds

s UNKNOWN MATERIAL	LIFE-GUARD RESPONDE	>480m		0	n.a.		77

cyanogen bromide
CAS Number: 506-68-3
Primary Class: 345   Inorganic Cyano Compounds

s BUTYL	TRELLEBORG TRELLCHE				n.a.	degrad.	71

cyclohexane (synonym: hexamethylene)
CAS Number: 110-82-7
Primary Class: 291   Hydrocarbons, Aliphatic and Alicyclic, Saturated

g BUTYL	BEST 878	44m	1420	4	0.75	degrad.	53
g BUTYL	BRUNSWICK BUTYL STD	4m	>230	4	0.63		89
g BUTYL	BRUNSWICK BUTYL-XTR	4m	>1000	5	0.63		89
g BUTYL	COMASEC BUTYL	35m	3780	4	0.55		40
g BUTYL	NORTH B-174	69m	1218	4	0.66	degrad.	34
g BUTYL/ECO	BRUNSWICK BUTYL-POL	>480m	0	0	0.63		89
g BUTYL/NEOPRENE	COMASEC BUTYL PLUS	75m	1200	4	0.50		40
s CPE	ILC DOVER CLOROPEL	>180m			n.a.		46
g HYPALON	COMASEC DIPCO	330m	306	3	0.60		40
g NATURAL RUBBER	BEST 65NFW				n.a.	degrad.	53
g NATURAL RUBBER	COMASEC FLEXIGUM	41m	13200	5	0.95		40
g NATURAL RUBBER	PIONEER N-190	2m	15000	5	0.31		5
g NATURAL+NEOPRENE	ANSELL OMNI 276	5m	14000	5	0.45		5
g NATURAL+NEOPRENE	PLAYTEX ARGUS 123	9m	16920	5	n.a.		72
g NEOPRENE	ANSELLEDMONT 29-870	57m	2	2	0.38		34
g NEOPRENE	BEST 32	36m	640	3	n.a.		53
g NEOPRENE	BEST 6780	220m	100	3	n.a.		53
g NEOPRENE	COMASEC COMAPRENE	6m	10800	5	n.a.		40

G A R M E Class & Number/ N test chemical/ T MATERIAL NAME	MANUFACTURER & PRODUCT IDENTIFICATION	Break- through Time in Minutes	Perm- eation Rate in mg/sq m /min.	I N D E X	Thick- ness in mm	degrad. and comments	R e f s
g NEOPRENE	MAPA-PIONEER N-44	>156m	420	3	0.56		36
g NEOPRENE	MAPA-PIONEER N-73	29m	1000	4	0.46		5
g NITRILE	ANSELLEDMONT 37-155	>360m	0	0	0.38		34
g NITRILE	ANSELLEDMONT 37-165	>60m			0.55		5
g NITRILE	ANSELLEDMONT 37-175	>60m			0.37		5
g NITRILE	BEST 22R	286m	10	2	n.a.		53
g NITRILE	BEST 727	>480m	0	0	0.38		53
g NITRILE	COMASEC COMATRIL	>480m	0	0	0.55		40
g NITRILE	COMASEC COMATRIL SU	>480m	0	0	0.60		40
g NITRILE	COMASEC FLEXITRIL	6m	840	4	n.a.		40
g NITRILE	MAPA-PIONEER A-14	>480m	0	0	0.54		36
g NITRILE	NORTH LA-142G	>360m	0	0	0.38		34
g NITRILE+PVC	COMASEC MULTIMAX	307m	0.6	0	n.a.		40
g NITRILE+PVC	COMASEC MULTIPLUS	222m	1	1	n.a.		40
g PE	ANSELLEDMONT 35-125	1m	1000	5	0.03		5
g PE/EVAL/PE	SAFETY4 4H	>240m	0	0	0.07	35°C	60
g PVAL	COMASEC SOLVATRIL	>480m	0	0	n.a.		40
g PVAL	EDMONT 25-950	47m	<1	2	0.25		34
g PVC	BEST 725R	88m	210	3	n.a.		53
g PVC	COMASEC MULTIPOST	102m	360	3	n.a.		40
g PVC	COMASEC MULTITOP	165m	60	2	n.a.		40
g PVC	COMASEC NORMAL	38m	2160	4	n.a.		40
g PVC	COMASEC OMNI	6m	4140	5	n.a.		40
g PVC	MAPA-PIONEER V-20	16m	1000	4	0.31		5
s TEFLON	CHEMFAB CHALL. 5100	>200m	0	1	n.a.		65
s UNKNOWN MATERIAL	CHEMRON CHEMREL	>1440m	0	0	n.a.		67
s UNKNOWN MATERIAL	CHEMRON CHEMREL MAX	>1440m	0	0	n.a.		67
s UNKNOWN MATERIAL	DU PONT BARRICADE	>480m	0	0	n.a.		8
g UNKNOWN MATERIAL	NORTH SILVERSHIELD	>360m	0	0	0.08		7
g VITON	NORTH F-091	>420m	0	0	0.23		34

cyclohexanol (synonym: hexahydrophenol)
CAS Number: 108-93-0
Primary Class: 312   Hydroxy Compounds, Aliphatic and Alicyclic, Secondary

g BUTYL	BRUNSWICK BUTYL STD	>480m	0	0	0.63		89
g BUTYL	BRUNSWICK BUTYL-XTR	>480m	0	0	0.63		89
g BUTYL	NORTH B-174	>660m	0	0	0.68		34
g BUTYL/NEOPRENE	COMASEC BUTYL PLUS	>480m	0	0	0.50		40
g NATURAL RUBBER	ACKWELL 5-109				0.15	degrad.	34
g NATURAL RUBBER	ANSELLEDMONT 36-124	15m	<900	3	0.46		6
g NATURAL RUBBER	ANSELLEDMONT 392	10m	<900	4	0.48		6
g NATURAL RUBBER	COMASEC FLEXIGUM	>480m	0	0	0.95		40
g NATURAL RUBBER	MARIGOLD BLACK HEVY	163m	30	2	0.65		79
g NATURAL RUBBER	MARIGOLD FEATHERLIT	1m	278	4	0.15		79
g NATURAL RUBBER	MARIGOLD MED WT EXT	170m	69	2	0.45		79
g NATURAL RUBBER	MARIGOLD MEDICAL	20m	191	3	0.28		79

G A R M E T	Class & Number/ test chemical/ MATERIAL NAME	MANUFACTURER & PRODUCT IDENTIFICATION	Break- through Time in Minutes	Perm- eation Rate in mg/sq m /min.	I N D E X	Thick- ness in mm	degrad. and comments	R e f s
g	NATURAL RUBBER	MARIGOLD ORANGE SUP	160m	26	2	0.73		79
g	NATURAL RUBBER	MARIGOLD R SURGEONS	20m	191	3	0.28		79
g	NATURAL RUBBER	MARIGOLD RED LT WT	149m	86	2	0.43		79
g	NATURAL RUBBER	MARIGOLD SENSOTECH	20m	191	3	0.28		79
g	NATURAL RUBBER	MARIGOLD SUREGRIP	165m	42	2	0.58		79
g	NATURAL RUBBER	MARIGOLD YELLOW	106m	121	3	0.38		79
g	NATURAL+NEOPRENE	MARIGOLD FEATHERWT	84m	139	3	0.35		79
g	NEOPRENE	ANSELLEDMONT 29-840	150m	<90	2	0.38		6
g	NEOPRENE	ANSELLEDMONT 29-870	>480m	0	0	0.48		34
g	NEOPRENE	ANSELLEDMONT NEOX	180m	<9	1	n.a.		6
g	NEOPRENE	BRUNSWICK NEOPRENE	97m	130	3	0.88		89
g	NEOPRENE	COMASEC COMAPRENE	180m	600	3	n.a.		40
g	NITRILE	ANSELLEDMONT 37-165	>360m	0	0	0.60		6
g	NITRILE	COMASEC COMATRIL	330m	90	2	0.55		40
g	NITRILE	COMASEC COMATRIL SU	>480m	0	0	0.60		40
g	NITRILE	COMASEC FLEXITRIL	75m	780	3	n.a.		40
g	NITRILE	MARIGOLD BLUE	>480m	0	0	0.45		79
g	NITRILE	MARIGOLD GREEN SUPA	>480m	0	0	0.30		79
g	NITRILE	MARIGOLD NITROSOLVE	>480m	0	0	0.75		79
g	NITRILE	NORTH LA-142G	>960m	0	0	0.33		34
g	NITRILE	NORTH LA-142G	5688m	0.4	1	0.38		63
g	NITRILE+PVC	COMASEC MULTIPLUS	>480m	0	0	0.60		40
g	PE/EVAL/PE	SAFETY4 4H	>240m	0	0	0.07	35°C	60
g	PVAL	ANSELLEDMONT PVA	>360m		0	n.a.		6
g	PVAL	EDMONT 15-552	360m	<9	1	n.a.		6
g	PVAL	EDMONT 25-545	>360m	0	0	n.a.		6
g	PVAL	EDMONT 25-950	>960m	0	0	0.27		34
g	PVC	ANSELLEDMONT M GRIP	360m	<9	1	n.a.		6
g	PVC	COMASEC MULTIPOST	>480m	0	0	n.a.		40
g	PVC	COMASEC MULTITOP	>480m	0	0	n.a.		40
g	PVC	COMASEC NORMAL	>480m	0	0	n.a.		40
g	PVC	COMASEC OMNI	294m	66	2	n.a.		40
g	UNKNOWN MATERIAL	NORTH SILVERSHIELD	>360m	0	0	0.08		7
g	VITON	NORTH F-091	>480m	0	0	0.27		34

cyclohexanone
CAS Number: 108-94-1
Primary Class: 391   Ketones, Aliphatic and Alicyclic

g	BUTYL	BEST 878	>480m	0	0	0.75		53
g	BUTYL	BRUNSWICK BUTYL STD	>480m	0	0	0.63		89
g	BUTYL	BRUNSWICK BUTYL-XTR	>480m	0	0	0.63		89
g	BUTYL	COMASEC BUTYL	>480m	0	0	0.55		40
g	BUTYL	NORTH B-161	842m	33	2	0.45		63
g	BUTYL	NORTH B-174	>960m	0	0	0.51		34
g	BUTYL/ECO	BRUNSWICK BUTYL-POL	408m	0.05	0	0.63		89
g	NATURAL RUBBER	BEST 65NFW				n.a.	degrad.	53

G A R M E Class & Number/ N test chemical/ T MATERIAL NAME	MANUFACTURER & PRODUCT IDENTIFICATION	Break- through Time in Minutes	Perm- eation Rate in mg/sq m /min.	I N D E X	Thick- ness in mm	R degrad. e and f comments s
g NATURAL RUBBER	MARIGOLD BLACK HEVY	15m	3837	4	0.65	degrad. 79
g NATURAL RUBBER	MARIGOLD FEATHERLIT	15m	14334	5	0.15	degrad. 79
g NATURAL RUBBER	MARIGOLD MED WT EXT	15m	5533	4	0.45	degrad. 79
g NATURAL RUBBER	MARIGOLD MEDICAL	15m	10667	5	0.28	degrad. 79
g NATURAL RUBBER	MARIGOLD ORANGE SUP	15m	3667	4	0.73	degrad. 79
g NATURAL RUBBER	MARIGOLD R SURGEONS	15m	10667	5	0.28	degrad. 79
g NATURAL RUBBER	MARIGOLD RED LT WT	15m	6266	4	0.43	degrad. 79
g NATURAL RUBBER	MARIGOLD SENSOTECH	15m	10667	5	0.28	degrad. 79
g NATURAL RUBBER	MARIGOLD SUREGRIP	15m	4346	4	0.58	degrad. 79
g NATURAL RUBBER	MARIGOLD YELLOW	15m	7733	4	0.38	degrad. 79
g NATURAL+NEOPRENE	MARIGOLD FEATHERWT	15m	8467	4	0.35	degrad. 79
g NATURAL+NEOPRENE	PLAYTEX ARGUS 123	17m	1320	4	n.a.	72
g NEOPRENE	ANSELLEDMONT 29-870	39m	6126	4	0.51	degrad. 34
g NEOPRENE	BEST 32	61m	760	3	n.a.	degrad. 53
g NEOPRENE	BEST 6780	151m	810	3	n.a.	degrad. 53
g NITRILE	BEST 22R				n.a.	degrad. 53
g NITRILE	BEST 727	60m	1480	4	0.38	degrad. 53
g NITRILE	MARIGOLD BLUE	71m	11664	5	0.45	degrad. 79
g NITRILE	MARIGOLD GREEN SUPA	30m	15333	5	0.30	degrad. 79
g NITRILE	MARIGOLD NITROSOLVE	79m	10930	5	0.75	degrad. 79
g NITRILE	NORTH LA-142G				0.38	degrad. 7
g NITRILE	NORTH LA-142G	30m	5760	4	0.34	63
g NITRILE+PVC	COMASEC MULTIMAX	86m	354	3	n.a.	40
g PE/EVAL/PE	SAFETY4 4H	>240m	0	0	0.07	35°C 60
g PE/EVAL/PE	SAFETY4 4H	>480m	0	0	0.07	60
g PVAL	COMASEC SOLVATRIL	>480m	0	0	n.a.	40
g PVAL	EDMONT 25-950	>420m	0	0	0.25	degrad. 34
g PVC	BEST 725R				n.a.	degrad. 53
g UNKNOWN MATERIAL	NORTH SILVERSHIELD	>360m	0	0	0.08	7
g VITON	NORTH F-091	29m	5178	4	0.25	degrad. 34

cyclohexylamine
CAS Number: 108-91-8
Primary Class: 141   Amines, Aliphatic and Alicyclic, Primary

g BUTYL	NORTH B-174	174m	1710	4	0.61	degrad. 34
g NATURAL RUBBER	ACKWELL 5-109	1m	53760	5	0.15	degrad. 34
g NEOPRENE	ANSELLEDMONT 29-870	24m	10920	5	0.50	degrad. 34
g NITRILE	ANSELLEDMONT 37-155	61m	11040	5	0.43	degrad. 34
g PVAL	EDMONT 25-950				0.40	degrad. 34
g PVC	ANSELLEDMONT 34-100				0.20	degrad. 34
g VITON	NORTH F-091				0.26	degrad. 34

cyclohexylamine, 32% & morpholine, 8% & water, 60%
Primary Class: 800   Multicomponent Mixtures With >2 Components

g PE/EVAL/PE	SAFETY4 4H	>240m		0	0.07	60

G A R M E Class & Number/ N test chemical/ T MATERIAL NAME	MANUFACTURER & PRODUCT IDENTIFICATION	Break-through Time in Minutes	Perm-eation Rate in mg/sq m /min.	I N D E X	Thick-ness in mm	degrad. and comments	R e f s

cyclopentane (synonym: pentamethylene)
CAS Number: 287-92-3
Primary Class: 291   Hydrocarbons, Aliphatic and Alicyclic, Saturated

| s BUTYL | TRELLEBORG TRELLCHE | | | | n.a. | degrad. | 71 |

cyclopentanone
CAS Number: 120-92-3
Primary Class: 391   Ketones, Aliphatic and Alicyclic

| g PE/EVAL/PE | SAFETY4 4H | >240m | 0 | 0 | 0.07 | 35°C | 60 |

Cymbush
Primary Class: 840   Pesticides, Mixtures and Formulations

| g PE/EVAL/PE | SAFETY4 4H | >240m | 0 | 0 | 0.07 | 35°C | 60 |

cypermethrin
CAS Number: 52315-07-8
Primary Class: 840   Pesticides, Mixtures and Formulations
Related Class: 224   Esters, Carboxylic, Aliphatic, Others
              261   Halogen Compounds, Aliphatic and Alicyclic

| g PE/EVAL/PE | SAFETY4 4H | >240m | 0 | 0 | 0.07 | 35°C | 60 |

d,l-limonene (synonyms: d,l-p-mentha-1,8-diene; dipentene; limonene)
CAS Number: 138-86-3
Primary Class: 294   Hydrocarbons, Aliphatic and Alicyclic, Unsaturated

g BUTYL	NORTH B-161				0.65	degrad.	34
g NATURAL RUBBER	ACKWELL 5-109				0.15	degrad.	34
g NEOPRENE	ANSELLEDMONT 29-870	65m	102	3	0.43	degrad.	34
g NITRILE	ANSELLEDMONT 37-155	>1200m	0	0	0.38		34
g PVAL	EDMONT 25-950	>480m	0	0	0.30		34
g VITON	NORTH F-091	>480m	0	0	0.25		34

decanal (synonym: caperaldehyde)
CAS Number: 1321-89-7
Primary Class: 121   Aldehydes, Aliphatic and Alicyclic

s BUTYL	MSA CHEMPRUF	>480m	0	0	n.a.		48
s BUTYL/NEOPRENE	MSA BETEX	>480m	0	0	n.a.		48
s PVC	MSA UPC	<120m			0.20		48
s VITON/NEOPRENE	MSA VAUTEX	>480m	0	0	n.a.		48

G A R M E Class & Number/ N test chemical/ T MATERIAL NAME	MANUFACTURER & PRODUCT IDENTIFICATION	Break- through Time in Minutes	Perm- eation Rate in mg/sq m /min.	I N D E X	Thick- ness in mm	degrad. and comments	R e f s
Deep Woods Off Primary Class: 900 Miscellaneous Unclassified Chemicals							
g PE/EVAL/PE	SAFETY4 4H	>240m		0	0.07		60
Degalan S309 (mixture) Primary Class: 900 Miscellaneous Unclassified Chemicals							
g PE/EVAL/PE	SAFETY4 4H	>240m		0	0.07		60
Degalan S696 Primary Class: 900 Miscellaneous Unclassified Chemicals							
g PE/EVAL/PE	SAFETY4 4H	>240m		0	0.07		60
DGEBA, 50% & methyl ethyl ketone, 50% Primary Class: 810 Epoxy Products							
di(2-ethylhexyl) phthalate (synonym: bis(2-ethylhexyl) phthalate) CAS Number: 117-81-7 Primary Class: 226 Esters, Carboxylic, Benzoates and Phthalates							
g BUTYL	NORTH B-174	>480m 0		0	0.66		34
g NITRILE	ANSELLEDMONT 37-155	260m	100	3	0.38		34
g PE/EVAL/PE	SAFETY4 4H	>240m 0		0	0.07	35°C	60
g PVC	ANSELLEDMONT 34-100	1m	94	4	0.21		34
g VITON	NORTH F-091	>480m 0		0	0.30		34
di-n-amylamine (synonym: di-n-pentylamine) CAS Number: 2050-92-2 Primary Class: 142 Amines, Aliphatic and Alicyclic, Secondary							
g BUTYL	NORTH B-174				0.66	degrad.	34
g NATURAL RUBBER	ACKWELL 5-109				0.15	degrad.	34
g NEOPRENE	ANSELLEDMONT 29-870	132m	643	3	0.33	degrad.	34
g NITRILE	ANSELLEDMONT 37-155	>480m 0		0	0.34		34
g PVC	ANSELLEDMONT 34-100	7m	1700	5	0.20		34
g VITON	NORTH F-091	>480m 0		0	0.24		34
di-n-butyl phthalate (synonym: DBP) CAS Number: 84-74-2 Primary Class: 226 Esters, Carboxylic, Benzoates and Phthalates							
g BUTYL	BEST 878	>480m 0		0	0.75		53

G A R M E N T	Class & Number/ test chemical/ MATERIAL NAME	MANUFACTURER & PRODUCT IDENTIFICATION	Break-through Time in Minutes	Perm-eation Rate in mg/sq m /min.	I N D E X	Thick-ness in mm	degrad. and comments	R e f s
g	BUTYL	BRUNSWICK BUTYL STD	>480m	0	0	0.63		89
g	BUTYL	BRUNSWICK BUTYL-XTR	>480m	0	0	0.63		89
g	BUTYL	NORTH B-161	>960m	0	0	0.42		34
g	BUTYL/ECO	BRUNSWICK BUTYL-POL	>480m	0	0	0.63		89
g	NATURAL RUBBER	ACKWELL 5-109				0.15	degrad.	34
g	NATURAL RUBBER	ANSELL CONFORM 4205	>360m	0	0	0.13		55
g	NATURAL RUBBER	ANSELL FL 200 254	>360m	0	0	0.51		55
g	NATURAL RUBBER	ANSELL ORANGE 208	>360m	0	0	0.76		55
g	NATURAL RUBBER	ANSELL PVL 040	>360m	0	0	0.46		55
g	NATURAL RUBBER	ANSELL STERILE 832	>360m	0	0	0.23		55
g	NATURAL RUBBER	ANSELLEDMONT 36-124	17m			0.46		6
g	NATURAL RUBBER	ANSELLEDMONT 392	20m	0	2	0.48		6
g	NATURAL RUBBER	MARIGOLD BLACK HEVY	96m	69	2	0.65		79
g	NATURAL RUBBER	MARIGOLD FEATHERLIT	1m	385	4	0.15	degrad.	79
g	NATURAL RUBBER	MARIGOLD MED WT EXT	55m	157	3	0.45	degrad.	79
g	NATURAL RUBBER	MARIGOLD MEDICAL	15m	290	3	0.28	degrad.	79
g	NATURAL RUBBER	MARIGOLD ORANGE SUP	100m	60	2	0.73		79
g	NATURAL RUBBER	MARIGOLD R SURGEONS	15m	290	3	0.28	degrad.	79
g	NATURAL RUBBER	MARIGOLD RED LT WT	49m	176	3	0.43	degrad.	79
g	NATURAL RUBBER	MARIGOLD SENSOTECH	15m	290	3	0.28	degrad.	79
g	NATURAL RUBBErR	MARIGOLD SUREGRIP	84m	95	2	0.58	degrad.	79
g	NATURAL RUBBER	MARIGOLD YELLOW	38m	214	3	0.38	degrad.	79
g	NATURAL+NEOPRENE	ANSELL OMNI 276	>360m	0	0	0.56		55
g	NATURAL+NEOPRENE	ANSELL TECHNICIANS	>360m	0	0	0.43		55
g	NATURAL+NEOPRENE	MARIGOLD FEATHERWT	32m	233	3	0.35	degrad.	79
g	NEOPRENE	ANSELL NEOPRENE 530	>360m	0	0	0.46		55
g	NEOPRENE	ANSELLEDMONT 29-840	120m	<9	1	0.38		6
g	NEOPRENE	ANSELLEDMONT NEOX	300m	<90	2	n.a.		6
g	NEOPRENE	BEST 32	>480m	0	0	n.a.		53
g	NEOPRENE	BEST 6780	>480m	0	0	n.a.		53
g	NEOPRENE/NATURAL	ANSELL CHEMI-PRO	>360m	0	0	0.72		55
g	NITRILE	ANSELL CHALLENGER	>360m	0	0	0.38		55
g	NITRILE	ANSELLEDMONT 37-165	>360m	0	0	0.60		6
g	NITRILE	BEST 22R	>480m	0	0	n.a.		53
g	NITRILE	MARIGOLD BLUE	>480m	0	0	0.45	degrad.	79
g	NITRILE	MARIGOLD GREEN SUPA	>480m	0	0	0.30	degrad.	79
g	NITRILE	MARIGOLD NITROSOLVE	>480m	0	0	0.75	degrad.	79
g	NITRILE	NORTH LA-142G	>960m	0	0	0.33		34
g	PE/EVAL/PE	SAFETY4 4H	>240m	0	0	0.07	35°C	60
g	PVAL	ANSELLEDMONT PVA	>360m		0	n.a.		6
g	PVAL	EDMONT 15-552	>360m	0	0	n.a.		6
g	PVAL	EDMONT 25-545	>360m	0	0	n.a.		6
g	PVAL	EDMONT 25-950	>960m	0	0	0.33		34
g	PVC	ANSELLEDMONT 34-100				0.20	degrad.	34
g	PVC	ANSELLEDMONT SNORKE				n.a.	degrad.	6
g	UNKNOWN MATERIAL	NORTH SILVERSHIELD	>360m	0	0	0.08		7
g	VITON	NORTH F-091	>480m	0	0	0.25		34

G A R M E Class & Number/ N test chemical/ T MATERIAL NAME	MANUFACTURER & PRODUCT IDENTIFICATION	Break- through Time in Minutes	Perm- eation Rate in mg/sq m /min.	I N D E X	Thick- ness in mm	degrad. and comments	R e f s

di-n-butylamine (synonym: dibutylamine)
CAS Number: 111-92-2
Primary Class: 142   Amines, Aliphatic and Alicyclic, Secondary

g BUTYL	NORTH B-174				0.63	degrad.	34
g NATURAL RUBBER	ACKWELL 5-109				0.15	degrad.	34
g NEOPRENE	ANSELLEDMONT 29-870				0.48	degrad.	34
g NITRILE	ANSELLEDMONT 37-155	>480m	0	0	0.40		34
g PVAL	EDMONT 25-950	>480m	0	0	0.78		34
g PVC	ANSELLEDMONT 34-100	3m	4440	5	0.20	degrad.	34
g VITON	NORTH F-091	>480m	0	0	0.27		34

di-n-octyl phthalate (synonym: DOP)
CAS Number: 117-84-0
Primary Class: 226   Esters, Carboxylic, Benzoates and Phthalates

g NATURAL RUBBER	ACKWELL 5-109				0.15	degrad.	34
g NATURAL RUBBER	ANSELL CONFORM 4205	>360m	0	0	0.13		55
g NATURAL RUBBER	ANSELL FL 200 254	>360m	0	0	0.51		55
g NATURAL RUBBER	ANSELL ORANGE 208	>360m	0	0	0.76		55
g NATURAL RUBBER	ANSELL PVL 040	>360m	0	0	0.46		55
g NATURAL RUBBER	ANSELL STERILE 832	>360m	0	0	0.23		55
g NATURAL RUBBER	ANSELLEDMONT 36-124	>360m	0	0	0.46		6
g NATURAL RUBBER	ANSELLEDMONT 392				0.48	degrad.	6
g NATURAL+NEOPRENE	ANSELL OMNI 276	>360m	0	0	0.56		55
g NATURAL+NEOPRENE	ANSELL TECHNICIANS	>360m	0	0	0.43		55
g NEOPRENE	ANSELL NEOPRENE 530	>360m	0	0	0.46		55
g NEOPRENE	ANSELLEDMONT 29-840	>360m	0	0	0.38		6
g NEOPRENE	ANSELLEDMONT NEOX	120m	<9	1	n.a.		6
g NEOPRENE/NATURAL	ANSELL CHEMI-PRO	>360m	0	0	0.72		55
g NITRILE	ANSELL CHALLENGER	>360m	0	0	0.38		55
g NITRILE	ANSELLEDMONT 37-165	>360m	0	0	0.60		6
g NITRILE	MARIGOLD NITROSOLVE	>480m	0	0	0.75		79
g PVAL	ANSELLEDMONT PVA	30m	<9000	4	n.a.		6
g PVAL	EDMONT 15-552	30m	<9000	4	n.a.		6
g PVAL	EDMONT 25-545	240m	<90	2	n.a.		6
g PVC	ANSELLEDMONT SNORKE				n.a.	degrad.	6

di-n-propylamine (synonym: dipropylamine)
CAS Number: 142-84-7
Primary Class: 142   Amines, Aliphatic and Alicyclic, Secondary

g BUTYL	NORTH B-174				0.65	degrad.	34
g NATURAL RUBBER	ACKWELL 5-109				0.15	degrad.	34
g NEOPRENE	ANSELLEDMONT 29-840	21m	8580	4	0.45	degrad.	34
g NITRILE	ANSELLEDMONT 37-155	81m	1380	4	0.35	degrad.	34
g PVAL	EDMONT 25-950				0.40	degrad.	34

G A R M E Class & Number/ N test chemical/ T MATERIAL NAME	MANUFACTURER & PRODUCT IDENTIFICATION	Break- through Time in Minutes	Perm- eation Rate in mg/sq m /min.	I N D E X	Thick- ness in mm	degrad. and comments	R e f s
g PVC	ANSELLEDMONT 34-100	2m	4020	5	0.18	degrad.	34
s TEFLON	CHEMFAB CHALL. 5100	>200m	0	1	n.a.		65
g VITON	NORTH F-091	257m	1320	4	0.30		34

diallylamine
CAS Number: 124-02-7
Primary Class: 142   Amines, Aliphatic and Alicyclic, Secondary

g BUTYL	NORTH B-174	198m	526	3	0.63	degrad.	34
g NATURAL RUBBER	ACKWELL 5-109				0.15	degrad.	34
g NEOPRENE	ANSELLEDMONT 29-870				0.48	degrad.	34
g NITRILE	ANSELLEDMONT 37-155				0.40	degrad.	34
g NITRILE+PVC	COMASEC MULTIMAX	3m			n.a.		40
g PVAL	EDMONT 25-950	426m	98	2	0.50		34
g PVC	ANSELLEDMONT 34-100	1m	14160	5	0.20	degrad.	34
g VITON	NORTH F-091	276m	0	0	0.23		34

diborane
CAS Number: 19287-45-7
Primary Class: 350   Inorganic Gases and Vapors

s BUTYL	TRELLEBORG TRELLCHE				n.a.	degrad.	71

dichloroacetyl chloride
CAS Number: 79-36-7
Primary Class: 111   Acid Halides, Carboxylic, Aliphatic and Alicyclic
Related Class: 261   Halogen Compounds, Aliphatic and Alicyclic

g BUTYL	NORTH B-174	245m	728	3	0.69	degrad.	34
g NATURAL RUBBER	ACKWELL 5-109				0.15	degrad.	34
g NEOPRENE	ANSELLEDMONT 29-870				0.48	degrad.	34
g NITRILE	ANSELLEDMONT 37-155				0.40	degrad.	34
g PVAL	EDMONT 25-950	210m			0.63		34
g PVC	ANSELLEDMONT 34-100	2m	4380	5	0.20	degrad.	34
g VITON	NORTH F-091	>480m	0	0	0.27		34

dichloroethane
CAS Number: 1300-21-6
Primary Class: 261   Halogen Compounds, Aliphatic and Alicyclic

u NEOPRENE	Unknown	10m	1200	5	0.54		30

dichlorotriazine, 80% & toluene, 20%
Primary Class: 274   Heterocyclic Compounds, Nitrogen, Others

s UNKNOWN MATERIAL	KAPPLER CPF III	>480m	0	0	n.a.		77

G A R M E Class & Number/ N test chemical/ T MATERIAL NAME	MANUFACTURER & PRODUCT IDENTIFICATION	Break- through Time in Minutes	Perm- eation Rate in mg/sq m /min.	I N D E X	Thick- ness in mm	degrad. and comments	R e f s

diesel fuel
CAS Number: 68334-30-5
Primary Class: 291   Hydrocarbons, Aliphatic and Alicyclic, Saturated

g	NATURAL RUBBER	BEST 65NFW				n.a.	degrad.	53
g	NITRILE	MARIGOLD BLUE	>480m	0	0	0.45		79
g	NITRILE	MARIGOLD GREEN SUPA	>480m	0	0	0.30		79
g	NITRILE	MARIGOLD NITROSOLVE	>480m	0	0	0.75		79
s	UNKNOWN MATERIAL	DU PONT BARRICADE	195m	0.9	1	n.a.		8
g	UNKNOWN MATERIAL	MARIGOLD R MEDIGLOV	15m	497	3	0.03	degrad.	79

diethanolamine
CAS Number: 111-42-2
Primary Class: 142   Amines, Aliphatic and Alicyclic, Secondary
Related Class: 311   Hydroxy Compounds, Aliphatic and Alicyclic, Primary

g	BUTYL	NORTH B-174	>480m	0	0	0.65		34
g	NATURAL RUBBER	ACKWELL 5-109				0.15	degrad.	34
g	NATURAL RUBBER	MAPA-PIONEER L-118	>480m	0	0	0.46		36
g	NATURAL+NEOPRENE	PLAYTEX ARGUS 123	>480m	0	0	n.a.		72
g	NEOPRENE	ANSELLEDMONT 29-870	>480m	0	0	0.46		34
g	NEOPRENE	MAPA-PIONEER N-44	>480m	0	0	0.74		36
g	NITRILE	ANSELLEDMONT 37-155	>480m	0	0	0.46		34
g	NITRILE	MAPA-PIONEER A-14	>480m	0	0	0.54		36
g	PE/EVAL/PE	SAFETY4 4H	>240m	0	0	0.07	50% in 2O	60
g	PVC	MAPA-PIONEER V-20	>480m	0	0	0.45		36
s	TEFLON	CHEMFAB CHALL. 5100	>180m	0	1	n.a.		65
g	VITON	NORTH F-091	>480m	0	0	0.26		34

diethyl carbonate
CAS Number: 105-58-8
Primary Class: 232   Esters, Non-Carboxylic, Carbonates

g	NATURAL RUBBER	ACKWELL 5-109				0.15	degrad.	34
g	NEOPRENE	ANSELLEDMONT 29-870				0.48	degrad.	34
g	NITRILE	ANSELLEDMONT 37-155				0.40	degrad.	34
g	VITON	NORTH F-091				0.25	degrad.	34

diethyl phthalate (synonyms: bis(ethyl) phthalate; ethyl phthalate)
CAS Number: 84-66-2
Primary Class: 226   Esters, Carboxylic, Benzoates and Phthalates

g	PE/EVAL/PE	SAFETY4 4H	>240m	0	0	0.07	35°C	60

diethylacetamide (synonym: N,N-diethylacetamide)
CAS Number: 685-91-6
Primary Class: 132   Amides, Aliphatic and Alicyclic

g	PE/EVAL/PE	SAFETY4 4H	>240m	0	0	0.07	35°C	60

	Class & Number/ test chemical/ MATERIAL NAME	MANUFACTURER & PRODUCT IDENTIFICATION	Break- through Time in Minutes	Perm- eation Rate in mg/sq m /min.	I N D E X	Thick- ness in mm	degrad. and comments	R e f s

diethylamine
CAS Number: 109-89-7
Primary Class: 142   Amines, Aliphatic and Alicyclic, Secondary

g	BUTYL	ANSELL LAMPRECHT	16m	9476	4	0.70		38
g	BUTYL	BEST 878	30m	600	3	0.75		53
g	BUTYL	BRUNSWICK BUTYL STD	27m	>5000	4	0.63		89
g	BUTYL	BRUNSWICK BUTYL-XTR	19m	>5000	4	0.63		89
s	BUTYL	DRAGER 500 OR 710	>30m			n.a.		91
s	BUTYL	LIFE-GUARD BUTYL	3m	53	4	n.a.		77
g	BUTYL	NORTH B-174	46m	2760	4	0.62	degrad.	34
s	BUTYL	WHEELER ACID KING	8m	<9000	5	n.a.		85
g	BUTYL/ECO	BRUNSWICK BUTYL-POL	28m	>5000	4	0.70		89
s	BUTYL/NEOPRENE	MSA BETEX	<5m	78	4	n.a.		48
s	CPE	STD. SAFETY	22m	679	3	n.a.		75
g	NATURAL RUBBER	ACKWELL 5-109				0.15	degrad.	34
g	NATURAL RUBBER	ANSELL ORANGE 208	4m	7	3	0.76		55
g	NATURAL RUBBER	ANSELLEDMONT 36-124				0.46	degrad.	6
g	NATURAL RUBBER	ANSELLEDMONT 392				0.48	degrad.	6
g	NATURAL RUBBER	BEST 65NFW	60m	20070	5	n.a.		53
g	NATURAL RUBBER	BEST 67NFW		5340	5	n.a.		53
u	NATURAL RUBBER	Unknown	3m	31205	5	0.64		38
g	NEOPRENE	ANSELL NEOPRENE 530	3m	5	3	0.46		55
g	NEOPRENE	ANSELLEDMONT 29-840				0.38	degrad.	6
g	NEOPRENE	BEST 32	30m	320	3	n.a.		53
g	NEOPRENE	BEST 6780	13m	1360	5	n.a.		53
g	NEOPRENE	BRUNSWICK NEOPRENE	15m	>5000	4	0.95		89
s	NEOPRENE	LIFE-GUARD NEOPRENE	16m	57	3	n.a.		77
b	NEOPRENE	RAINFAIR	36m	>75	3	n.a.		90
b	NEOPRENE	RANGER	24m	>100	3	n.a.		90
b	NEOPRENE	SERVUS NO 22204	45m	>150	3	n.a.		90
g	NEOPRENE/NATURAL	ANSELL CHEMI-PRO	2m	8	3	0.72		55
g	NITRILE	ANSELL CHALLENGER	15m	0.2		0.38		55
g	NITRILE	ANSELLEDMONT 37-155	12m	33780	5	0.43	degrad.	34
g	NITRILE	ANSELLEDMONT 37-165	45m	<9000	4	0.60		6
g	NITRILE	ANSELLEDMONT 37-175	24m	2926	4	0.42		38
g	NITRILE	BEST 727	60m	20070	5	0.38		53
b	NITRILE+PUR+PVC	BATA HAZMAX	166m	3	1	n.a.		86
b	NITRILE+PVC	SERVUS NO 73101	115m	95	2	n.a.		90
b	NITRILE+PVC	TINGLEY	96m	25	2	n.a.		90
c	PE	DU PONT TYVEK QC	1m	1410	5	n.a.		8
g	PE/EVAL/PE	SAFETY4 4H	6m	0	3	0.07	35°C	60
g	PE/EVAL/PE	SAFETY4 4H	60m			0.07		60
g	PVAL	ANSELLEDMONT PVA				n.a.	degrad.	6
g	PVAL	EDMONT 15-552				n.a.	degrad.	6
g	PVAL	EDMONT 25-950				0.40	degrad.	34
g	PVC	ANSELLEDMONT 34-100	2m	22200	5	0.20	degrad.	34
g	PVC	ANSELLEDMONT SNORKE				n.a.	degrad.	6
b	PVC	BATA STANDARD	<1m	0.7	3	n.a.		86
g	PVC	BEST 812		4140	5	n.a.		53
b	PVC	JORDAN DAVID	69m	>280	3	n.a.		90

G A R M E Class & Number/ N test chemical/ T MATERIAL NAME	MANUFACTURER & PRODUCT IDENTIFICATION	Break-through Time in Minutes	Perm-eation Rate in mg/sq m /min.	I N D E X	Thick-ness in mm	degrad. and comments	R e f s
b PVC	STD. SAFETY GL-20	4m	>4000	5	n.a.		75
s PVC	STD. SAFETY WG-20	8m	>4000	5	n.a.		75
s PVC	WHEELER ACID KING	6m	<9000	5	n.a.	degrad.	85
b PVC+POLYURETHANE	BATA POLYBLEND	4m	6	3	n.a.		86
b PVC+POLYURETHANE	BATA POLYBLEND	152m	37	2	n.a.		90
b PVC+POLYURETHANE	BATA POLYMAX	196m	1	1	n.a.		86
b PVC+POLYURETHANE	BATA SUPER POLY	155m	3	1	n.a.		86
s SARANEX-23	DU PONT TYVEK SARAN	6m	3000	5	n.a.		8
s SARANEX-23 2-PLY	DU PONT TYVEK SARAN	145m	54	2	n.a.		8
u TEFLON	CHEMFAB	2m	179	4	0.10		65
s TEFLON	CHEMFAB CHALL. 5100	>480m	0	0	n.a.		56
s TEFLON	CHEMFAB CHALL. 5100	>270m	0	0	n.a.		65
s TEFLON	CHEMFAB CHALL. 5200	>300m	0	0	n.a.		80
s TEFLON	CHEMFAB CHALL. 6000	>180m			n.a.		80
s TEFLON	LIFE-GUARD TEFGUARD	>480m	0	0	n.a.		77
s TEFLON	WHEELER ACID KING	>480m	0	0	n.a.		85
v TEFLON-FEP	CHEMFAB CHALL.	>180m			0.25		65
s UNKNOWN MATERIAL	CHEMRON CHEMREL	110m	120	3	n.a.		67
s UNKNOWN MATERIAL	DU PONT BARRICADE	>480m	0	0	n.a.		8
s UNKNOWN MATERIAL	LIFE-GUARD RESPONDE	>480m	0	0	n.a.		77
g UNKNOWN MATERIAL	NORTH SILVERSHIELD	>480m	0	0	0.08		7
g VITON	NORTH F-091	34m	21600	5	0.30	degrad.	34
g VITON	NORTH F-091	31m	2090	4	0.26		38
s VITON/BUTYL	DRAGER 500 OR 710	>30m			n.a.		91
s VITON/BUTYL/UNKN.	TRELLEBORG HPS	>180m			n.a.		71
s VITON/CHLOROBUTYL	LIFE-GUARD VC-100	32m			n.a.		77
s VITON/CHLOROBUTYL	LIFE-GUARD VNC 200	13m	14	4	n.a.		77
g VITON/NEOPRENE	ERISTA VITRIC	29m	3205	4	0.60		38
s VITON/NEOPRENE	MSA VAUTEX	35m	5200	4	n.a.		48

diethylbenzene
CAS Number: 25340-17-4
Primary Class: 292   Hydrocarbons, Aromatic

s BUTYL	TRELLEBORG TRELLCHE				n.a.	degrad.	71

diethyldichlorosilane (synonym: dichlorodiethylsilane)
CAS Number: 1719-53-5
Primary Class: 480   Organosilicon Compounds

g BUTYL	NORTH B-161				0.65	degrad.	34
g NATURAL RUBBER	ACKWELL 5-109				0.15	degrad.	34
g NEOPRENE	ANSELLEDMONT 29-870	<1m	4260	5	0.43	degrad.	34
g NITRILE	ANSELLEDMONT 37-155	>480m	0	0	0.36		34
g PVAL	EDMONT 25-950				0.35	degrad.	34
g PVC	ANSELLEDMONT 34-100	<1m	15180	5	0.18		34
g VITON	NORTH F-091	>480m	0	0	0.30		34

G A R M E Class & Number/ N test chemical/ T MATERIAL NAME	MANUFACTURER & PRODUCT IDENTIFICATION	Break- through Time in Minutes	Perm- eation Rate in mg/sq m /min.	I N D E X	Thick- ness in mm	R degrad. e and f comments s

---

diethylene glycol (synonym: 2-hydroxyethyl ether)
CAS Number: 111-46-6
Primary Class: 314   Hydroxy Compounds, Aliphatic and Alicyclic, Polyols

g BUTYL	NORTH B-161	>480m	0	0	0.43	34
g NITRILE	ANSELLEDMONT 37-155	>480m	0	0	0.38	34

---

diethylenetriamine
CAS Number: 111-40-0
Primary Class: 148   Amines, Poly, Aliphatic and Alicyclic
Related Class: 141   Amines, Aliphatic and Alicyclic, Primary
                142   Amines, Aliphatic and Alicyclic, Secondary

g BUTYL	NORTH B-174	>480m	0	0	0.68	34
g NATURAL RUBBER	ACKWELL 5-109				0.15	degrad. 34
g NEOPRENE	ANSELLEDMONT 29-870	>480m	0	0	0.48	34
g NITRILE	ANSELLEDMONT 37-155				0.40	degrad. 34
g PE/EVAL/PE	SAFETY4 4H	>240m	0		0.07	60
g PVC	ANSELLEDMONT 34-100	38m	17	3	0.20	34
g VITON	NORTH F-091	>480m	0	0	0.25	34

---

diglycidyl ether with bisphenol A, 50% & methyl ethyl ketone, 50%
Primary Class: 275   Heterocyclic Compounds, Oxygen, Epoxides

---

diisobutyl ketone (synonyms: 2,6-dimethyl-4-heptanone; DIBK)
CAS Number: 108-83-8
Primary Class: 391   Ketones, Aliphatic and Alicyclic

g BUTYL	BEST 878	>480m	0	0	0.75	53
g BUTYL	NORTH B-161	197m	2472	4	0.43	degrad. 34
g BUTYL/NEOPRENE	COMASEC BUTYL PLUS	>480m	0	0	0.50	40
g NATURAL RUBBER	ACKWELL 5-109				0.15	degrad. 34
g NATURAL RUBBER	ANSELL CONFORM 4205	1m	68	4	0.13	55
g NATURAL RUBBER	ANSELL FL 200 254	13m	5	3	0.51	55
g NATURAL RUBBER	ANSELL ORANGE 208	18m	3	2	0.76	55
g NATURAL RUBBER	ANSELL PVL 040	21m	2	2	0.46	55
g NATURAL RUBBER	ANSELL STERILE 832	2m	21	4	0.23	55
g NATURAL RUBBER	ANSELLEDMONT 36-124				0.46	degrad. 6
g NATURAL RUBBER	ANSELLEDMONT 392				0.48	degrad. 6
g NATURAL RUBBER	COMASEC FLEXIGUM	45m	1980	4	0.95	40
g NATURAL RUBBER	MARIGOLD BLACK HEVY	14m	5018	5	0.65	degrad. 79
g NATURAL RUBBER	MARIGOLD FEATHERLIT	3m	2200	5	0.15	degrad. 79
g NATURAL RUBBER	MARIGOLD MED WT EXT	1m	9200	5	0.45	degrad. 79
g NATURAL RUBBER	MARIGOLD MEDICAL	2m	2200	5	0.28	degrad. 79

G A R M E Class & Number/ N test chemical/ T MATERIAL NAME	MANUFACTURER & PRODUCT IDENTIFICATION	Break- through Time in Minutes	Perm- eation Rate in mg/sq m /min.	I N D E X	Thick- ness in mm	degrad. and comments	R e f s
g NATURAL RUBBER	MARIGOLD ORANGE SUP	15m	4600	4	0.73	degrad.	79
g NATURAL RUBBER	MARIGOLD R SURGEONS	2m	2200	5	0.28	degrad.	79
g NATURAL RUBBER	MARIGOLD RED LT WT	1m	8200	5	0.43	degrad.	79
g NATURAL RUBBER	MARIGOLD SENSOTECH	2m	2200	5	0.28	degrad.	79
g NATURAL RUBBER	MARIGOLD SUREGRIP	10m	6273	5	0.58	degrad.	79
g NATURAL RUBBER	MARIGOLD YELLOW	1m	6200	5	0.38	degrad.	79
g NATURAL+NEOPRENE	ANSELL OMNI 276	16m	4	2	0.56		55
g NATURAL+NEOPRENE	ANSELL TECHNICIANS	8m	6	3	0.43		55
g NATURAL+NEOPRENE	MARIGOLD FEATHERWT	2m	5200	5	0.35	degrad.	79
g NEOPRENE	ANSELL NEOPRENE 530	33m	2	2	0.46		55
g NEOPRENE	ANSELLEDMONT 29-840				0.38	degrad.	6
g NEOPRENE	ANSELLEDMONT 29-870				0.48	degrad.	34
g NEOPRENE	BEST 6780	>480m	0	0	n.a.		53
g NEOPRENE	COMASEC COMAPRENE	15m	4500	4	n.a.		40
g NEOPRENE/NATURAL	ANSELL CHEMI-PRO	33m	2	2	0.72		55
g NITRILE	ANSELL CHALLENGER	240m			0.2	0.38	55
g NITRILE	ANSELLEDMONT 37-165	120m	<9000	4	0.60		6
g NITRILE	COMASEC COMATRIL	85m	30	2	0.55		40
g NITRILE	COMASEC COMATRIL SU	282m	60	2	0.60		40
g NITRILE	COMASEC FLEXITRIL	50m	1080	4	n.a.		40
g NITRILE	MARIGOLD BLUE	380m	47	2	0.45	degrad.	79
g NITRILE	MARIGOLD GREEN SUPA	320m	76	2	0.30	degrad.	79
g NITRILE	MARIGOLD NITROSOLVE	392m	42	2	0.75	degrad.	79
g NITRILE	NORTH LA-142G	173m	2940	4	0.33	80% in H_2O	34
g NITRILE+PVC	COMASEC MULTIMAX	92m	420	3	n.a.		40
g NITRILE+PVC	COMASEC MULTIPLUS	85m	12	2	n.a.		40
g PE/EVAL/PE	SAFETY4 4H	>240m	0	0	0.07	35°C	60
g PVAL	ANSELLEDMONT PVA	>360m		0	n.a.		6
g PVAL	EDMONT 15-552	>360m	0	0	n.a.		6
g PVAL	EDMONT 25-545	120m	<9	1	n.a.		6
g PVAL	EDMONT 25-950	>960m	0	0	0.33	80% in H_2O	34
g PVC	ANSELLEDMONT 34-100				0.20	degrad.	34
g PVC	ANSELLEDMONT SNORKE				n.a.	degrad.	6
g PVC	COMASEC MULTIPOST	40m	120	3	n.a.		40
g PVC	COMASEC MULTITOP	60m	48	2	n.a.		40
g PVC	COMASEC NORMAL	64m	120	3	n.a.		40
g PVC	COMASEC OMNI	31m	168	3	n.a.		40
g VITON	NORTH F-091	68m	5436	4	0.25	degr 80%	34

diisobutylamine
CAS Number: 110-96-3
Primary Class: 142   Amines, Aliphatic and Alicyclic, Secondary

g BUTYL	NORTH B-174				0.63	degrad.	34
g NATURAL RUBBER	ACKWELL 5-109				0.15	degrad.	34
g NEOPRENE	ANSELLEDMONT 29-870	52m	1374	4	0.48	degrad.	34
g NITRILE	ANSELLEDMONT 37-155	>480m	0	0	0.31		34

G A R M E Class & Number/ N test chemical/ T MATERIAL NAME	MANUFACTURER & PRODUCT IDENTIFICATION	Break-through Time in Minutes	Perm-eation Rate in mg/sq m /min.	I N D E X	Thick-ness in mm	degrad. and comments	R e f s
g PVAL	EDMONT 25-950	>480m	0	0	0.50		34
g PVC	ANSELLEDMONT 34-100				0.20	degrad.	34
g VITON	NORTH F-091	>480m	0	0	0.25		34

diisooctyl phthalate
CAS Number: 27554-26-3
Primary Class: 226   Esters, Carboxylic, Benzoates and Phthalates

g NATURAL RUBBER	MARIGOLD BLACK HEVY	98m	10	2	0.65		79
g NATURAL RUBBER	MARIGOLD FEATHERLIT	34m	136	3	0.15	degrad.	79
g NATURAL RUBBER	MARIGOLD MED WT EXT	25m	0	3	0.45	degrad.	79
g NATURAL RUBBER	MARIGOLD MEDICAL	30m	83	3	0.28	degrad.	79
g NATURAL RUBBER	MARIGOLD ORANGE SUP	105m	10	2	0.73		79
g NATURAL RUBBER	MARIGOLD R SURGEONS	30m	83	3	0.28	degrad.	79
g NATURAL RUBBER	ARIGOLD RED LT WT	25m	21	3	0.43		79
g NATURAL RUBBER	ARIGOLD SENSOTECH	3m	83	3	0.28	degrad.	79
g NATURAL RUBBER	MARIGOLD SUREGRIP	76m	10	2	0.58		79
g NATURAL RUBBER	MARIGOLD YELLOW	27m	42	3	0.38	degrad.	79
g NATURAL+NEOPRENE	MARIGOLD FEATHERWT	28m	52	3	0.35	degrad.	79
g NITRILE	MARIGOLD BLUE	>480m	0	0	0.45		79
g NITRILE	MARIGOLD GREEN SUPA	>480m	0	0	0.30		79
g UNKNOWN MATERIAL	MARIGOLD R MEDIGLOV	15m	15	3	0.03	degrad.	79

diisopropylamine
CAS Number: 108-18-9
Primary Class: 142   Amines, Aliphatic and Alicyclic, Secondary

g BUTYL	NORTH B-174				0.66	degrad.	34
g NATURAL RUBBER	ACKWELL 5-109				0.15	degrad.	34
g NEOPRENE	ANSELLEDMONT 29-870	40m	2722	4	0.41	degrad.	34
g NITRILE	ANSELLEDMONT 37-155	192m	554	3	0.36		34
g PVC	ANSELLEDMONT 34-100	2m	7942	5	0.20		34
s TEFLON	CHEMFAB CHALL. 5100	>670m	0	0	n.a.		65
g VITON	NORTH F-091	>480m	0	0	0.24		34

dimercaptothiodiazole, 10% in butyldioxothol & methyl ethyl ketone, 1:1
Primary Class: 800   Multicomponent Mixtures With > 2 Components

g PE/EVAL/PE	SAFETY4 4H	>240m	0	0	0.07	35°C	60

dimethyl acetamide (synonyms: DMAC; N,N-dimethyl acetamide)
CAS Number: 127-19-5
Primary Class: 132   Amides, Aliphatic and Alicyclic

g BUTYL	NORTH B-174	480m	0	0	0.62		34
u CPE	ILC DOVER	40m			n.a.		29
g NATURAL RUBBER	ACKWELL 5-109				0.15	degrad.	34

G A R M E Class & Number/ N test chemical/ T MATERIAL NAME	MANUFACTURER & PRODUCT IDENTIFICATION	Break-through Time in Minutes	Perm-eation Rate in mg/sq m /min.	I N D E X	Thick-ness in mm	degrad. and comments	R e f s
g NATURAL RUBBER	ANSELLEDMONT 36-124	6m	< 900	4	0.46		6
g NATURAL RUBBER	ANSELLEDMONT 392	15m	< 900	3	0.48		6
g NATURAL RUBBER	MAPA-PIONEER L-118				0.46	degrad.	36
g NATURAL RUBBER	MARIGOLD BLACK HEVY	29m	670	3	0.65	degrad.	79
g NATURAL RUBBER	MARIGOLD FEATHERLIT	15m	3652	4	0.15		79
g NATURAL RUBBER	MARIGOLD MED WT EXT	15m	1767	4	0.45	degrad.	79
g NATURAL RUBBER	MARIGOLD MEDICAL	15m	2867	4	0.28		79
g NATURAL RUBBER	MARIGOLD ORANGE SUP	30m	560	3	0.73		79
g NATURAL RUBBER	MARIGOLD R SURGEONS	15m	2867	4	0.28		79
g NATURAL RUBBER	MARIGOLD RED LT WT	15m	1924	4	0.43	degrad.	79
g NATURAL RUBBER	MARIGOLD SENSOTECH	15m	2867	4	0.28		79
g NATURAL RUBBER	MARIGOLD SUREGRIP	25m	999	3	0.58	degrad.	79
g NATURAL RUBBER	MARIGOLD YELLOW	15m	2238	4	0.38	degrad.	79
g NATURAL+NEOPRENE	MARIGOLD FEATHERWT	15m	2395	4	0.35	degrad.	79
g NATURAL+NEOPRENE	PLAYTEX ARGUS 123	36m	840	3	n.a.		72
g NEOPRENE	ANSELLEDMONT 29-840				0.38	degrad.	6
g NEOPRENE	ANSELLEDMONT 29-870	100m	300	3	0.50	degrad.	34
g NEOPRENE	ANSELLEDMONT NEOX				n.a.		6
g NEOPRENE	MAPA-PIONEER GF-N	<15m	12000	5	0.43		36
g NEOPRENE	MAPA-PIONEER N-44				0.56	degrad.	36
g NITRILE	ANSELLEDMONT 37-155				0.40	degrad.	34
g NITRILE	ANSELLEDMONT 37-165				0.60	degrad.	6
g NITRILE	BEST 727				0.38	degrad.	53
g NITRILE	MAPA-PIONEER A-14	28m	>10000	5	0.56	degrad.	36
g NITRILE	MARIGOLD BLUE	26m	2224	4	0.45	degrad.	79
g NITRILE	MARIGOLD GREEN SUPA	15m	293	3	0.30	degrad.	79
g NITRILE	MARIGOLD NITROSOLVE'	28m	2610	4	0.75	degrad.	79
g PE/EVAL/PE	SAFETY4 4H	>240m	0	0	0.07	35°C	60
g PVAL	ANSELLEDMONT PVA				n.a.	degrad.	6
g PVAL	EDMONT 15-552				n.a.	degrad.	6
g PVAL	EDMONT 25-950	54m	660	3	0.30		34
g PVC	ANSELLEDMONT 34-100				0.20	degrad.	34
g PVC	ANSELLEDMONT SNORKE				n.a.	degrad.	6
g PVC	MAPA-PIONEER V-20	20m	>10000	5	0.51	degrad.	36
s SARANEX-23	DU PONT TYVEK SARAN	64m	20	2	n.a.		8
s UNKNOWN MATERIAL	DU PONT BARRICADE	>480m	0	0	n.a.		8
s UNKNOWN MATERIAL	LIFE-GUARD RESPONDE	>480m	0	0	n.a.		77
g VITON	NORTH F-091	25m	180	3	0.26	degrad.	34
u VITON/CHLOROBUTYL	Unknown	>180m			0.36		29

dimethyl ether
CAS Number: 115-10-6
Primary Class: 241　　Ethers, Aliphatic and Alicyclic

g BUTYL	NORTH B-174	>480m	0	0	0.62		34
g NATURAL RUBBER	ACKWELL 5-109	<1m	125000	5	0.15		34
g NEOPRENE	ANSELLEDMONT 29-870	>480m	0	0	0.62		34
g PVC	ANSELLEDMONT 34-100	<1m	84200	5	0.18		34

G A R M E Class & Number/ N test chemical/ T MATERIAL NAME	MANUFACTURER & PRODUCT IDENTIFICATION	Break- through Time in Minutes	Perm- eation Rate in mg/sq m /min.	I N D E X	Thick- ness in mm	degrad. and comments	R e f s

dimethyl formamide (synonyms: DMF; N,N-dimethyl formamide)
CAS Number: 68-12-2
Primary Class: 132   Amides, Aliphatic and Alicyclic

g BUTYL	BEST 878	>480m	0	0	0.75		53
g BUTYL	BRUNSWICK BUTYL STD	>480m	0	0	0.78		89
g BUTYL	BRUNSWICK BUTYL-XTR	>480m	0	0	0.88		89
g BUTYL	COMASEC BUTYL	>480m	0	0	0.47		40
s BUTYL	LIFE-GUARD BUTYL	>480m	0	0	n.a.		77
g BUTYL	NORTH B-161	>480m	0	0	0.41		34
s BUTYL	WHEELER ACID KING	>480m	0	0	n.a.		8
g BUTYL/ECO	BRUNSWICK BUTYL-POL	>480m	0	0	0.63		89
g BUTYL/NEOPRENE	COMASEC BUTYL PLUS	>480m	0	0	0.60		40
s BUTYL/NEOPRENE	MSA BETEX	>240m	0	0	n.a.		48
s CPE	STD. SAFETY	112m	184	3	n.a.		75
g HYPALON	COMASEC DIPCO	61m			0.55		40
g NAT+NEOPR+NITRILE	MAPA-PIONEER TRIONI	8m	>5000	5	0.48		36
g NATURAL RUBBER	ACKWELL 5-109				0.15	degrad.	34
g NATURAL RUBBER	ANSELLEDMONT 30-139	30m	<900	3	0.25		42
g NATURAL RUBBE	ANSELLEDMONT 36-124	30m	<90	3	0.46		6
g NATURAL RUBBER	ANSELLEDMONT 392	25m	<90	3	0.48		6
g NATURAL RUBBER	COMASEC FLEXIGUM	126m	3600	4	0.95		40
g NATURAL RUBBER	MAPA-PIONEER 258	30m	2.5	2	0.60	30°C	87
g NATURAL RUBBER	MAPA-PIONEER 300	38m	2	2	n.a.	30°C	87
g NATURAL RUBBER	MAPA-PIONEER 730	142m	3	1	1.10	30°C	87
g NATURAL RUBBER	MAPA-PIONEER L-118	67m	2460	4	0.46		36
g NATURAL RUBBER	MARIGOLD BLACK HEVY	29m	1015	4	0.65		79
g NATURAL RUBBER	MARIGOLD FEATHERLIT	15m	1216	4	0.15		79
g NATURAL RUBBER	MARIGOLD MED WT EXT	15m	2169	4	0.45	degrad.	79
g NATURAL RUBBER	MARIGOLD MEDICAL	15m	1613	4	0.28		79
g NATURAL RUBBER	MARIGOLD ORANGE SUP	30m	900	3	0.73		79
g NATURAL RUBBER	MARIGOLD R SURGEONS	15m	1613	4	0.28		79
g NATURAL RUBBER	MARIGOLD RED LT WT	15m	2090	4	0.43	degrad.	79
g NATURAL RUBBER	MARIGOLD SENSOTECH	15m	1613	4	0.28		79
g NATURAL RUBBER	MARIGOLD SUREGRIP	25m	1361	4	0.58	degrad.	79
g NATURAL RUBBER	MARIGOLD YELLOW	15m	1931	4	0.38		79
g NATURAL+NEOPRENE	MAPA-PIONEER 424	40m	3.5	2	0.63	30°C	87
g NATURAL+NEOPRENE	MAPA-PIONEER NS-53	15m	1200	4	0.85		36
g NATURAL+NEOPRENE	MARIGOLD FEATHERWT	15m	1852	4	0.35		79
g NATURAL+NEOPRENE	PLAYTEX ARGUS 123	37m	660	3	n.a.		72
g NEOPRENE	ANSELLEDMONT 29-840	10m	<900	4	0.38		6
g NEOPRENE	ANSELLEDMONT 29-870	51m	660	3	0.50	degrad.	34
g NEOPRENE	ANSELLEDMONT NEOX	60m	<900	3	n.a.		6
g NEOPRENE	ANSELLEDMONT NEOX	10m	570	4	n.a.		31
g NEOPRENE	ANSELLEDMONT NEOX	3m	84	4	n.a.	preexp. 15h	31
g NEOPRENE	BEST 32	97m	220	3	n.a.		53
g NEOPRENE	BEST 6780	61m	1420	4	n.a.		53

G A R M E N T	Class & Number/ test chemical/ MATERIAL NAME	MANUFACTURER & PRODUCT IDENTIFICATION	Break- through Time in Minutes	Perm- eation Rate in mg/sq m /min.	I N D E X	Thick- ness in mm	degrad. and comments	R e f s
g	NEOPRENE	BRUNSWICK NEOPRENE	96m	>5000	4	0.95		89
g	NEOPRENE	COMASEC COMAPRENE	8m	960	4	n.a.		40
s	NEOPRENE	LIFE-GUARD NEOPRENE	60m	11	2	n.a.		77
g	NEOPRENE	MAPA-PIONEER 360	60m	3.5	1	n.a.	30°C	87
g	NEOPRENE	MAPA-PIONEER 420	48m	4.5	2	0.75	30°C	87
g	NEOPRENE	MAPA-PIONEER N-44	110m	2460	4	0.56		36
b	NEOPRENE	RAINFAIR	120m	2	1	n.a.		90
b	NEOPRENE	RANGER	164m	1.5	1	n.a.		90
b	NEOPRENE	SERVUS NO 22204	>180m			n.a.		90
u	NEOPRENE	Unknown	34m	470	3	0.38		1
g	NEOPRENE/NATURAL	ANSELL CHEMI-PRO	46m	0.2		0.72		55
g	NITRILE	ANSELLEDMONT 37-165				0.60	degrad.	6
g	NITRILE	ANSELLEDMONT 37-175	10m	16500	5	0.42		38
g	NITRILE	BEST 22R		660	5	n.a.		53
g	NITRILE	BEST 727		1140	5	0.38	degrad.	53
g	NITRILE	COMASEC COMATRI	45m	720	3	0.56		40
g	NITRILE	COMASEC COMATRIL SU	55m	600	3	0.60		40
g	NITRILE	COMASEC FLEXITRIL	20m	2100	4	n.a.		40
g	NITRILE	MAPA-PIONEER 490	10m	110	4	0.38	30°C	87
g	NITRILE	MAPA-PIONEER A-14	35m	2460	4	0.56		36
g	NITRILE	MARIGOLD BLUE	28m	2144	4	0.45	degrad.	79
g	NITRILE	MARIGOLD GREEN SUPA	15m	398	3	0.30	degrad.	79
g	NITRILE	MARIGOLD NITROSOLVE	31m	2493	4	0.75	degrad.	79
g	NITRILE	NORTH LA-142G	9m	900	4	0.33	degrad.	34
b	NITRILE+PUR+PVC	BATA HAZMAX	>180m			n.a.		86
g	NITRILE+PVC	COMASEC MULTIMAX	90m	3	1	n.a.		40
g	NITRILE+PVC	COMASEC MULTIPLUS	90m	1320	4	n.a.		40
b	NITRILE+PVC	SERVUS NO 73101	>180m			n.a.		90
b	NITRILE+PVC	TINGLEY	>180m			n.a.		90
c	PE	DU PONT TYVEK QC	45m	12	3	n.a.		
g	PE/EVAL/PE	SAFETY4 4H	86m	0.24	1	0.07	35°C	60
g	PE/EVAL/PE	SAFETY4 4H	>1440m	0	0	0.07		60
g	PVAL	ANSELLEDMONT PVA				n.a.	degrad.	6
g	PVAL	COMASEC SOLVATRIL	18m	1260	4	n.a.		40
g	PVAL	EDMONT 15-552				n.a.	degrad.	6
g	PVAL	EDMONT 25-950	20m	780	3	0.38		34
g	PVAL	EDMONT 25-950	12m	246	4	0.28		34
u	PVAL	Unknown	20m	480	3	0.68		1
g	PVC	ANSELLEDMONT SNORKE				n.a.	degrad.	6
b	PVC	BATA STANDARD	12m	0.3	3	n.a.		86
g	PVC	COMASEC MULTIPOST	75m	1320	4	n.a.		40
g	PVC	COMASEC MULTITOP	80m	1380	4	n.a.		40
g	PVC	COMASEC NORMAL	60m	1380	4	n.a.		40
g	PVC	COMASEC OMNI	39m	1200	4	n.a.		40
b	PVC	JORDAN DAVID	>180m			n.a.		90
b	PVC	STD. SAFETY GL-20	20m	>4000	4	n.a.		75
s	PVC	STD. SAFETY WG-20	12m	>4000	5	n.a.		75

G A R M E Class & Number/ N test chemical/ T MATERIAL NAME	MANUFACTURER & PRODUCT IDENTIFICATION	Break- through Time in Minutes	Perm- eation Rate in mg/sq m /min.	I N D E X	Thick- ness in mm	degrad. and comments	R e f s
s PVC	WHEELER ACID KING	14m	<9000	5	n.a.	degrad.	85
b PVC+POLYURETHANE	BATA POLYBLEND	4m	0.6	3	n.a.		86
b PVC+POLYURETHANE	BATA POLYBLEND	>180m			n.a.		90
b PVC+POLYURETHANE	BATA POLYMAX	234m	3	1	n.a.		86
b PVC+POLYURETHANE	BATA SUPER POLY	6m	0.6	3	n.a.		86
s SARANEX-23	DU PONT TYVEK SARAN	120m	180	3	n.a.		8
s SARANEX-23 2-PLY	DU PONT TYVEK SARAN	>480m	0	0	n.a.		8
u TEFLON	CHEMFAB	2m	44	4	0.10		65
s TEFLON	CHEMFAB CHALL. 5100	>480m	0	0	n.a.		56
s TEFLON	CHEMFAB CHALL. 5100	>190m			n.a.		65
s TEFLON	CHEMFAB CHALL. 5200	>300m	0	0	n.a.		80
s TEFLON	CHEMFAB CHALL. 6000	>180m			n.a.		80
s TEFLON	LIFE-GUARD TEFGUARD	>480m	0	0	n.a.		77
s TEFLON	WHEELER ACID KING	>480m	0	0	n.a.		85
v TEFLON-FEP	CHEMFAB CHALL.	>180m			0.25		65
s UNKNOWN MATERIAL	CHEMRON CHEMREL	>1440m	0	0	n.a.		67
s UNKNOWN MATERIAL	CHEMRON CHEMREL MAX	>1440m	0	0	n.a.		67
s UNKNOWN MATERIAL	DU PONT BARRICADE	226m	25	2	n.a.		8
s UNKNOWN MATERIAL	KAPPLER CPF III	79m	6	1	n.a.		77
s UNKNOWN MATERIAL	LIFE-GUARD RESPONDE	>480m	0	0	n.a.		77
g UNKNOWN MATERIAL	NORTH SILVERSHIELD	>480m	0	0	0.10		7
g VITON	NORTH F-091	8m	390	4	0.33	degrad.	34
g VITON	NORTH F-091	7m	16500	5	0.26		38
s VITON/BUTYL/UNKN.	TRELLEBORG HPS	>180m			n.a.		71
s VITON/CHLOROBUTYL	LIFE-GUARD VC-100	176m			n.a.		77
s VITON/CHLOROBUTYL	LIFE-GUARD VNC 200	>480m	0	0	n.a.		77
g VITON/NEOPRENE	ERISTA VITRIC	12m	11300	5	0.60		38
s VITON/NEOPRENE	MSA VAUTEX	30m	77	3	n.a.		4

dimethyl sulfate (synonyms: DMS; methyl sulfate; sulfuric acid dimethyl ester)
CAS Number: 77-78-1
Primary Class: 507   Sulfur Compounds, Sulfonates, Sulfates, and Sulfites

g PE/EVAL/PE	SAFETY4 4H	>240m	0	0	0.07	35°C	60

dimethyl sulfoxide (synonyms: DMSO; methyl sulfoxide)
CAS Number: 67-68-5
Primary Class: 503   Sulfur Compounds, Sulfones and Sulfoxides

g BUTYL	ANSELL LAMPRECHT	>60m			0.70		38
g BUTYL	BEST 878	>480m	0	0	0.75		53
g BUTYL	NORTH B-174	>480m	0	0	0.67	34	
s CPE	ILC DOVER CLOROPEL	>180m			n.a.		46
g HYPALON	COMASEC DIPCO	>480m	0		0.59		40
g NATURAL RUBBER	ACKWELL 5-109	40m	276	3	0.15		34
g NATURAL RUBBER	ANSELL CONFORM 4205	1m	15	4	0.13		55
g NATURAL RUBBER	ANSELL FL 200 254	74m	0.4		0.51		55

G A R M E Class & Number/ N test chemical/ T MATERIAL NAME	MANUFACTURER & PRODUCT IDENTIFICATION	Break-through Time in Minutes	Perm-eation Rate in mg/sq m /min.	I N D E X	Thick-ness in mm	degrad. and comments	R e f s
g NATURAL RUBBER	ANSELL ORANGE 208	>360m	0	0	0.76		55
g NATURAL RUBBER	ANSELL PVL 040	240m	0.2		0.46		55
g NATURAL RUBBER	ANSELL STERILE 832	15m	2	2	0.23		55
g NATURAL RUBBER	ANSELLEDMONT 36-124	60m	<90	2	0.46		6
g NATURAL RUBBER	ANSELLEDMONT 392	180m	<9	1	0.48		6
g NATURAL RUBBER	ANSELLEDMONT 46-320	>60m			0.31		5
g NATURAL RUBBER	COMASEC FLEXIGUM	126m	2340	4	0.95		40
g NATURAL RUBBER	MAPA-PIONEER L-118	240m	0		0.46		36
g NATURAL RUBBER	MARIGOLD BLACK HEVY	92m	117	3	0.65		79
g NATURAL RUBBER	MARIGOLD FEATHERLIT	9m	264	4	0.15		79
g NATURAL RUBBER	MARIGOLD MED WT EXT	60m	288	3	0.45	degrad.	79
g NATURAL RUBBER	MARIGOLD MEDICAL	30m	274	3	0.28		79
g NATURAL RUBBER	MARIGOLD ORANGE SUP	95m	100	3	0.73		79
g NATURAL RUBBER	MARIGOLD R SURGEONS	30m	274	3	0.28		79
g NATURAL RUBBER	MARIGOLD RED LT WT	56m	286	3	0.43	degrad.	79
g NATURAL RUBBER	MARIGOLD SENSOTECH	30m	274	3	0.28		79
g NATURAL RUBBER	MARIGOLD SUREGRIP	82m	168	3	0.58	degrad.	79
g NATURAL RUBBER	MARIGOLD YELLOW	47m	282	3	0.38		79
g NATURAL RUBBER	PHARMASEAL LAB 8070	120m	50000	5	0.20		2
g NATURAL+NEOPRENE	ANSELL TECHNICIANS	150m	0.2		0.43		55
g NATURAL+NEOPRENE	MARIGOLD FEATHERWT	43m	280	3	0.35		79
g NATURAL+NEOPRENE	PLAYTEX 835	240m	50000	5	0.50		2
g NEOPRENE	ANSELL NEOPRENE 530	>360m	0	0	0.46		55
g NEOPRENE	ANSELLEDMONT 29-840	>360m	0	0	0.38		6
g NEOPRENE	ANSELLEDMONT 29-870	>480m	0	0	0.50		2
g NEOPRENE	ANSELLEDMONT 29-870	>480m	0	0	0.46		34
g NEOPRENE	ANSELLEDMONT NEOX	>180m			n.a.		6
g NEOPRENE	BEST 6780	>480m	0	0	n.a.		53
g NEOPRENE	COMASEC COMAPRENE	>360m	0	0	n.a.		40
g NEOPRENE	MAPA-PIONEER GF-N-	>60m			0.47		5
g NEOPRENE	MAPA-PIONEER N-44	<1m			0.56		36
g NEOPRENE/NATURAL	ANSELL CHEMI-PRO	>360m	0	0	0.72		55
g NITRILE	ANSELL CHALLENGER	30m	17	3	0.38		55
g NITRILE	ANSELLEDMONT 37-155	80m	335000	5	0.40		2
g NITRILE	ANSELLEDMONT 37-155	>60m			0.35		5
g NITRILE	ANSELLEDMONT 37-155				0.40	degrad.	34
g NITRILE	ANSELLEDMONT 37-165	>240m	0	0	0.60		6
g NITRILE	ANSELLEDMONT 37-175	>60m			0.42		38
g NITRILE	COMASEC COMATRIL	260m	54	2	0.55		40
g NITRILE	COMASEC FLEXITRIL	12m	7500	5	n.a.		40
g NITRILE	MAPA-PIONEER A-14	<1m	66	4	0.56	degrad.	36
g NITRILE	MARIGOLD BLUE	118m	25	2	0.45	degrad.	79
g NITRILE	MARIGOLD GREEN SUPA	45m	12	3	0.30	degrad.	79
g NITRILE	MARIGOLD NITROSOLVE	132m	28	2	0.75	degrad.	79
g NITRILE+PVC	COMASEC MULTIMAX	240m	0	0	n.a.		40
g NITRILE+PVC	COMASEC MULTIPLUS	80m	48	2	n.a.		40
g PE/EVAL/PE	SAFETY4 4H	103m	22	2	0.07	35°C	60
g PE/EVAL/PE	SAFETY4 4H	>480m	0	0	0.07		60

G A R M E Class & Number/ N test chemical/ T MATERIAL NAME	MANUFACTURER & PRODUCT IDENTIFICATION	Break- through Time in Minutes	Perm- eation Rate in mg/sq m /min.	I N D E X	Thick- ness in mm	degrad. and comments	R e f s
g PVAL	ANSELLEDMONT PVA				n.a.	degrad.	6
g PVAL	COMASEC SOLVATRIL	10m			n.a.		40
g PVAL	EDMONT 15-552				n.a.	degrad.	6
g PVAL	EDMONT 25-950				0.40	degrad.	34
g PVC	ANSELLEDMONT 34-100				0.20	degrad.	34
g PVC	ANSELLEDMONT M GRIP	70m	<90	2	n.a.		6
g PVC	COMASEC MULTIPOST	45m	72	3	n.a.		40
g PVC	COMASEC MULTITOP	60m	60	2	n.a.		40
g PVC	COMASEC NORMAL	50m	60	3	n.a.		40
g PVC	COMASEC OMNI	35m	84	3	n.a.		40
g PVC	MAPA-PIONEER V-20	60m			0.51		36
g VITON	NORTH F-091	95m	135	3	0.30	degrad.	34
g VITON	NORTH F-091	>60m			0.26		38
s VITON/CHLOROBUTYL	LIFE-GUARD VC-100	>240m		0	n.a.		77
u VITON/CHLOROBUTYL	Unknown	>180m			0.36		46
g VITON/NEOPRENE	ERISTA VITRIC	> 60m			0.60		38

dimethyl sulfoxide, 50% & toluene, 50%
Primary Class: 503   Sulfur Compounds, Sulfones and Sulfoxides

u NEOPRENE	Unknown	32/26m			0.40	37°C TL	16
u NITRILE	Unknown	21/11m			0.50	37°C TL	16
u PVC	Unknown	18/16m			0.30	37°C TL	16

dimethyl-d6 sulfoxide (synonym: hexadeuterodimethyl sulfoxide)
CAS Number: 2206-27-1
Primary Class: 503   Sulfur Compounds, Sulfones and Sulfoxides

s UNKNOWN MATERIAL	LIFE-GUARD RESPONDE	>240m	0	0	n.a.		77

dimethylamine
CAS Number: 124-40-3
Primary Class: 142   Amines, Aliphatic and Alicyclic, Secondary

g BUTYL	NORTH B-174	>480m	0	0	0.45		34
g NATURAL RUBBER	ACKWELL 5-109	2m	485	4	0.15		34
g NEOPRENE	ANSELLEDMONT 29-870	>480m	0	0	0.38		34
g PE/EVAL/PE	SAFETY4 4H	>240m	0		0.07	35°C	60
g PVAL	EDMONT 25-950	16m	221	3	0.57		34
g PVC	ANSELLEDMONT 34-100	5m	119	4	0.20		34

dimethyldichlorosilane (synonym: dichlorodimethylsilane)
CAS Number: 75-78-5
Primary Class: 480   Organosilicon Compounds

s BUTYL	TRELLEBORG TRELLCHE				n.a.	degrad.	71

G A R M E Class & Number/ N test chemical/ T MATERIAL NAME	MANUFACTURER & PRODUCT IDENTIFICATION	Break-through Time in Minutes	Perm-eation Rate in mg/sq m /min.	I N D E X	Thick-ness in mm	degrad. and comments	R e f s

dimethylvinyl chloride (synonym: 2-methyl-1-chloropropene)
CAS Number: 513-37-1
Primary Class: 263   Halogen Compounds, Vinylic
Related Class: 261   Halogen Compounds, Aliphatic and Alicyclic

g BUTYL	NORTH B-174				0.66	degrad.	34
g NATURAL RUBBER	ACKWELL 5-109				0.15	degrad.	34
g NEOPRENE	ANSELLEDMONT 29-870				0.48	degrad.	34
g NITRILE	ANSELLEDMONT 37-155	9m	3516	5	0.43	degrad.	34
g PVAL	EDMONT 25-950	70m	72	2	0.71		34
g PVC	ANSELLEDMONT 34-100	1m	4200	5	0.20	degrad.	34
g VITON	NORTH F-091	130m	267	3	0.30		34

Dinol
Primary Class: 900   Miscellaneous Unclassified Chemicals

g PE/EVAL/PE	SAFETY4 4H	>240m	0		0.07		60

dinoseb, 48% & xylene, 52%
Primary Class: 316   Hydroxy Compounds, Aromatic (Phenols)
Related Class: 840   Pesticides, Mixtures and Formulations

g PE/EVAL/PE	SAFETY4 4H	>240m	0	0	0.07	35°C	60

dipropylene glycol (synonym: 2,2'-dihydroxyisopropyl ether)
CAS Number: 110-98-5
Primary Class: 314   Hydroxy Compounds, Aliphatic and Alicyclic, Polyols

g BUTYL	NORTH B-161	>480m	0	0	0.38		34
g NITRILE	ANSELLEDMONT 37-175	>480m	0	0	0.38		34
g PVC	ANSELLEDMONT 34-100	115m	0.1	1	0.19		34

diquat dibromide (synonym: Reglone)
CAS Number: 85-00-7
Primary Class: 274   Heterocyclic Compounds, Nitrogen, Others
Related Class: 840   Pesticides, Mixtures and Formulations

g PE/EVAL/PE	SAFETY4 4H	>240m	0	0	0.07	35°C	60

divinylbenzene (synonym: vinylstyrene)
CAS Number: 1321-74-0
Primary Class: 292   Hydrocarbons, Aromatic
Related Class: 294   Hydrocarbons, Aliphatic and Alicyclic, Unsaturated

g BUTYL	BRUNSWICK BUTYL STD	59m	>640	3	0.63		89

G A R M E Class & Number/ N test chemical/ T MATERIAL NAME	MANUFACTURER & PRODUCT IDENTIFICATION	Break- through Time in Minutes	Perm- eation Rate in mg/sq m /min.	I N D E X	Thick- ness in mm	degrad. and comments	R e f s
g BUTYL	BRUNSWICK BUTYL-XTR	95m	>400	3	0.63		89
g BUTYL	NORTH B-174	132m	14160	5	0.50	degrad.	34
g BUTYL/ECO	BRUNSWICK BUTYL-POL	85m	>950	3	0.63		89
g NEOPRENE	ANSELLEDMONT 29-870	43m	2340	4	0.51		34
g NITRILE	ANSELLEDMONT 37-155	60m	27000	5	0.38	degrad.	34
g NITRILE	NORTH LA-142G				0.38	degrad.	7
g PVAL	EDMONT 15-552	>1080m	0	0	0.25		34
g UNKNOWN MATERIAL	NORTH SILVERSHIELD	>480m	0	0	0.10		7
g VITON	NORTH F-091	>1020m	0	0	0.23		34

dodecane (synonym: n-dodecane)
CAS Number: 112-40-3
Primary Class: 291   Hydrocarbons, Aliphatic and Alicyclic, Saturated

g PE/EVAL/PE	SAFETY4 4H	>480m	0	0	0.07		60
g PE/EVAL/PE	SAFETY4 4H	>240m		0	0.07	35°C	60

doxorubicin
CAS Number: 25316-40-9
Primary Class: 870   Antineoplastic Drugs and Other Pharmaceuticals

g EMA	Unknown	90m			0.07		59
u NATURAL RUBBER	Unknown	>1440m	0	0	0.18		59
g NEOPRENE	ANSELL NEOPRENE 530	>1440m	0	0	0.28		59
u PVC	Unknown	1440m		0	0.19		59

dynamite (synonym: ethylene glycol dinitrate, 70% & nitroglycerin, 30%)
Primary Class: 510   Nitrates and Nitrites

g NEOPRENE	ANSELLEDMONT 29-865	210m			0.48		82
g NITRILE	ANSELLEDMONT 37-145	282m		0	0.30		82
g NITRILE	ANSELLEDMONT 37-155	276m		0	0.36		82
g NITRILE	ANSELLEDMONT 37-165	1170m		0	0.60		82
g NITRILE	ANSELLEDMONT 37-175	360m		0	0.41		82
g NITRILE	ANSELLEDMONT HYNIT	65m			n.a.		82
g NITRILE	BEST 735-L	30m			n.a.		82
g NITRILE	COMASEC FLEXITRIL	90m			n.a.		82
g PVC	BEMAC 789 POLARFLEX	318m		0	n.a.		82
g PVC	TASKALL SPIT FIRE	120m			n.a.		82

electroless copper
Primary Class: 900   Miscellaneous Unclassified Chemicals

g NATURAL RUBBER	ANSELL CONFORM 4205	>360m	0	0	0.13		55
g NATURAL RUBBER	ANSELL FL 200 254	>360m	0	0	0.51		55
g NATURAL RUBBER	ANSELL ORANGE 208	>360m	0	0	0.76		55
g NATURAL RUBBER	ANSELL PVL 040	>360m	0	0	0.46		55

G A R M E Class & Number/ N test chemical/ T MATERIAL NAME	MANUFACTURER & PRODUCT IDENTIFICATION	Break- through Time in Minutes	Perm- eation Rate in mg/sq m /min.	I N D E X	Thick- ness in mm	degrad. and comments	R e f s
g NATURAL RUBBER	ANSELL STERILE 832	>360m	0	0	0.23		55
g NATURAL RUBBER	ANSELLEDMONT 392	>360m	0	0	0.48		6
g NATURAL+NEOPRENE	ANSELL OMNI 276	>360m	0	0	0.56		55
g NATURAL+NEOPRENE	ANSELL TECHNICIANS	>360m	0	0	0.43		55
g NEOPRENE	ANSELL NEOPRENE 530	>360m	0	0	0.46		55
g NEOPRENE	ANSELLEDMONT 29-840	>360m		0	0.38		6
g NEOPRENE	ANSELLEDMONT NEOX	>360m		0	n.a.		6
g NEOPRENE/NATURAL	ANSELL CHEMI-PRO	>360m	0	0	0.72		55
g NITRILE	ANSELL CHALLENGER	>360m	0	0	0.38		55
g NITRILE	ANSELLEDMONT 37-165	>360m		0	0.54		6
g PVAL	ANSELLEDMONT PVA				n.a.	degrad.	6
g PVC	ANSELLEDMONT M GRIP	>360m		0	n.a.		6

electroless nickel
Primary Class: 900   Miscellaneous Unclassified Chemicals

g NATURAL RUBBER	ANSELL CONFORM 4205	>360m	0	0	0.13		55
g NATURAL RUBBER	ANSELL FL 200 254	>360m	0	0	0.51		55
g NATURAL RUBBER	ANSELL ORANGE 208	>360m	0	0	0.76		55
g NATURAL RUBBER	ANSELL PVL 040	>360m	0	0	0.46		55
g NATURAL RUBBER	ANSELL STERILE 832	>360m	0	0	0.23		55
g NATURAL RUBBER	ANSELLEDMONT 392	>360m	0	0	0.48		6
g NATURAL+NEOPRENE	ANSELL OMNI 276	>360m	0	0	0.56		55
g NATURAL+NEOPRENE	ANSELL TECHNICIANS	>360m	0	0	0.43		55
g NEOPRENE	ANSELL NEOPRENE 530	>360m	0	0	0.46		55
g NEOPRENE	ANSELLEDMONT 29-840	>360m	0		0.38		6
g NEOPRENE	ANSELLEDMONT NEOX	>360m	0		n.a.		6
g NEOPRENE/NATURAL	ANSELL CHEMI-PRO	>360m	0	0	0.72		55
g NITRILE	ANSELL CHALLENGER	>360m	0	0	0.38		55
g NITRILE	ANSELLEDMONT 37-165	>360m		0	0.54		6
g PVAL	ANSELLEDMONT PVA				n.a.	degrad.	6
g PVC	ANSELLEDMONT M GRIP	>360m		0	n.a.		6

emulsions with metals
Primary Class: 850   Cutting Fluids

g NATURAL RUBBER	MARIGOLD ORANGE SUP	120m			0.74		12
g NEOPRENE	NOLATO 1505	>150m			0.40		12
g NITRILE	ERISTA SPECIAL	150m			0.37		12
u NITRILE	NOLATO	>170m			0.38		12
g PVC	EDMONT 34-590	150m			0.17		12
g PVC	EDMONT 34-590	77m			0.17	concentr.	12

epibromohydrin (synonym: 1,2-epoxy-3-bromopropane)
CAS Number: 3132-64-7
Primary Class: 275   Heterocyclic Compounds, Oxygen, Epoxides
Related Class: 261   Halogen Compounds, Aliphatic and Alicyclic

G A R M E Class & Number/ N test chemical/ T MATERIAL NAME	MANUFACTURER & PRODUCT IDENTIFICATION	Break- through Time in Minutes	Perm- eation Rate in mg/sq m /min.	I N D E X	Thick- ness in mm	degrad. and comments	R e f s
g BUTYL	NORTH B-174	>480m	0	0	0.62		34
g NAT+NEOPR+NITRILE	MAPA-PIONEER TRIONI	4m	>5000	5	0.46		36
g NATURAL RUBBER	ACKWELL 5-109	<1m	1380	5	0.15		34
g NEOPRENE	ANSELLEDMONT 29-870				0.48	degrad.	34
g NITRILE	ANSELLEDMONT 37-155				0.40	degrad.	34
g PVAL	EDMONT 25-950	>480m	0	0	0.35		34
g PVC	ANSELLEDMONT 34-100				0.20	degrad.	34
s UNKNOWN MATERIAL	KAPPLER CPF III	67m	21	2	n.a.		77
g VITON	NORTH F-091	>480m	0	0	0.27		34

epichlorohydrin (synonym: 1,2-epoxy-3-chloropropane)
CAS Number: 106-89-8
Primary Class: 275   Heterocyclic Compounds, Oxygen, Epoxides
Related Class: 261   Halogen Compounds, Aliphatic and Alicyclic

Class & Number/ test chemical/ MATERIAL NAME	MANUFACTURER & PRODUCT IDENTIFICATION	Break- through Time	Perm- eation Rate	INDEX	Thick- ness	degrad. and comments	Refs
g BUTYL	NORTH B-161	1440m	0	0	0.38		25
g BUTYL	NORTH B-161	5m	6	3	0.38	pre-exp.24h	25
g BUTYL	NORTH B-174	>480m	0	0	0.65		34
u BUTYL	Unknown	4740m	20	2	0.84		3
g NAT+NEOPR+NITRILE	MAPA-PIONEER TRIONI	4m	>5000	5	0.46		36
g NATURAL RUBBER	ACKWELL 5-109	<1m	3780	5	0.15		34
g NATURAL RUBBER	ANSELLEDMONT 36-124	3m	<9000	5	0.46		6
g NATURAL RUBBER	ANSELLEDMONT 392	5m	<9000	5	0.48		6
u NATURAL RUBBER	Unknown	3m	1340	5	0.23		25
g NEOPRENE	ANSELLEDMONT 29-840				0.38	degrad.	6
g NEOPRENE	ANSELLEDMONT 29-870	17m	3500	4	0.43		25
g NEOPRENE	ANSELLEDMONT 29-870				0.61	degrad.	34
g NEOPRENE	ANSELLEDMONT NEOX	10m	<9000	5	n.a.		6
u NEOPRENE	Unknown	60m	1100	4	0.76		3
g NITRILE	ANSELLEDMONT 37-155				0.40	degrad.	34
g NITRILE	ANSELLEDMONT 37-165				0.60	degrad.	6
g NITRILE	SURETY	22m	12000	5	0.38		25
g PE	ANSELLEDMONT 35-125	3m	96	4	0.03		25
g PE/EVAL/PE	SAFETY4 4H	>240m	0	0	0.07	35°C	60
g PVAL	ANSELLEDMONT PVA	300m	<9	1	n.a.		6
g PVA	EDMONT 15-552	16m	<90	3	n.a.		6
g PVAL	EDMONT 15-552	2m	1100	5	n.a.		25
g PVAL	EDMONT 25-950	233m	2	1	0.30		34
u PVAL	Unknown	<5m	1300	5	0.08		
g PVC	ANSELLEDMONT 34-100				0.20	degrad.	34
g PVC	ANSELLEDMONT SNORKE				n.a.	degrad.	6
s SARANEX-23	DU PONT TYVEK SARAN	57m	522	3	n.a.		8
s SARANEX-23	DU PONT TYVEK SARAN	60m	34	2	n.a.		25
s TEFLON	CHEMFAB CHALL. 5100	>180m	0	1	n.a.		65
g TEFLON	CLEAN ROOM PRODUCTS	420m	0.1	1	0.05		25
g TEFLON	CLEAN ROOM PRODUCTS	450m	0.1	1	0.05	pre-exp.24h	25
c UNKNOWN MATERIAL	CHEMRON CHEMTUFF	60m	2	1	n.a.		67
s UNKNOWN MATERIAL	DU PONT BARRICADE	>480m		0	n.a.		8
s UNKNOWN MATERIAL	KAPPLER CPF III	67m	21	2	n.a.		77

G A R M E Class & Number/ N test chemical/ T MATERIAL NAME	MANUFACTURER & PRODUCT IDENTIFICATION	Break- through Time in Minutes	Perm- eation Rate in mg/sq m /min.	I N D E X	Thick- ness degrad. in and mm comments	R e f s
s UNKNOWN MATERIAL	LIFE-GUARD RESPONDE	>240m	0	0	n.a.	77
g VITON	NORTH F-091	60m	510	3	0.23	25
g VITON	NORTH F-091	112m	480	3	0.27	34
g VITON	NORTH SF	50m	600	3	0.23	25
g VITON/NITRILE	NORTH VITRILE	80m	1000	4	0.20	25

epoxy (accelerator)
Primary Class: 810   Epoxy Products

g PE/EVAL/PE	SAFETY4 4H	2280m	0	0	0.06 40°C	47

epoxy (base & accelerator), & acetone, methoxyethanol, methyl alcohol
Primary Class: 810   Epoxy Products

g BUTYL	ERISTA BX	>240m	0	0	0.65	33
g BUTYL	ERISTA BX	>240m	0	0	0.65 preexp. 20h	33
g PVAL	EDMONT 15-552	>240m	0	0	n.a.	33
g PVAL	EDMONT 15-552	>240m	0	0	n.a. preexp. 20h	33
g PVC	KID VINYLPRODUKTER	1m	14880	5	0.52	33

epoxy (base)
Primary Class: 810   Epoxy Products

g PE/EVAL/PE	SAFETY4 4H	>1440m	0	0	0.06 40°C TL	47

epoxy (base) & toluene, 30-70% & methyl isobutyl ketone, <30% & xylene, <30%
Primary Class: 810   Epoxy Products

g PVAL	EDMONT 15-552	>240m	0	0	n.a.	33
g PVAL	EDMONT 15-552	>240m	0	0	n.a. preexp. 20h	33

epoxy (DGEBA), 50% & n-butanol, <30% & methyl ethyl ketone, <30% & xylene, <30%
Primary Class: 810   Epoxy Products

g PE/EVAL/PE	SAFETY4 4H	>300m	0	0	0.06 40°C TL	47

epoxytrichloropropane (synonym: trichloroepoxypropane)
CAS Number: 67664-94-2
Primary Class: 275   Heterocyclic Compounds, Oxygen, Epoxides

g NATURAL RUBBER	ACKWELL 5-109				0.15 degrad.	34
g NEOPRENE	ANSELLEDMONT 29-870				0.48 degrad.	34
g NITRILE	ANSELLEDMONT 37-155				0.40 degrad.	34
g PVC	ANSELLEDMONT 34-100				0.20 degrad.	34

G A R M E Class & Number/ N test chemical/ T MATERIAL NAME	MANUFACTURER & PRODUCT IDENTIFICATION	Break-through Time in Minutes	Perm-eation Rate in mg/sq m /min.	I N D E X	Thick-ness in mm	degrad. and comments	R e f s

ethanolamine (synonyms: 2-aminoethanol; MEA; monoethanolamine)
CAS Number: 141-43-5
Primary Class: 141   Amines, Aliphatic and Alicyclic, Primary
Related Class: 311   Hydroxy Compounds, Aliphatic and Alicyclic, Primary

	Class & Number/Material	Manufacturer & Product	Breakthrough	Perm Rate	Index	Thickness	degrad./comments	Refs
g	BUTYL	NORTH B-174	480m	0	0	0.48		34
g	BUTYL/NEOPRENE	COMASEC BUTYL PLUS	>480m	0	0	0.50		40
g	HYPALON	COMASEC DIPCO	>480m	0	0	0.58		40
g	NATURAL RUBBER	ACKWELL 5-109				0.15	degrad.	34
g	NATURAL RUBBER	ANSELL CONFORM 4205	>360m	0	0	0.13		55
g	NATURAL RUBBER	ANSELL FL 200 254	>360m	0	0	0.51		55
g	NATURAL RUBBER	ANSELL ORANGE 208	>360m	0	0	0.76		55
g	NATURAL RUBBER	ANSELL PVL 040	>360m	0	0	0.46		5
g	NATURAL RUBBER	ANSELL STERILE 832	>360m	0	0	0.23		55
g	NATURAL RUBBER	ANSELLEDMONT 36-124	210m	<90	2	0.46		6
g	NATURAL RUBBER	ANSELLEDMONT 392	50m	<9	2	0.48		6
g	NATURAL RUBBER	COMASEC FLEXIGUM	480m	66	2	0.95		40
g	NATURAL RUBBER	MARIGOLD BLACK HEVY	465m	8	1	0.65		79
g	NATURAL RUBBER	MARIGOLD FEATHERLIT	257m	35	2	0.15		79
g	NATURAL RUBBER	MARIGOLD MED WT EXT	317m	90	2	0.45		79
g	NATURAL RUBBER	MARIGOLD MEDICAL	282m	58	2	0.28		79
g	NATURAL RUBBER	MARIGOLD ORANGE SUP	>480m	0	0	0.73		79
g	NATURAL RUBBER	MARIGOLD R SURGEONS	282m	58	2	0.28		79
g	NATURAL RUBBER	MARIGOLD RED LT WT	312m	86	2	0.43		79
g	NATURAL RUBBER	MARIGOLD SENSOTECH	421m	33	2	0.28		79
g	NATURAL RUBBER	MARIGOLD SUREGRIP	391m	49	2	0.58		79
g	NATURAL RUBBER	MARIGOLD YELLOW	302m	76	2	0.38		79
g	NATURAL+NEOPRENE	ANSELL OMNI 276	360m	0	0	0.56		55
g	NATURAL+NEOPRENE	ANSELL TECHNICIANS	>360m	0	0	0.43		55
g	NATURAL+NEOPRENE	MARIGOLD FEATHERWT	297m	72	2	0.35		79
g	NEOPRENE	ANSELL NEOPRENE 530	>360m	0	0	0.46		55
g	NEOPRENE	ANSELLEDMONT 29-840	>360m	0	0	0.38		6
g	NEOPRENE	ANSELLEDMONT 29-870	>480m	0	0	0.41		34
g	NEOPRENE	ANSELLEDMONT NEOX	>360m	0	0	n.a.		6
g	NEOPRENE	COMASEC COMAPRENE	>360m	0	0	n.a.		40
g	NEOPRENE/NATURAL	ANSELL CHEMI-PRO	>360m	0	0	0.72		55
g	NITRILE	ANSELL CHALLENGER	>360m	0	0	0.38		55
g	NITRILE	ANSELLEDMONT 37-165	>360m	0	0	0.60		6
g	NITRILE	COMASEC COMATRIL	>260m	180	3	0.56		40
g	NITRILE	COMASEC COMATRIL SU	>480m	0	0	0.60		40
g	NITRILE	COMASEC FLEXITRIL	90m	90	2	n.a.		40
g	NITRILE	MARIGOLD BLUE	>480m	0	0	0.45		79
g	NITRILE	MARIGOLD GREEN SUPA	>480m	0	0	0.30		79
g	NITRILE	MARIGOLD NITROSOLVE	>480m	0	0	0.75		79
g	NITRILE	NORTH LA-142G	2450m	<1	1	0.35		63
g	NITRILE+PVC	COMASEC MULTIPLUS	300m	42	2	n.a.		40

G A R M E Class & Number/ N test chemical/ T MATERIAL NAME	MANUFACTURER & PRODUCT IDENTIFICATION	Break- through Time in Minutes	Perm- eation Rate in mg/sq m /min.	I N D E X	Thick- ness in mm	degrad. and comments	R e f s
g PE/EVAL/PE	SAFETY4 4H	>240m	0	0	0.07	35°C	60
g PE/EVAL/PE	SAFETY4 4H	>480m	0	0	0.07		60
g PVAL	ANSELLEDMONT PVA	>360m		0	n.a.		6
g PVAL	EDMONT 15-552	150m	<90	2	n.a.		6
g PVAL	EDMONT 25-545	300m	0	0	n.a.		6
g PVC	ANSELLEDMONT 34-100	>480m	0	0	0.20		34
g PVC	ANSELLEDMONT M GRIP	>360m	0	0	n.a.		6
g PVC	COMASEC MULTIPOST	110m	72	2	n.a.		40
g PVC	COMASEC MULTITOP	210m	66	2	n.a.		40
g PVC	COMASEC NORMAL	120m	78	2	n.a.		40
g PVC	COMASEC OMNI	80m	114	3	n.a.		40
g VITON	NORTH F-091	>480m	0	0	0.30		34

ethion 4
CAS Number: 563-12-2
Primary Class: 462   Organophosphorus Compounds and Derivatives of Phosphorus-based Acids
Related Class: 233   Esters, Non-Carboxylic, Carbamates and Others

s TEFLON	CHEMFAB CHALL. 5100	>288m	0	0	n.a.		65

ethyl acetate
CAS Number: 141-78-6
Primary Class: 222   Esters, Carboxylic, Acetates

g BUTYL	BEST 878	212m	80	2	0.75		53
g BUTYL	BRUNSWICK BUTYL STD	253m	6600	4	0.73		89
g BUTYL	BRUNSWICK BUTYL-XTR	191m	29000	5	0.63		89
s BUTYL	LIFE-GUARD BUTYL	28m	2	2	n.a.		77
g BUTYL	NORTH B-161	455m	204	3	0.45		7
s BUTYL	WHEELER ACID KING	59m	<9	2	n.a.		85
g BUTYL/ECO	BRUNSWICK BUTYL-POL	165m	>200	3	0.63		89
g BUTYL/NEOPRENE	COMASEC BUTYL PLUS	230m	372	3	0.50		40
s BUTYL/NEOPRENE	MSA BETEX	30m	580	3	n.a.		4
s CPE	ILC DOVER CLOROPEL	64m			n.a.		46
s CPE	STD. SAFETY	30m	37	3	n.a.		7
g HYPALON	COMASEC DIPCO	42m	84	3	0.60		40
g NATURAL RUBBER	ANSELL CONFORM 4205	1m	>118	4	0.13		55
g NATURAL RUBBER	ANSELL FL 200 254	4m	9	3	0.51		55
g NATURAL RUBBER	ANSELL ORANGE 208	7m	4	3	0.76		55
g NATURAL RUBBER	ANSELL PVL 040	4m	5	3	0.46		55
g NATURAL RUBBER	ANSELL STERILE 832	1m	30	4	0.23		55
g NATURAL RUBBER	ANSELLEDMONT 36-124	5m	<900	4	0.46		6
g NATURAL RUBBER	ANSELLEDMONT 392	5m	<900	4	0.48		6
g NATURAL RUBBER	COMASEC FLEXIGUM	29m	480	3	0.95		40
g NATURAL RUBBER	MARIGOLD BLACK HEVY	4m	4404	5	0.65	degrad.	79
g NATURAL RUBBER	MARIGOLD FEATHERLIT	1m	9921	5	0.15	degrad.	79

G A R M E N T	Class & Number/ test chemical/ MATERIAL NAME	MANUFACTURER & PRODUCT IDENTIFICATION	Break- through Time in Minutes	Perm- eation Rate in mg/sq m /min.	I N D E X	Thick- ness in mm	degrad. and comments	R e f s
g	NATURAL RUBBER	MARIGOLD MED WT EXT	2m	9110	5	0.45	degrad.	79
g	NATURAL RUBBER	MARIGOLD MEDICAL	1m	9583	5	0.28	degrad.	79
g	NATURAL RUBBER	MARIGOLD ORANGE SUP	4m	3933	5	0.73	degrad.	79
g	NATURAL RUBBER	MARIGOLD R SURGEONS	1m	9583	5	0.28	degrad.	79
g	NATURAL RUBBER	MARIGOLD RED LT WT	2m	9178	5	0.43	degrad.	79
g	NATURAL RUBBER	MARIGOLD SENSOTECH	1m	9583	5	0.28	degrad.	79
g	NATURAL RUBBER	MARIGOLD SUREGRIP	3m	5816	5	0.58	degrad.	79
g	NATURAL RUBBER	MARIGOLD YELLOW	2m	9313	5	0.38	degrad.	79
u	NATURAL RUBBER	Unknown	6m	7943	5	0.64		38
g	NATURAL+NEOPRENE	ANSELL OMNI 276	4m	5	3	0.56		55
g	NATURAL+NEOPRENE	ANSELL TECHNICIANS	1m	11	4	0.43		55
g	NATURAL+NEOPRENE	MARIGOLD FEATHERWT	2m	9380	5	0.35	degrad.	79
g	NATURAL+NEOPRENE	PLAYTEX ARGUS 123	5m	4860	5	n.a.		72
g	NEOPRENE	ANSELL NEOPRENE 530	4m	4	3	0.46		55
g	NEOPRENE	ANSELLEDMONT 29-840	15m	<900	3	0.38		6
g	NEOPRENE	ANSELLEDMONT NEOX	20m	<900	3	n.a.		6
g	NEOPRENE	BEST 32	8m	1560	5	n.a.		53
g	NEOPRENE	BEST 6780	17m	1830	4	n.a.		53
g	NEOPRENE	BRUNSWICK NEOPRENE	24m	>5000	4	0.91		89
g	NEOPRENE	COMASEC COMAPRENE	12m	480	4	n.a.		40
s	NEOPRENE	LIFE-GUARD NEOPRENE	17m	21	3	n.a.		77
g	NEOPRENE	MAPA-PIONEER N-44	34m	10680	5	0.56		36
b	NEOPRENE	RAINFAIR	40m	98	3	n.a.		90
b	NEOPRENE	RANGER	41m	130	3	n.a.		90
b	NEOPRENE	SERVUS NO 22204	59m	90	3	n.a.		90
g	NEOPRENE/NATURAL	ANSELL CHEMI-PRO	5m	4	3	0.72		55
g	NITRILE	ANSELLEDMONT 37-165				0.60	degrad.	6
g	NITRILE	ANSELLEDMONT 37-175	12m	9476	5	0.42		38
g	NITRILE	BEST 727	30m	2180	4	0.38	degrad.	53
g	NITRILE	COMASEC COMATRIL	25m	660	3	0.56		40
g	NITRILE	COMASEC COMATRIL SU	37m	660	3	0.60		40
g	NITRILE	COMASEC FLEXITRIL	3m	660	4	n.a.		40
g	NITRILE	MARIGOLD BLUE	21m	6022	4	0.45	degrad.	79
g	NITRILE	MARIGOLD GREEN SUPA	9m	6441	5	0.30	degrad.	79
g	NITRILE	MARIGOLD NITROSOLVE	23m	5938	4	0.75	degrad.	79
g	NITRILE	NORTH LA-142G	8m	8700	5	0.36		7
b	NITRILE+PUR+PVC	BATA HAZMAX	128m	70	2	n.a.		86
g	NITRILE+PVC	COMASEC MULTIMAX	37m	84	3	n.a.		40
g	NITRILE+PVC	COMASEC MULTIPLUS	29m	540	3	n.a.		40
b	NITRILE+PVC	SERVUS NO 73101	>180m			n.a.		90
b	NITRILE+PVC	TINGLEY	119m	>5	1	n.a.		90
c	PE	DU PONT TYVEK QC	<1m	19900	5	n.a.		8
c	PE	MOLNLYCKE H D	2m	1925	5	n.a.		54
g	PE/EVAL/PE	SAFETY4 4H	>240m	0	0	0.07	35°C	60
g	PE/EVAL/PE	SAFETY4 4H	>1440m	0	0	0.07		60
g	PVAL	ANSELLEDMONT PVA	>360m		0	n.a.		6
g	PVAL	COMASEC SOLVATRIL	>480m	0	0	n.a.		40

G A R M E N T	Class & Number/ test chemical/ MATERIAL NAME	MANUFACTURER & PRODUCT IDENTIFICATION	Break-through Time in Minutes	Perm-eation Rate in mg/sq m /min.	I N D E X	Thick-ness in mm	degrad. and comments	R e f s
g	PVAL	EDMONT 15-552	>360m	0	0	n.a.		6
g	PVAL	EDMONT 25-545	>360m	0	0	n.a.		6
g	PVC	ANSELLEDMONT SNORKE				n.a.	degrad.	6
b	PVC	BATA STANDARD	30m	62	3	n.a.	degrad.	86
g	PVC	COMASEC MULTIPOST	18m	780	3	n.a.		40
g	PVC	COMASEC MULTITOP	25m	600	3	n.a.		40
g	PVC	COMASEC NORMAL	20m	780	3	n.a.		40
g	PVC	COMASEC OMNI	14m	480	4	n.a.		40
b	PVC	JORDAN DAVID	124m	> 9	1	n.a.		90
b	PVC	STD. SAFETY GL-20	4m	>4000	5	n.a.		75
s	PVC	STD. SAFETY WG-20	4m	>4000	5	n.a.		75
s	PVC	WHEELER ACID KING	6m	<90000	5	n.a.	degrad.	85
b	PVC+POLYURETHANE	BATA POLYBLEND	37m	53	3	n.a.	degrad.	86
b	PVC+POLYURETHANE	BATA POLYBLEND	157m	26	2	n.a.		90
b	PVC+POLYURETHANE	BATA POLYMAX	107m	379	3	n.a.		86
b	PVC+POLYURETHANE	BATA SUPER POLY	143m	45	2	n.a.	degrad.	86
s	SARANEX-23	DU PONT TYVEK SARAN	36m	66	3	n.a.		8
s	SARANEX-23 2-PLY	DU PONT TYVEK SARAN	130m	5	1	n.a.		8
u	TEFLON	CHEMFAB	2m	45	4	0.10		65
s	TEFLON	CHEMFAB CHALL. 5100	>280m	0	0	n.a.		65
s	TEFLON	CHEMFAB CHALL. 5200	>300m	0	0	n.a.		80
s	TEFLON	CHEMFAB CHALL. 6000	>180m			n.a.		80
s	TEFLON	LIFE-GUARD TEFGUARD	>480m	0	0	n.a.		77
s	TEFLON	WHEELER ACID KING	>480m	0	0	n.a.		85
v	TEFLON-FEP	CHEMFAB CHALL.	>180m			0.25		65
s	UNKNOWN MATERIAL	CHEMRON CHEMREL	52m	<1	2	n.a.		67
s	UNKNOWN MATERIAL	CHEMRON CHEMREL MAX	>1440m	0	0	n.a.		67
c	UNKNOWN MATERIAL	CHEMRON CHEMTUFF	23m	5	2	n.a.		67
s	UNKNOWN MATERIAL	DU PONT BARRICADE	>480m	0	0	n.a.		8
s	UNKNOWN MATERIAL	KAPPLER CPF III	>480m	0	0	n.a.		77
s	UNKNOWN MATERIAL	LIFE-GUARD RESPONDE	>480m	0	0	n.a.		77
g	UNKNOWN MATERIAL	NORTH SILVERSHIELD	>360m	0	0	0.08		7
g	VITON	NORTH F-091				0.25	degrad.	7
g	VITON	NORTH F-091	3m	45000	5	0.26		38
s	VITON/BUTYL/UNKN.	TRELLEBORG HPS	>180m			n.a.		71
s	VITON/CHLOROBUTYL	LIFE-GUARD VNC 200	49m	2	2	n.a.		77
u	VITON/CHLOROBUTYL	Unknown	30m			n.a.		46
g	VITON/NEOPRENE	ERISTA VITRIC	13m	5156	5	0.60		38
s	VITON/NEOPRENE	MSA VAUTEX	15m	465	3	n.a.		48

ethyl acetate, >70% & acetone, ethyl alcohol and methyl alcohol
Primary Class: 800   Multicomponent Mixtures With >2 Components

g	BUTYL	ERISTA BX	>240m	0	0	0.67		33
g	BUTYL	ERISTA BX	>240m	0	0	0.67	preexp. 20h	33
g	PE/EVAL/PE	SAFETY4 4H	>240m	0	0	0.07	35°C	60
g	PVAL	EDMONT 15-552	>240m	0	0	n.a.	degrad.	33

G A R M E Class & Number/ N test chemical/ T MATERIAL NAME	MANUFACTURER & PRODUCT IDENTIFICATION	Break- through Time in Minutes	Perm- eation Rate in mg/sq m /min.	I N D E X	Thick- ness in mm	degrad. and comments	R e f s

ethyl acetate, >70% & ethyl alcohol, <30%
Primary Class: 222   Esters, Carboxylic, Acetates

g BUTYL	ERISTA BX	>240m	0	0	0.68		33
g BUTYL	ERISTA BX	>240m	0	0	0.68	pre-exp.20h	33
g PVC	KID VINYLPRODUKTER	2m	11000	5	0.50		33
g VITON/NEOPRENE	ERISTA VITRIC	8m	2800	5	0.50		33

ethyl acrylate (synonym: ethyl 2-propenoate)
CAS Number: 140-88-5
Primary Class: 223   Esters, Carboxylic, Acrylates and Methacrylates

g BUTYL	NORTH B-131	124m			0.35		83
g BUTYL	NORTH B-174	>480m	0	0	0.68		34
m BUTYL	PLYMOUTH RUBBER	39m			n.a.		46
s BUTYL/NEOPRENE	MSA BETEX	25m	45	3	n.a.		48
u CPE	ILC DOVER	24m			n.a.		29
s CPE	ILC DOVER CLOROPEL	67m			n.a.		46
g NATURAL RUBBER	ACKWELL 5-109	1m	10380	5	0.15	degrad.	34
g NEOPRENE	ANSELLEDMONT 29-870				0.48	degrad.	34
g NITRILE	ANSELLEDMONT 37-155				0.40	degrad.	34
g PE/EVAL/PE	SAFETY4 4H	>240m	0	0	0.07	35°C	60
g PVAL	EDMONT 25-950	>480m	0	0	0.58		34
g PVC	ANSELLEDMONT 34-100	2m	10380	5	0.20	degrad.	34
s TEFLON	CHEMFAB CHALL. 5100	>480m	0	0	n.a.		56
s TEFLON	CHEMFAB CHALL. 5100	>1020m	0	0	n.a.		65
s UNKNOWN MATERIAL	LIFE-GUARD RESPONDE	>480m	0	0	n.a.		77
g UNKNOWN MATERIAL	NORTH SILVERSHIELD	5m			0.08		83
g VITON	NORTH F-091				0.26	degrad.	34
g VITON	NORTH F-091	4m			0.25		83
u VITON/CHLOROBUTYL	Unknown	22m			n.a.		46
s VITON/NEOPRENE	MSA VAUTEX	<5m	465	4	n.a.		48

ethyl alcohol (synonyms: ethanol; grain alcohol)
CAS Number: 64-17-5
Primary Class: 311   Hydroxy Compounds, Aliphatic and Alicyclic, Primary

g BUTYL	BEST 878	>480m	0	0	0.75		53
g BUTYL	BRUNSWICK BUTYL STD	>480m	0	0	0.63		89
g BUTYL	BRUNSWICK BUTYL-XTR	>480m	0	0	0.63		89
g BUTYL	NORTH B-174	>480m	0	0	0.58		34
g BUTYL/ECO	BRUNSWICK BUTYL-POL	>480m	0	0	0.63		89
g NATURAL RUBBER	ACKWELL 5-109	<1m	1140	5	0.15		34
g NATURAL RUBBER	ANSELL CONFORM 4205	1m	8	3	0.13		55
g NATURAL RUBBER	ANSELL FL 200 254	16m	0.2		0.51		55
g NATURAL RUBBER	ANSELL ORANGE 208	21m	0.1		0.76		55
g NATURAL RUBBER	ANSELL PVL 040	24m	0.1		0.46		55

G A R M E Class & Number/ N test chemical/ T MATERIAL NAME	MANUFACTURER & PRODUCT IDENTIFICATION	Break- through Time in Minutes	Perm- eation Rate in mg/sq m /min.	I N D E X	Thick- ness in mm	degrad. and comments	R e f s
g NATURAL RUBBER	ANSELL STERILE 832	2m	1	3	0.23		55
g NATURAL RUBBER	ANSELLEDMONT 36-124	30m	<90	3	0.46		6
g NATURAL RUBBER	ANSELLEDMONT 392	15m	<90	3	0.48		6
g NATURAL RUBBER	ANSELLEDMONT 46-320	28m	40	3	0.31		5
g NATURAL RUBBER	BEST 65NFW	>480m	0	0	n.a.		53
g NATURAL RUBBER	COMASEC FLEXIGUM	90m	540	3	0.95		40
g NATURAL RUBBER	MAPA-PIONEER L-118	>480m	0	0	0.46		36
g NATURAL RUBBER	MARIGOLD BLACK HEVY	15m	117	3	0.65		79
g NATURAL RUBBER	MARIGOLD FEATHERLIT	1m	217	4	0.15		79
g NATURAL RUBBER	MARIGOLD MED WT EXT	6m	120	4	0.45		79
g NATURAL RUBBER	MARIGOLD MEDICAL	2m	177	4	0.28		79
g NATURAL RUBBER	MARIGOLD ORANGE SUP	16m	117	3	0.73		79
g NATURAL RUBBER	MARIGOLD R SURGEONS	2m	177	4	0.28		79
g NATURAL RUBBER	MARIGOLD RED LT WT	5m	128	4	0.43		79
g NATURAL RUBBER	MARIGOLD SENSOTECH	2m	177	4	0.28		79
g NATURAL RUBBER	MARIGOLD SUREGRIP	12m	118	4	0.58		79
g NATURAL RUBBER	MARIGOLD YELLOW	4m	144	4	0.38		79
g NATURAL+NEOPRENE	ANSELL OMNI 276	18m	0.2		0.56		55
g NATURAL+NEOPRENE	ANSELL TECHNICIANS	10m	0.3		0.43		55
g NATURAL+NEOPRENE	MARIGOLD FEATHERWT	4m	152	4	0.35		79
g NATURAL+NEOPRENE	PLAYTEX ARGUS 123	22m	<60	3	n.a.		72
g NEOPRENE	ANSELL NEOPRENE 520	550m	0	0	0.38		58
g NEOPRENE	ANSELL NEOPRENE 530	90m	0.06		0.46		55
g NEOPRENE	ANSELLEDMONT 29-840	90m	<90	2	0.31		6
g NEOPRENE	ANSELLEDMONT 29-870	317m	6	1	0.50		34
g NEOPRENE	ANSELLEDMONT 29-870	520m	0	0	0.46		58
g NEOPRENE	ANSELLEDMONT NEOX	180m	<90	2	n.a.		6
g NEOPRENE	BEST 32	71m	230	3	n.a.		53
g NEOPRENE	BEST 6780	400m	610	3	n.a.		53
g NEOPRENE	BEST 723	125m			0.64		58
g NEOPRENE	BRUNSWICK NEOPRENE	>480m	0	0	0.88		89
g NEOPRENE	COMASEC COMAPRENE	120m	30	2	n.a.		40
g NEOPRENE	GRANET 2001	95m			0.55		58
g NEOPRENE	MAPA-PIONEER N-36	750m	0	0	0.54		58
g NEOPRENE	MAPA-PIONEER N-44	>480m	0	0	0.56		36
g NEOPRENE	MAPA-PIONEER N-44	>480m	0	0	0.74		36
g NEOPRENE/NATURAL	ANSELL CHEMI-PRO	21m	0.1		0.72		55
g NITRILE	ANSELL CHALLENGER	120m	0.1		0.38		55
g NITRILE	ANSELLEDMONT 37-155	>60m			0.35		5
g NITRILE	ANSELLEDMONT 37-165	240m	<90	2	0.60		6
g NITRILE	BEST 22R	>480m	0	0	n.a.		53
g NITRILE	BEST 727	225m	2160	4	0.38		53
g NITRILE	COMASEC COMATRIL	>480m	0	0	0.55		40
g NITRILE	COMASEC COMATRIL SU	>480m	0	0	0.60		40
g NITRILE	COMASEC FLEXITRIL	120m	96	2	n.a.		40
g NITRILE	MAPA-PIONEER A-14	>480m	0	0	0.56		36
g NITRILE	MARIGOLD BLUE	260m	42	2	0.45	degrad.	79
g NITRILE	MARIGOLD GREEN SUPA	169m	31	2	0.30	degrad.	79

G A R M E N T Class & Number/ test chemical/ MATERIAL NAME	MANUFACTURER & PRODUCT IDENTIFICATION	Break- through Time in Minutes	Perm- eation Rate in mg/sq m /min.	I N D E X	Thick- ness in mm	degrad. and comments	R e f s
g NITRILE	MARIGOLD NITROSOLVE	278m	44	2	0.75	degrad.	79
g NITRILE+PVC	COMASEC MULTIPLUS	>480m	0	0	n.a.		40
g PE	HANDGARDS	>60m			0.07		5
g PE/EVAL/PE	SAFETY4 4H	>240m	0	0	0.07	35°C	60
g PE/EVAL/PE	SAFETY4 4H	>480m	0	0	0.07		60
g PVAL	ANSELLEDMONT PVA				n.a.	degrad.	6
g PVAL	EDMONT 15-552				n.a.	degrad.	6
g PVAL	EDMONT 15-552	100m	55	2	n.a.	degrad.	33
g PVAL	EDMONT 25-950				0.35	degrad.	34
g PVC	ANSELLEDMONT 34-100	4m	480	4	0.15		34
g PVC	ANSELLEDMONT SNORKE	60m	<90	2	n.a.		6
g PVC	BEST 725R	66m	40	2	n.a.		53
g PVC	COMASEC MULTIPOST	>480m	0	0	n.a.		40
g PVC	COMASEC MULTITOP	>480m	0	0	n.a.		40
g PVC	COMASEC NORMAL	150m	6	1	n.a.		40
g PVC	COMASEC OMNI	75m	60	2	n.a.		40
g PVC	MAPA-PIONEER V-20	20m	280	3	0.31		5
s UNKNOWN MATERIAL	CHEMRON CHEMREL MAX	>1440m	0	0	n.a.		67
g UNKNOWN MATERIAL	MARIGOLD R MEDIGLOV	1m	193	4	0.03		79

ethyl alcohol, 50% & methyl acetate, 50%
Primary Class: 311   Hydroxy Compounds, Aliphatic and Alicyclic, Primary

ethyl bromide (synonym: bromoethane)
CAS Number: 74-96-4
Primary Class: 261   Halogen Compounds, Aliphatic and Alicyclic

g BUTYL	NORTH B-174				0.63	degrad.	34
g NATURAL RUBBER	ACKWELL 5-109				0.15	degrad.	34
g NEOPRENE	ANSELLEDMONT 29-870	4m	13200	5	0.41	degrad.	34
g NITRILE	ANSELLEDMONT 37-155				0.40	degrad.	34
g PVAL	EDMONT 25-950	64m	5	1	0.58		34
g PVC	ANSELLEDMONT 34-100	1m	8890	5	0.20	degrad.	34
g VITON	NORTH F-091	86m	314	3	0.26		34

ethyl ether (synonym: diethyl ether)
CAS Number: 60-29-7
Primary Class: 241   Ethers, Aliphatic and Alicyclic

g BUTYL	ANSELL LAMPRECHT	19m	5853	4	0.70		38
g BUTYL	BEST 878	19m	2140	4	0.75		53
g BUTYL	NORTH B-161	8m	5532	5	0.45		7
g NATURAL RUBBER	ANSELLEDMONT 36-124				0.46	degrad.	6
g NATURAL RUBBER	ANSELLEDMONT 392				0.48	degrad.	6
g NATURAL RUBBER	COMASEC FLEXIGUM	10m	15600	5	0.95		40
g NATURAL RUBBER	MARIGOLD BLACK HEVY	3m	24273	5	0.65	degrad.	79

G A R M E N T	Class & Number/ test chemical/ MATERIAL NAME	MANUFACTURER & PRODUCT IDENTIFICATION	Break-through Time in Minutes	Perm-eation Rate in mg/sq m /min.	I N D E X	Thick-ness in mm	degrad. and comments	R e f s
g	NATURAL RUBBER	MARIGOLD FEATHERLIT	1m	5571	5	0.15	degrad.	79
g	NATURAL RUBBER	MARIGOLD MED WT EXT	1m	57000	5	0.45	degrad.	79
g	NATURAL RUBBER	MARIGOLD MEDICAL	1m	27000	5	0.28	degrad.	79
g	NATURAL RUBBER	MARIGOLD MEDICAL	1m	27000	5	0.28	degrad.	79
g	NATURAL RUBBER	MARIGOLD ORANGE SUP	3m	21000	5	0.73	degrad.	79
g	NATURAL RUBBER	MARIGOLD RED LT WT	1m	52714	5	0.43	degrad.	79
g	NATURAL RUBBER	MARIGOLD SENSOTECH	1m	27000	5	0.28	degrad.	79
g	NATURAL RUBBER	MARIGOLD SUREGRIP	2m	34091	5	0.58	degrad.	79
g	NATURAL RUBBER	MARIGOLD YELLOW	1m	44143	5	0.38	degrad.	79
u	NATURAL RUBBER	Unknown	2m	33720	5	0.64		38
g	NATURAL+NEOPRENE	MARIGOLD FEATHERWT	1m	39857	5	0.35	degrad.	79
g	NEOPRENE	ANSELL NEOPRENE 530	2m	5	3	0.46		55
g	NEOPRENE	ANSELLEDMONT 29-840	10m	<900	4	0.38		6
g	NEOPRENE	ANSELLEDMONT NEOX	10m	<900	4	n.a.		6
g	NEOPRENE	BEST 32	9m	2180	5	n.a.		53
g	NEOPRENE	BEST 6780	14m	1750	5	n.a.		53
g	NEOPRENE	COMASEC COMAPRENE	2m	12300	5	n.a.		40
g	NEOPRENE	MAPA-PIONEER N-44	18m	4860	4	0.74		36
g	NEOPRENE/NATURAL	ANSELL CHEMI-PRO	2m	9	3	0.72		55
g	NITRILE	ANSELL CHALLENGER	21m	2	2	0.38		55
g	NITRILE	ANSELLEDMONT 37-165	120m	<900	3	0.60		6
g	NITRILE	ANSELLEDMONT 37-175	39m	965	3	0.42		38
g	NITRILE	BEST 22R	8m	1650	5	n.a.		53
g	NITRILE	BEST 727	33m	360	3	0.38		53
g	NITRILE	COMASEC COMATRIL	95m	960	3	0.55		40
g	NITRILE	COMASEC COMATRIL SU	150m	840	3	0.57		40
g	NITRILE	COMASEC FLEXITRIL	8m	14400	5	n.a.		40
g	NITRILE	MAPA-PIONEER A-14	64m	780	3	0.54		36
g	NITRILE	MARIGOLD BLUE	62m	409	3	0.45	degrad.	79
g	NITRILE	MARIGOLD GREEN SUPA	30m	210	3	0.30	degrad.	79
g	NITRILE	MARIGOLD NITROSOLVE	69m	499	3	0.75	degrad.	79
g	NITRILE	NORTH LA-142G	14m	1308	5	0.36		7
g	NITRILE+PVC	COMASEC MULTIPLUS	25m	19800	5	n.a.		40
g	PE/EVAL/PE	SAFETY4 4H	>240m		0	0.07	35°C	60
g	PE/EVAL/PE	SAFETY4 4H	>480m	0	0	0.07		60
g	PVAL	ANSELLEDMONT PVA	>360m		0	n.a.		6
g	PVAL	COMASEC SOLVATRIL	>480m	0	0	n.a.		40
g	PVAL	EDMONT 15-552	>360m	0	0	n.a.		6
g	PVAL	EDMONT 25-545	150m	<9	1	n.a.		6
g	PVAL	EDMONT 25-950	>480m	0	0	0.33		34
g	PVC	ANSELLEDMONT SNORKE				n.a.	degrad.	6
g	PVC	COMASEC MULTIPOST	25m	19800	5	n.a.		40
g	PVC	COMASEC MULTITOP	20m	19320	5	n.a.		40
g	PVC	COMASEC NORMAL	20m	21000	5	n.a.		40
g	PVC	COMASEC OMNI	10m	21780	5	n.a.		40
s	SARANEX-23	DU PONT TYVEK SARAN	1m	18	4	n.a.		8
s	TEFLON	CHEMFAB CHALL. 5100	>480m	0	0	n.a.		56
s	TEFLON	CHEMFAB CHALL. 5100	>180m	0	1	n.a.		65

G A R M E Class & Number/ N test chemical/ T MATERIAL NAME	MANUFACTURER & PRODUCT IDENTIFICATION	Break-through Time in Minutes	Perm-eation Rate in mg/sq m /min.	I N D E X	Thick-ness in mm	degrad. and comments	R e f s
s UNKNOWN MATERIAL	CHEMRON CHEMREL	1m	3	3	n.a.		67
s UNKNOWN MATERIAL	DU PONT BARRICADE	>480m	0	0	n.a.		8
s UNKNOWN MATERIAL	LIFE-GUARD RESPONDE	>240m	0	0	n.a.		77
g UNKNOWN MATERIAL	NORTH SILVERSHIELD	>360m	0	0	0.08		7
g VITON	NORTH F-091	12m	1290	5	0.28		34
g VITON	NORTH F-091	16m	2285	4	0.26		38
u VITON/CHLOROBUTYL	Unknown	5m			n.a.		46
g VITON/NEOPRENE	ERISTA VITRIC	16m	5978	4	0.60		38

ethyl methacrylate (synonym: ethyl 2-methylpropenoate)
CAS Number: 97-63-2
Primary Class: 223   Esters, Carboxylic, Acrylates and Methacrylates

g BUTYL	NORTH B-174	395m	120	3	0.70	degrad.	34
u CPE	ILC DOVER	32m			0.05		29
g NATURAL RUBBER	ACKWELL 5-109				0.15	degrad.	34
g NEOPRENE	ANSELLEDMONT 29-870				0.48	degrad.	34
g NITRILE	ANSELLEDMONT 37-155	23m	1860	4	0.43	degrad.	34
g PVAL	EDMONT 25-950	>480m	0	0	0.58		34
g PVC	ANSELLEDMONT 34-100	2m	840	4	0.20	degrad.	34
s UNKNOWN MATERIAL	LIFE-GUARD RESPONDE	>240m	0	0	n.a.		77
g VITON	NORTH F-091				0.25	degrad.	34
u VITON/CHLOROBUTYL	Unknown	30m			0.36		29

ethyl parathion, 30-70% (synonym: parathion)
CAS Number: 56-38-2
Primary Class: 462   Organophosphorus Compounds and Derivatives of Phosphorus-based Acids
Related Class: 233   Esters, Non-Carboxylic, Carbamates and Others
              840   Pesticides, Mixtures and Formulations

s TEFLON	CHEMFAB CHALL. 5100	>180m	0	1	n.a.		65

ethyl parathion, <30% (synonym: parathion <30%)
Primary Class: 462   Organophosphorus Compounds and Derivatives of Phosphorus-based Acids

g NATURAL RUBBER	ANSELL 9070-5	<15m	44	3	0.10	3%/corn oil	34
g NEOPRENE	ANSELLEDMONT 29-870	>480m	0	0	0.48	3%/corn oil	34
g NITRILE	ANSELLEDMONT 37-175	>480m	0	0	0.65	3%/corn oil	34
g PVC	ANSELLEDMONT 34-100	90m	9	1	0.19	3%/corn oil	34

ethyl vinyl ether (synonym: vinyl ethyl ether)
CAS Number: 109-92-2
Primary Class: 246   Ethers, Vinyl

s UNKNOWN MATERIAL	LIFE-GUARD RESPONDE	>180m	0	1	n.a.		77

G A R M E Class & Number/ N test chemical/ T MATERIAL NAME	MANUFACTURER & PRODUCT IDENTIFICATION	Break- through Time in Minutes	Perm- eation Rate in mg/sq m /min.	I N D E X	Thick- ness in mm	degrad. and comments	R e f s
ethyl-n-butylamine (synonym: butylethylamine)							
CAS Number: 617-79-8							
Primary Class: 142 Amines, Aliphatic and Alicyclic, Secondary							
g BUTYL	NORTH B-174				0.66	degrad.	34
g NATURAL RUBBER	ACKWELL 5-109				0.15	degrad.	34
g NEOPRENE	ANSELLEDMONT 29-870				0.48	degrad.	34
g NITRILE	ANSELLEDMONT 37-155	72m	1289	4	0.36	degrad.	34
g PVAL	EDMONT 25-950	402m	135	3	0.56		34
g PVC	ANSELLEDMONT 34-100	4m	15834	5	0.20	degrad.	34
g VITON	NORTH F-091	228m	8880	4	0.25		34
ethylamine (synonym: monoethylamine)							
CAS Number: 75-04-7							
Primary Class: 141 Amines, Aliphatic and Alicyclic, Primary							
g BUTYL	BRUNSWICK BUTYL STD	>480m	0	0	0.63		89
g BUTYL	BRUNSWICK BUTYL-XTR	>480m	0	0	0.63		89
g BUTYL	NORTH B-174	>480m	0	0	0.59		34
g BUTYL/ECO	BRUNSWICK BUTYL-POL	>480m	0	0	0.63		89
g NATURAL RUBBER	ACKWELL 5-109	<1m	2640	5	0.15		34
g NEOPRENE	ANSELLEDMONT 29-870	104m	180	3	0.56		34
g PVAL	EDMONT 25-950				0.35	degrad.	34
g PVC	ANSELLEDMONT 34-100	1m	1260	5	0.18		34
g VITON	NORTH F-091				0.24	degrad.	34
ethylamine, 30-70% (synonym: monoethylamine, 30-70%)							
Primary Class: 141 Amines, Aliphatic and Alicyclic, Primary							
g BUTYL	NORTH B-161	>720m	0	0	0.45	70% soln.	7
g NITRILE	NORTH LA-142G	66m	1806	4	0.36	70% soln.	7
s TEFLON	CHEMFAB CHALL. 5100	>180m	0	1	n.a.		65
s UNKNOWN MATERIAL	LIFE-GUARD RESPONDE	>240m	0	0	n.a.	70% in H_2O	77
g UNKNOWN MATERIAL	NORTH SILVERSHIELD	28m	360	3	0.08	70% soln.	7
s VITON/BUTYL	DRAGER 500 OR 710	>30m			n.a.	70% in H_2O	91
ethylbenzene							
CAS Number: 100-41-4							
Primary Class: 292 Hydrocarbons, Aromatic							
g BUTYL	BEST 878				0.75	degrad.	53
g BUTYL	NORTH B-174	34m	4931	4	0.51	degrad.	34
g BUTYL/ECO	BRUNSWICK BUTYL-POL	36m	>200	3	0.63		89
g NATURAL RUBBER	ACKWELL 5-109				0.15	degrad.	34
g NATURAL RUBBER	BEST 65NFW				n.a.	degrad.	53
g NEOPRENE	ANSELLEDMONT 29-870				0.48	degrad.	34

G A R M E N T	Class & Number/ test chemical/ MATERIAL NAME	MANUFACTURER & PRODUCT IDENTIFICATION	Break- through Time in Minutes	Perm- eation Rate in mg/sq m /min.	I N D E X	Thick- ness in mm	degrad. and comments	R e f s
g	NEOPRENE	BEST 32				n.a.	degrad.	53
g	NEOPRENE	BEST 6780				n.a.	degrad.	53
g	NITRILE	ANSELLEDMONT 37-155	24m	1915	4	0.38	degrad.	34
g	NITRILE	BEST 727	43m	1240	4	0.38	degrad.	53
g	PVAL	EDMONT 25-950	33m			0.50		34
g	PVC	ANSELLEDMONT 34-100				0.20	degrad.	34
s	SARANEX-23	DU PONT TYVEK SARAN	45m	180	3	n.a.		8
s	TEFLON	CHEMFAB CHALL. 5100	>180m	0	1	n.a.		65
s	UNKNOWN MATERIAL	CHEMRON CHEMREL MAX	>1440m	0	0	n.a.		67
s	UNKNOWN MATERIAL	DU PONT BARRICADE	>480m	0	0	n.a.		8
s	UNKNOWN MATERIAL	LIFE-GUARD RESPONDE	>480m	0	0	n.a.		77
g	VITON	NORTH F-091	>480m	0	0	0.25		34

ethylene dibromide (synonym: 1,2-dibromoethane)
CAS Number: 106-93-4
Primary Class: 261   Halogen Compounds, Aliphatic and Alicyclic

	Class & Number/ test chemical/ MATERIAL NAME	MANUFACTURER & PRODUCT IDENTIFICATION	Break- through Time	Perm- eation Rate	I N D E X	Thick- ness	degrad. and comments	Refs
s	BUTYL	MSA CHEMPRUF	<60m			n.a.		48
g	BUTYL	NORTH B-161	115m	770	3	0.38		25
g	BUTYL	NORTH B-174	210m	384	3	0.50	degrad.	34
s	BUTYL/NEOPRENE	MSA BETEX	15m	233	3	n.a.		4
u	CPE	ILC DOVER	44m			0.05		29
g	NATURAL RUBBER	ACKWELL 5-109				0.15	degrad.	34
u	NATURAL RUBBER	Unknown	<1m	>7300	5	0.23		25
g	NEOPRENE	ANSELLEDMONT 29-870	10m	>7300	5	0.43		25
g	NEOPRENE	ANSELLEDMONT 29-870				0.48	degrad.	34
g	NEOPRENE	BEST 6780		3540	5	n.a.		53
g	NITRILE	ANSELLEDMONT 37-155				0.40	degrad.	34
g	NITRILE	BEST 22R		5820	5	n.a.		53
g	NITRILE	BEST 727		4080	5	n.a.		53
g	NITRILE	SURETY	31m	>7300	4	0.38		25
g	PE	ANSELLEDMONT 35-125	<2m	1500	5	0.05		25
g	PVAL	EDMONT 15-552	>1440m	0	0	0.46		25
g	PVAL	EDMONT 15-552	>1440m	0	0	0.46	preexp. 24h	25
g	PVAL	EDMONT 25-950	>480m	0	0	0.50		34
g	PVC	ANSELLEDMONT 34-100	2m	14040	5	0.20	degrad.	34
g	PVC	BEST 812		2940	5	n.a.		53
s	PVC	MSA UPC	<3m			0.20		48
s	SARANEX-23	DU PONT TYVEK SARAN	9m	490	4	0.15		25
s	TEFLON	CHEMFAB CHALL. 5100	>200m	0	1	n.a.		65
g	TEFLON	CLEAN ROOM PRODUCTS	>1440m	0	0	0.05		25
g	VITON	NORTH F-091	>1440m	0	0	0.23		25
g	VITON	NORTH SF	>1440m	0	0	0.23		25
u	VITON/CHLOROBUTYL	Unknown	>180m			0.36		29
s	VITON/NEOPRENE	MSA VAUTEX	>240m	0	0	n.a.		48
g	VITON/NITRILE	NORTH VITRILE	>1440m	0	0	0.20		25
g	VITON/NITRILE	NORTH VITRILE	>1440m	0	0	0.20	preexp. 24h	25

G A R M E Class & Number/ N test chemical/ T MATERIAL NAME	MANUFACTURER & PRODUCT IDENTIFICATION	Break- through Time in Minutes	Perm- eation Rate in mg/sq m /min.	I N D E X	Thick- ness in mm	degrad. and comments	R e f s
ethylene dichloride (synonyms: 1,2-dichloroethane; EDC)							
CAS Number: 107-06-2							
Primary Class: 261 Halogen Compounds, Aliphatic and Alicyclic							
g BUTYL	BEST 878	69m	320	3	0.75		53
g BUTYL	NORTH B-161	140m			0.56	degrad.	4
g BUTYL	NORTH B-161	175m	3180	4	0.43		7
g BUTYL	NORTH B-161	67m	520	3	0.46		92
g BUTYL	NORTH B-161	30m	1200	4	0.46	35°C	92
g BUTYL	NORTH B-161	16m	2850	4	0.46	50°C	92
g BUTYL	NORTH B-174	180m	3204	4	0.62	degrad.	34
u CPE	ILC DOVER	15m			n.a.		29
g NATURAL RUBBER	ACKWELL 5-109	<1m	100	5	0.15	degrad.	34
g NATURAL RUBBER	ANSELL PVL 040				0.46	degrad.	55
g NATURAL RUBBER	ANSELLEDMONT 36-124				0.46	degrad.	6
g NATURAL RUBBER	ANSELLEDMONT 392				0.48	degrad.	6
g NATURAL+NEOPRENE	ANSELL OMNI 276	5m	13000	5	0.45		5
g NATURAL+NEOPRENE	ANSELL OMNI 276				0.56	degrad.	55
g NEOPRENE	ANSELLEDMONT 29-840				0.38	degrad.	6
g NEOPRENE	ANSELLEDMONT 29-870				0.48	degrad.	34
g NEOPRENE	ANSELLEDMONT NEOX	2m	7000	5	n.a.		5
g NEOPRENE	BEST 32	16m	4400	4	n.a.	degrad.	53
g NEOPRENE	BEST 6780	11m	3120	5	n.a.	degrad.	53
g NEOPRENE	MAPA-PIONEER N-44	33m	14820	5	0.74	degrad.	36
g NEOPRENE	MAPA-PIONEER N-73	8m	10000	5	0.46		5
u NEOPRENE	Unknown	20m			0.58	degrad.	4
g NEOPRENE/NATURAL	ANSELL CHEMI-PRO				0.76	degrad.	55
g NITRILE	ANSELL CHALLENGER				0.38	degrad.	55
g NITRILE	ANSELLEDMONT 37-155				0.40	degrad.	34
g NITRILE	ANSELLEDMONT 37-165				0.60	degrad.	6
g NITRILE	ANSELLEDMONT 37-175	7m	39000	5	0.37		5
g NITRILE	BEST 727	6m	3310	5	0.38	degrad.	53
g NITRILE	MAPA-PIONEER A-14	16m	17520	5	0.54	degrad.	36
g NITRILE	NORTH LA-142G	8m	18660	5	0.36		7
u NITRILE	Unknown	2m			0.20		4
u NITRILE	Unknown	2m			0.30	degrad.	4
g PE	ANSELLEDMONT 35-125	1m	100	4	0.03		5
u PE	Unknown	2m			0.05		4
g PE/EVAL/PE	SAFETY4 4H	>240m		0	0.07	35°C	60
g PVAL	ANSELLEDMONT PVA	>360m		0	n.a.		6
g PVAL	EDMONT 15-552	22m			n.a.		4
g PVAL	EDMONT 15-552	>180m	<90	2	n.a.		6
g PVAL	EDMONT 25-545	60m	<9	1	n.a.		6
g PVAL	EDMONT 25-950	>480m	0	0	0.38		34
g PVC	ANSELLEDMONT 34-100				0.20	degrad.	34
g PVC	ANSELLEDMONT SNORKE				1.40	degrad.	6
g PVC	MAPA-PIONEER V-20	1m	69000	5	0.31		5
s TEFLON	CHEMFAB CHALL. 5100	>340m	0	0	n.a.		65

G A R M E Class & Number/ N test chemical/ T MATERIAL NAME	MANUFACTURER & PRODUCT IDENTIFICATION	Break-through Time in Minutes	Perm-eation Rate in mg/sq m /min.	I N D E X	Thick-ness in mm	degrad. and comments	R e f s
g TEFLON	CLEAN ROOM PRODUCTS	>1440m		0	0.05		4
g TEFLON	CLEAN ROOM PRODUCTS	90m			0.05	crumpled	4
s UNKNOWN MATERIAL	DU PONT BARRICADE	>480m	0	0	n.a.		8
s UNKNOWN MATERIAL	LIFE-GUARD RESPONDE	>480m	0	0	n.a.		77
g UNKNOWN MATERIAL	NORTH SILVERSHIELD	>360m	0	0	0.08		7
g VITON	NORTH F-091	820m	0	0	0.25		4
g VITON	NORTH F-091	415m	49	2	0.24		7
g VITON	NORTH F-091	>480m	0	0	0.24		34

ethylene glycol (synonym: 1,2-ethanediol)
CAS Number: 107-21-1
Primary Class: 314   Hydroxy Compounds, Aliphatic and Alicyclic, Polyols

g BUTYL	BEST 878	>480m	0	0	0.75		53
g NATURAL RUBBER	ANSELL CONFORM 4205	>360m	0	0	0.13		55
g NATURAL RUBBER	ANSELL FL 200 254	>360m	0	0	0.51		55
g NATURAL RUBBER	ANSELL ORANGE 208	>360m	0	0	0.76		55
g NATURAL RUBBER	ANSELL PVL 040	>360m	0	0	0.46		55
g NATURAL RUBBER	ANSELL STERILE 832	>360m	0	0	0.23		55
g NATURAL RUBBER	ANSELLEDMONT 36-124	>360m	0	0	0.46		6
g NATURAL RUBBER	ANSELLEDMONT 392	>360m	0	0	0.48		6
g NATURAL RUBBER	COMASEC FLEXIGUM	>480m	0	0	0.95		40
g NATURAL RUBBER	MAPA-PIONEER L-118	>480m	0	0	0.46		36
g NATURAL RUBBER	MARIGOLD BLACK HEVY	>480m	0	0	0.65		79
g NATURAL RUBBER	MARIGOLD FEATHERLI	>480m	0	0	0.15		79
g NATURAL RUBBER	MARIGOLD MED WT EXT	>480m	0	0	0.45		79
g NATURAL RUBBER	MARIGOLD MEDICAL	>480m	0	0	0.28		79
g NATURAL RUBBER	MARIGOLD ORANGE SUP	>480m	0	0	0.73		79
g NATURAL RUBBER	MARIGOLD R SURGEONS	>480m	0	0	0.28		79
g NATURAL RUBBER	MARIGOLD RED LT WT	>480m	0	0	0.43		79
g NATURAL RUBBER	MARIGOLD SENSOTECH	>480m	0	0	0.28		79
g NATURAL RUBBER	MARIGOLD SUREGRIP	>480m	0	0	0.58		79
g NATURAL RUBBER	MARIGOLD YELLOW	>480m	0	0	0.38		79
g NATURAL+NEOPRENE	ANSELL OMNI 276	>360m	0	0	0.56		55
g NATURAL+NEOPRENE	ANSELL TECHNICIANS	>60m			0.45		5
g NATURAL+NEOPRENE	ANSELL TECHNICIANS	>360m	0	0	0.43		55
g NATURAL+NEOPRENE	MARIGOLD FEATHERWT	>480m	0	0	0.35		79
g NEOPRENE	ANSELL NEOPRENE 530	>360m	0	0	0.46		55
g NEOPRENE	ANSELLEDMONT 29-840	>360m	0	0	0.38		6
g NEOPRENE	ANSELLEDMONT NEOX	>360m	0	0	n.a.		6
g NEOPRENE	BEST 32	>480m	0	0	n.a.		53
g NEOPRENE	BEST 6780	>480m	0	0	n.a.		53
g NEOPRENE	COMASEC COMAPRENE	>360m	0	0	n.a		40
g NEOPRENE	MAPA-PIONEER N-44	>480m	0	0	0.56		36
g NEOPRENE/NATURAL	ANSELL CHEMI-PRO	>360m	0	0	0.72		55
g NITRILE	ANSELL CHALLENGER	>360m	0	0	0.38		55
g NITRILE	ANSELLEDMONT 37-155	>60m			0.35		5
g NITRILE	ANSELLEDMONT 37-165	>360m	0	0	0.60		6

G A R M E Class & Number/ N test chemical/ T MATERIAL NAME	MANUFACTURER & PRODUCT IDENTIFICATION	Break-through Time in Minutes	Perm-eation Rate in mg/sq m /min.	I N D E X	Thick-ness in mm	degrad. and comments	R e f s
g NITRILE	BEST 22R	>480m	0	0	n.a.		53
g NITRILE	BEST 727	>480m	0	0	0.38		53
g NITRILE	COMASEC COMATRIL	>480m	0	0	0.55		40
g NITRILE	COMASEC COMATRIL SU	>480m	0	0	0.60		40
g NITRILE	COMASEC FLEXITRIL	>480m	0	0	n.a.		40
g NITRILE	MAPA-PIONEER A-14	>480m	0	0	0.56		36
g NITRILE	MARIGOLD BLUE	>480m	0	0	0.45		79
g NITRILE	MARIGOLD GREEN SUPA	>480m	0	0	0.30		79
g NITRILE	MARIGOLD NITROSOLVE	>480m	0	0	0.75		79
g NITRILE+PVC	COMASEC MULTIPLUS	>480m	0	0	n.a.		40
g PE	ANSELLEDMONT 35-125	>60m			0.03		5
c PE	DU PONT TYVEK QC	>480m	0	0	n.a.		8
g PE/EVAL/PE	SAFETY4 4H	>240m	0	0	0.07	35°C	60
g PVAL	ANSELLEDMONT PVA	120m	<90	2	n.a.		6
g PVAL	EDMONT 15-552	120m	<90	2	n.a.		6
g PVAL	EDMONT 25-545	150m	<90	2	n.a.		6
g PVC	ANSELLEDMONT SNORKE	>360m	0	0	n.a.		6
g PVC	COMASEC MULTIPOST	>480m	0	0	n.a.		40
g PVC	COMASEC MULTITOP	>480m	0	0	n.a.		40
g PVC	COMASEC NORMAL	>480m	0	0	n.a.		40
g PVC	COMASEC OMNI	>480m	0	0	n.a.		40
g PVC	MAPA-PIONEER V-20	>60m			0.31		5
g PVC	MAPA-PIONEER V-20	>480m	0	0	0.51		36
s SARANEX-23	DU PONT TYVEK SARAN	>480m	0	0	n.a.		8
s TEFLON	CHEMFAB CHALL. 5100	>1000m	0	0	n.a.		65
s UNKNOWN MATERIAL	DU PONT BARRICADE	>480m	0	0	n.a.		8
s UNKNOWN MATERIAL	LIFE-GUARD RESPONDE	>240m	0	0	n.a.		77

ethylene glycol dimethyl ether (synonym: 1,2-dimethoxyethane)
CAS Number: 110-71-4
Primary Class: 245   Ethers, Glycols

g BUTYL	NORTH B-161	125m		0.51		78
g BUTYL	NORTH B-161	155m		0.56	pre-exp 16h	78
g BUTYL	NORTH B-161	105m		0.56	exp 2 x 16h	78
g BUTYL	NORTH B-161	120m		0.56	exp 16h 50C	78
g BUTYL	NORTH B-161	145m		0.56	exp 2 x 16h	78
g BUTYL	NORTH B-174	375m	0	0.77		78
g HYPALON	NORTH Y-1532	<15m		0.43		78
g NEOPRENE	ANSELLEDMONT NEOX	20m		n.a.		78
u NEOPREN	Unknown	<15m		0.43		78
u NEOPRENE	Unknown	30m		0.81		78
g NITRILE	ANSELLEDMONT 37-165	<30m		0.51		78
g PVAL	EDMONT 15-554	<120m		n.a.		78
g PVC	AMERICAN SCIENTIFIC			0.17	degrad.	78
g VITON	NORTH F-091	<30m		0.25		78

G A R M E N T	Class & Number/ test chemical/ MATERIAL NAME	MANUFACTURER & PRODUCT IDENTIFICATION	Break- through Time in Minutes	Perm- eation Rate in mg/sq m /min.	I N D E X	Thick- ness in mm	degrad. and comments	R e f s

ethylene oxide (synonym: oxirane)
CAS Number: 75-21-8
Primary Class: 275   Heterocyclic Compounds, Oxygen, Epoxides

G A R M E N T	Class & Number/ test chemical/ MATERIAL NAME	MANUFACTURER & PRODUCT IDENTIFICATION	Break- through Time in Minutes	Perm- eation Rate in mg/sq m /min.	I N D E X	Thick- ness in mm	degrad. and comments	R e f s
g	BUTYL	BRUNSWICK BUTYL STD	173m	35	2	0.63		89
g	BUTYL	BRUNSWICK BUTYL-XTR	346m	180	3	0.70		89
s	BUTYL	LIFE-GUARD BUTYL	48m	4	2	n.a.		84
g	BUTYL	NORTH B-131	>480m	0	0	0.35		84
s	BUTYL	WHEELER ACID KING	55m	<0.9	2	n.a.		85
s	BUTYL	WHEELER ACID KING	85m	0.9	1	n.a.	100%	88
s	BUTYL	WHEELER ACID KING	400m	0	0	n.a.	0.2%	88
g	BUTYL/ECO	BRUNSWICK BUTYL-POL	268m	380	3	0.63		89
s	BUTYL/NEOPRENE	MSA BETEX	165m	1.4	1	n.a.	100%	88
s	BUTYL/NEOPRENE	MSA BETEX	>400m	0	0	n.a.	0.2%	88
s	CPE	ILC DOVER CLOROPEL	118m	201	3	n.a.	100%	88
s	CPE	ILC DOVER CLOROPEL	>375m	0	0	n.a.	0.2%	88
g	NATURAL RUBBER	DAYTON SURGICAL	3m	1347	5	0.16	100%	88
g	NATURAL RUBBER	DAYTON SURGICAL	5m	301	4	0.16	0.2%	88
g	NEOPRENE	BRUNSWICK NEOPRENE	40m	>5000	4	0.90		89
s	NEOPRENE	FAIRPRENE	51m	128	3	n.a.	100%	88
s	NEOPRENE	FAIRPRENE	158m	5.6	1	n.a.	0.2%	88
g	NEOPRENE	MAPA-PIONEER N-44	31m	600	3	0.74		36
g	NITRILE	BEST 727		4	4	n.a.		53
g	NITRILE	MAPA-PIONEER A-14	32m	1260	4	0.54		36
g	NITRILE	MAPA-PIONEER A-15	195m	115	3	0.35	100%	88
g	NITRILE	MAPA-PIONEER A-15	>315m	0	0	0.35	0.2%	88
c	PE	DU PONT TYVEK QC	<1m	180	4	n.a.		8
m	POLYURETHANE	ILC DOVER	29m	494	3	0.43	100%	88
m	POLYURETHANE	ILC DOVER	65m	0.7	1	0.43	0.2%	88
v	PVC	LIFE-GUARD VISOR	158m	17	2	n.a.		84
s	PVC	WHEELER ACID KING	44m	<9000	4	n.a.		85
s	PVC	WHEELER ACID KING	13m	1080	5	n.a.	100%	88
s	PVC	WHEELER ACID KING	31m	348	3	n.a.	0.2%	88
s	SARANEX-23	DU PONT TYVEK SARAN	6m	84	4	n.a.		8
s	SARANEX-23	DU PONT TYVEK SARAN	121m	60	2	n.a.	100%	88
s	SARANEX-23	DU PONT TYVEK SARAN	>400m	0	0	n.a.	0.2%	88
s	SARANEX-23 2-PLY	DU PONT TYVEK SARAN	55m	79	3	n.a.		8
s	TEFLON	CHEMFAB CHALL. 5100	>950m	0	0	n.a.	100%	88
s	TEFLON	CHEMFAB CHALL. 5200	64m	7.6	1	n.a.		80
s	TEFLON	CHEMFAB CHALL. 5200	31m	0.33	2	n.a.		84
s	TEFLON	LIFE-GUARD TEFGUARD	34m	0.04	2	n.a.		77
s	TEFLON	WHEELER ACID KING	71m	<9	1	n.a.		85
v	TEFLON-FEP	DU PONT	160m	0.04	1	n.a.		84
v	TEFLON-FEP/PC	CHEMRON	353m	0.03	0	n.a.		84
s	UNKNOWN MATERIAL	CHEMRON CHEMREL	>1440m	0	0	n.a.		67
s	UNKNOWN MATERIAL	CHEMRON CHEMREL	>1050m	0	0	n.a.	100%	88
s	UNKNOWN MATERIAL	CHEMRON CHEMREL MAX	>480m	0	0	n.a.		84
s	UNKNOWN MATERIAL	DU PONT BARRICADE	>480m	0	0	n.a.		8

G A R M E Class & Number/ N test chemical/ T MATERIAL NAME	MANUFACTURER & PRODUCT IDENTIFICATION	Break- through Time in Minutes	Perm- eation Rate in mg/sq m /min.	I N D E X	Thick- ness in mm	degrad. and comments	R e f s
s UNKNOWN MATERIAL	LIFE-GUARD RESPONDE	>480m	0	0	n.a.		77
s UNKNOWN MATERIAL	LIFE-GUARD RESPONDE	>180m	0	1	n.a.	liquid	77
s UNKNOWN MATERIAL	LIFE-GUARD RESPONDE	>480m	0	0	n.a.		84
g UNKNOWN MATERIAL	NORTH SILVERSHIELD	>480m	0	0	0.08		84
s VITON/BUTYL	DRAGER 500 OR 710	>60m			n.a.		91

ethylenediamine (synonym: 1,2-diaminoethane)
CAS Number: 107-15-3
Primary Class: 148   Amines, Poly, Aliphatic and Alicyclic
Related Class: 141   Amines, Aliphatic and Alicyclic, Primary

g BUTYL	NORTH B-174	>480m	0	0	0.65		34
g NATURAL RUBBER	ACKWELL 5-109	5m	3000	5	0.15		34
g NEOPRENE	ANSELLEDMONT 29-870	396m	147	3	0.48		34
g NITRILE	ANSELLEDMONT 37-155				0.40	degrad.	34
c PE	DU PONT TYVEK QC	195m	32	2	n.a.		8
g PE/EVAL/PE	SAFETY4 4H	47m			0.07	35°C	60
g PE/EVAL/PE	SAFETY4 4H	92m			0.07	21°C	60
g PVAL	EDMONT 25-950				0.50	degrad.	34
g PVC	ANSELLEDMONT 34-100	9m	508	4	0.20		34
s SARANEX-23	DU PONT TYVEK SARAN	>480m	0	0	n.a.		8
s TEFLON	CHEMFAB CHALL. 5100	>200m	0	1	n.a.		65
g VITON	NORTH F-091				0.26	degrad.	34

ethyleneimine (synonym: aziridine)
CAS Number: 151-56-4
Primary Class: 274   Heterocyclic Compounds, Nitrogen, Others
Related Class: 142   Amines, Aliphatic and Alicyclic, Secondary

u BUTYL	Unknown	600m	45	2	0.84		3
u NEOPRENE	Unknown	<5m			0.20		3
s UNKNOWN MATERIAL	LIFE-GUARD RESPONDE	>357m		0	n.a.		77

ethylhydroxynol (synonym: methoxypropanol acetate)
CAS Number: 84540-57-8
Primary Class: 222   Esters, Carboxylic, Acetates
Related Class: 241   Ethers, Aliphatic and Alicyclic
              245   Ethers, Glycols

ferric chloride (synonym: ferric trichloride)
CAS Number: 7705-08-0
Primary Class: 340   Inorganic Salts

s UNKNOWN MATERIAL	LIFE-GUARD RESPONDE	>480m	0	0	n.a.		77

G A R M E Class & Number/ N test chemical/ T MATERIAL NAME	MANUFACTURER & PRODUCT IDENTIFICATION	Break- through Time in Minutes	Perm- eation Rate in mg/sq m /min.	I N D E X	Thick- ness in mm	degrad. and comments	R e f s

ferrous chloride
CAS Number: 7758-94-3
Primary Class: 340   Inorganic Salts

s UNKNOWN MATERIAL	LIFE-GUARD RESPONDE	>480m	0	0	n.a.		77

fluorine
CAS Number: 7782-41-4
Primary Class: 350   Inorganic Gases and Vapors
Related Class: 330   Elements

s BUTYL	TRELLEBORG TRELLCHE				n.a.	degrad.	71
c PE	DU PONT TYVEK QC	<1m	>1000	5	n.a.		8
s PVC	TRELLEBORG TRELLCHE				n.a.	degrad.	71
s SARANEX-23	DU PONT TYVEK SARAN	>480m	0	0	n.a.		8
s UNKNOWN MATERIAL	DU PONT BARRICADE	>480m	0	0	n.a.		8
s UNKNOWN MATERIAL	LIFE-GUARD RESPONDE	>480m	0	0	n.a.		77

fluorosilic acid (synonym: fluosilic acid)
CAS Number: 16961-83-4
Primary Class: 370   Inorganic Acids

s UNKNOWN MATERIAL	LIFE-GUARD RESPONDE	>480m		0	n.a.		77

fluorosulfonic acid
CAS Number: 7789-21-1
Primary Class: 370   Inorganic Acids
Related Class: 504   Sulfur Compounds, Sulfonic Acids

c PE	DU PONT TYVEK QC	10m			n.a.		8
s SARANEX-23	DU PONT TYVEK SARAN	>360m	0	0	n.a.		8

formaldehyde, 30-70% (synonym: formalin)
CAS Number: 50-00-0
Primary Class: 121   Aldehydes, Aliphatic and Alicyclic

g BUTYL	NORTH B-161	>960m	0	0	0.43	37% soln.	34
s BUTYL/NEOPRENE	MSA BETEX	>240m	0	0	n.a.		48
u CPE	ILC DOVER	>180m			0.05		29
g NATURAL RUBBER	ACKWELL 5-109	4m	12	4	0.16	37% soln.	34
g NATURAL RUBBER	ANSELL CONFORM 4205	>360m	0	0	0.13		55
g NATURAL RUBBER	ANSELL FL 200 254	>360m	0	0	0.51		55
g NATURAL RUBBER	ANSELL ORANGE 208	>360m	0	0	0.76		55
g NATURAL RUBBER	ANSELL PVL 040	>360m	0	0	0.46		55
g NATURAL RUBBER	ANSELL STERILE 832	>360m	0	0	0.23		55
g NATURAL RUBBER	ANSELLEDMONT 36-124	60m	<90	2	0.46		6

G A R M E N T	Class & Number/ test chemical/ MATERIAL NAME	MANUFACTURER & PRODUCT IDENTIFICATION	Break- through Time in Minutes	Perm- eation Rate in mg/sq m /min.	I N D E X	Thick- ness in mm	degrad. and comments	R e f s
g	NATURAL RUBBER	ANSELLEDMONT 392	105m	<90	2	0.48		6
g	NATURAL RUBBER	MAPA-PIONEER L-118	>480m	0	0	0.46	37% soln.	36
g	NATURAL RUBBER	MAPA-PIONEER L-118	>480m	0	0	0.46		36
g	NATURAL RUBBER	MARIGOLD BLACK HEVY	30m	8	2	0.65		79
g	NATURAL RUBBER	MARIGOLD FEATHERLIT	30m	24	3	0.15		79
g	NATURAL RUBBER	MARIGOLD MED WT EXT	30m	17	3	0.45		79
g	NATURAL RUBBER	MARIGOLD MEDICAL	30m	21	3	0.28		79
g	NATURAL RUBBER	MARIGOLD ORANGE SUP	30m	8	2	0.73		79
g	NATURAL RUBBER	MARIGOLD R SURGEONS	30m	21	3	0.28		79
g	NATURAL RUBBER	MARIGOLD RED LT WT	30m	18	3	0.43		79
g	NATURAL RUBBER	MARIGOLD SENSOTECH	30m	21	3	0.28		79
g	NATURAL RUBBER	MARIGOLD SUREGRIP	30m	11	3	0.58		79
g	NATURAL RUBBER	MARIGOLD YELLOW	30m	19	3	0.38		79
u	NATURAL RUBBER	Unknown	6m	33	4	0.15	37% soln.	61
g	NATURAL+NEOPRENE	ANSELL OMNI 276	>360m	0	0	0.56		55
g	NATURAL+NEOPRENE	ANSELL TECHNICIANS	>360m	0	0	0.43		55
g	NATURAL+NEOPRENE	MARIGOLD FEATHERWT	30m	19	3	0.35		79
g	NATURAL+NEOPRENE	PLAYTEX ARGUS 123	35m	<60	3	n.a.		72
g	NEOPRENE	ANSELL NEOPRENE 530	>360m	0	0	0.46		55
g	NEOPRENE	ANSELLEDMONT 29-840	120m	<9	1	0.38		6
g	NEOPRENE	ANSELLEDMONT NEOX	120m	<90	2	n.a.		6
g	NEOPRENE	BEST 6780	>480m	0	0	n.a.	37% soln.	53
g	NEOPRENE	MAPA-PIONEER N-44	>480m	0	0	0.56	37% soln.	36
g	NEOPRENE/NATURAL	ANSELL CHEMI-PRO	>360m	0	0	0.72		55
g	NITRILE	ANSELL CHALLENGER	>360m	0	0	0.38		55
g	NITRILE	ANSELLEDMONT 37-165	>360m	0	0	0.60		6
g	NITRILE	BEST 22R	>480m	0	0	n.a.	37% soln.	53
g	NITRILE	BEST 727	>480m	0	0	n.a.	37% soln.	53
g	NITRILE	MAPA-PIONEER A-14	>480m	0	0	0.56	37% soln.	36
g	NITRILE	MARIGOLD BLUE	>480m	0	0	0.45		79
g	NITRILE	MARIGOLD GREEN SUPA	>480m	0	0	0.30		79
g	NITRILE ·	MARIGOLD NITROSOLVE	>480m	0	0	0.75		79
g	NITRILE	NORTH LA-142G	>1260m	0	0	0.36	37% soln.	34
u	NITRILE	Unknown	>360m	0	0	0.28	37% soln.	61
c	PE	DU PONT TYVEK QC	<1m	3	3	n.a.	37% soln.	8
g	PE/EVAL/PE	SAFETY4 4H	>240m	0	0	0.07	37% 35°C	60
g	PVAL	ANSELLEDMONT PVA				n.a.	degrad.	6
g	PVAL	EDMONT 15-552				n.a.	degrad.	6
g	PVAL	EDMONT 25-950				0.50	degrad.	34
g	PVC	ANSELLEDMONT M GRIP	180m	<90	2	n.a.		6
g	PVC	BEST 812	>480m	0	0	n.a.	37% soln.	53
g	PVC	MAPA-PIONEER V-20	>480m	0	0	0.51	37% soln.	36
s	SARANEX-23	DU PONT TYVEK SARAN	>480m	0	0	n.a.	37% soln.	8
s	TEFLON	CHEMFAB CHALL. 5100	>180m	0	1	n.a.		65
s	UNKNOWN MATERIAL	CHEMRON CHEMREL	>480m	0	0	n.a.		67
s	UNKNOWN MATERIAL	CHEMRON CHEMREL MAX	>1440m	0	0	n.a		67
s	UNKNOWN MATERIAL	DU PONT BARRICADE	>480m	0	0	n.a.	37% soln.	8
s	UNKNOWN MATERIAL	KAPPLER CPF III	>480m	0	0	n.a.		77

	G A R M E N T			Perm-	I N D E X		R e f
	Class & Number/ test chemical/ MATERIAL NAME	MANUFACTURER & PRODUCT IDENTIFICATION	Break- through Time in Minutes	eation Rate in mg/sq m /min.		Thick- ness degrad. in and mm comments	
s	UNKNOWN MATERIAL	LIFE-GUARD RESPONDE	>240m		0	n.a.	77
g	UNKNOWN MATERIAL	NORTH SILVERSHIELD	>360m 0		0	0.08 37% soln.	7
g	VITON	NORTH F-091	>960m 0		0	0.23 37% soln.	34
s	VITON/CHLOROBUTYL	LIFE-GUARD VC-100	>240m		0	n.a.	77
g	VITON/NEOPRENE	ERISTA VITRIC	> 60m			0.60	38
s	VITON/NEOPRENE	MSA VAUTEX	>240m 0		0	n.a.	48

formaldehyde, 37% & methyl alcohol, 10% & water 53%
Primary Class: 121   Aldehydes, Aliphatic and Alicyclic
Related Class: 311   Hydroxy Compounds, Aliphatic and Alicyclic, Primary

g	PE/EVAL/PE	SAFETY4 4H	>240m 0		0	0.07 35°C	60

formaldehyde, <30%
Primary Class: 121   Aldehydes, Aliphatic and Alicyclic

g	NATURAL RUBBER	ACKWELL 5-109	12m	<1	3	0.20 10% soln.	34
g	PVC	ANSELLEDMONT 34-100	4m	<1	3	0.15 10% soln.	34

formic acid
CAS Number: 64-18-6
Primary Class: 102   Acids, Carboxylic, Aliphatic and Alicyclic, Unsubstituted

g	BUTYL	BEST 878	>480m	0	0	0.75 90%	53
s	BUTYL	MSA CHEMPRUF	>480m	0	0	n.a.	48
g	NATURAL RUBBER	ANSELL CONFORM 4205	120m			0.13 90%	55
g	NATURAL RUBBER	ANSELL FL 200 254	120m			0.51 90%	55
g	NATURAL RUBBER	ANSELL ORANGE 208	>360m	0	0	0.76 90%	55
g	NATURAL RUBBER	ANSELL PVL 040	>360m	0	0	0.46 90%	55
g	NATURAL RUBBER	ANSELL STERILE 832	60m			0.23 90%	55
g	NATURAL RUBBER	ANSELLEDMONT 36-124	120m			0.46	6
g	NATURAL RUBBER	ANSELLEDMONT 392	150m	0	1	0.48 90%	6
g	NATURAL RUBBER	BEST 65NFW	>480m	0	0	n.a. 90%	53
g	NATURAL RUBBER	MARIGOLD BLACK HEVY	69m	502	3	0.65 degrad.	79
g	NATURAL RUBBER	MARIGOLD FEATHERLIT	21m	1545	4	0.15 degrad.	79
g	NATURAL RUBBER	MARIGOLD MED WT EXT	30m	1618	4	0.45 degrad.	79
g	NATURAL RUBBER	MARIGOLD MEDICAL	25m	1576	4	0.28 degrad.	79
g	NATURAL RUBBER	MARIGOLD ORANGE SUP	73m	391	3	0.73 degrad.	79
g	NATURAL RUBBER	MARIGOLD R SURGEONS	25m	1576	4	0.28 degrad.	79
g	NATURAL RUBBER	MARIGOLD RED LT WT	29m	1612	4	0.43 degrad.	79
g	NATURAL RUBBER	MARIGOLD SENSOTECH	25m	1576	4	0.28 degrad.	79
g	NATURAL RUBBER	MARIGOLD SUREGRIP	57m	837	3	0.58 degrad.	79
g	NATURAL RUBBER	MARIGOLD YELLOW	28m	1600	4	0.38 degrad.	79
g	NATURAL+NEOPRENE	ANSELL OMNI 276	150m			0.56 90%	55
g	NATURAL+NEOPRENE	ANSELL TECHNICIANS	150m			0.43 90%	55
g	NATURAL+NEOPRENE	MARIGOLD FEATHERWT	27m	1594	4	0.35 degrad.	79
g	NATURAL+NEOPRENE	PLAYTEX ARGUS 123	190m	120	3	n.a.	72

G A R M E N T	Class & Number/ test chemical/ MATERIAL NAME	MANUFACTURER & PRODUCT IDENTIFICATION	Break- through Time in Minutes	Perm- eation Rate in mg/sq m /min.	I N D E X	Thick- ness in mm	degrad. and comments	R e f s
g	NEOPRENE	ANSELL NEOPRENE 530	>360m	0	0	0.46	90%	55
g	NEOPRENE	ANSELLEDMONT 29-840	>360m	0	0	0.38		6
g	NEOPRENE	ANSELLEDMONT NEOX	>360m	0	0	n.a.		6
g	NEOPRENE	BEST 32	>480m	0	0	n.a.	90%	53
g	NEOPRENE	BEST 6780	>480m	0	0	n.a.	90%	53
g	NEOPRENE/NATURAL	ANSELL CHEMI-PRO	270m		0	0.72	90%	55
g	NITRILE	ANSELL CHALLENGER	180m			0.38	90%	55
g	NITRILE	ANSELLEDMONT 37-165	240m		0	0.60		6
g	NITRILE	BEST 22R	420m	41	2	n.a.	90%	53
g	NITRILE	BEST 727	75m	81	2	0.38	90%, degrad.	53
g	NITRILE	MARIGOLD BLUE	78m	1653	4	0.45	degrad.	79
g	NITRILE	MARIGOLD GREEN SUPA	26m	1771	4	0.30	degrad.	79
g	NITRILE	MARIGOLD NITROSOLVE	88m	1630	4	0.75	degrad.	79
c	PE	DU PONT TYVEK QC	4m	3	3	n.a.		8
g	PE/EVAL/PE	SAFETY4 4H	120m			0.07		60
g	PE/EVAL/PE	SAFETY4 4H	60m			0.07	35°C	60
g	PVAL	ANSELLEDMONT PVA				n.a.	degrad.	6
g	PVAL	EDMONT 15-552				n.a.	degrad.	6
g	PVC	ANSELLEDMONT M GRIP	>360m	0	0	n.a.		6
g	PVC	BEST 725R	>480m	0	0	n.a.	90%	53
s	PVC	MSA UPC	>480m	0	0	0.20		48
s	SARANEX-23	DU PONT TYVEK SARAN	>480m	0	0	n.a.		8
s	UNKNOWN MATERIAL	CHEMRON CHEMREL	>1440m	0	0	n.a.		67
s	UNKNOWN MATERIAL	CHEMRON CHEMREL MAX	>1440m	0	0	n.a		67
s	UNKNOWN MATERIAL	DU PONT BARRICADE	>480m	0	0	n.a.		8
s	UNKNOWN MATERIAL	LIFE-GUARD RESPONDE	>480m	0	0	n.a.		77
s	VITON/CHLOROBUTYL	LIFE-GUARD VC-100	>240m		0	n.a.		77

freon 113 (TF) (synonym: trichlorotrifluoroethane)
CAS Number: 76-13-1
Primary Class: 261   Halogen Compounds, Aliphatic and Alicyclic

m	BUTYL	PLYMOUTH RUBBER	37m			n.a.		46
s	CPE	ILC DOVER CLOROPEL	>180m			n.a.		46
g	HYPALON/NEOPRENE	NORTH		20	5	0.88		50
g	NAT+NEOPR+NITRILE	MAPA-PIONEER TRIONI	20m	4098	4	0.48		36
g	NATURAL RUBBER	ANSELLEDMONT 30-139				0.25	degrad.	42
g	NATURAL RUBBER	ANSELLEDMONT 36-124				0.46	degrad.	6
g	NATURAL RUBBER	ANSELLEDMONT 392				0.48	degrad.	6
g	NATURAL RUBBER	ANSELLEDMONT 46-320	9m	10000	5	0.31		5
g	NATURAL+NEOPRENE	ANSELL OMNI 276	16m	7000	4	0.45		5
g	NATURAL+NEOPRENE	ANSELL OMNI 276				0.56	degrad.	55
g	NATURAL+NEOPRENE	PLAYTEX ARGUS 123	16m	4740	4	n.a.		72
g	NEOPRENE	ANSELL NEOPRENE 530	240m	0.06		0.46		55
g	NEOPRENE	ANSELLEDMONT 29-840	240m	<9	1	0.38		6
g	NEOPRENE	ANSELLEDMONT NEOX	>60m			0.77		5
g	NEOPRENE	ANSELLEDMONT NEOX	120m	<90	2	n.a.		6
g	NEOPRENE	MAPA-PIONEER N-44	>480m	0	0	0.56		36

G A R M E Class & Number/ N test chemical/ T MATERIAL NAME	MANUFACTURER & PRODUCT IDENTIFICATION	Break- through Time in Minutes	Perm- eation Rate in mg/sq m /min.	I N D E X	Thick- ness in mm	degrad. and comments	R e f s
g NEOPRENE	NORTH		180	5	0.38		50
g NEOPRENE	NORTH		70	5	0.76		50
g NEOPRENE	PIONEER		40	5	0.84		50
g NEOPRENE/NATURAL	ANSELL CHEMI-PRO	14m	3	3	0.72		55
g NITRILE	ANSELL CHALLENGER	>360m	0	0	0.38		55
g NITRILE	ANSELLEDMONT 37-165	>360m	0	0	0.60		6
g NITRILE	ANSELLEDMONT 37-175	>60m			0.37		5
g NITRILE	ANSELLEDMONT 49-155	>360m	0	0	0.38		42
g NITRILE	MAPA-PIONEER A-14	>480m	0	0	0.56		36
g PE	ANSELLEDMONT 35-125	5m	100	4	0.03		5
g PE/EVAL/PE	SAFETY4 4H	>240m	0	0	0.07	35°C	60
g PVAL	ANSELLEDMONT PVA	>360m		0	n.a.		6
g PVAL	EDMONT 15-552	30m	<90	3	n.a.		6
g PVAL	EDMONT 25-545	>360m	0	0	n.a.		6
g PVC	ANSELLEDMONT SNORKE				n.a.	degrad.	6
g PVC	MAPA-PIONEER V-20	11m	1900	5	0.31		5
s TEFLON	CHEMFAB CHALL. 5100	>480m	0	0	n.a.		56
s UNKNOWN MATERIAL	CHEMRON CHEMREL	>480m	0	0	n.a.		67
s UNKNOWN MATERIAL	CHEMRON CHEMREL	>1440m	0	0	n.a.		67
s UNKNOWN MATERIAL	CHEMRON CHEMREL MAX	>1440m	0	0	n.a.		67
s UNKNOWN MATERIAL	DU PONT BARRICADE	>480m	0	0	n.a.		8
s UNKNOWN MATERIAL	KAPPLER CPF III	>480m	0	0	n.a		77
s UNKNOWN MATERIAL	LIFE-GUARD RESPONDE	>480m	0	0	n.a.		77
s VITON/CHLOROBUTYL	LIFE-GUARD VC-100	>240m		0	n.a.		77
u VITON/CHLOROBUTYL	Unknown	>180m			n.a.		46

freon TMC (synonym: trichlorotrifluoroethane, 50% & methylene chloride, 50%)
Primary Class: 261   Halogen Compounds, Aliphatic and Alicyclic

g NATURAL RUBBER	ANSELLEDMONT 36-124	3m	<90000	5	0.46		6
g NATURAL RUBBER	ANSELLEDMONT 392				0.48	degrad.	6
g NEOPRENE	ANSELLEDMONT 29-840	10m	<90000	5	0.31		6
g NEOPRENE	ANSELLEDMONT NEOX	3m	<90000	5	n.a.		6
g NITRILE	ANSELLEDMONT 37-165	10m	<90000	5	0.60		6
g PVAL	ANSELLEDMONT PVA	>360m		0	n.a.		6
g PVAL	EDMONT 15-552	>360m	0	0	n.a.		6
g PVAL	EDMONT 25-545	>360m	0	0	n.a.		6
g PVC	ANSELLEDMONT SNORKE				n.a.	degrad.	6

furan (synonym: furfurane)
CAS Number: 110-00-9
Primary Class: 277   Heterocyclic Compounds, Oxygen, Furans

g BUTYL	NORTH B-174	81m	576	3	0.70	degrad.	34
g NATURAL RUBBER	ACKWELL 5-109				0.15	degrad.	34
g NEOPRENE	ANSELLEDMONT 29-870				0.48	degrad.	34
g NITRILE	ANSELLEDMONT 37-155				0.40	degrad.	34

G A R M E Class & Number/ N test chemical/ T MATERIAL NAME	MANUFACTURER & PRODUCT IDENTIFICATION	Break-through Time in Minutes	Perm-eation Rate in mg/sq m /min.	I N D E X	Thick-ness in mm	degrad. and comments	R e f s
g PVAL	EDMONT 25-950	86m	2	1	0.63		34
g PVC	ANSELLEDMONT 34-100	1m	29400	5	0.20	degrad.	34
g VITON	NORTH F-091	20m	1380	4	0.26	degrad.	34

furfural (synonym: 2-furaldehyde)
CAS Number: 98-01-1
Primary Class: 122   Aldehydes, Aromatic
Related Class: 277   Heterocyclic Compounds, Oxygen, Furans

g BUTYL	BEST 878	>480m		0	0.75		53
g BUTYL	COMASEC BUTYL	>480m	0	0	0.53		40
g BUTYL	NORTH B-161	>960m	0	0	0.43		34
g BUTYL	NORTH B-161	700m	11	2	0.44		63
g BUTYL/NEOPRENE	COMASEC BUTYL PLUS	>480m	0	0	0.60		40
g NATURAL RUBBER	ANSELL CONFORM 4205	1m	7	3	0.13		55
g NATURAL RUBBER	ANSELLEDMONT 36-124	15m	<900	3	0.46		6
g NATURAL RUBBER	ANSELLEDMONT 392	15m	<90	3	0.48		6
g NATURAL RUBBER	COMASEC FLEXIGUM	35m	180	3	0.95		40
g NATURAL RUBBER	MARIGOLD BLACK HEVY	15m	111	3	0.65		79
g NATURAL RUBBER	MARIGOLD FEATHERLIT	15m	319	3	0.15		79
g NATURAL RUBBER	MARIGOLD MED WT EXT	15m	153	3	0.45		79
g NATURAL RUBBER	MARIGOLD MEDICAL	15m	250	3	0.28		79
g NATURAL RUBBER	MARIGOLD ORANGE SUP	15m	107	3	0.73		79
g NATURAL RUBBER	MARIGOLD R SURGEONS	15m	250	3	0.28		79
g NATURAL RUBBER	MARIGOLD RED LT WT	15m	167	3	0.43		79
g NATURAL RUBBER	MARIGOLD SENSOTECH	15m	250	3	0.28		79
g NATURAL RUBBER	MARIGOLD SUREGRIP	15m	124	3	0.58		79
g NATURAL RUBBER	MARIGOLD YELLOW	15m	195	3	0.38		79
g NATURAL+NEOPRENE	MARIGOLD FEATHERWT	15m	209	3	0.35		79
g NEOPRENE	ANSELL NEOPRENE 530	45m	2	2	0.46		55
g NEOPRENE	ANSELLEDMONT 29-840	20m	<900	3	0.38		6
g NEOPRENE	ANSELLEDMONT 29-870	126m	60	2	0.51		34
g NEOPRENE	ANSELLEDMONT NEOX	120m	<900	3	n.a.		6
g NEOPRENE	BEST 32	81m	150	3	n.a.		53
g NEOPRENE	BEST 6780	116m	80	2	n.a.		53
g NEOPRENE	COMASEC COMAPRENE	30m	180	3	n.a.		40
g NEOPRENE/NATURAL	ANSELL CHEMI-PRO	17m	1	2	0.72		55
g NITRILE	ANSELLEDMONT 37-165				0.60	degrad.	6
g NITRILE	BEST 22R				n.a.	degrad.	53
g NITRILE	BEST 727				0.38	degrad.	53
g NITRILE	COMASEC COMATRIL	55m	1620	4	0.55		40
g NITRILE	COMASEC COMATRIL SU	57m	1560	4	0.60		40
g NITRILE	COMASEC FLEXITRIL	6m	1680	5	n.a.		40
g NITRILE	MARIGOLD BLUE	36m	10154	5	0.45	degrad.	79
g NITRILE	MARIGOLD GREEN SUPA	15m	12667	5	0.30	degrad.	79
g NITRILE	MARIGOLD NITROSOLVE	40m	9652	4	0.75	degrad.	79
g NITRILE	NORTH LA-142G	24m	15880	5	0.33	degrad.	34
g NITRILE+PVC	COMASEC MULTIPLUS	100m	1440	4	n.a.		40
g PE/EVAL/PE	SAFETY4 4H	>240m	0	0	0.07	35°C	60

G A R M E Class & Number/ N test chemical/ T MATERIAL NAME	MANUFACTURER & PRODUCT IDENTIFICATION	Break- through Time in Minutes	Perm- eation Rate in mg/sq m /min.	I N D E X	Thick- ness in mm	degrad. and comments	R e f s
g PE/EVAL/PE	SAFETY4 4H	>480m	0	0	0.07		60
g PVAL	ANSELLEDMONT PVA	>360m		0	n.a.		6
g PVAL	EDMONT 15-552	>360m	0	0	n.a.		6
g PVAL	EDMONT 25-545	>360m	0	0	n.a.		6
g PVAL	EDMONT 25-950	>960m	0	0	0.30		34
g PVC	ANSELLEDMONT 34-100				0.20	degrad.	34
g PVC	ANSELLEDMONT SNORKE				n.a.	degrad.	6
g PVC	COMASEC MULTIPOST	57m	1140	4	n.a.		40
g PVC	COMASEC MULTITOP	75m	1080	4	n.a.		40
g PVC	COMASEC NORMAL	70m	1080	4	n.a.		40
g PVC	COMASEC OMNI	57m	1740	4	n.a.		40
s SARANEX-23	DU PONT TYVEK SARAN	245m	2	1	n.a.		8
s TEFLON	CHEMFAB CHALL. 5100	>60m			n.a.		65
s UNKNOWN MATERIAL	DU PONT BARRICADE	>480m	0	0	n.a.		8
g UNKNOWN MATERIAL	NORTH SILVERSHIELD	>480m	0	0	0.08		7
g VITON	NORTH F-091	210m	888	3	0.27		34
g VITON	NORTH F-091	157m	40	2	0.35		92
g VITON	NORTH F-091	71m	90	2	0.35	37°C	92
g VITON	NORTH F-091	5m	150	3	0.35	50°C	92

furfuryl alcohol
CAS Number: 98-00-0
Primary Class: 311   Hydroxy Compounds, Aliphatic and Alicyclic, Primary

g PE/EVAL/PE	SAFETY4 4H	>240m		0	0.07	35°C	60
g PE/EVAL/PE	SAFETY4 4H	>480m	0	0	0.07		60

fusilade 250EC (synonym: fluazifop-butyl)
CAS Number: 69806-50-4
Primary Class: 840   Pesticides, Mixtures and Formulations

g NATURAL RUBBER	ANSELLEDMONT 36-124	<30m	1440	4	0.51		74
g NEOPRENE	ANSELLEDMONT 29-865	<30m	1320	4	0.51		74
g NITRILE	ANSELLEDMONT 37-175	270m	18	2	0.46		74
g PE/EVAL/PE	SAFETY4 4H	>240m	0	0	0.07	35°C	60
g PVC	EDMONT CANADA 14112	<30m	240	3	n.a.		74

gasoline with 40-55% aromatics (b.p. 35-210°C)
CAS Number: 8006-61-9
Primary Class: 292   Hydrocarbons, Aromatic

g BUTYL	ANSELL LAMPRECHT	20m	10312	5	0.70		38
s BUTYL/NEOPRENE	MSA BETEX	25m	287	3	n.a.		48
u NATURAL RUBBER	Unknown	4m	20350	5	0.64		38
g NEOPRENE	MAPA-PIONEER N-44	96m	960	3	0.56		36
g NITRILE	ANSELLEDMONT 37-175	>60m			0.42		38
g NITRILE	MAPA-PIONEER A-14	>480m	0	0	0.56		36
g NITRILE	MARIGOLD BLUE	>480m	0	0	0.45	degrad.	79

G A R M E Class & Number/ N test chemical/ T MATERIAL NAME	MANUFACTURER & PRODUCT IDENTIFICATION	Break- through Time in Minutes	Perm- eation Rate in mg/sq m /min.	I N D E X	Thick- ness in mm	degrad. and comments	R e f s
g NITRILE	MARIGOLD GREEN SUPA	>480m	0	0	0.30	degrad.	79
g NITRILE	MARIGOLD NITROSOLVE	>480m	0	0	0.75	degrad.	79
g PE/EVAL/PE	SAFETY4 4H	>240m	0	0	0.07	35°C	60
g PVC	KID VINYLPRODUKTER	5m	2655	5	0.60	degrad.	33
s TEFLON	CHEMFAB CHALL. 5100	>890m	0	0	n.a.		65
s UNKNOWN MATERIAL	DU PONT BARRICADE	>480m	0	0	n.a.		8
s UNKNOWN MATERIAL	LIFE-GUARD RESPONDE	>480m		0	n.a.		77
g VITON	NORTH F-091	>60m			0.26		38
s VITON/CHLOROBUTYL	LIFE-GUARD VC-100	>240m		0	n.a.		77
g VITON/NEOPRENE	ERISTA VITRIC	>60m			0.60		38
s VITON/NEOPRENE	MSA VAUTEX	>480m	0	0	n.a.		48

gasoline, 50% & acetone, 50%
Primary Class: 292   Hydrocarbons, Aromatic

g PE/EVAL/PE	SAFETY4 4H	>240m	0	0	0.07		60
g PE/EVAL/PE	SAFETY4 4H	3m			0.07	35°C	60

gasoline, 7% & ethanol, 60% & methanol, 33% (synonym: gashol)
Primary Class: 800   Multicomponent Mixtures With > 2   Components

c PE	DU PONT TYVEK QC	<1m	8	3	n.a.		8
s SARANEX-23	DU PONT TYVEK SARAN	>480m		0	n.a.		8
s UNKNOWN MATERIAL	DU PONT BARRICADE	170m	2.4	1	n.a.		8

gasoline, 80% & N-methylpyrrolidone, 20%
Primary Class: 292   Hydrocarbons, Aromatic

g PE/EVAL/PE	SAFETY4 4H	>240m	0	0	0.07	35°C	60

gasoline, unleaded
Primary Class: 292   Hydrocarbons, Aromatic

g BUTYL	BEST 878				0.75	degrad.	53
s BUTYL	MSA CHEMPRUF	<60m			n.a.		48
g NATURAL RUBBER	ANSELLEDMONT 36-124				0.46	degrad.	6
g NATURAL RUBBER	ANSELLEDMONT 392				0.48	degrad.	6
g NATURAL RUBBER	BEST 65NFW				n.a.	degrad.	53
g NEOPRENE	ANSELLEDMONT 29-840				0.38	degrad.	6
g NITRILE	ANSELLEDMONT 37-165	>360m	0	0	0.60		6
g PE/EVAL/PE	SAFETY4 4H	>240m	0	0	0.07	35°C	60
g PVAL	ANSELLEDMONT PVA	>360m		0	n.a.		6
g PVAL	EDMONT 15-552	>360m	0	0	n.a.		6
g PVAL	EDMONT 25-545	>360m	0	0	n.a.		6
g PVC	ANSELLEDMONT SNORKE				n.a.	degrad.	6
s UNKNOWN MATERIAL	KAPPLER CPF III	>480m	0	0	n.a.		77
s UNKNOWN MATERIAL	LIFE-GUARD RESPONDE	>480m	0	0	n.a.		77

G A R M E Class & Number/ N test chemical/ T MATERIAL NAME	MANUFACTURER & PRODUCT IDENTIFICATION	Break- through Time in Minutes	Perm- eation Rate in mg/sq m /min.	I N D E X	Thick- ness in mm	degrad. and comments	R e f s

Glance (mixture)
Primary Class: 900   Miscellaneous Unclassified Chemicals

g PE/EVAL/PE	SAFETY4 4H	>240m	0		0.07		60

glutaraldehyde (synonym: 1,5-pentanedial)
CAS Number: 111-30-8
Primary Class: 121   Aldehydes, Aliphatic and Alicyclic

g BUTYL	BEST 878	>480m	0	0	0.75		53
g BUTYL	NORTH B-174	>480m	0	0	0.68		34
g NEOPRENE	ANSELLEDMONT 29-870	>480m	0	0	0.46		34
g NEOPRENE	BEST 6780	>480m	0	0	n.a.		53
g PE/EVAL/PE	SAFETY4 4H	>240m	0	0	0.07	2% 35°C	60
g PE/EVAL/PE	SAFETY4 4H	>240m	0	0	0.07	25% 35°C	60
g PVAL	EDMONT 25-950				0.50	degrad.	34
g PVC	ANSELLEDMONT 34-100	70m	75	2	0.20		34
g VITON	NORTH F-091	>480m	0	0	0.40		34

glycerol (synonym: glycerin)
CAS Number: 56-81-5
Primary Class: 314   Hydroxy Compounds, Aliphatic and Alicyclic, Polyols

g NATURAL RUBBER	MARIGOLD BLACK HEVY	>480m	0	0	0.65		79
g NATURAL RUBBER	MARIGOLD FEATHERLIT	>480m	0	0	0.15		79
g NATURAL RUBBER	MARIGOLD MED WT EXT	>480m	0	0	0.45		79
g NATURAL RUBBER	MARIGOLD MEDICAL	>480m	0	0	0.28		79
g NATURAL RUBBER	MARIGOLD ORANGE SUP	>480m	0	0	0.73		79
g NATURAL RUBBER	MARIGOLD R SURGEONS	>480m	0	0	0.28		79
g NATURAL RUBBER	MARIGOLD RED LT WT	>480m	0	0	0.43		79
g NATURAL RUBBER	MARIGOLD SENSOTECH	>480m	0	0	0.28		79
g NATURAL RUBBER	MARIGOLD SUREGRIP	>480m	0	0	0.58		79
g NATURAL RUBBER	MARIGOLD YELLOW	>480m	0	0	0.38		79
g NATURAL+NEOPRENE	MARIGOLD FEATHERWT	>480m	0	0	0.35		79
g NATURAL+NEOPRENE	PLAYTEX ARGUS 123	>480m	0	0	n.a.		72
g NITRILE	MARIGOLD BLUE	>480m	0	0	0.45		79
g NITRILE	MARIGOLD GREEN SUPA	>480m	0	0	0.30		79
g NITRILE	MARIGOLD NITROSOLVE	>480m	0	0	0.75		79
g PE/EVAL/PE	SAFETY4 4H	>240m	0	0	0.07	35°C	60
g UNKNOWN MATERIAL	MARIGOLD R MEDIGLOV	>480m	0	0	0.03		79

glycerol monothioglycolate, >70%
CAS Number: 68148-42-5
Primary Class: 501   Sulfur Compounds, Thiols
Related Class: 224   Esters, Carboxylic, Aliphatic, Others

g PE/EVAL/PE	SAFETY4 4H	>240m	0	0	0.07	35°C	60

G A R M E Class & Number/ N test chemical/ T MATERIAL NAME	MANUFACTURER & PRODUCT IDENTIFICATION	Break- through Time in Minutes	Perm- eation Rate in mg/sq m /min.	I N D E X	Thick- ness in mm	degrad. and comments	R e f s

glycerolpropoxy acrylate
Primary Class: 223   Esters, Carboxylic, Acrylates and Methacrylates

| g PE/EVAL/PE | SAFETY4 4H | >240m | 0 | 0 | 0.07 | 35°C | 60 |

glycidyl methacrylate (synonym: methacrylic acid 2,3-epoxypropyl ester)
CAS Number: 106-91-2
Primary Class: 223   Esters, Carboxylic, Acrylates and Methacrylates

| g PE/EVAL/PE | SAFETY4 4H | >240m | | 0 | 0.07 | | 60 |

guthion (synonym: azinphos-methyl)
CAS Number: 86-50-0
Primary Class: 462   Organophosphorus Compounds and Derivatives of Phosphorus-based Acids
Related Class: 233   Esters, Non-Carboxylic, Carbamates and Others

g NATURAL RUBBER	ANSELLEDMONT 36-124	190m	5	1	0.51		74
g NEOPRENE	ANSELLEDMONT 29-865	>510m	0	0	0.51		74
g NITRILE	ANSELLEDMONT 37-175	>510m	0	0	0.46		74
g PVC	EDMONT CANADA 14112	250m	12	2	n.a.		74

halothane (synonym: 2-bromo-2-chloro-1,1,1-trifluoroethane)
CAS Number: 151-67-7
Primary Class: 261   Halogen Compounds, Aliphatic and Alicyclic

g BUTYL	NORTH B-174	180m	1357	4	0.63	degrad.	34
g NATURAL RUBBER	ACKWELL 5-109				0.15	degrad.	34
g NEOPRENE	ANSELLEDMONT 29-870				0.48	degrad.	34
g NITRILE	ANSELLEDMONT 37-155				0.40	degrad.	34
g PVAL	EDMONT 25-950	>480m	0	0	0.49		34
g PVC	ANSELLEDMONT 34-100	2m	40950	5	0.20	degrad.	34
g VITON	NORTH F-091	37m	4348	4	0.37	degrad.	34

heptane (synonym: n-heptane)
CAS Number: 142-82-5
Primary Class: 291   Hydrocarbons, Aliphatic and Alicyclic, Saturated

g BUTYL	BEST 878	23m	2210	4	0.75	degrad.	53
g BUTYL/NEOPRENE	COMASEC BUTYL PLUS	4m	1920	5	0.50		40
g HYPALON	COMASEC DIPCO	414m	54	2	0.58		40
g NATURAL RUBBER	COMASEC FLEXIGUM	15m	5220	4	0.95		40
u NATURAL RUBBER	Unknown	2m	>5000	5	n.a.		10
u NATURAL RUBBER	Unknown	3m	>5000	5	0.42		10
u NATURAL+NEOPRENE	Unknown	5m	>5000	5	0.50		10
g NEOPRENE	BEST 32	48m			n.a.		5
g NEOPRENE	BEST 6780	114m			n.a.		53

G A R M E Class & Number/ N test chemical/ T MATERIAL NAME	MANUFACTURER & PRODUCT IDENTIFICATION	Break- through Time in Minutes	Perm- eation Rate in mg/sq m /min.	I N D E X	Thick- ness in mm	degrad. and comments	R e f s
g NEOPRENE	COMASEC COMAPRENE	45m	4980	4	n.a.		40
g NEOPRENE	MAPA-PIONEER N-44	124m	120	3	0.74		36
g NEOPRENE	MAPA-PIONEER N-44	97m	210	3	0.55		92
g NEOPRENE	MAPA-PIONEER N-44	47m	390	3	0.55	37°C	92
g NEOPRENE	MAPA-PIONEER N-44	24m	750	3	0.55	50°C	92
u NEOPRENE	Unknown	>60m			0.60		10
g NITRILE	BEST 22R	>480m	0	0	n.a.		53
g NITRILE	BEST 727	>480m	0	0	0.38		53
g NITRILE	COMASEC COMATRIL	>480m	0	0	0.55		40
g NITRILE	COMASEC COMATRIL SU	>480m	0	0	0.60		40
g NITRILE	COMASEC FLEXITRIL	60m	6180	4	n.a.		40
g NITRILE	MAPA-PIONEER A-14	2m	0.18	3	0.54		36
g NITRILE+PVC	COMASEC MULTIMAX	206m	18	2	n.a.		40
g NITRILE+PVC	COMASEC MULTIPLUS	180m	6	1	n.a.		40
g PE/EVAL/PE	SAFETY4 4H	>480m	0	0	0.38		53
g PE/EVAL/PE	SAFETY4 4H	>240m	0	0	0.07	35°C	60
g PVAL	EDMONT 15-552	>60m			n.a.		10
g PVC	COMASEC MULTIPOST	78m	300	3	n.a.		40
g PVC	COMASEC MULTITOP	144m	66	2	n.a.		40
g PVC	COMASEC NORMAL	32m	1740	4	n.a.		40
g PVC	COMASEC OMNI	26m	9060	4	n.a.		40
u PVC	Unknown	15m	>5000	4	n.a.		10
g VITON	NORTH F-121	>60m			0.31		10

hexachloro-1,3-butadiene (synonym: HCBD)
CAS Number: 87-68-3
Primary Class: 263   Halogen Compounds, Vinylic

s UNKNOWN MATERIAL	LIFE-GUARD RESPONDE	>480m	0	0	n.a.		77

hexachlorocyclopentadiene
CAS Number: 77-47-4
Primary Class: 262   Halogen Compounds, Allylic and Benzylic

g BUTYL	NORTH B-174	>480m	0	0	0.46		34
g NATURAL RUBBER	ACKWELL 5-109				0.15	degrad.	34
g NEOPRENE	ANSELLEDMONT 29-870				0.48	degrad.	34
g NITRILE	ANSELLEDMONT 37-155	>480m	0	0	0.36		34
g PVAL	EDMONT 25-950	>480m	0	0	0.56		34
g PVC	ANSELLEDMONT 34-100				0.20	degrad.	34
g VITON	NORTH F-091	>480m	0	0	0.25		34

hexamethylene-1,6-diisocyanate (synonym: HMDI)
CAS Number: 822-06-0
Primary Class: 211   Isocyanates, Aliphatic and Alicyclic

s SARANEX-23	DU PONT TYVEK SARAN	>480m	0	0	n.a.		8

G A R M E Class & Number/ N test chemical/ T MATERIAL NAME	MANUFACTURER & PRODUCT IDENTIFICATION	Break- through Time in Minutes	Perm- eation Rate in mg/sq m /min.	I N D E X	Thick- ness in mm	degrad. and comments	R e f s
s UNKNOWN MATERIAL	DU PONT BARRICADE	>480m	0		0	n.a.	8
s UNKNOWN MATERIAL	LIFE-GUARD RESPONDE	>480m	0		0	n.a.	77

hexamethylphosphoramide (synonym: HMPA)
CAS Number: 680-31-9
Primary Class: 462   Organophosphorus Compounds and Derivatives of Phosphorus-based Acids

u BUTYL	Unknown	60m	1	1	0.84		3
u NITRILE	Unknown	60m	130	3	0.94		3
u PE	Unknown	15m	40	3	0.06		3

hexane (synonym: n-hexane)
CAS Number: 110-54-3
Primary Class: 291   Hydrocarbons, Aliphatic and Alicyclic, Saturated

g BUTYL	BEST 878	13m	2650	5	0.75	degrad.	53
g BUTYL	BRUNSWICK BUTYL STD	4m	>5000	5	0.73		89
g BUTYL	BRUNSWICK BUTYL-XTR	21m	>5000	4	0.65		89
g BUTYL	COMASEC BUTYL	3m	2100	5	0.50		40
s BUTYL	LIFE-GUARD BUTYL	4m	49	4	n.a.		77
g BUTYL	NORTH B-161				0.43	degrad.	7
g BUTYL	NORTH B-161	2m	2556	5	n.a.		19
m BUTYL	PLYMOUTH RUBBER	14m			n.a.		46
u BUTYL	Unknown	21m	57	3	0.40		52
u BUTYL	Unknown	10m	222	4	0.40	45°C	52
s BUTYL	WHEELER ACID KING	1m	<9000	5	n.a.		85
g BUTYL/ECO	BRUNSWICK BUTYL-POL	223m	>5000	4	0.63		89
g BUTYL/NEOPRENE	COMASEC BUTYL PLUS	3m	2100	5	0.50		40
s BUTYL/NEOPRENE	MSA BETEX	<5m	366	4	n.a.		48
s CPE	ILC DOVER CLOROPEL	>180m			n.a.		46
s CPE	STD. SAFETY	239m	0.4	1	n.a.		75
g NATURAL RUBBER	ANSELL PVL 040				0.46	degrad.	55
g NATURAL RUBBER	ANSELLEDMONT 30-139				0.25	degrad.	42
g NATURAL RUBBER	ANSELLEDMONT 36-124				0.46	degrad.	6
g NATURAL RUBBER	ANSELLEDMONT 392				0.48	degrad.	6
g NATURAL RUBBER	COMASEC FLEXIGUM	10m	6180	5	0.95		40
g NATURAL+NEOPRENE	ANSELL TECHNICIANS				0.43	degrad.	55
g NEOPRENE	ANSELL NEOPRENE 520	116m			0.38		58
g NEOPRENE	ANSELL NEOPRENE 530	16m	0.4		0.46		55
g NEOPRENE	ANSELLEDMONT 29-840	5m	<9000	4	0.38		6
g NEOPRENE	ANSELLEDMONT 29-870	137m	276	3	0.50		19
g NEOPRENE	ANSELLEDMONT 29-870	45m	<1	2	0.51		34
g NEOPRENE	ANSELLEDMONT 29-870	90m			0.48		58
g NEOPRENE	ANSELLEDMONT NEOX	60m	1260	4	n.a.		31
g NEOPRENE	ANSELLEDMONT NEOX	18m	8	2	n.a.	preexp. 15h	31

G A R M E N T	Class & Number/ test chemical/ MATERIAL NAME	MANUFACTURER & PRODUCT IDENTIFICATION	Break- through Time in Minutes	Perm- eation Rate in mg/sq m /min.	I N D E X	Thick- ness in mm	degrad. and comments	R e f s
g	NEOPRENE	ANSELLEDMONT NEOX	90m	450	3	n.a.		31
g	NEOPRENE	ANSELLEDMONT NEOX	40m	1	2	n.a.	pre-exposed	31
g	NEOPRENE	ANSELLEDMONT NEOX	90m	<900	3	n.a.		6
g	NEOPRENE	BEST 32	31m	240	3	n.a.		53
g	NEOPRENE	BEST 6780	149m	480	3	n.a.		53
g	NEOPRENE	BEST 723	24m			0.64		58
g	NEOPRENE	BRUNSWICK NEOPRENE	7m	>5000	4	0.88		89
g	NEOPRENE	COMASEC COMAPRENE	40m	5760	4	n.a.		40
g	NEOPRENE	GRANET 2001	26m			0.55		58
g	NEOPRENE	GRANET 2714	221m	2	1	n.a.		73
s	NEOPRENE	LIFE-GUARD NEOPRENE	20m	8	2	n.a.		77
g	NEOPRENE	MAPA-PIONEER N-36	79m			0.54		58
g	NEOPRENE	MAPA-PIONEER N-44	39m	360	3	0.56		36
b	NEOPRENE	RAINFAIR	95m	68	2	n.a.		90
b	NEOPRENE	RANGER	96m	95	2	n.a.		90
b	NEOPRENE	SERVUS NO 22204	44m	>190	3	1.50		90
u	NEOPRENE	Unknown	20m			0.40	37°C TL	16
g	NEOPRENE/NATURAL	ANSELL CHEMI-PRO	5m	3	3	0.72		55
g	NITRILE	ANSELL CHALLENGER	>360m	0	0	0.38		55
g	NITRILE	ANSELLEDMONT 37-155	>240m	0	0	0.38		34
g	NITRILE	ANSELLEDMONT 37-165	>360m	0	0	0.60		6
g	NITRILE	ANSELLEDMONT 49-125	360m	<9	1	0.23		42
g	NITRILE	ANSELLEDMONT 49-155	5m	<9	3	0.38		42
g	NITRILE	BEST 22R	234m	20	2	n.a.		53
g	NITRILE	BEST 727	>480m	0	0	0.38		53
g	NITRILE	COMASEC COMATRIL SU	>360m	0	0	0.60		40
g	NITRILE	COMASEC FLEXITRIL	15m	3300	4	n.a.		40
g	NITRILE	MAPA-PIONEER A-14	>480m	0	0	0.56		36
g	NITRILE	MARIGOLD BLUE	>480m	0	0	0.45		79
g	NITRILE	MARIGOLD GREEN SUPA	>480m	0	0	0.30		79
g	NITRILE	MARIGOLD NITROSOLVE	>480m	0	0	0.75		79
g	NITRILE	NORTH LA-142G	1559m	0.08	1	0.38		63
u	NITRILE	Unknown	78m			0.50	37°C TL	16
b	NITRILE+PUR+PVC	BATA HAZMAX	>180m			n.a.		86
g	NITRILE+PVC	COMASEC MULTIMAX	274m	2	1	n.a.		40
g	NITRILE+PVC	COMASEC MULTIPLUS	90m	420	3	n.a.		40
b	NITRILE+PVC	SERVUS NO 73101	>180m			n.a.		90
b	NITRILE+PVC	TINGLEY	169m	2	1	n.a.		90
c	PE	DU PONT TYVEK QC	<1m	4100	5	n.a.		8
u	PE	Unknown	4m			0.12	37°C TL	16
g	PE/EVAL/PE	SAFETY4 4H	>240m	0	0	0.07	35°C TL	60
g	PE/EVAL/PE	SAFETY4 4H	>1440m	0	0	0.07		60
g	PVAL	ANSELLEDMONT PVA	>360m		0	n.a.		6
g	PVAL	EDMONT 15-552	>360m	0	0	n.a.		6
g	PVAL	EDMONT 25-545	>491m	0	0	n.a.		19
g	PVAL	EDMONT 25-545	>360m	0	0	n.a.		6

G A R M E N T Class & Number/ test chemical/ MATERIAL NAME	MANUFACTURER & PRODUCT IDENTIFICATION	Break-through Time in Minutes	Perm-eation Rate in mg/sq m /min.	I N D E X	Thick-ness in mm	degrad. and comments	R e f s
g PVAL	EDMONT 25-950	>2100m	0	0	0.25		34
g PVC	ANSELLEDMONT SNORKE				n.a.	degrad.	6
b PVC	BATA STANDARD	188m	37	2	n.a.		86
g PVC	BEST 812		900	5	n.a.		53
g PVC	COMASEC MULTIPOST	23m	2700	4	n.a.		40
g PVC	COMASEC MULTITOP	55m	720	3	n.a.		40
g PVC	COMASEC NORMAL	25m	2700	4	n.a.		40
g PVC	COMASEC OMNI	18m	2880	4	n.a.		40
b PVC	JORDAN DAVID	152m	110	3	n.a.		90
b PVC	STD. SAFETY GL-20	20m	171	3	n.a.		75
s PVC	STD. SAFETY WG-20	12m	221	4	n.a.		75
u PVC	Unknown	29m			0.30	37°C TL	16
s PVC	WHEELER ACID KING	12m	<9000	5	n.a.		85
b PVC+POLYURETHANE	BATA POLYBLEND	173m	25	2	n.a.		86
b PVC+POLYURETHANE	BATA POLYBLEND	>180m			n.a.		90
b PVC+POLYURETHANE	BATA POLYMAX	145m	70	2	n.a.		86
b PVC+POLYURETHANE	BATA SUPER POLY	209m	20	2	n.a.		86
s SARANEX-23	DU PONT TYVEK SARAN	2m	0.3		n.a.		8
s SARANEX-23 2-PLY	DU PONT TYVEK SARAN	>480m	0	0	n.a.		8
u TEFLON	CHEMFAB	2m	308	4	0.10		65
s TEFLON	CHEMFAB CHALL. 5100	>480m	0	0	n.a.		56
s TEFLON	CHEMFAB CHALL. 5100	>300m	0	0	n.a.		65
s TEFLON	CHEMFAB CHALL. 5200	>300m	0	0	n.a.		80
s TEFLON	CHEMFAB CHALL. 6000	>180m			n.a.		80
s TEFLON	LIFE-GUARD TEFGUARD	>480m	0	0	n.a.		77
s TEFLON	WHEELER ACID KING	>480m	0	0	n.a.		85
v TEFLON-FEP	CHEMFAB CHALL.	>180m			0.25		65
s UNKNOWN MATERIAL	CHEMRON CHEMREL	>1440m	0	0	n.a.		67
s UNKNOWN MATERIAL	CHEMRON CHEMREL MAX	>1440m	0	0	n.a.		67
s UNKNOWN MATERIAL	DU PONT BARRICADE	>456m	0.1	0	n.a.		8
s UNKNOWN MATERIAL	KAPPLER CPF III	>480m		0	n.a.		77
s UNKNOWN MATERIAL	LIFE-GUARD RESPONDE	>480m	0	0	n.a.		77
g UNKNOWN MATERIAL	MARIGOLD R MEDIGLOV	1m	19113	5	0.03	degrad.	79
g UNKNOWN MATERIAL	NORTH SILVERSHIELD	>360m	0	0	0.08		7
g VITON	NORTH F-091	>660m	0	0	0.23		34
g VITON	NORTH F-091	>360m	0	0	0.24		41
s VITON/BUTYL/UNKN.	TRELLEBORG HPS	>180m			n.a.		71
s VITON/CHLOROBUTYL	LIFE-GUARD VC-100	>480m			n.a.		77
s VITON/CHLOROBUTYL	LIFE-GUARD VNC 200	>480m	0	0	n.a.		77
u VITON/CHLOROBUTYL	Unknown	>180m			n.a.		46
s VITON/NEOPRENE	MSA VAUTEX	>240m	0	0	n.a.		48

G A R M E Class & Number/ N test chemical/ T MATERIAL NAME	MANUFACTURER & PRODUCT IDENTIFICATION	Break- through Time in Minutes	Perm- eation Rate in mg/sq m /min.	I N D E X	Thick- ness in mm	R degrad. and comments	R e f s
hexane, 50% & acetone, 50% Primary Class: 291 Hydrocarbons, Aliphatic and Alicyclic, Saturated							
u VITON/CHLOROBUTYL	Unknown	4m			0.36		43
hexane, 50% & methyl alcohol, 50% Primary Class: 291 Hydrocarbons, Aliphatic and Alicyclic, Saturated							
u NEOPRENE	Unknown	19/20m			0.40	37°C TL	16
u NITRILE	Unknown	44/42m			0.50	37°C TL	16
u PE	Unknown	6/6m			0.12	37°C TL	16
u PVC	Unknown	24/20m			0.30	37°C TL	16
hexane, 50% & methyl ethyl ketone, 50% Primary Class: 291 Hydrocarbons, Aliphatic and Alicyclic, Saturated							
g VITON	NORTH F-091	3/3m	/19920	5	0.24		41
hexane, 50% & methylene chloride, 50% Primary Class: 291 Hydrocarbons, Aliphatic and Alicyclic, Saturated							
u VITON/CHLOROBUTYL	Unknown	59/44m			0.36		43
hexane, 50% & toluene, 50% Primary Class: 291 Hydrocarbons, Aliphatic and Alicyclic, Saturated							
u NEOPRENE	Unknown	16/18m			0.40	37°C TL	16
u NITRILE	Unknown	23/22m			0.50	37°C TL	16
u PE	Unknown	8/8m			0.12	37°C TL	16
u PVC	Unknown	29/22m			0.30	37°C TL	16
hexane, 90% & benzene, 10% Primary Class: 291 Hydrocarbons, Aliphatic and Alicyclic, Saturated							
g PE/EVAL/P	SAFETY4 4H	>240m	0	0	0.07	35°C	60
hexane, 90% & methyl ethyl ketone, 10% Primary Class: 291 Hydrocarbons, Aliphatic and Alicyclic, Saturated							
g VITON	NORTH F-091	36/40m	420/	3	0.24		41

G A R M E Class & Number/ N test chemical/ T MATERIAL NAME	MANUFACTURER & PRODUCT IDENTIFICATION	Break-through Time in Minutes	Perm-eation Rate in mg/sq m /min.	I N D E X	Thick-ness in mm	degrad. and comments	R e f s
hydraulic oil Primary Class: 860 Coals, Charcoals, Oils							
s BUTYL	MSA CHEMPRUF	<30m			n.a.		48
g NITRILE	ANSELLEDMONT 37-145	>240m	0	0	0.33		33
g NITRILE	ANSELLEDMONT 37-145	>240m	0	0	0.33	preexp. 20h	33
g PE/EVAL/PE	SAFETY4 4H	>240m	0	0	0.07	35°C	60
g PVC	KID 490	>240m	0	0	n.a.		33
g PVC	KID 490	>240m	0	0	n.a.	preexp. 20h	33
g PVC	KID VINYLPRODUKTER	>240m	0	0	0.45		33
g PVC	KID VINYLPRODUKTER	>240m	0	0	0.45	preexp. 20h	33
s PVC	MSA UPC	<60m			0.20		48
hydrazine (synonym: diamine) CAS Number: 302-01-2 Primary Class: 280 Hydrazines							
g BUTYL	BRUNSWICK BUTYL-XTR	>480m	0	0	0.63		89
g BUTYL	NORTH B-161	>480m	0	0	0.38		34
s BUTYL/NEOPRENE	MSA BETEX	>240m	0	0	n.a.		48
s CPE	STD. SAFETY	>180m			n.a.		75
g NATURAL RUBBER	ANSELL FL 200 254	>360m	0	0	051		55
g NATURAL RUBBER	ANSELL ORANGE 208	>360m	0	0	0.76		55
g NATURAL RUBBER	ANSELL PVL 040	>360m	0	0	0.46		55
g NATURAL RUBBER	MAPA-PIONEER L-118	218m	120	3	0.46		36
g NATURAL+NEOPRENE	ANSELL OMNI 276	>360m	0	0	0.56		55
g NATURAL+NEOPRENE	ANSELL TECHNICIANS	>360m	0	0	0.43		55
g NEOPRENE	ANSELL NEOPRENE 530	>360m	0	0	0.46		55
g NEOPRENE	ANSELLEDMONT 29-870	>960m	0	0	0.51		34
g NEOPRENE	MAPA-PIONEER N-44	>480m	0	0	0.74		36
g NEOPRENE/NATURAL	ANSELL CHEMI-PRO	>360m	0	0	0.72		55
g NITRILE	ANSELL CHALLENGER	>360m	0	0	0.38		55
g NITRILE	MAPA-PIONEER A-14	>480m	0	0	0.54		36
g NITRILE	NORTH LA-142G	>480m	0	0	0.36		34
g PE/EVAL/PE	SAFETY4 4H	>240m	0	0	0.07	35°C 80%	60
g PVAL	EDMONT 25-950				0.40	degrad.	34
g PVC	ANSELLEDMONT 34-100	>480m	0	0	0.25		34
g PVC	MAPA-PIONEER V-20	>480m	0	0	0.45		36
s SARANEX-23	DU PONT TYVEK SARAN	>480m	0	0	n.a.		8
s SARANEX-23 2-PLY	DU PONT TYVEK SARAN	>480m	0	0	n.a.		8
s TEFLON	CHEMFAB CHALL. 5200	>480m	0	0	n.a.		80
s UNKNOWN MATERIAL	CHEMRON CHEMREL	>1440m	0	0	n.a.		67
s UNKNOWN MATERIAL	CHEMRON CHEMREL MAX	>1440m	0	0	n.a.		67
s UNKNOWN MATERIAL	DU PONT BARRICADE	>480m	0	0	n.a.		8
s UNKNOWN MATERIAL	LIFE-GUARD RESPONDE	>480m	0	0	n.a.		77
g VITON	NORTH F-091				0.26	degrad.	34
s VITON/NEOPRENE	MSA VAUTEX	>240m	0	0	n.a.		48

G A R M E Class & Number/ N test chemical/ T MATERIAL NAME	MANUFACTURER & PRODUCT IDENTIFICATION	Break- through Time in Minutes	Perm- eation Rate in mg/sq m /min.	I N D E X	Thick- ness in mm	degrad. and comments	R e f s
hydrazine hydrate CAS Number: 10217-52-4 Primary Class: 280 Hydrazines							
s UNKNOWN MATERIAL	LIFE-GUARD RESPONDE	>240m		0	n.a.		77
s VITON/CHLOROBUTYL	LIFE-GUARD VC-100	>240m		0	n.a.		77
hydrazine, 30-70% Primary Class: 280 Hydrazines							
g BUTYL	NORTH B-161	>480m	0	0	0.45		7
g NATURAL RUBBER	ANSELLEDMONT 36-124	>360m	0	0	0.46		6
g NATURAL RUBBER	ANSELLEDMONT 392	150m	<90	2	0.48		6
g NATURAL RUBBER	ANSELLEDMONT 46-320	>360m	0	0	0.31		6
g NATURAL RUBBER	ANSELLEDMONT 46-320	>360m	0	0	0.31		6
g NATURAL+NEOPRENE	PLAYTEX ARGUS 123	>480m	0	0	n.a.		72
g NEOPRENE	ANSELLEDMONT NEOX	>360m	0	0	n.a.		6
g NITRILE	ANSELLEDMONT 37-165	>360m	0	0	0.60		6
g NITRILE	NORTH LA-142G	>480m	0	0	0.36		7
g PVAL	ANSELLEDMONT PVA				n.a.	degrad.	6
g PVAL	EDMONT 15-552				n.a.	degrad.	6
g PVC	ANSELLEDMONT M GRIP	>360m	0	0	n.a.		6
g UNKNOWN MATERIAL	NORTH SILVERSHIELD	126m	71	2	0.10		7
g VITON	NORTH F-091				0.25	degrad.	7
hydrobromic acid, 30-70% CAS Number: 10035-10-6 Primary Class: 370 Inorganic Acids							
s UNKNOWN MATERIAL	LIFE-GUARD RESPONDE	>480m	0	0	n.a.		77
hydrochloric acid, 30-70% CAS Number: 7647-01-0 Primary Class: 370 Inorganic Acids							
g BUTYL	BEST 878	>480m	0	0	0.75		53
g BUTYL	BRUNSWICK BUTYL STD	>480m	0	0	0.63		89
g BUTYL	BRUNSWICK BUTYL-XTR	>480m	0	0	0.63		89
s BUTYL	MSA CHEMPRUF	>480m	0	0	n.a.		48
g BUTYL/ECO	BRUNSWICK BUTYL-POL	>480m	0	0	0.63		89
s BUTYL/NEOPRENE	MSA BETEX	>480m	0	0	n.a.		48
u CPE	ILC DOVER	>180m			n.a.		29
g HYPALON	COMASEC DIPCO	>480m		0	0.53		40
g NAT+NEOPR+NITRILE	MAPA-PIONEER TRIONI	101m	228	3	0.48		36
g NATURAL RUBBER	ANSELL CONFORM 4205	180m			0.13		55

G A R M E N T	Class & Number/ test chemical/ MATERIAL NAME	MANUFACTURER & PRODUCT IDENTIFICATION	Break-through Time in Minutes	Perm-eation Rate in mg/sq m /min.	I N D E X	Thick-ness in mm	degrad. and comments	R e f s
g	NATURAL RUBBER	ANSELL CONT ENVIRON	>360m	0	0	0.55		55
g	NATURAL RUBBER	ANSELL FL 200 254	360m	0	0	0.51		55
g	NATURAL RUBBER	ANSELL ORANGE 208	>360m	0	0	0.76		55
g	NATURAL RUBBER	ANSELL PVL 040	>360m	0	0	0.46		55
g	NATURAL RUBBER	ANSELL STERILE 832	>360m		0	0.23		55
g	NATURAL RUBBER	ANSELLEDMONT 36-124	>300m	0	0	0.46		6
g	NATURAL RUBBER	ANSELLEDMONT 392	299m	0	0	0.48		6
g	NATURAL RUBBER	COMASEC FLEXIGUM	>480m	0	0	0.95		40
g	NATURAL RUBBER	MAPA-PIONEER L-118	210m	13080	5	0.46		36
g	NATURAL RUBBER	MARIGOLD BLACK HEVY	447m	7	1	0.65		79
g	NATURAL RUBBER	MARIGOLD FEATHERLIT	1m	652	4	0.15	degrad.	79
g	NATURAL RUBBER	MARIGOLD MED WT EXT	113m	76	2	0.45		79
g	NATURAL RUBBER	MARIGOLD MEDICAL	40m	412	3	0.28	degrad.	79
g	NATURAL RUBBER	MARIGOLD ORANGE SUP	>480m	0	0	0.73		79
g	NATURAL RUBBER	MARIGOLD R SURGEONS	40m	412	3	0.28	degrad.	79
g	NATURAL RUBBER	MARIGOLD RED LT WT	103m	124	3	0.43		79
g	NATURAL RUBBER	MARIGOLD SENSOTECH	40m	412	3	0.28	degrad.	79
g	NATURAL RUBBER	MARIGOLD SUREGRIP	347m	28	2	0.58		79
g	NATURAL RUBBER	MARIGOLD YELLOW	82m	220	3	0.38		79
u	NATURAL RUBBER	Unknown	>60m			n.a.	degrad.	9
u	NATURAL RUBBER	Unknown	> 60m			0.43		9
g	NATURAL+NEOPRENE	ANSELL OMNI 276	>180m			0.56		55
g	NATURAL+NEOPRENE	ANSELL TECHNICIANS	300m	0	0	0.43		55
g	NATURAL+NEOPRENE	MARIGOLD FEATHERWT	72m	268	3	0.35	degrad.	79
g	NATURAL+NEOPRENE	PLAYTEX ARGUS 123	265m	120	3	n.a.		72
u	NATURAL+NEOPRENE	Unknown	> 60m			n.a.		9
g	NEOPRENE	ANSELL NEOPRENE 520	>360m	0	0	0.46		55
g	NEOPRENE	ANSELLEDMONT 29-840	>360m	0	0	0.31		6
g	NEOPRENE	ANSELLEDMONT NEOX	>360m	0	0	n.a.		6
g	NEOPRENE	BEST 32	>480m	0	0	n.a.		53
g	NEOPRENE	BEST 6780	>480m	0	0	n.a.		53
g	NEOPRENE	BRUNSWICK NEOPRENE	>480m	0	0	0.63		89
g	NEOPRENE	COMASEC COMAPRENE	>360m	0	0	n.a.		40
g	NEOPRENE	MAPA-PIONEER N-44	>480m	0	0	0.56		36
u	NEOPRENE	Unknown	>60m			0.61		9
g	NEOPRENE/NATURAL	ANSELL CHEMI-PRO	>360m	0	0	0.72		55
g	NITRILE	ANSELL CHALLENGER	>360m	0	0	0.38		55
g	NITRILE	ANSELLEDMONT 37-165	>360m	0	0	0.60		6
g	NITRILE	BEST 727	>480m	0	0	0.38		53
g	NITRILE	COMASEC COMATRIL	330m	6600	4	0.55		40
g	NITRILE	COMASEC COMATRIL SU	>480m	0	0	0.60		40
g	NITRILE	COMASEC FLEXITRIL	18m	2400	4	n.a.		40
g	NITRILE	MAPA-PIONEER A-14	>480m	0	0	0.56		36
g	NITRILE	MARIGOLD BLUE	>480m	0	0	0.45		79
g	NITRILE	MARIGOLD GREEN SUPA	>480m	0	0	0.30		79
g	NITRILE	MARIGOLD NITROSOLVE	>480m	0	0	0.75		79

G A R M E Class & Number/ N test chemical/ T MATERIAL NAME	MANUFACTURER & PRODUCT IDENTIFICATION	Break- through Time in Minutes	Perm- eation Rate in mg/sq m /min.	I N D E X	Thick- ness in mm	degrad. and comments	R e f s
g NITRILE	NORTH LA-142G				0.38	degrad.	7
u NITRILE	Unknown	>60m			0.45		9
g NITRILE+PVC	COMASEC MULTIPLUS	>480m	0	0	n.a.		40
c PE	DU PONT TYVEK QC	81m	28	2	n.a.		8
g PE/EVAL/PE	SAFETY4 4H	>240m	0	0	0.07	35° C	60
g PVAL	ANSELLEDMONT PVA				n.a.	degrad.	6
g PVAL	EDMONT 15-552				n.a.	degrad.	6
g PVC	ANSELLEDMONT M GRIP	>300m	0	0	n.a.		6
g PVC	COMASEC MULTIPOST	>480m	0	0	n.a.		40
g PVC	COMASEC MULTITOP	>480m	0	0	n.a.		40
g PVC	COMASEC NORMAL	>480m	0	0	n.a.		40
g PVC	COMASEC OMNI	>300m	48	2	n.a.		40
g PVC	MAPA-PIONEER V-20	>480m	0	0	0.51		36
s PVC	MSA UPC	>480m	0	0	0.20		48
u PVC	Unknown	24m			0.22		9
u PVC	Unknown	>60m			n.a.		9
s SARANEX-23	DU PONT TYVEK SARAN	>2880m	0	0	0.15		8
s UNKNOWN MATERIAL	CHEMRON CHEMREL MAX	>1440m	0	0	n.a.		67
c UNKNOWN MATERIAL	CHEMRON CHEMTUFF	>480m	0	0	n.a.		67
s UNKNOWN MATERIAL	DU PONT BARRICADE	>480m	0	0	n.a.		8
s UNKNOWN MATERIAL	LIFE-GUARD RESPONDE	>480m		0	n.a.		77
g VITON	NORTH F-121	>60m			0.31		9
s VITON/CHLOROBUTYL	LIFE-GUARD VC-100	>240m		0	n.a.		77
u VITON/CHLOROBUTYL	Unknown	>180m			0.36		29
s VITON/NEOPRENE	MSA VAUTEX	>480m	0	0	0.58		48

hydrochloric acid, <30%
Primary Class: 370   Inorganic Acids

g NATURAL RUBBER	ANSELL CONFORM 4205	>360m	0	0	0.13		55
g NATURAL RUBBER	ANSELL CONT ENVIRON	>480m	0	0	0.55		55
g NATURAL RUBBER	ANSELL FL 200 254	>360m		0	0.51		55
g NATURAL RUBBER	ANSELL ORANGE 208	>360m	0	0	0.76		55
g NATURAL RUBBER	ANSELL PVL 040	>360m	0	0	0.40	10%	55
g NATURAL RUBBER	ANSELL STERILE 832	>360m	0	0	0.23		55
g NATURAL RUBBER	ANSELLEDMONT 36-124	>360m	0	0	0.46		6
g NATURAL RUBBER	ANSELLEDMONT 392	>360m	0	0	0.48		6
g NATURAL+NEOPRENE	ANSELL OMNI 276	>360m	0	0	0.56	10%	55
g NATURAL+NEOPRENE	ANSELL TECHNICIANS	>360m	0	0	0.43	10%	55
g NEOPRENE	ANSELL NEOPRENE 530	>360m	0	N	0.46	10%	55
g NEOPRENE	ANSELLEDMONT 29-840	>360m	0	0	0.31		6
g NEOPRENE	ANSELLEDMONT NEOX	>360m	0	0	n.a.		6
g NEOPRENE/NATURAL	ANSELL CHEMI-PRO	>360m	0	0	0.72	10%	55
g NITRILE	ANSELL CHALLENGER	>360m	0	0	0.38		55
g NITRILE	ANSELLEDMONT 37-165	>360m	0	0	0.60		6
g PE/EVAL/PE	SAFETY4 4H	>240m	0	0	0.07	35°C 7%	60

G A R M E Class & Number/ N test chemical/ T MATERIAL NAME	MANUFACTURER & PRODUCT IDENTIFICATION	Break- through Time in Minutes	Perm- eation Rate in mg/sq m /min.	I N D E X	Thick- ness in mm	degrad. and comments	R e f s
g PVAL	ANSELLEDMONT PVA				n.a.	degrad.	6
g PVAL	EDMONT 15-552				n.a.	degrad.	6
g PVC	ANSELLEDMONT M GRIP	>360m	0	0	n.a.		6

hydrofluoric acid & ammonium trifluoride (synonym: oxide etch)
Primary Class: 820   Etching Products

g NAT+NEOPR+NITRILE	MAPA-PIONEER TRIONI	>480m	0	0	0.46		36
g NATURAL RUBBER	ANSELL CONT ENVIRON	>240m	0	0	0.55		55
g NATURAL RUBBER	ANSELL FL 200 254	>240m	0	0	0.51		55
g NATURAL RUBBER	ANSELL PVL 040	>240m	0	0	0.46		55
g NATURAL+NEOPRENE	ANSELL OMNI 276	>240m	0	0	0.56		55
g NEOPRENE	ANSELL NEOPRENE 530	>240m	0	0	0.46		55
g NEOPRENE/NATURAL	ANSELL CHEMI-PRO	>240m	0	0	0.72		55

hydrofluoric acid & nitric acid & acetic acid (synonym: silicon etch)
Primary Class: 820   Etching Products

g NATURAL RUBBER	ANSELLEDMONT 36-124				0.46	degrad.	6
g NEOPRENE	ANSELLEDMONT 29-840	>360m	0	0	0.31		6
g NEOPRENE	ANSELLEDMONT NEOX	>360m	0	0	n.a.		6
g NITRILE	ANSELL 650				n.a.	degrad.	55
g NITRILE	ANSELL CHALLENGER				0.38	degrad.	55
g NITRILE	ANSELLEDMONT 37-165				0.60	degrad.	6
g PVAL	EDMONT 15-552				n.a.	degrad.	6
g PVC	ANSELLEDMONT SNORKE	150m			n.a.		6

hydrofluoric acid, 30-70%
CAS Number: 7664-39-3
Primary Class: 370   Inorganic Acids

g BUTYL	BRUNSWICK BUTYL STD	>480m	0	0	0.63	47%	89
g BUTYL	BRUNSWICK BUTYL-XTR	>480m	0	0	0.63	47%	89
g BUTYL/ECO	BRUNSWICK BUTYL-POL	>480m	0	0	0.63	47%	89
g NAT+NEOPR+NITRILE	MAPA-PIONEER TRIONI	91m	48	2	0.48	48%	36
g NATURAL RUBBER	ANSELL CONFORM 4205	150m			0.13	48%	55
g NATURAL RUBBER	ANSELL CONT ENVIRON	>480m	0	0	0.51	48%	55
g NATURAL RUBBER	ANSELL FL 200 254	90m			0.51	48%	55
g NATURAL RUBBER	ANSELL ORANGE 208	90m			0.76	48%	55
g NATURAL RUBBER	ANSELL STERILE 832	240m	0	0	0.23	48%	55
g NATURAL RUBBER	ANSELLEDMONT 30-139	210m			0.25	48%	42
g NATURAL RUBBER	ANSELLEDMONT 36-124	210m			0.46	48%	6
g NATURAL RUBBER	ANSELLEDMONT 392	190m			0.48	48%	6
g NATURAL RUBBER	MAPA-PIONEER L-118	>480m	0	0	0.46	48%	36
g NATURAL RUBBER	MARIGOLD BLACK HEVY	137m	143	3	0.82		79

G A R M E Class & Number/ N test chemical/ T MATERIAL NAME	MANUFACTURER & PRODUCT IDENTIFICATION	Break-through Time in Minutes	Perm-eation Rate in mg/sq m /min.	I N D E X	Thick-ness in mm	degrad. and comments	R e f s
g NATURAL RUBBER	MARIGOLD FEATHERLIT	69m	63	2	0.15		79
g NATURAL RUBBER	MARIGOLD MED WT EXT	50m	370	3	0.45		79
g NATURAL RUBBER	MARIGOLD MEDICAL	61m	191	3	0.28		79
g NATURAL RUBBER	MARIGOLD ORANGE SUP	170m	58	2	0.73		79
g NATURAL RUBBER	MARIGOLD R SURGEONS	61m	191	3	0.28		79
g NATURAL RUBBER	MARIGOLD RED LT WT	52m	344	3	0.43		79
g NATURAL RUBBER	MARIGOLD SENSOTECH	61m	191	3	0.28		79
g NATURAL RUBBER	MARIGOLD SUREGRIP	105m	228	3	0.89		79
g NATURAL RUBBER	MARIGOLD YELLOW	55m	293	3	0.38		79
g NATURAL+NEOPRENE	ANSELL OMNI 276	180m			0.56	48%	55
g NATURAL+NEOPRENE	ANSELL TECHNICIANS	180m			0.48	48%	55
g NATURAL+NEOPRENE	MARIGOLD FEATHERWT	56m	268	3	0.35		79
g NATURAL+NEOPRENE	PLAYTEX ARGUS 123	>480m	0	0	n.a.		72
g NEOPRENE	ANSELL NEOPRENE 520	>360m	0	0	0.46	48%	55
g NEOPRENE	ANSELLEDMONT 29-840	60m			0.31	48%	6
g NEOPRENE	ANSELLEDMONT NEOX	75m			n.a.	48%	6
g NEOPRENE	MAPA-PIONEER N-44	>480m	0	0	0.56	48%	36
g NEOPRENE/NATURAL	ANSELL CHEMI-PRO	>360m	0	0	0.72	48%	55
g NITRILE	ANSELL CHALLENGER	>180m			0.38	48%	55
g NITRILE	ANSELLEDMONT 37-165	120m			0.60	48%	6
g NITRILE	ANSELLEDMONT 49-125	30m			0.23		42
g NITRILE	ANSELLEDMONT 49-155	180m			0.38		42
g NITRILE	MAPA-PIONEER A-14	132m	282	3	0.56		36
g NITRILE	MARIGOLD BLUE	131m	811	3	0.43		79
g NITRILE	MARIGOLD GREEN SUPA	46m	889	3	0.40		79
g NITRILE	MARIGOLD NITROSOLVE	149m	796	3	0.75	degrad.	79
b NITRILE+PUR+PVC	BATA HAZMAX	>480m	0	0	n.a.		86
c PE	DU PONT TYVEK QC	180m	0.8	1	n.a.	50%	8
g PE/EVAL/PE	SAFETY4 4H	30m			0.07	49% 35°C	60
g PE/EVAL/PE	SAFETY4 4H	120m			0.07	40% 35°C	60
g PE/EVAL/PE	SAFETY4 4H	>240m	0	0	0.07	30%	60
g PE/EVAL/PE	SAFETY4 4H	120m			0.07	30% 35°C	60
g PE/EVAL/PE	SAFETY4 4H	>240m	0	0	0.07	40%	60
g PE/EVAL/PE	SAFETY4 4H	120m			0.07	40% 35°C	60
g PE/EVAL/PE	SAFETY4 4H	>240m	0	0	0.07	49%	60
g PE/EVAL/PE	SAFETY4 4H	30m			0.07	49% 35°C	60
g PVAL	ANSELLEDMONT PVA				n.a.	degrad.	6
g PVAL	EDMONT 15-552				n.a.	degrad.	6
g PVC	ANSELLEDMONT SNORKE	40m			n.a.	48%	6
g PVC	MAPA-PIONEER V-20	110m	22	2	0.51		36
s SARANEX-23	DU PONT TYVEK SARAN	>400m	0	0	n.a.	50%	8
s TEFLON	CHEMFAB CHALL. 5200	>480m	0	0	n.a.	50%	80
s UNKNOWN MATERIAL	CHEMRON CHEMREL	>1440m	0	0	n.a.	48%	67
s UNKNOWN MATERIAL	CHEMRON CHEMREL MAX	>1440m	0	0	n.a.	48%	67
c UNKNOWN MATERIAL	CHEMRON CHEMTUFF	15m	0.2	2	n.a.	48%	67
s UNKNOWN MATERIAL	DU PONT BARRICADE	>480m	0	0	n.a.	50%	8
s UNKNOWN MATERIAL	LIFE-GUARD RESPONDE	>480m		0	n.a.	48%	77

G A R M E Class & Number/ N test chemical/ T MATERIAL NAME	MANUFACTURER & PRODUCT IDENTIFICATION	Break- through Time in Minutes	Perm- eation Rate in mg/sq m /min.	I N D E X	Thick- ness in mm	degrad. and comments	R e f s
hydrofluoric acid, <30% Primary Class: 370 Inorganic Acids							
g NAT+NEOPR+NITRILE	MAPA-PIONEER TRIONI	>720m	0	0	0.48		36
g PE/EVAL/PE	SAFETY4 4H	>240m	0	0	0.07	10%	60
g PE/EVAL/PE	SAFETY4 4H	120m			0.07	10% 35°C	60
hydrofluoric acid, >70% Primary Class: 370 Inorganic Acids							
s BUTYL/NEOPRENE	MSA BETEX	>480m	0	0	n.a.		48
s UNKNOWN MATERIAL	DU PONT BARRICADE	67m	28	2	n.a.	92%, 90°	8
s VITON/NEOPRENE	MSA VAUTEX	>480m	0	0	n.a.		48
hydrogen chloride (synonym: hydrochloric acid, anhydrous) Primary Class: 350 Inorganic Gases and Vapors							
g BUTYL	BRUNSWICK BUTYL STD	>480m	0	0	0.71		89
g BUTYL	BRUNSWICK BUTYL-XTR	>480m	0	0	0.70		89
s BUTYL	WHEELER ACID KING	>480m	0	0	n.a.		85
g BUTYL/ECO	BRUNSWICK BUTYL-POL	>480m	0	0	0.63		89
g NEOPRENE	BRUNSWICK NEOPRENE	>480m	0	0	0.81		89
c PE	DU PONT TYVEK QC	<1m	>1000	5	n.a.		8
s PVC	WHEELER ACID KING	>480m	0	0	n.a.		85
s SARANEX-23	DU PONT TYVEK SARAN	>480m	0		n.a.		8
s TEFLON	CHEMFAB CHALL. 5200	>480m	0	0	n.a.		80
s TEFLON	LIFE-GUARD TEFGUARD	>480m	0	0	n.a.		77
s TEFLON	WHEELER ACID KING	>480m	0	0	n.a.		85
s UNKNOWN MATERIAL	DU PONT BARRICADE	>480m	0	0	n.a.		8
s UNKNOWN MATERIAL	LIFE-GUARD RESPONDE	>480m	0	0	n.a.		77
s VITON/CHLOROBUTYL	LIFE-GUARD VC-100	>480m	0	0	n.a.		77
hydrogen cyanide CAS Number: 74-90-8 Primary Class: 345 Inorganic Cyano Compounds							
u BUTYL	Unknown	60m	0.15	1	0.38		61
s BUTYL/NEOPRENE	MSA BETEX	>240m	0	0	n.a.		48
c PE	DU PONT TYVEK QC	60m	1100	4	n.a.		8
c PE	DU PONT TYVEK QC	60m	1	1	n.a.		61
g PE/EVAL/PE	SAFETY4 4H	>240m	0	0	0.07		60
u PVC	Unknown	30m	3	2	0.79		61
s TEFLON	CHEMFAB CHALL. 5200	>480m	0	0	n.a.		80
s UNKNOWN MATERIAL	DU PONT BARRICADE	108m	5	1	n.a.		8

G A R M E Class & Number/ N test chemical/ T MATERIAL NAME	MANUFACTURER & PRODUCT IDENTIFICATION	Break- through Time in Minutes	Perm- eation Rate in mg/sq m /min.	I N D E X	Thick- ness in mm	degrad. and comments	R e f s
s UNKNOWN MATERIAL	LIFE-GUARD RESPONDE	>180m	0	1	n.a.		77
s VITON/NEOPRENE	MSA VAUTEX	80m	7	1	n.a.		48

hydrogen fluoride (synonym: hydrofluoric acid, anhydrous)
Primary Class: 350   Inorganic Gases and Vapors

s CPE	STD. SAFETY	>180m	0	1	n.a.		75
m NEOPRENE	STD. SAFETY BOOTIE	173m	21	2	n.a.		75
c PE	DU PONT TYVEK QC	13m	0.06	3	n.a.	10°C	8
c PE	DU PONT TYVEK QC	7m	60	4	n.a.	25°C	8
v PVC	STD. SAFETY	25m	25	3	n.a.		75
b PVC	STD. SAFETY GL-20	30m	9060	4	n.a.		75
g PVC	STD. SAFETY SD-6100	13m	22980	5	n.a.		75
s PVC	STD. SAFETY WG-20	30m	>4000	4	n.a.		75
s SARANEX-23	DU PONT TYVEK SARAN	>30m			n.a.	10°C	8
s SARANEX-23	DU PONT TYVEK SARAN	20m	30	3	n.a.	25°C	8
s TEFLON	CHEMFAB CHALL. 5200	>480m	0	0	n.a.		80
s UNKNOWN MATERIAL	DU PONT BARRICADE	91m	3570	4	n.a.		8
s UNKNOWN MATERIAL	LIFE-GUARD RESPONDE	>180m	0	1	n.a.		77

hydrogen peroxide, 30-70%
CAS Number: 7722-84-1
Primary Class: 300   Peroxides

g NAT+NEOPR+NITRILE	MAPA-PIONEER TRIONI	>960m	0	0	0.48		36
g NATURAL RUBBER	ANSELL CONFORM 4205	150m			0.13		55
g NATURAL RUBBER	ANSELL CONT ENVIRON	>480m	0	0	0.55		55
g NATURAL RUBBER	ANSELL FL 200 254	>360m	0	0	0.51		55
g NATURAL RUBBER	ANSELL ORANGE 208	>360m	0	0	0.76		55
g NATURAL RUBBER	ANSELL PVL 040	>360m	0	0	0.46		55
g NATURAL RUBBER	ANSELL STERILE 832	>360m		0	0.23		55
g NATURAL RUBBER	ANSELLEDMONT 36-124	>360m	0	0	0.46		6
g NATURAL RUBBER	ANSELLEDMONT 392	>360m		0	0.48		6
g NATURAL RUBBER	MARIGOLD BLACK HEVY	>480m	0	0	0.65		79
g NATURAL RUBBER	MARIGOLD FEATHERLIT	>480m	0	0	0.15		79
g NATURAL RUBBER	MARIGOLD MED WT EXT	>480m	0	0	0.45		79
g NATURAL RUBBER	MARIGOLD MEDICAL	>480m	0	0	0.28		79
g NATURAL RUBBER	MARIGOLD ORANGE SUP	>480m	0	0	0.73		79
g NATURAL RUBBER	MARIGOLD R SURGEON	>480m	0	0	0.28		79
g NATURAL RUBBER	MARIGOLD RED LT WT	>480m	0	0	0.43		79
g NATURAL RUBBER	MARIGOLD SENSOTECH	>480m	0	0	0.28		79
g NATURAL RUBBER	MARIGOLD SUREGRIP	>480m	0	0	0.58		79
g NATURAL RUBBER	MARIGOLD YELLOW	>480m	0	0	0.38		79
g NATURAL+NEOPRENE	ANSELL OMNI 276	>360m	0	0	0.56		55
g NATURAL+NEOPRENE	ANSELL TECHNICIANS	>360m	0	0	0.43		55

G A R M E Class & Number/ N test chemical/ T MATERIAL NAME	MANUFACTURER & PRODUCT IDENTIFICATION	Break- through Time in Minutes	Perm- eation Rate in mg/sq m /min.	I N D E X	Thick- ness in mm	degrad. and comments	R e f s
g NATURAL+NEOPRENE	MARIGOLD FEATHERWT	>480m	0	0	0.35		79
g NEOPRENE	ANSELL NEOPRENE 530	>360m	0	0	0.46		55
g NEOPRENE	ANSELLEDMONT 29-840	5m			0.31		6
g NEOPRENE	ANSELLEDMONT NEOX	7m			n.a.		6
g NEOPRENE/NATURAL	ANSELL CHEMI-PRO	>360m	0	0	0.72		55
g NITRILE	ANSELL CHALLENGER	>360m	0	0	0.38		55
g NITRILE	ANSELLEDMONT 37-165	>360m	0	0	0.60		6
g NITRILE	MARIGOLD BLUE	>480m	0	0	0.45		79
g NITRILE	MARIGOLD GREEN SUPA	>480m	0	0	0.30		79
g NITRILE	MARIGOLD NITROSOLVE	>480m	0	0	0.75		79
g PE/EVAL/PE	SAFETY4 4H	>240m		0	0.07		60
g PVAL	ANSELLEDMONT PVA				n.a.	degrad.	6
g PVAL	EDMONT 15-552				n.a.	degrad.	6
g PVC	ANSELLEDMONT M GRIP	>360m	0	0	n.a.		6
s UNKNOWN MATERIAL	LIFE-GUARD RESPONDE	>480m	0	0	n.a.		77
g UNKNOWN MATERIAL	MARIGOLD R MEDIGLOV	>480m	0	0	0.03		79

hydrogen sulfide
CAS Number: 7783-06-4
Primary Class: 350   Inorganic Gases and Vapors

s UNKNOWN MATERIAL	LIFE-GUARD RESPONDE	>180m	0	1	n.a.		77

hydroquinone (synonym: p-benzenediol)
CAS Number: 123-31-9
Primary Class: 316   Hydroxy Compounds, Aromatic (Phenols)

g NATURAL RUBBER	ANSELL CONFORM 4205	>360m	0	0	0.13		55
g NATURAL RUBBER	ANSELL FL 200 254	>360m	0	0	0.51		55
g NATURAL RUBBER	ANSELL ORANGE 208	>360m	0	0	0.76		55
g NATURAL RUBBER	ANSELL PVL 040	>360m	0	0	0.46		55
g NATURAL RUBBER	ANSELL STERILE 832	>360m	0	0	0.23		55
g NATURAL RUBBER	ANSELLEDMONT 36-124	>360m	0	0	0.46		6
g NATURAL RUBBER	ANSELLEDMONT 392	>360m	0	0	0.48		6
g NATURAL+NEOPRENE	ANSELL OMNI 276	>360m	0	0	0.56		55
g NATURAL+NEOPRENE	ANSELL TECHNICIANS	>360m	0	0	0.43		55
g NEOPRENE	ANSELL NEOPRENE 530	>360m	0	0	0.46		55
g NEOPRENE	ANSELLEDMONT 29-840	>360m	0	0	0.31		6
g NEOPRENE	ANSELLEDMONT NEOX	>360m	0	0	n.a.		6
g NEOPRENE/NATURAL	ANSELL CHEMI-PRO	>360m	0	0	0.72		55
g NITRILE	ANSELL CHALLENGER	>360m	0	0	0.38		55
g NITRILE	ANSELLEDMONT 37-165	>360m	0	0	0.60		6
g PE/EVAL/PE	SAFETY4 4H	>240m	0	0	0.07	33%/ethanol	60
g PVAL	ANSELLEDMONT PVA				n.a.	degrad.	6
g PVAL	EDMONT 15-552				n.a.	degrad.	6

G A R M E Class & Number/ N test chemical/ T MATERIAL NAME	MANUFACTURER & PRODUCT IDENTIFICATION	Break- through Time in Minutes	Perm- eation Rate in mg/sq m /min.	I N D E X	Thick- ness in mm	R e f degrad. and s comments

iodine, solid
CAS Number: 7553-56-2
Primary Class: 330   Elements

c PE	DU PONT TYVEK QC	440m	300	3	n.a.	8
s SARANEX-23	DU PONT TYVEK SARAN	>480m	0	0	n.a.	8

---

isoamyl acetate (synonym: isopentyl acetate)
CAS Number: 123-92-2
Primary Class: 222   Esters, Carboxylic, Acetates

g NATURAL RUBBER	ACKWELL 5-109				0.15	degrad. 34
g NATURAL RUBBER	ANSELLEDMONT 46-320	5m	11000	5	0.31	5
g NEOPRENE	ANSELLEDMONT 29-870				0.46	degrad. 34
g NEOPRENE	ANSELLEDMONT NEOX	12m	1400	5	n.a.	5
g NEOPRENE	MAPA-PIONEER N-73	30m	3100	4	0.46	5
g NITRILE	ANSELLEDMONT 37-155				0.40	degrad. 34
g NITRILE	ANSELLEDMONT 37-175	65m	630	3	0.37	5
g PE	ANSELLEDMONT 35-125	2m	200	4	0.03	5
g PVC	ANSELLEDMONT 34-100				0.20	degrad. 34
g PVC	MAPA-PIONEER V-20	5m	16000	5	0.31	5
g VITON	NORTH F-091				0.25	degrad. 34

---

isoamyl nitrite (synonyms: 3-methylbutyl nitrite; isopentyl nitrite)
CAS Number: 110-46-3
Primary Class: 510   Nitrates and Nitrites

g BUTYL	NORTH B-161				0.65	degrad. 34
g NATURAL RUBBER	ACKWELL 5-109				0.15	degrad. 34
g NEOPRENE	ANSELLEDMONT 29-870	47m	2238	4	0.51	degrad. 34
g NITRILE	ANSELLEDMONT 37-155	173m	96	2	0.38	34
g PVAL	EDMONT 25-950	>480m	0	0	0.25	34
g PVC	ANSELLEDMONT 34-100				0.20	degrad. 34
g VITON	NORTH F-091	68m	552	3	0.23	34

---

isobutyl acrylate (synonym: isobutyl 2-propenoate)
CAS Number: 106-63-8
Primary Class: 223   Esters, Carboxylic, Acrylates and Methacrylates

g BUTYL	NORTH B-174	>480m	0	0	0.66	34
g NATURAL RUBBER	ACKWELL 5-109				0.15	degrad. 34
g NEOPRENE	ANSELLEDMONT 29-870				0.48	degrad. 34
g NITRILE	ANSELLEDMONT 37-155	67m	1281	4	0.37	degrad. 34
g PVAL	EDMONT 25-950	>480m	0	0	0.57	34
g PVC	ANSELLEDMONT 34-100	1m	2052	5	0.20	degrad. 34
g VITON	NORTH F-091				0.26	degrad. 34

		Break- through Time in Minutes	Perm- eation Rate in mg/sq m /min.	I N D E X	Thick- ness in mm	degrad. and comments	R e f s
MANUFACTURER & PRODUCT IDENTIFICATION							

isobutyl alcohol　(synonyms: 2-methyl-1-propanol; isobutanol)
CAS Number: 78-83-1
Primary Class: 311　　Hydroxy Compounds, Aliphatic and Alicyclic, Primary

g BUTYL	BEST 878	>480m	0	0	0.75		53
g BUTYL	NORTH B-174	>480m	0	0	0.71		34
g BUTYL/NEOPRENE	COMASEC BUTYL PLUS	>480m	0	0	0.50		40
g NATURAL RUBBER	ANSELL CONFORM 4205	1m	3	3	0.13		55
g NATURAL RUBBER	ANSELL FL 200 254	26m	0.3		0.51		55
g NATURAL RUBBER	ANSELL ORANGE 208	69m	0.1		0.76		55
g NATURAL RUBBER	ANSELL PVL 040	84m	0.1		0.46		55
g NATURAL RUBBER	ANSELL STERILE 832	6m	1	3	0.23		55
g NATURAL RUBBER	ANSELLEDMONT 36-124	25m	<900	3	0.46		6
g NATURAL RUBBER	ANSELLEDMONT 392	15m	<90	3	0.48		6
g NATURAL RUBBER	COMASEC FLEXIGUM	>480m	0	0	0.95		40
g NATURAL RUBBER	MARIGOLD BLACK HEVY	27m	93	3	0.65		79
g NATURAL RUBBER	MARIGOLD FEATHERLIT	1m	172	4	0.15		79
g NATURAL RUBBER	MARIGOLD MED WT EXT	22m	71	3	0.45		79
g NATURAL RUBBER	MARIGOLD MEDICAL	10m	130	4	0.28		79
g NATURAL RUBBER	MARIGOLD ORANGE SUP	27m	96	3	0.73		79
g NATURAL RUBBER	MARIGOLD R SURGEONS	10m	130	4	0.28		79
g NATURAL RUBBER	MARIGOLD RED LT WT	20m	80	3	0.43		79
g NATURAL RUBBER	MARIGOLD SENSOTECH	10m	130	4	0.28		79
g NATURAL RUBBER	MARIGOLD SUREGRIP	25m	87	3	0.58		79
g NATURAL RUBBER	MARIGOLD YELLOW	17m	96	3	0.38		79
g NATURAL+NEOPRENE	ANSELL OMNI 276	29m	0.2		0.56		55
g NATURAL+NEOPRENE	ANSELL TECHNICIANS	27m	0.1		0.43		55
g NATURAL+NEOPRENE	MARIGOLD FEATHERWT	15m	105	3	0.35		79
g NATURAL+NEOPRENE	PLAYTEX ARGUS 123	79m	<60	2	n.a.		72
g NEOPRENE	ANSELL NEOPRENE 530	>360m	0	0	0.46		55
g NEOPRENE	ANSELLEDMONT 29-840	10m	<9	3	0.31		6
g NEOPRENE	ANSELLEDMONT 29-870	>480m	0	0	0.46		34
g NEOPRENE	ANSELLEDMONT NEOX	>360m	0	0	n.a.		6
g NEOPRENE	BEST 32	>480m	0	0	n.a.		53
g NEOPRENE	BEST 6780	>480m	0	0	n.a.		53
g NEOPRENE	COMASEC COMAPRENE	>360m	0	0	n.a.		40
g NEOPRENE/NATURAL	ANSELL CHEMI-PRO	44m	0.1		0.72		55
g NITRILE	ANSELL CHALLENGER	>360m	0	0	0.38		55
g NITRILE	ANSELLEDMONT 37-155	>480m	0	0	0.46		34
g NITRILE	ANSELLEDMONT 37-165	>360m	0	0	0.60		6
g NITRILE	BEST 22R	>480m	0	0	n.a.		53
g NITRILE	BEST 727	>480m	0	0	0.38		53
g NITRILE	COMASEC COMATRIL	306m	18	2	0.55		40
g NITRILE	COMASEC COMATRIL SU	>480m	0	0	0.60		40
g NITRILE	COMASEC FLEXITRIL	70m	1320	4	n.a.		40
g NITRILE	MARIGOLD BLUE	>480m	0	0	0.45		79
g NITRILE	MARIGOLD GREEN SUPA	>480m	0	0	0.30		79
g NITRILE	MARIGOLD NITROSOLVE	>480m	0	0	0.75		79

G A R M E Class & Number/ N test chemical/ T MATERIAL NAME	MANUFACTURER & PRODUCT IDENTIFICATION	Break- through Time in Minutes	Perm- eation Rate in mg/sq m /min.	I N D E X	Thick- ness in mm	degrad. and comments	R e f s
g NITRILE+PVC	COMASEC MULTIPLUS	>480m	0	0	n.a.		40
g PE/EVAL/PE	SAFETY4 4H	>240m	0	0	0.07	35°C	60
g PVAL	ANSELLEDMONT PVA				n.a.	degrad.	6
g PVAL	EDMONT 15-552				n.a.	degrad.	6
g PVAL	EDMONT 25-950				0.35	degrad.	34
g PVC	ANSELLEDMONT 34-100				0.20	degrad.	34
g PVC	ANSELLEDMONT SNORKE	10m	<90	4	n.a.		6
g PVC	COMASEC MULTIPOST	190m	132	3	n.a.		40
g PVC	COMASEC MULTITOP	294m	108	3	n.a.		40
g PVC	COMASEC NORMAL	120m	150	3	n.a.		40
g PVC	COMASEC OMNI	80m	240	3	n.a.		40
g VITON	NORTH F-091	>480m	0	0	0.30		34

isobutyl nitrite
CAS Number: 542-56-3
Primary Class: 510   Nitrates and Nitrites

g BUTYL	NORTH B-174	78m	1320	4	0.70	degrad.	34
g NITRILE	ANSELLEDMONT 37-155	98m	60	2	0.61	degrad.	34
g PVC	ANSELLEDMONT 34-100	2m	14520	5	0.20	degrad.	34
g VITON	NORTH F-091	20m	6180	4	0.41	degrad.	34

isobutylamine
CAS Number: 78-81-9
Primary Class: 141   Amines, Aliphatic and Alicyclic, Primary

g BUTYL	NORTH B-174	222m	575	3	0.63	degrad.	34
s BUTYL/NEOPRENE	MSA BETEX	50m	0.7	2	0.48		48
g NATURAL RUBBER	ACKWELL 5-109				0.10	degrad.	34
g NEOPRENE	ANSELLEDMONT 29-870	19m	8910	4	0.41	degrad.	34
g NITRILE	ANSELLEDMONT 37-155				0.40	degrad.	34
g PE/EVAL/PE	SAFETY4 4H	9m			0.07		60
g PE/EVAL/PE	SAFETY4 4H	2m			0.07	35°C	60
g PVAL	EDMONT 25-950	19m	8349	4	0.57		34
g PVC	ANSELLEDMONT 34-100	1m	34260	5	0.20		34
g VITON	NORTH F-091				0.25	degrad.	34
s VITON/NEOPRENE	MSA VAUTEX	45m	1.5	2	0.58		48

isobutyraldehyde
CAS Number: 78-84-2
Primary Class: 121   Aldehydes, Aliphatic and Alicyclic

g BUTYL	NORTH B-174	>480m	0	0	0.64		34
g NEOPRENE	ANSELLEDMONT 29-870	25m	486	3	0.46		34
g NITRILE	ANSELLEDMONT 37-155	19m	276	3	0.38		34
g PVAL	EDMONT 25-950	1m	156	4	0.38		34
g VITON	NORTH F-091	4m	690	4	0.24		34

G A R M E Class & Number/ N test chemical/ T MATERIAL NAME	MANUFACTURER & PRODUCT IDENTIFICATION	Break- through Time in Minutes	Perm- eation Rate in mg/sq m /min.	I N D E X	Thick- ness in mm	R degrad. e and f comments s
isooctane CAS Number: 540-84-1 Primary Class: 291 Hydrocarbons, Aliphatic and Alicyclic, Saturated						
g BUTYL	BEST 878	56m	1550	4	0.75	53
g NATURAL RUBBER	ANSELL CONT ENVIRON	15m			0.55	55
g NATURAL RUBBER	ANSELLEDMONT 36-124				0.46	degrad. 6
g NATURAL RUBBER	ANSELLEDMONT 392				0.48	degrad. 6
g NATURAL RUBBER	BEST 65NFW		600	5	n.a.	53
g NATURAL RUBBER	BEST 67NFW		2940	5	n.a.	53
g NEOPRENE	ANSELL NEOPRENE 530	17m	0.5	2	0.46	55
g NEOPRENE	ANSELLEDMONT 29-840	60m	<900	3	0.38	6
g NEOPRENE	ANSELLEDMONT NEOX	360m	<9	1	n.a.	6
g NEOPRENE	BEST 32	>480m	0	0	n.a.	53
g NEOPRENE	BEST 6780	>480m	0	0	n.a.	53
g NEOPRENE/NATURAL	ANSELL CHEMI-PRO	13m	0.9	3	0.72	55
g NITRILE	ANSELL CHALLENGER	>360m	0	0	0.38	55
g NITRILE	ANSELLEDMONT 37-165	360m	<9	1	0.60	6
g NITRILE	BEST 22R	>480m	0	0	n.a.	53
g NITRILE	BEST 727	>480m	0	0	n.a.	53
g PVAL	ANSELLEDMONT PVA	>360m	0	0	n.a.	6
g PVAL	EDMONT 15-552	40m	<90	3	n.a.	6
g PVAL	EDMONT 25-545	>360m	0	0	n.a.	6
g PVC	ANSELLEDMONT SNORKE				n.a.	degrad. 6
g PVC	BEST 812		30	5	n.a.	53
isophorone CAS Number: 78-59-1 Primary Class: 391 Ketones, Aliphatic and Alicyclic						
g NATURAL RUBBER	ACKWELL 5-109				0.15	degrad. 34
g NEOPRENE	ANSELLEDMONT 29-870	173m	240	3	0.48	degrad. 34
g NITRILE	ANSELLEDMONT 37-155				0.40	degrad. 34
g PVAL	EDMONT 25-950	>480m	0	0	0.43	degrad. 34
g PVC	ANSELLEDMONT 34-100	10m	300	4	0.20	degrad. 34
s UNKNOWN MATERIAL	LIFE-GUARD RESPONDE	>180m			n.a.	77
g VITON	NORTH F-091	73m	180	3	0.31	degrad. 34
isophoronediisocyanate (synonym: IDI) CAS Number: 4098-71-9 Primary Class: 211 Isocyanates, Aliphatic and Alicyclic						
g BUTYL	NORTH B-174	>480m	0	0	0.74	34
g NATURAL RUBBER	ACKWELL 5-109	>480m	0	0	0.15	degrad. 34
g NITRILE	MAPA-PIONEER A-15	>480m	0	0	0.41	36
g PVAL	EDMONT 25-950	>480m	0	0	0.33	34

G A R M E Class & Number/ N test chemical/ T MATERIAL NAME	MANUFACTURER & PRODUCT IDENTIFICATION	Break- through Time in Minutes	Perm- eation Rate in mg/sq m /min.	I N D E X	Thick- ness in mm	degrad. and comments	R e f s
s UNKNOWN MATERIAL	LIFE-GUARD RESPONDE	>480m	0	0	n.a.		77
g VITON	NORTH F-091	>480m	0	0	0.33		34

isoprene (synonym: 2-methyl-1,3-butadiene)
CAS Number: 78-79-5
Primary Class: 294   Hydrocarbons, Aliphatic and Alicyclic, Unsaturated

g NEOPRENE	ANSELLEDMONT 29-870	16m	813	3	0.50		34
g NITRILE	ANSELLEDMONT 37-155	52m	117	3	0.36		34
g PVAL	EDMONT 25-950	>720m	0	0	0.30		34
s UNKNOWN MATERIAL	LIFE-GUARD RESPONDE	>180m	0	1	n.a.		77
g VITON	NORTH F-091	372m	11	2	0.26		34

isopropanolamine (synonym: 1-amino-2-propanol)
CAS Number: 78-96-6
Primary Class: 141   Amines, Aliphatic and Alicyclic, Primary
Related Class: 312   Hydroxy Compounds, Aliphatic and Alicyclic, Secondary

g BUTYL	NORTH B-174	>480m	0	0	0.48		34
g NEOPRENE	ANSELLEDMONT 29-870	>480m	0	0	0.39		34
g PVC	ANSELLEDMONT 34-100	>480m	0	0	0.20		34
g VITON	NORTH F-091	>480m	0	0	0.34		34

isopropyl alcohol (synonyms: 2-propanol; isopropanol)
CAS Number: 67-63-0
Primary Class: 32   Hydroxy Compounds, Aliphatic and Alicyclic, Secondary

g BUTYL	BEST 878	>480m	0	0	0.75		53
u CPE	ILC DOVER	>180m			n.a.		46
g NAT+NEOPR+NITRILE	MAPA-PIONEER TRIONI	38m	35	3	0.46		36
g NATURAL RUBBER	ANSELL CONFORM 4205	1m	0.8	3	0.13		55
g NATURAL RUBBER	ANSELL FL 200 254	6m	0.18	3	0.51		55
g NATURAL RUBBER	ANSELL ORANGE 208	26m	0.1		0.76		55
g NATURAL RUBBER	ANSELL PVL 040	15m	0.1	2	0.46		55
g NATURAL RUBBER	ANSELL STERILE 832	15m	0.1		0.23		55
g NATURAL RUBBER	ANSELLEDMONT 30-139	7m	<9	3	0.25		42
g NATURAL RUBBER	ANSELLEDMONT 36-124	7m	<9	3	0.46		6
g NATURAL RUBBER	ANSELLEDMONT 392	20m	<90	3	0.48		6
g NATURAL RUBBER	COMASEC FLEXIGUM	186m	60	2	0.95		40
g NATURAL RUBBER	MAPA-PIONEER L-118	>60m			0.46		36
g NATURAL RUBBER	MARIGOLD BLACK HEVY	15m	12	3	0.65		79
g NATURAL RUBBER	MARIGOLD FEATHERLIT	1m	173	4	0.15		79
g NATURAL RUBBER	MARIGOLD MED WT EXT	9m	30	4	0.45		79
g NATURAL RUBBER	MARIGOLD MEDICAL	4m	113	4	0.28		79
g NATURAL RUBBER	MARIGOLD ORANGE SUP	16m	10	3	0.73		79
g NATURAL RUBBER	MARIGOLD R SURGEONS	4m	113	4	0.28		79

G A R M E Class & Number/ N test chemical/ T MATERIAL NAME	MANUFACTURER & PRODUCT IDENTIFICATION	Break- through Time in Minutes	Perm- eation Rate in mg/sq m /min.	I N D E X	Thick- ness in mm	degrad. and comments	R e f s
g NATURAL RUBBER	MARIGOLD RED LT WT	8m	42	4	0.43		79
g NATURAL RUBBER	MARIGOLD SENSOTECH	4m	113	4	0.28		79
g NATURAL RUBBER	MARIGOLD SUREGRIP	13m	17	4	0.58		79
g NATURAL RUBBER	MARIGOLD YELLOW	7m	65	4	0.38		79
g NATURAL+NEOPRENE	ANSELL OMNI 276	40m	0.1	2	0.56		55
g NATURAL+NEOPRENE	ANSELL TECHNICIANS	14m	0.1	3	0.43		55
g NATURAL+NEOPRENE	MARIGOLD FEATHERWT	6m	77	4	0.35		79
g NEOPRENE	ANSELL NEOPRENE 530	>360m	0	0	0.46		55
g NEOPRENE	ANSELLEDMONT 29-840	>360m	0	0	0.31		6
g NEOPRENE	ANSELLEDMONT NEOX	>360m	0	0	n.a.		6
g NEOPRENE	BEST 32	1m	80	4	n.a.		53
g NEOPRENE	BEST 6780	>480m	0	0	n.a.		53
g NEOPRENE	COMASEC COMAPRENE	120m	48	2	n.a.		40
g NEOPRENE	MAPA-PIONEER N-44	>60m			0.56		36
g NEOPRENE/NATURAL	ANSELL CHEMI-PRO	50m	0.1	2	0.72		55
g NITRILE	ANSELL CHALLENGER	>360m	0	0	0.38		55
g NITRILE	ANSELLEDMONT 37-165	>360m	0	0	0.60		6
g NITRILE	ANSELLEDMONT 49-125	150m	<9	1	0.23		42
g NITRILE	BEST 22R	>480m	0	0	n.a.		53
g NITRILE	BEST 727	>480m	0	0	0.38		53
g NITRILE	COMASEC COMATRIL	>480m	0	0	0.55		40
g NITRILE	COMASEC COMATRIL SU	>480m	0	0	0.60		40
g NITRILE	COMASEC FLEXITRIL	80m	234	3	n.a.		40
g NITRILE	MAPA-PIONEER A-14	>480m	0	0	0.56		36
g NITRILE	MARIGOLD BLUE	>480m	0	0	0.30		79
g NITRILE	MARIGOLD NITROSOLVE	>480m	0	0	0.75		79
g NITRILE+PVC	COMASEC MULTIPLUS	>480m	0	0	n.a.		40
g PE/EVAL/PE	SAFETY4 4H	>240m	0	0	0.07	35°C	60
g PE/EVAL/PE	SAFETY4 4H	>480m	0	0	0.07		60
g PVAL	ANSELLEDMONT PVA				n.a.	degrad.	6
g PVAL	EDMONT 15-552				n.a.	degrad.	6
g PVC	ANSELLEDMONT SNORKE	150m	<9	1	n.a.		6
g PVC	COMASEC MULTIPOST	>480m	0	0	n.a.		40
g PVC	COMASEC MULTITOP	>480m	0	0	n.a.		40
g PVC	COMASEC NORMAL	120m	90	2	n.a.		40
g PVC	COMASEC OMNI	85m	210	3	n.a.		40
g PVC	MAPA-PIONEER V-20	210m	20	2	0.51		36
s SARANEX-23	DU PONT TYVEK SARAN	114m			n.a.		8
s TEFLON	CHEMFAB CHALL. 5100	>180m			n.a.		65
g UNKNOWN MATERIAL	MARIGOLD R MEDIGLOV	1m	160	4	0.03		79

isopropyl alcohol, 75% & 1-napthylamine, 25%
Primary Class: 312   Hydroxy Compounds, Aliphatic and Alicyclic, Secondary

| g PE/EVAL/PE | SAFETY4 4H | >240m | 0 | | 0.07 | 35°C | 60 |

G A R M E Class & Number/ N test chemical/ T MATERIAL NAME	MANUFACTURER & PRODUCT IDENTIFICATION	Break-through Time in Minutes	Perm-eation Rate in mg/sq m /min.	I N D E X	Thick-ness in mm	R degrad. e and f comments s
isopropyl ether (synonym: diisopropyl ether) CAS Number: 108-20-3 Primary Class: 241 Ethers, Aliphatic and Alicyclic						
u CPE	ILC DOVER	>180m			n.a.	29
u NATURAL RUBBER	Unknown	3m	>5000	5	n.a.	10
u NATURAL RUBBER	Unknown	3m	>5000	5	0.42	10
u NATURAL+NEOPRENE	Unknown	7m	>5000	5	0.50	10
u NEOPRENE	Unknown	55m			0.60	10
u NEOPRENE	Unknown	>60m			0.93	10
u PVC	Unknown	15m	>5000	4	n.a.	10
u VITON/CHLOROBUTYL	Unknown	>180m			0.36	29
isopropyl methacrylate (synonym: isopropyl 2-methylpropenoate) CAS Number: 4655-34-9 Primary Class: 223 Esters, Carboxylic, Acrylates and Methacrylates						
g BUTYL	NORTH B-174	>480m	0	0	0.69	34
g NATURAL RUBBER	ACKWELL 5-109				0.15	degrad. 34
g NEOPRENE	ANSELLEDMONT 29-870				0.48	degrad. 34
g NITRILE	ANSELLEDMONT 37-155	112m	381	3	0.45	degrad. 34
g PVAL	EDMONT 25-950	>480m	0	0	0.60	34
g PVC	ANSELLEDMONT 34-100	1m	3540	5	0.20	degrad. 34
g VITON	NORTH F-091				0.26	degrad. 34
isopropyl nitrite CAS Number: 1712-64-7 Primary Class: 510 Nitrates and Nitrites						
g PE/EVAL/PE	SAFETY4 4H	>240m	0	0	0.07	60
isopropylamine (synonym: 2-aminopropane) CAS Number: 75-31-0 Primary Class: 141 Amines, Aliphatic and Alicyclic, Primary						
g BUTYL	NORTH B-174	246m	381	3	0.63	34
g NATURAL RUBBER	ACKWELL 5-109				0.15	degrad. 34
g NEOPRENE	ANSELLEDMONT 29-870	14m	9136	5	0.37	degrad. 34
g NITRILE	ANSELLEDMONT 37-155				0.40	degrad. 34
g PVC	ANSELLEDMONT 34-100	2m	46620	5	0.20	34
s TEFLON	CHEMFAB CHALL. 5100	>180m	0	1	n.a.	65
g VITON	NORTH F-091	11m	33360	5	0.30	degrad. 34

G A R M E Class & Number/ N test chemical/ T MATERIAL NAME	MANUFACTURER & PRODUCT IDENTIFICATION	Break- through Time in Minutes	Perm- eation Rate in mg/sq m /min.	I N D E X	Thick- ness in mm	degrad. and comments	R e f s

jet fuel with <30% aromatics (synonym: JP-4)
Primary Class: 291   Hydrocarbons, Aliphatic and Alicyclic, Saturated

g BUTYL	ANSELL LAMPRECHT	52m	5020	4	0.70		38
g BUTYL	ERISTA BX	52m	20630	5	0.67		33
u NATURAL RUBBER	Unknown	7m	10730	5	0.64		38
g NITRILE	ANSELLEDMONT 37-145	>240m	0	0	0.33		33
g NITRILE	ANSELLEDMONT 37-145	>240m	0	0	0.33	preexp. 20h	33
g PE/EVAL/PE	SAFETY4 4H	>240m	0	0	0.07	35°C	60
g PVC	KID 490	>240m	0	0	n.a.		33
g PVC	KID VINYLPRODUKTER	37m	2520	4	0.55	degrad.	33
s SARANEX-23	DU PONT TYVEK SARAN	458m	30	2	n.a.		8
s UNKNOWN MATERIAL	LIFE-GUARD RESPONDE	>240m	0	0	n.a.		77
g VITON	NORTH F-091	>240m	0	0	0.25		33
s VITON/CHLOROBUTYL	LIFE-GUARD VC-100	>240m		0	n.a.		77
g VITON/NEOPRENE	ERISTA VITRIC	> 60m			0.60		38

Kovac's indol reagent
Primary Class: 900   Miscellaneous Unclassified Chemicals

g PE/EVAL/PE	SAFETY4 4H	>240m	0	0	0.07	35 °C	60

KVK parathion
Primary Class: 462   Organophosphorus Compounds and Derivatives of Phosphorus-based Acids

g PE/EVAL/PE	SAFETY4 4H	>240m	0	0	0.07		60

lactic acid
CAS Number: 50-21-5
Primary Class: 103   Acids, Carboxylic, Aliphatic and Alicyclic, Substituted
Related Class: 312   Hydroxy Compounds, Aliphatic and Alicyclic, Secondary

g BUTYL	BEST 878	>480m	0	0	0.75	85%	53
g NATURAL RUBBER	ANSELL CONFORM 4205	>360m	0	0	0.13	85%	55
g NATURAL RUBBER	ANSELL FL 200 254	>360m	0	0	0.51	85%	55
g NATURAL RUBBER	ANSELL ORANGE 208	>360m	0	0	0.76	85%	55
g NATURAL RUBBER	ANSELL PVL 040	>360m	0	0	0.46	85%	55
g NATURAL RUBBER	ANSELL STERILE 832	>360m	0	0	0.23	85%	55
g NATURAL RUBBER	ANSELLEDMONT 36-124	>360m	0	0	0.46	85%	6
g NATURAL RUBBER	ANSELLEDMONT 392	>360m	0	0	0.48	85%	6
g NATURAL RUBBER	BEST 65NFW	>480m	0	0	n.a.	85%	53
g NATURAL RUBBER	MARIGOLD BLACK HEVY	>480m	0	0	0.65		79
g NATURAL RUBBER	MARIGOLD FEATHERLIT	>480m	0	0	0.15		79
g NATURAL RUBBER	MARIGOLD MED WT EXT	>480m	0	0	0.45		79
g NATURAL RUBBER	MARIGOLD MEDICAL	>480m	0	0	0.28		79
g NATURAL RUBBER	MARIGOLD ORANGE SUP	>480m	0	0	0.73		79

G A R M E Class & Number/ N test chemical/ T MATERIAL NAME	MANUFACTURER & PRODUCT IDENTIFICATION	Break-through Time in Minutes	Perm-eation Rate in mg/sq m /min.	I N D E X	Thick-ness in mm	degrad. and comments	R e f s
g NATURAL RUBBER	MARIGOLD R SURGEON	>480m	0	0	0.28		79
g NATURAL RUBBER	MARIGOLD RED LT WT	>480m	0	0	0.43		79
g NATURAL RUBBER	MARIGOLD SENSOTECH	>480m	0	0	0.28		79
g NATURAL RUBBER	MARIGOLD SUREGRIP	>480m	0	0	0.58		79
g NATURAL RUBBER	MARIGOLD YELLOW	>480m	0	0	0.38		79
g NATURAL+NEOPRENE	ANSELL OMNI 276	>360m	0	0	0.56		55
g NATURAL+NEOPRENE	ANSELL TECHNICIANS	>360m	0	0	0.43	85%	55
g NATURAL+NEOPRENE	MARIGOLD FEATHERWT	>480m	0	0	0.35		79
g NEOPRENE	ANSELL NEOPRENE 530	>360m	0	0	0.46	85%	55
g NEOPRENE	ANSELLEDMONT 29-840	>360m	0	0	0.38	85%	6
g NEOPRENE	ANSELLEDMONT NEOX	>360m	0	0	n.a.	85%	6
g NEOPRENE	BEST 32	>480m	0	0	n.a.	85%	53
g NEOPRENE	BEST 6780	>480m	0	0	n.a.	85%	53
g NEOPRENE/NATURAL	ANSELL CHEMI-PRO	>360m	0	0	0.72	85%	55
g NITRILE	ANSELL CHALLENGER	>360m	0	0	0.38	85%	55
g NITRILE	ANSELLEDMONT 37-165	>360m	0	0	0.60	85%	6
g NITRILE	BEST 22R	>480m	0	0	n.a.	85%	53
g NITRILE	BEST 727	>480m	0	0	0.38	85%	53
g NITRILE	MARIGOLD BLUE	>480m	0	0	0.45		79
g NITRILE	MARIGOLD GREEN SUPA	>480m	0	0	0.30		79
g NITRILE	MARIGOLD NITROSOLVE	>480m	0	0	0.75		79
g PVAL	ANSELLEDMONT PVA	>360m		0	n.a.	85%	6
g PVAL	EDMONT 15-552	>360m	0	0	n.a	85%	6
g PVAL	EDMONT 25-545	20m			n.a.	85%	6
g PVC	ANSELLEDMONT M GRIP	>360m	0	0	n.a.	85%	6
g PVC	BEST 725R	>480m	0	0	n.a.	85%	53
g UNKNOWN MATERIAL	MARIGOLD R MEDIGLOV	>480m	0	0	0.03		79

lauric acid, 30-70% (synonym: dodecanoic acid, 30-70%)
CAS Number: 1430-70-7
Primary Class: 102   Acids, Carboxylic, Aliphatic and Alicyclic, Unsubstituted

g NATURAL RUBBER	ANSELL CONFORM 4205	60m			0.13	ETOH	55
g NATURAL RUBBER	ANSELL ORANGE 208	>360m		0	0.76	ETOH	55
g NATURAL RUBBER	ANSELLEDMONT 36-124	>360m	0	0	0.46	ETOH	6
g NATURAL RUBBER	ANSELLEDMONT 392	>360m		0	0.48	ETOH	6
g NATURAL+NEOPRENE	ANSELL OMNI 276	>360m	0	0	0.56	ETOH	55
g NEOPRENE	ANSELL NEOPRENE 530	>360m	0	0	0.46	ETOH	55
g NEOPRENE	ANSELLEDMONT 29-840	>360m	0	0	0.38	ETOH	6
g NEOPRENE	ANSELLEDMONT NEOX	>360m	0	0	n.a.	ETOH	6
g NEOPRENE/NATURAL	ANSELL CHEMI-PRO	>360m	0	0	0.72	ETOH	55
g NITRILE	ANSELL CHALLENGER	>360m	0	0	0.38	ETOH	55
g NITRILE	ANSELLEDMONT 37-165	>360m	0	0	0.60	ETOH	6
g PVAL	ANSELLEDMONT PVA				n.a.	degrad.	6
g PVAL	EDMONT 15-552				n.a.	degrad.	6
g PVC	ANSELLEDMONT M GRIP	15m			n.a.	ETOH	6

G A R M E Class & Number/ N test chemical/ T MATERIAL NAME	MANUFACTURER & PRODUCT IDENTIFICATION	Break- through Time in Minutes	Perm- eation Rate in mg/sq m /min.	I N D E X	Thick- ness in mm	degrad. and comments	R e f s
lauric acid, >70% (synonym: dodecanoic acid, >70%) Primary Class: 102 Acids, Carboxylic, Aliphatic and Alicyclic, Unsubstituted							
lindane & chloroform Primary Class: 840 Pesticides, Mixtures and Formulations							
m BUTYL	PLYMOUTH RUBBER	5m			n.a.		46
u VITON/CHLOROBUTYL	Unknown	>180m			n.a.		46
lindane & xylene Primary Class: 840 Pesticides, Mixtures and Formulations							
m BUTYL	PLYMOUTH RUBBER	85m			n.a.		46
u VITON/CHLOROBUTYL	Unknown	>180m			n.a.		46
linoleic acid (synonym: linolic acid) CAS Number: 60-33-3 Primary Class: 102 Acids, Carboxylic, Aliphatic and Alicyclic, Unsubstituted							
g NATURAL RUBBER	ACKWELL 5-109				0.15	degrad.	34
liquified coal Primary Class: 860 Coals, Charcoals, Oils							
u NATURAL RUBBER	Unknown	280m		0	n.a.		18
u NITRILE	Unknown	720m		0	0.36		18
u PVC	Unknown	360m	0	0	n.a.		18
lubricating oil Primary Class: 860 Coals, Charcoals, Oils							
g NATURAL RUBBER	MARIGOLD BLACK HEVY	>480m	0	0	0.65		79
g NATURAL RUBBER	MARIGOLD FEATHERLIT	146m	36	2	0.15	degrad.	79
g NATURAL RUBBER	MARIGOLD MED WT EXT	>480m	0	0	0.45		79
g NATURAL RUBBER	MARIGOLD MEDICAL	285m	21	2	0.28	degrad.	79
g NATURAL RUBBER	MARIGOLD ORANGE SUP	>480m	0	0	0.73		79
g NATURAL RUBBER	MARIGOLD R SURGEONS	285m	21	2	0.28	degrad.	79
g NATURAL RUBBER	MARIGOLD RED LT WT	452m	3	1	0.43		79
g NATURAL RUBBER	MARIGOLD SENSOTECH	285m	21	2	0.28	degrad.	79
g NATURAL RUBBER	MARIGOLD SUREGRIP	>480m	0	0	0.58		79
g NATURAL RUBBER	MARIGOLD YELLOW	396m	9	1	0.38		79
g NATURAL+NEOPRENE	MARIGOLD FEATHERWT	369m	12	2	0.35	degrad.	79
g NITRILE	MARIGOLD BLUE	>480m	0	0	0.45		79
g NITRILE	MARIGOLD GREEN SUPA	>480m	0	0	0.30		79

G A R M E Class & Number/ N test chemical/ T MATERIAL NAME	MANUFACTURER & PRODUCT IDENTIFICATION	Break- through Time in Minutes	Perm- eation Rate in mg/sq m /min.	I N D E X	Thick- ness in mm	degrad. and comments	R e f s
g NITRILE	MARIGOLD NITROSOLVE	>480m	0	0	0.75		79
g PE/EVAL/PE	SAFETY4 4H	>240m	0	0	0.07	35°C	60
g UNKNOWN MATERIAL	MARIGOLD R MEDIGLOV	320m	105	3	0.03		79

m-cresol (synonyms: 3-hydroxytoluene; 3-methylphenol)
CAS Number: 108-39-4
Primary Class: 316   Hydroxy Compounds, Aromatic (Phenols)
Related Class: 292   Hydrocarbons, Aromatic

g NATURAL RUBBER	ANSELLEDMONT 46-320	36m<	10	3	0.31		5
g NATURAL RUBBER	MAPA-PIONEER L-118	150m	120	3	0.46		36
g NATURAL+NEOPRENE	ANSELL OMNI 276	30m	20	3	0.45		5
g NEOPRENE	ANSELLEDMONT NEOX	>60m			n.a.		5
g NEOPRENE	MAPA-PIONEER N-44	>480m	0	0	0.74		36
g NEOPRENE	MAPA-PIONEER N-73	>60m			0.46		5
g NITRILE	ANSELLEDMONT 37-175	>60m			0.37		5
g NITRILE	MAPA-PIONEER A-14	210m	1260	4	0.54		36
g PE	ANSELLEDMONT 35-125	>60m			0.03		5
g PVC	MAPA-PIONEER V-10	14m	630	4	0.20		5
g PVC	MAPA-PIONEER V-20	>60m			0.31		5
g PVC	MAPA-PIONEER V-20	150m	360	3	0.45		36
s TEFLON	CHEMFAB CHALL. 5100	>240m	0	0	n.a.		65

m-cresol, 50% & methyl ethyl ketone, 50%
Primary Class: 316   Hydroxy Compounds, Aromatic (Phenols)

g PE/EVAL/PE	SAFETY4 4H	>240m	0	0	0.07	35°C	60

m-xylene (synonym: 1,3-dimethylbenzene)
CAS Number: 108-38-3
Primary Class: 292   Hydrocarbons, Aromatic

g BUTYL	NORTH B-174	39m	876	3	0.67	degrad.	34
g NATURAL RUBBER	ACKWELL 5-109				0.15	degrad.	34
g NEOPRENE	ANSELLEDMONT 29-870				0.48	degrad.	34
g NITRILE	ANSELLEDMONT 37-155	62m	1884	4	0.36	degrad.	34
g PVAL	EDMONT 25-950	>761m	0	0	0.30		34
g VITON	NORTH F-091	>960m	0	0	0.30		34

malathion
CAS Number: 121-75-5
Primary Class: 462   Organophosphorus Compounds and Derivatives of Phosphorus-based Acids
Related Class: 233   Esters, Non-Carboxylic, Carbamates and Others

g PE/EVAL/PE	SAFETY4 4H	>240m	0	0	0.07		60
s TEFLON	CHEMFAB CHALL. 5100	>186m	1	1	n.a.		65

G A R M E Class & Number/ N test chemical/ T MATERIAL NAME	MANUFACTURER & PRODUCT IDENTIFICATION	Break- through Time in Minutes	Perm- eation Rate in mg/sq m /min.	I N D E X	Thick- ness in mm	degrad. and comments	R e f s
maleic acid CAS Number: 110-16-7 Primary Class: 104 Acids, Carboxylic, Aliphatic and Alicyclic, Polybasic							
g NATURAL RUBBER	ANSELL CONFORM 4205	>360m	0	0	0.13		55
g NATURAL RUBBER	ANSELL FL 200 254	>360m	0	0	0.51		55
g NATURAL RUBBER	ANSELL ORANGE 208	>360m	0	0	0.76		55
g NATURAL RUBBER	ANSELL PVL 040	>360m	0	0	0.46		55
g NATURAL RUBBER	ANSELL STERILE 832	>360m	0	0	0.23		55
g NATURAL RUBBER	ANSELLEDMONT 36-124	>360m	0	0	0.46		6
g NATURAL RUBBER	ANSELLEDMONT 392	>360m	0	0	0.48		6
g NATURAL+NEOPRENE	ANSELL OMNI 276	>360m	0	0	0.56		55
g NATURAL+NEOPRENE	ANSELL TECHNICIANS	>360m	0	0	0.43		55
g NEOPRENE	ANSELL NEOPRENE 530	>360m	0	0	0.46		55
g NEOPRENE	ANSELLEDMONT 29-840	>360m	0	0	0.38		6
g NEOPRENE	ANSELLEDMONT NEOX	>360m	0	0	n.a.		6
g NEOPRENE/NATURAL	ANSELL CHEMI-PRO	>360m	0	0	0.72		55
g NITRILE	ANSELL CHALLENGER	>360m	0	0	0.38		55
g NITRILE	ANSELLEDMONT 37-165	>360m	0	0	0.60		6
g PVAL	ANSELLEDMONT PVA				n.a.	degrad.	6
g PVAL	EDMONT 15-552				n.a.	degrad.	6
g PVC	ANSELLEDMONT M GRIP	>360m	0	0	n.a.		6
s UNKNOWN MATERIAL	LIFE-GUARD RESPONDE	>480m		0	n.a.		77
maleic anhydride CAS Number: 108-31-6 Primary Class: 161 Anhydrides, Aliphatic and Alicyclic							
s UNKNOWN MATERIAL	LIFE-GUARD RESPONDE	>480m		0	n.a.		77
melamine-formaldehyde resin in solution (butyl alcohol, 1% & water, 50%) Primary Class: 830 Lacquer Products							
g BUTYL	ERISTA BX	>240m	0	0	0.67		33
g BUTYL	ERISTA BX	>240m	0	0	0.67	preexp. 20h	33
g PVAL	EDMONT 15-552	6m			n.a.		33
mercaptoacetic acid (synonym: thioglycolic acid) CAS Number: 68-11-1 Primary Class: 103 Acids, Carboxylic, Aliphatic and Alicyclic, Substituted							
g BUTYL	NORTH B-174	>480m	0	0	0.64		34
g NATURAL RUBBER	ACKWELL 5-109				0.15	degrad.	34
g NEOPRENE	ANSELLEDMONT 29-870	>480m	0	0	0.52		34
g NITRILE	ANSELLEDMONT 37-155	150m	1380	4	0.38		34
g PE/EVAL/PE	SAFETY4 4H	>240m	0	0	0.07	35°C	60
g PVAL	EDMONT 25-950				n.a.	degrad.	34

G				Perm-	I			
A								
R								R
M			Break-	eation	N	Thick-		
E Class & Number/	MANUFACTURER		through	Rate in	D	ness	degrad.	e
N test chemical/	& PRODUCT		Time in	mg/sq m	E	in	and	f
T MATERIAL NAME	IDENTIFICATION		Minutes	/min.	X	mm	comments	s

g PVC	ANSELLEDMONT 34-100					0.20	degrad.	34
g VITON	NORTH F-091		>480m	0	0	0.28		34

mercuric chloride (saturated solution)
CAS Number: 7487-94-7
Primary Class: 340   Inorganic Salts

s SARANEX-23	DU PONT TYVEK SARAN	>480m		0	n.a.		8
s UNKNOWN MATERIAL	DU PONT BARRICADE	>480m		0	n.a.		8

mercury (synonym: quick silver)
CAS Number: 7439-97-6
Primary Class: 330   Elements

s SARANEX-23	DU PONT TYVEK SARAN	390m	0.1	0	n.a.		8
s SARANEX-23 2-PLY	DU PONT TYVEK SARAN	>480m	0	0	n.a.		8
s UNKNOWN MATERIAL	DU PONT BARRICADE	>480m	0	0	n.a.		8

methacrylic acid (synonym: 2-methylpropenoic acid)
CAS Number: 79-41-4
Primary Class: 102   Acids, Carboxylic, Aliphatic and Alicyclic, Unsubstituted

g BUTYL	NORTH B-174	>480m	0	0	0.61		34
g NATURAL RUBBER	ACKWELL 5-109				0.15	degrad.	34
g NEOPRENE	ANSELLEDMONT 29-870				0.48	degrad.	34
g NITRILE	ANSELLEDMONT 37-155	107m	1380	4	0.36	degrad.	34
g PE/EVAL/PE	SAFETY4 4H	>240m	0	0	0.07	35°C	60
g PE/EVAL/PE	SAFETY4 4H	>480m	0	0	0.07		60
g PVAL	EDMONT 25-950				0.35	degrad.	34
g PVC	ANSELLEDMONT 34-100	<1m	600	4	0.18		34
s UNKNOWN MATERIAL	LIFE-GUARD RESPONDE	>480m	0	0	n.a.		77
g VITON	NORTH F-091	>480m	0	0	0.33		34

methane
CAS Number: 74-82-8
Primary Class: 291   Hydrocarbons, Aliphatic and Alicyclic, Saturated

s UNKNOWN MATERIAL	LIFE-GUARD RESPONDE	>480m		0	n.a.		77

methanesulfonic acid
CAS Number: 75-75-2
Primary Class: 504   Sulfur Compounds, Sulfonic Acids

g NEOPRENE	ANSELLEDMONT 29-840	>240m	0	0	0.45		33
g PVC	KID VINYLPRODUKTER	>240m	0	0	0.45		33

G A R M E Class & Number/ N test chemical/ T MATERIAL NAME	MANUFACTURER & PRODUCT IDENTIFICATION	Break- through Time in Minutes	Perm- eation Rate in mg/sq m /min.	I N D E X	Thick- ness in mm	degrad. and comments	R e f s

methomyl (synonym: lannate)
CAS Number: 16752-77-5
Primary Class: 233   Esters, Non-Carboxylic, Carbamates and Others

Related Class: 840	Pesticides, Mixtures and Formulations						
c PE	DU PONT TYVEK QC	<15m	<1	2	n.a.		8
s SARANEX-23	DU PONT TYVEK SARAN	<15m	<1	2	n.a.		8

methotrexate
CAS Number: 59-05-2
Primary Class: 870   Antineoplastic Drugs and Other Pharmaceuticals

g EMA	Unknown	>1440m	0	0	0.07		59
u NATURAL RUBBER	Unknown	>1440m	0	0	0.18		59
g NEOPRENE	ANSELL NEOPRENE 530	>1440m	0	0	0.28		59
u PVC	Unknown	>1440m	0	0	0.19		59

methyl acetate
CAS Number: 79-20-9
Primary Class: 222   Esters, Carboxylic, Acetates

g BUTYL	NORTH B-174	>480m	0	0	0.69		34
g NATURAL RUBBER	ACKWELL 5-109	1m	6000	5	0.15		34
g NATURAL RUBBER	KACHELE 706	5m	4000	5	0.71		15
g NEOPRENE	KACHELE 722	7m	3000	5	0.65		15
g NITRILE	ANSELLEDMONT 37-155				0.40	degrad.	34
g NITRILE	KURSAAL 85/62	7m	20000	5	0.41		15
g PE/EVAL/PE	SAFETY4 4H	>240m	0	0	0.07	35°C	60
g PVAL	EDMONT 25-950	41m	98	3	0.71		34
g PVC	ANSELLEDMONT 34-100	1m	60000	5	0.20		34
g PVC	NORTH 800	3m	16000	5	n.a.		15
g VITON	NORTH F-091				0.25	degrad.	34
s VITON/BUTYL	DRAGER 500 OR 710	> 60m			n.a.		91
g VITON/NEOPRENE	ERISTA VITRIC	4m	4000	5	0.45		15

methyl acetate, 50% & ethyl alcohol, 50%
Primary Class: 222   Esters, Carboxylic, Acetates

g BUTYL	NORTH B-161	>240m	0	0	0.40		33
g NITRILE	NORTH LA-102G	7m	1050	5	0.25		30
g VITON	NORTH F-091	4m	2240	5	0.25		30

methyl acrylate (synonym: methyl 2-propenoate)
CAS Number: 96-33-3
Primary Class: 223   Esters, Carboxylic, Acrylates and Methacrylates

g BUTYL	NORTH B-174	>480m	0	0	0.66		34

G A R M E Class & Number/ N test chemical/ T MATERIAL NAME	MANUFACTURER & PRODUCT IDENTIFICATION	Break- through Time in Minutes	Perm- eation Rate in mg/sq m /min.	I N D E X	Thick- ness in mm	degrad. and comments	R e f s
g NATURAL RUBBER	ACKWELL 5-109	1m	6240	5	0.15	degrad.	34
g NEOPRENE	ANSELLEDMONT 29-870	15m	31340	5	0.50	degrad.	34
g NITRILE	ANSELLEDMONT 37-155				0.40	degrad.	34
g PVAL	EDMONT 25-950	90m	16	2	0.65		34
g PVC	ANSELLEDMONT 34-100				0.20	degrad.	34
s TEFLON	CHEMFAB CHALL. 5100	>180m	0	1	n.a.		65
g VITON	NORTH F-091				0.25	degrad.	34

methyl alcohol (synonyms: methanol; wood alcohol)
CAS Number: 67-56-1
Primary Class: 311   Hydroxy Compounds, Aliphatic and Alicyclic, Primary

g BUTYL	ANSELL LAMPRECHT	>60m			0.70		38
g BUTYL	BEST 878	>480m	0	0	0.75		53
g BUTYL	BRUNSWICK BUTYL STD	>480m	0	0	0.63		89
g BUTYL	BRUNSWICK BUTYL-XTR	>480m	0	0	0.80		89
g BUTYL	COMASEC BUTYL	>480m	0	0	0.55		40
s BUTYL	LIFE-GUARD BUTYL	303m	0.1		n.a.		77
s BUTYL	MSA CHEMPRUF	>480m	0	0	n.a.		48
m BUTYL	PLYMOUTH RUBBER	>180m			n.a.		46
u BUTYL	Unknown	>480m	0	0	0.30		30
s BUTYL	WHEELER ACID KING	>480m	0	0	n.a.		85
g BUTYL/ECO	BRUNSWICK BUTYL-POL	>480m	0	0	0.63		89
g BUTYL/NEOPRENE	COMASEC BUTYL PLUS	>480m	0	0	0.60		40
s BUTYL/NEOPRENE	MSA BETEX	>480m	0	0	n.a.		48
s CPE	ILC DOVER CLOROPEL	>180m			n.a.		46
s CPE	STD. SAFETY	>180m			n.a.		75
g NAT+NEOPR+NITRILE	MAPA-PIONEER TRIONI	19m	72	3	0.48		36
g NATURAL RUBBER	ANSELL CONFORM 4205	1m	>30	4	0.13		55
g NATURAL RUBBER	ANSELL FL 200 254	10m	1	3	0.51		55
g NATURAL RUBBER	ANSELL ORANGE 208	17m	0.9	2	0.76		55
g NATURAL RUBBER	ANSELL PVL 040	18m	0.9	2	0.46		55
g NATURAL RUBBER	ANSELL STERILE 832	2m	6	3	0.23		55
g NATURAL RUBBER	ANSELLEDMONT 36-124	13m	<9	3	0.46		6
g NATURAL RUBBER	ANSELLEDMONT 392	20m	<90	3	0.48		6
g NATURAL RUBBER	ANSELLEDMONT 46-320	20m	80	3	0.31		5
g NATURAL RUBBER	KACHELE 706	2m			0.71		15
g NATURAL RUBBER	MAPA-PIONEER 258	27m	1	2	0.64	30°C	87
g NATURAL RUBBER	MAPA-PIONEER 300	22m	1	2	n.a.	30°C	87
g NATURAL RUBBER	MAPA-PIONEER 730	92m	0.3	1	1.10	30°C	87
g NATURAL RUBBER	MAPA-PIONEER L-118	>60m			0.46		36
g NATURAL RUBBER	MARIGOLD BLACK HEVY	9m	51	4	0.65		79
g NATURAL RUBBER	MARIGOLD FEATHERLIT	2m	301	4	0.15		79
g NATURAL RUBBER	MARIGOLD MED WT EXT	4m	54	4	0.45		79
g NATURAL RUBBER	MARIGOLD MEDICAL	3m	198	4	0.28		79
g NATURAL RUBBER	MARIGOLD ORANGE SUP	10m	50	4	0.73		79
g NATURAL RUBBER	MARIGOLD R SURGEONS	3m	198	4	0.28		79
g NATURAL RUBBER	MARIGOLD RED LT WT	4m	75	4	0.45		79
g NATURAL RUBBER	MARIGOLD SENSOTECH	3m	198	4	0.28		79

G A R M E N T Class & Number/ test chemical/ MATERIAL NAME	MANUFACTURER & PRODUCT IDENTIFICATION	Break-through Time in Minutes	Perm-eation Rate in mg/sq m /min.	I N D E X	Thick-ness in mm	degrad. and comments	R e f s
g NATURAL RUBBER	MARIGOLD SUREGRIP	8m	52	4	0.58		79
g NATURAL RUBBER	MARIGOLD YELLOW	3m	116	4	0.38		79
g NATURAL+NEOPRENE	ANSELL OMNI 276	15m	0.12	2	0.56		55
g NATURAL+NEOPRENE	ANSELL TECHNICIANS	6m	2	3	0.43		55
g NATURAL+NEOPRENE	MAPA-PIONEER 424	25m	1	2	0.65	30°C	87
g NATURAL+NEOPRENE	MARIGOLD FEATHERWT	3m	136	4	0.35		79
g NATURAL+NEOPRENE	PLAYTEX ARGUS 123	18m	<60	3	n.a.		72
g NEOPRENE	ANSELL NEOPRENE 520	15m	0.12	2	0.46		55
g NEOPRENE	ANSELLEDMONT 29-840	120m	<9	1	0.38		6
g NEOPRENE	ANSELLEDMONT NEOX	15m	<9	2	n.a.		6
g NEOPRENE	BEST 32	49m	10	3	n.a.		53
g NEOPRENE	BEST 6780	64m	80	2	n.a.		53
g NEOPRENE	BRUNSWICK NEOPRENE	239m	7	1	0.88		89
g NEOPRENE	GRANET 2714	>95m	14	2	n.a.		73
s NEOPRENE	LIFE-GUARD NEOPRENE	210m	0.3		n.a.		77
g NEOPRENE	MAPA-PIONEER 360	114m	0.3	1	n.a.	30°C	87
g NEOPRENE	MAPA-PIONEER 420	70m	0.6	1	0.80	30°C	87
g NEOPRENE	MAPA-PIONEER N-44	>60m			0.56		36
b NEOPRENE	RAINFAIR	>180m			n.a.		90
b NEOPRENE	RANGER	>180m			n.a.		90
b NEOPRENE	SERVUS NO 22204	>180m			n.a.		90
u NEOPRENE	Unknown	>60m			0.60		10
u NEOPRENE	Unknown	62m			0.40	37°C TL	16
g NEOPRENE/NATURAL	ANSELL CHEMI-PRO	5m	2	3	0.72		55
g NITRILE	ANSELL CHALLENGER	88m	0.7	1	0.38		55
g NITRILE	ANSELLEDMONT 37-155	69m	230	3	0.35		5
g NITRILE	ANSELLEDMONT 37-155	110m	300	3	0.35		41
g NITRILE	ANSELLEDMONT 37-165	11m	<9000	5	0.60		6
g NITRILE	ANSELLEDMONT 37-175	54m	360	3	0.37		5
g NITRILE	ANSELLEDMONT 37-185	>60m			0.55		5
g NITRILE	BEST 22R	30m	30	3	n.a.		53
g NITRILE	BEST 727	28m	410	3	0.38		53
g NITRILE	COMASEC FLEXITRIL	82m	516	3	n.a.		40
g NITRILE	MAPA-PIONEER 370	17m	22	3	n.a.	30°C	87
g NITRILE	MAPA-PIONEER 490	33m	22	3	0.36	30°C	87
g NITRILE	MAPA-PIONEER 492	50m	23	3	0.48	30°C	87
g NITRILE	MAPA-PIONEER 493	60m	6	1	0.55	30°C	87
g NITRILE	MAPA-PIONEER A-14	118m	180	3	0.56		36
g NITRILE	MARIGOLD BLUE	93m	375	3	0.45	degrad.	79
g NITRILE	MARIGOLD GREEN SUPA	34m	468	3	0.30	degrad.	79
g NITRILE	MARIGOLD NITROSOLVE	105m	357	3	0.75	degrad.	79
u NITRILE	Unknown	54m			0.50	37°C TL	16
b NITRILE+PUR+PVC	BATA HAZMAX	>180m			n.a.		86
g NITRILE+PVC	COMASEC MULTIMAX	80m			n.a.		40
b NITRILE+PVC	SERVUS NO 73101	>180m			n.a.		90
b NITRILE+PVC	TINGLEY	>180m			n.a.		90
g PE	ANSELLEDMONT 35-125	>60m			0.03		3
c PE	DU PONT TYVEK QC	1m	22	4	n.a.		8
g PE/EVAL/PE	SAFETY4 4H	30m	16	3	0.07	35°C	60
g PE/EVAL/PE	SAFETY4 4H	>1440m	0	0	0.07		60

G A R M E N T						
Class & Number/ test chemical/ MATERIAL NAME	**MANUFACTURER & PRODUCT IDENTIFICATION**	**Break- through Time in Minutes**	**Perm- eation Rate in mg/sq m /min.**	**I N D E X**	**Thick- ness in mm**	**degrad. and comments**	**R e f s**
g PVAL	COMASEC SOLVATRIL	18m	1038	4	n.a.		40
g PVAL	EDMONT 15-552				n.a.	degrad.	6
g PVAL	EDMONT 15-552	3m	<50	4	n.a.		10
g PVAL	EDMONT 15-552	1m	1245	5	n.a.	degrad.	33
u PVAL	Unknown	19m			0.08	37°C TL	16
g PVC	ANSELLEDMONT M GRIP	45m	<900	3	n.a.		6
g PVC	ANSELLEDMONT M GRIP	60m	80	2	1.40		31
g PVC	ANSELLEDMONT M GRIP	30m	<1	2	1.40	preexp. 15h	31
b PVC	BATA STANDARD	39m	5	2	n.a.		86
b PVC	JORDAN DAVID	>180m			n.a.		90
g PVC	MAPA-PIONEER V-20	3m	180	4	0.31		5
s PVC	MSA UPC	<30m			n.a.		48
g PVC	NORTH 800	2m	300	4	n.a.		15
b PVC	STD. SAFETY GL-20	24m	66	3	n.a.		75
s PVC	STD. SAFETY WG-20	16m	87	3	n.a.		75
u PVC	Unknown	40m			0.30	37°C TL	16
s PVC	WHEELER ACID KING	32m	<90	3	n.a.	degrad.	85
b PVC+POLYURETHANE	BATA POLYBLEND	237m	1	1	n.a.		86
b PVC+POLYURETHANE	BATA POLYBLEND	>180m			n.a.		90
b PVC+POLYURETHANE	BATA POLYMAX	78m	5	1	n.a.		86
b PVC+POLYURETHANE	BATA SUPER POLY	277m	1	1	n.a.		86
u PVDC/PE/PVDC	MOLNLYCKE	>240m	0	0	0.07		54
s SARANEX-23	DU PONT TYVEK SARAN	>480m	0	0	n.a.		8
s SARANEX-23 2-PLY	DU PONT TYVEK SARAN	>480m	0	0	n.a.		8
u TEFLON	CHEMFAB	2m	3	3	0.10		65
s TEFLON	CHEMFAB CHALL. 5100	>300m	0	0	n.a.		80
s TEFLON	CHEMFAB CHALL. 5200	>480m	0	0	n.a.		56
s TEFLON	CHEMFAB CHALL. 5200	>850m	0	0	n.a.		65
s TEFLON	CHEMFAB CHALL. 6000	>180m			n.a.		80
s TEFLON	LIFE-GUARD TEFGUARD	>480m	0	0	n.a.		77
s TEFLON	WHEELER ACID KING	>480m	0	0	n.a.		85
v TEFLON-FEP	CHEMFAB CHALL.	>180m			0.25		65
s UNKNOWN MATERIAL	CHEMRON CHEMREL	136m	5	1	n.a.		67
s UNKNOWN MATERIAL	DU PONT BARRICADE	119m	11	2	n.a.		8
s UNKNOWN MATERIAL	KAPPLER CPF III	8m	10	4	n.a.		77
s UNKNOWN MATERIAL	LIFE-GUARD RESPONDE	>480m	0	0	n.a.		77
g VITRON	NORTH F-091	> 60m			0.26		38
g VITRON	NORTH F-121	> 60m			0.31		10
s VITON/BUTYL/UNKN.	TRELLEBORG HPS	>180m			n.a.		71
s VITON/CHLOROBUTYL	LIFE-GUARD VC-100	101m			n.a.		77
s VITON/CHLOROBUTYL	LIFE-GUARD VNC 200	392m	0.1		n.a.		77
u VITON/CHLOROBUTYL	Unknown	>180m			n.a.		46
g VITON/NEOPRENE	ERISTA VITRIC	> 60m			0.60		38
s VITON/NEOPRENE	MSA VAUTEX	>480m	0	0	n.a.		48

methyl alcohol, 50% & butyl acetate, 50%
Primary Class: 311   Hydroxy Compounds, Aliphatic and Alicyclic, Primary

G A R M E Class & Number/ N test chemical/ T MATERIAL NAME	MANUFACTURER & PRODUCT IDENTIFICATION	Break- through Time in Minutes	Perm- eation Rate in mg/sq m /min.	I N D E X	Thick- ness in mm	R degrad. e and f comments s

methyl alcohol, 50% & hexane, 50%
Primary Class: 311   Hydroxy Compounds, Aliphatic and Alicyclic, Primary

methyl alcohol, 50% & toluene, 50%
Primary Class: 311   Hydroxy Compounds, Aliphatic and Alicyclic, Primary

methyl alcohol, 50% & water, 50%
Primary Class: 311   Hydroxy Compounds, Aliphatic and Alicyclic, Primary

| u PVAL | Unknown | 19/17m | | | 0.07 | 37°C TL 16 |

methyl alcohol, 90% & butyl acetate, 10%
Primary Class: 311   Hydroxy Compounds, Aliphatic and Alicyclic, Primary

| g NITRILE | ANSELLEDMONT 37-155 | 106/m | 600/ | 4 | 0.35 | 41 |

methyl alcohol, 95% & 4,4'-methylenedianiline, 5%
Primary Class: 311   Hydroxy Compounds, Aliphatic and Alicyclic, Primary

| g NAT+NEOPR+NITRILE | MAPA-PIONEER TRIONI | 430m | 0.03 | 0 | 0.50 | 36 |

methyl bromide (synonym: bromomethane)
CAS Number: 74-83-9
Primary Class: 261   Halogen Compounds, Aliphatic and Alicyclic

g BUTYL	NORTH B-174	>480m	0	0	0.63	34
g NATURAL RUBBER	ACKWELL 5-109	>480m	0	0	0.15	34
g NEOPRENE	ANSELLEDMONT 29-870	>480m	0	0	0.48	34
g PVC	ANSELLEDMONT 34-100	<1m	300000	5	0.20	34
s SARANEX-23	DU PONT TYVEK SARAN	47m	0.1	2	n.a.	8

methyl chloride (synonym: chloromethane)
CAS Number: 74-87-3
Primary Class: 261   Halogen Compounds, Aliphatic and Alicyclic

g BUTYL	BRUNSWICK BUTYL STD	176m	9	1	0.63	89
g BUTYL	BRUNSWICK BUTYL-XTR	223m	81	2	0.70	89
s BUTYL	WHEELER ACID KING	51m	<9	2	n.a.	85
g BUTYL/ECO	BRUNSWICK BUTYL-POL	136m	2250	4	0.63	89
g NEOPRENE	BRUNSWICK NEOPRENE	39m	680	3	0.80	89
c PE	DU PONT TYVEK QC	<1m	>1000	5	n.a.	8
s PVC	WHEELER ACID KING	18m	<900	3	n.a.	85
s SARANEX-23	DU PONT TYVEK SARAN	>480m	0	0	n.a.	8

G A R M E Class & Number/ N test chemical/ T MATERIAL NAME	MANUFACTURER & PRODUCT IDENTIFICATION	Break- through Time in Minutes	Perm- eation Rate in mg/sq m /min.	I N D E X	Thick- ness in mm	degrad. and comments	R e f s
s TEFLON	CHEMFAB CHALL. 5200	21m	7.6	2	n.a.		80
s TEFLON	LIFE-GUARD TEFGUARD	27m	<0.03	2	n.a.		77
s TEFLON	WHEELER ACID KING	14m	<9	3	n.a.		85
s UNKNOWN MATERIAL	DU PONT BARRICADE	>480m	0	0	n.a.		8
s UNKNOWN MATERIAL	LIFE-GUARD RESPONDE	>480m	0	0	n.a.		77
s VITON/CHLOROBUTYL	LIFE-GUARD VC-100	>480m	0	0	n.a.		77

methyl chloroacetate
CAS Number: 96-34-4
Primary Class: 224   Esters, Carboxylic, Aliphatic, Others
Related Class: 222   Esters, Carboxylic, Acetates
              261   Halogen Compounds, Aliphatic and Alicyclic

s SARANEX-23	DU PONT TYVEK SARAN	>480m	<180	3	n.a.		8

methyl ethyl ketone (synonyms: 2-butanone; MEK)
CAS Number: 78-93-3
Primary Class: 391   Ketones, Aliphatic and Alicyclic

g BUTYL	BEST 878	232m	94	2	0.75		53
g BUTYL	COMASEC BUTYL	>480m	0	0	0.45		40
g BUTYL	ERISTA BX	>240m	0	0	0.65		33
s BUTYL	MSA CHEMPRUF	<120m			n.a.		48
g BUTYL	NORTH B-174	>480m	0	0	0.65		34
g BUTYL/NEOPRENE	COMASEC BUTYL PLUS	>480m	0	0	0.50		40
s BUTYL/NEOPRENE	MSA BETEX	30m	124	3	n.a.		48
s CPE	ILC DOVER CLOROPEL	31m			n.a.		46
s CPE	STD. SAFETY				n.a.	degrad.	75
g HYPALON	COMASEC DIPCO	24m	4868	4	0.50		40
g NAT+NEOPR+NITRILE	MAPA-PIONEER TRIONI	4m	>5000	5	0.48		36
g NATURAL RUBBER	ACKWELL 5-109				0.15	degrad.	34
g NATURAL RUBBER	ANSELL CONFORM 4205	1m	119	4	0.13		55
g NATURAL RUBBER	ANSELL FL 200 254	4m	7	3	0.51		55
g NATURAL RUBBER	ANSELL ORANGE 208	5m	5	3	0.76		55
g NATURAL RUBBER	ANSELL PVL 040	7m	7	3	0.46		55
g NATURAL RUBBER	ANSELL STERILE 832	1m	23	4	0.23		55
g NATURAL RUBBER	ANSELLEDMONT 36-124	10m	<90000	5	0.46		6
g NATURAL RUBBER	ANSELLEDMONT 392	5m	<9000	5	0.48		6
g NATURAL RUBBER	ANSELLEDMONT 46-320	2m	6000	5	0.31		5
g NATURAL RUBBER	BEST 65NFW		5220	5	n.a.		53
g NATURAL RUBBER	BEST 67NFW		9240	5	n.a.		53
g NATURAL RUBBER	COMASEC FLEXIGUM	45m	2880	4	0.95		40
g NATURAL RUBBER	MAPA-PIONEER L-118	6m	5220	5	0.46		36
g NATURAL RUBBER	MARIGOLD BLACK HEVY	3m	3931	5	0.65	degrad.	79
g NATURAL RUBBER	MARIGOLD FEATHERLIT	1m	10766	5	0.15	degrad.	79
g NATURAL RUBBER	MARIGOLD MED WT EXT	2m	5567	5	0.45	degrad.	79
g NATURAL RUBBER	MARIGOLD MEDICAL	1m	8600	5	0.28	degrad.	79
g NATURAL RUBBER	MARIGOLD ORANGE SUP	3m	3767	5	0.73	degrad.	79

G A R M E N T		Break-through Time in Minutes	Perm-eation Rate in mg/sq m /min.	I N D E X	Thick-ness in mm	degrad. and comments	R e f s
E Class & Number/ N test chemical/ T MATERIAL NAME	MANUFACTURER & PRODUCT IDENTIFICATION						
g NATURAL RUBBER	MARIGOLD R SURGEONS	1m	8600	5	0.28	degrad.	79
g NATURAL RUBBER	MARIGOLD RED LT WT	2m	6000	5	0.43	degrad.	79
g NATURAL RUBBER	MARIGOLD SENSOTECH	1m	8600	5	0.28	degrad.	79
g NATURAL RUBBER	MARIGOLD SUREGRIP	3m	4422	5	0.58	degrad.	79
g NATURAL RUBBER	MARIGOLD YELLOW	2m	6867	5	0.38	degrad.	79
g NATURAL+NEOPRENE	ANSELL OMNI 276	6m	3100	5	0.45		5
g NATURAL+NEOPRENE	ANSELL OMNI 276	5m	5	3	0.56		55
g NATURAL+NEOPRENE	ANSELL TECHNICIANS	6m	7	3	0.43		55
g NATURAL+NEOPRENE	MARIGOLD FEATHERWT	1m	7300	5	0.35	degrad.	79
g NATURAL+NEOPRENE	PLAYTEX ARGUS 123	5m	10020	5	n.a.		72
g NEOPRENE	ANSELL NEOPRENE 530	2m	10	4	0.46		55
g NEOPRENE	ANSELLEDMONT 29-840				0.38	degrad.	6
g NEOPRENE	ANSELLEDMONT 29-870				0.48	degrad.	34
g NEOPRENE	BEST 32	9m	1350	5	n.a.		53
g NEOPRENE	BEST 6780	8m	1040	5	n.a.		53
g NEOPRENE	COMASEC COMAPRENE	7m	7200	5	n.a.		40
g NEOPRENE	MAPA-PIONEER N-44	22m	9300	4	0.56		36
g NEOPRENE	MAPA-PIONEER N-73	10m	6000	5	0.46		5
g NEOPRENE/NATURAL	ANSELL CHEMI-PRO	6m	4	3	0.72		55
g NITRILE	ANSELLEDMONT 37-155	6m	31000	5	0.35		5
g NITRILE	ANSELLEDMONT 37-155				0.40	degrad.	34
g NITRILE	ANSELLEDMONT 37-165	12m	19000	5	0.55		5
g NITRILE	ANSELLEDMONT 37-165				0.60	degrad.	6
g NITRILE	ANSELLEDMONT 37-175	6m	22000	5	0.37		5
g NITRILE	BEST 727				0.38	degrad.	53
g NITRILE	COMASEC COMATRIL	15m	5880	4	0.55		40
g NITRILE	COMASEC COMATRIL SU	20m	4920	4	0.60		40
g NITRILE	COMASEC FLEXITRIL	5m	11280	5	n.a.		40
g NITRILE	MARIGOLD BLUE	10m	9785	5	0.45	degrad.	79
g NITRILE	MARIGOLD GREEN SUPA	4m	10000	5	0.30	degrad.	79
g NITRILE	MARIGOLD NITROSOLVE	11m	9742	5	0.75	degrad.	79
g NITRILE+PVC	COMASEC MULTIMAX	23m	132	3	n.a.		40
g NITRILE+PVC	COMASEC MULTIPLUS	15m	6060	4	n.a.		40
G PE	ANSELLEDMONT 35-125	<1m	100	4	0.03		5
g PE/EVAL/PE	SAFETY4 4H	>240m	0	0	0.07	35°C	60
g PE/EVAL/PE	SAFETY4 4H	>1440m	0	0	0.07		60
g PVAL	ANSELLEDMONT PVA	90m	<90	2	n.a.		6
g PVAL	COMASEC SOLVATRIL	>480m	0	0	n.a.		40
g PVAL	EDMONT 15-552	30m	<900	3	n.a.		6
g PVAL	EDMONT 25-545	>360m	0	0	n.a.		6
g PVAL	EDMONT 25-950	325m	1	1	0.74		34
g PVC	ANSELLEDMONT 34-100				0.20	degrad.	34
g PVC	ANSELLEDMONT SNORKE				n.a.	degrad.	6
g PVC	COMASEC MULTIPOST	15m	6600	4	n.a.		40
g PVC	COMASEC MULTITOP	15m	6900	4	n.a.		40
g PVC	COMASEC NORMAL	16m	7200	4	n.a.		40
g PVC	COMASEC OMNI	12m	7980	5	n.a.		40

G A R M E Class & Number/ N test chemical/ T MATERIAL NAME	MANUFACTURER & PRODUCT IDENTIFICATION	Break-through Time in Minutes	Perm-eation Rate in mg/sq m /min.	I N D E X	Thick-ness in mm	degrad. and comments	R e f s
g PVC	MAPA-PIONEER V-20	<1m	>90000	5	0.31		5
s PVC	MSA UPC	<3m			0.20		48
s SARANEX-23	DU PONT TYVEK SARAN	29m	78	3	n.a.		8
s TEFLON	CHEMFAB CHALL. 5100	>480m	0	0	n.a.		56
s TEFLON	CHEMFAB CHALL. 5100	>180m	0	1	n.a.		65
s UNKNOWN MATERIAL	CHEMRON CHEMREL	>1440m	0	0	n.a.		67
s UNKNOWN MATERIAL	CHEMRON CHEMREL MAX	>1440m	0	0	n.a.		67
c UNKNOWN MATERIAL	CHEMRON CHEMTUFF	32m	3.1	2	n.a.		67
s UNKNOWN MATERIAL	DU PONT BARRICADE	>480m	0	0	n.a.		8
s UNKNOWN MATERIAL	KAPPLER CPF III	>480m	0	0	n.a.		77
s UNKNOWN MATERIAL	LIFE-GUARD RESPONDE	>240m		0	n.a.		77
g VITRON	NORTH F-091				0.26	degrad.	34
g VITRON	NORTH F-091	<2m	65520	5	0.24		41
s VITON/CHLOROBUTYL	LIFE-GUARD VC-100	26m			n.a.		77
u VITON/CHLOROBUTYL	Unknown	32m			n.a.		46
s VITON/NEOPRENE	MSA VAUTEX	4m	646	4	n.a.		48

methyl ethyl ketone, 30-70% & ethylene glycol acetate, MIBK, isopropyl alcohol
Primary Class: 800   Multicomponent Mixtures With > 2 Components

g BUTYL	ERISTA BX	>240m	0	0	0.70		33
g BUTYL	ERISTA BX	>1020m	0	0	0.70	preexp. 20h	33
g NATURAL RUBBER	ANSELLEDMONT 36-755	20m	240	3	0.50		33

methyl ethyl ketone, 30-70% & MIBK, isopropyl alcohol, binding material
Primary Class: 800   Multicomponent Mixtures With > 2 Components

g BUTYL	ERISTA BX	>240m	0	0	0.69		33
g NITRILE	ANSELLEDMONT 37-175	19m	2600	4	0.39		33
g PVC	KID VINYLPRODUKTER	9m	3140	5	0.55		33

methyl ethyl ketone, 50% & 1,4-butanediolglycidyl ether, 50%
Primary Class: 391   Ketones, Aliphatic and Alicyclic

| g PE/EVAL/PE | SAFETY4 4H | >240m | 0 | 0 | 0.07 | 35°C | 60 |

methyl ethyl ketone, 50% & diglycidyl ether with bisphenol A, 50%
Primary Class: 391   Ketones, Aliphatic and Alicyclic

| g PE/EVAL/PE | SAFETY4 4H | >240m | 0 | 0 | 0.07 | 35°C | 60 |
| g PE/EVAL/PE | SAFETY4 4H | >480m | 0 | 0 | 0.07 | | 60 |

methyl ethyl ketone, 50% & hexane, 50%
Primary Class: 391   Ketones, Aliphatic and Alicyclic

G A R M E Class & Number/ N test chemical/ T MATERIAL NAME	MANUFACTURER & PRODUCT IDENTIFICATION	Break- through Time in Minutes	Perm- eation Rate in mg/sq m /min.	I N D E X	Thick- ness in mm	R degrad. and comments	R e f s
methyl ethyl ketone, 50% & p-cresol, 50% Primary Class: 391 Ketones, Aliphatic and Alicyclic							
g PE/EVAL/PE	SAFETY4 4H	>240m	0	0	0.07	35°C	60
methyl ethyl ketone, 50% & phenol, 50% Primary Class: 391 Ketones, Aliphatic and Alicyclic							
g PE/EVAL/PE	SAFETY4 4H	>240m	0	0	0.07	35°C	60
methyl ethyl ketone, 50% & toluene, 50% Primary Class: 391 Ketones, Aliphatic and Alicyclic							
g PE/EVAL/PE	SAFETY4 4H	9m			0.07	35°C	60
g PE/EVAL/PE	SAFETY4 4H	114m			0.07		60
s UNKNOWN MATERIAL	CHEMRON CHEMREL	32m	1	2	n.a.		67
methyl ethyl ketone, 50% & triethylenetetraamine, 50% Primary Class: 391 Ketones, Aliphatic and Alicyclic							
g PE/EVAL/PE	SAFETY4 4H	>240m	0	0	0.07	35°C	60
methyl ethyl ketone, 85% & acrylamide, 15% Primary Class: 391 Ketones, Aliphatic and Alicyclic							
g PE/EVAL/PE	SAFETY4 4H	>240m	0	0	0	07 35°C	60
methyl ethyl ketone, 90% & hexane, 10% Primary Class: 391 Ketones, Aliphatic and Alicyclic							
g VITON	NORTH F-091	<2m	67560	5	0.24		41
methyl eugenol CAS Number: 93-15-2							
Primary Class: 242	Ethers, Aromatic						
g BUTYL	NORTH B-174	>480m	0	0	0.69		34
g NATURAL RUBBER	ACKWELL 5-109	15m	540	3	0.18	degrad.	34
g NEOPRENE	ANSELLEDMONT 29-870	132m	420	3	0.51	degrad.	34
g VITON	NORTH F-091	>480m	0	0	0.41		34

G A R M E Class & Number/ N test chemical/ T MATERIAL NAME	MANUFACTURER & PRODUCT IDENTIFICATION	Break- through Time in Minutes	Perm- eation Rate in mg/sq m /min.	I N D E X	Thick- ness in mm	R degrad. e and f comments s

methyl iodide (synonym: iodomethane)
CAS Number: 74-88-4
Primary Class: 261   Halogen Compounds, Aliphatic and Alicyclic

g BUTYL	NORTH B-174	55m	4900	4	0.70	degrad. 34
g NATURAL RUBBER	ACKWELL 5-109				0.15	degrad. 34
g NATURAL RUBBER	ANSELLEDMONT 36-124				0.46	degrad. 6
g NATURAL RUBBER	ANSELLEDMONT 392				0.48	6
g NATURAL RUBBER	ANSELLEDMONT 46-320	1m	130000	5	0.31	5
g NATURAL+NEOPRENE	ANSELL OMNI 276	2m	89000	5	0.45	5
g NEOPRENE	ANSELLEDMONT 29-840				0.38	degrad. 6
g NEOPRENE	ANSELLEDMONT 29-870	6m	78800	5	0.50	degrad. 34
g NEOPRENE	ANSELLEDMONT NEOX	15m	14000	5	0.77	5
g NEOPRENE	ANSELLEDMONT NEOX				n.a.	degrad. 6
g NEOPRENE	MAPA-PIONEER N-54	12m	37000	5	0.70	5
g NITRILE	ANSELLEDMONT 37-155				0.40	degrad. 34
g NITRILE	ANSELLEDMONT 37-165				0.60	degrad. 6
g NITRILE	ANSELLEDMONT 37-175	6m	60000	5	0.37	5
g NITRILE	ANSELLEDMONT 37-185	8m	80000	5	0.55	5
g PE	ANSELLEDMONT 35-125	1m	11000	5	0.03	5
g PVAL	ANSELLEDMONT PVA	>360m		0	n.a.	6
g PVAL	EDMONT 15-552	50m	<9	2	n.a.	6
g PVAL	EDMONT 25-950				0.60	degrad. 34
g PVC	ANSELLEDMONT SNORKE				n.a.	degrad. 6
g PVC	MAPA-PIONEER V-20	<1m	>90000	5	0.31	5
g VITON	NORTH F-091	400m	40	2	0.30	34

methyl isobutyl ketone (synonyms: 4-methyl-2-pentanone; MIBK)
CAS Number: 108-10-1
Primary Class: 391   Ketones, Aliphatic and Alicyclic

g BUTYL	BEST 878	226m	31	2	0.75	53
g BUTYL	COMASEC BUTYL	140m			0.50	40
g BUTYL	NORTH B-161	294m	70	2	0.45	30
g BUTYL/NEOPRENE	COMASEC BUTYL PLUS	>480m	0	0	0.50	40
g NATURAL RUBBER	ANSELL FL 200 254	6m	11	4	0.51	55
g NATURAL RUBBER	ANSELL ORANGE 208	12m	5	3	0.76	55
g NATURAL RUBBER	ANSELL PVL 040	12m	5	3	0.46	55
g NATURAL RUBBER	ANSELL STERILE 832	1m	100	4	0.23	55
g NATURAL RUBBER	ANSELLEDMONT 30-139	6m	<900	4	0.25	42
g NATURAL RUBBER	ANSELLEDMONT 36-124	6m	<9000	5	0.46	6
g NATURAL RUBBER	ANSELLEDMONT 392				0.48	degrad. 6
g NATURAL RUBBER	COMASEC FLEXIGUM	55m	3300	4	0.95	40
g NATURAL RUBBER	MARIGOLD BLACK HEVY	9m	3794	5	0.65	degrad. 79
g NATURAL RUBBER	MARIGOLD FEATHERLIT	1m	4228	5	0.03	degrad. 79

G A R M E Class & Number/ N test chemical/ T MATERIAL NAME	MANUFACTURER & PRODUCT IDENTIFICATION	Break-through Time in Minutes	Perm-eation Rate in mg/sq m /min.	I N D E X	Thick-ness in mm	R degrad. and comments	R e f s
g NATURAL RUBBER	MARIGOLD MED WT EXT	5m	6400	5	0.45	degrad.	79
g NATURAL RUBBER	MARIGOLD MEDICAL	2m	5133	5	0.28	degrad.	79
g NATURAL RUBBER	MARIGOLD ORANGE SUP	9m	3533	5	0.73	degrad.	79
g NATURAL RUBBER	MARIGOLD R SURGEONS	2m	5133	5	0.28	degrad.	79
g NATURAL RUBBER	MARIGOLD RED LT WT	5m	6219	5	0.43	degrad.	79
g NATURAL RUBBER	MARIGOLD SENSOTECH	2m	5133	5	0.28	degrad.	79
g NATURAL RUBBER	MARIGOLD SUREGRIP	8m	4576	5	0.58	degrad.	79
g NATURAL RUBBER	MARIGOLD YELLOW	4m	5857	5	0.38	degrad.	79
g NATURAL+NEOPRENE	ANSELL OMNI 276	12m	6	3	0.56		55
g NATURAL+NEOPRENE	ANSELL TECHNICIANS	1m	41	4	0.43		55
g NATURAL+NEOPRENE	MARIGOLD FEATHERWT	4m	5676	5	0.35	degrad.	79
g NATURAL+NEOPRENE	PLAYTEX ARGUS 123	8m	3420	5	n.a.		72
g NEOPRENE	ANSELLEDMONT 29-840				0.38	degrad.	6
g NEOPRENE	BEST 32	23m	2220	4	n.a.	degrad.	53
g NEOPRENE	BEST 6780	57m	1450	4	n.a.		53
g NEOPRENE	COMASEC COMAPRENE	15m	540	4	n.a.		40
g NITRILE	ANSELLEDMONT 37-165				0.60	degrad.	6
g NITRILE	BEST 727				0.38	degrad.	53
g NITRILE	COMASEC COMATRIL	37m	9060	4	0.55		40
g NITRILE	COMASEC COMATRIL SU	100m	8400	4	0.60		40
g NITRILE	COMASEC FLEXITRIL	5m	6240	5	n.a.		40
g NITRILE	MARIGOLD BLUE	37m	6845	4	0.45	degrad.	79
g NITRILE	MARIGOLD GREEN SUPA	19m	4967	4	0.30	degrad.	79
g NITRILE	MARIGOLD NITROSOLVE	41m	7221	4	0.75	degrad.	79
g NITRILE	NORTH LA-132G	27m	38000	5	0.32		30
g NITRILE+PVC	COMASEC MULTIPLUS	30m	7800	4	n.a.		40
g PE/EVAL/PE	SAFETY4 4H	>240m	0	0	0.07	35°C	60
g PVAL	ANSELLEDMONT PVA	>360m		0	n.a.		6
g PVAL	EDMONT 15-552	>360m	0	0	n.a.		6
g PVAL	EDMONT 25-545	>360m	0	0	n.a.		6
g PVC	ANSELLEDMONT SNORKE				n.a.	degrad.	6
g PVC	COMASEC MULTIPOST	30m	9000	4	n.a.		40
g PVC	COMASEC MULTITOP	25m	8280	4	n.a.		40
g PVC	COMASEC NORMAL	27m	9900	4	n.a.		40
g PVC	COMASEC OMNI	11m	8160	5	n.a.		40
s TEFLON	CHEMFAB CHALL. 5100	>180m	0	1	n.a.		65
s UNKNOWN MATERIAL	LIFE-GUARD RESPONDE	>480m	0	0	n.a.		77
g VITON	NORTH F-091	19m	35000	5	0.27		30

methyl isobutyl ketone, 50% & toluene, 50%
Primary Class: 391   Ketones, Aliphatic and Alicyclic

g PVAL	EDMONT 15-552	>240m	0	0	n.a.		33
g PVAL	EDMONT 15-552	>240m	0	0	n.a.	preexp. 20h	33

methyl isobutyl ketone, 50% & xylenes, 50%
Primary Class: 391   Ketones, Aliphatic and Alicyclic

g NITRILE	NORTH LA-132G	12m	27000	5	0.32		30

G A R M E Class & Number/ N test chemical/ T MATERIAL NAME	MANUFACTURER & PRODUCT IDENTIFICATION	Break- through Time in Minutes	Perm- eation Rate in mg/sq m /min.	I N D E X	Thick- ness in mm	R degrad. and e comments	R e f s
g VITRON	NORTH F-091	20m	30000	5	0.27		30

methyl isocyanate (synonym: MIC)
CAS Number: 624-83-9
Primary Class: 211   Isocyanates, Aliphatic and Alicyclic

s BUTYL	FYREPEL	7m			0.36		69
s BUTYL	MSA CHEMPRUF	13m			n.a.		69
g BUTYL	NORTH B-174	43m			0.62		34
m BUTYL	PAIGE	8m			0.37		69
s BUTYL	WHEELER ACID KING	5m			n.a.		69
s BUTYL/NEOPRENE	MSA BETEX	7m			n.a.		69
s CPE	ILC DOVER CLOROPEL	6m			n.a.		69
g NATURAL RUBBER	ACKWELL 5-109	<1m			0.15		34
g NATURAL RUBBER	TRAVENOL	<1m			0.18		69
g NEOPRENE	ANSELLEDMONT 29-870	<1m			0.46	degrad.	34
g NITRILE	ANSELLEDMONT 37-155				0.40	degrad.	34
g NITRILE	ANSELLEDMONT 37-155	2m			0.36		69
c PE	ANSELLEDMONT 55-530	<1m			n.a.		69
m POLYURETHANE	ILC DOVER	4m			0.44		69
g PVAL	EDMONT 25-950	>480m	0	0	0.50		34
g PVC	ANSELLEDMONT 34-100				0.20	degrad.	34
s PVC	FYREPEL	<1m			0.36		69
s PVC	WHEELER ACID KING	1m			n.a.		69
s SARANEX-23	DU PONT TYVEK SARAN	2m	2100	5	n.a.		8
s TEFLON	CHEMFAB CHALL. 5100	30m			n.a.		69
s TEFLON	WHEELER ACID KING	1m			0.23		69
s UNKNOWN MATERIAL	CHEMRON CHEMREL	6m	3	3	n.a.		67
s UNKNOWN MATERIAL	DU PONT BARRICADE	>480m	0	0	n.a.		8
s UNKNOWN MATERIAL	LIFE-GUARD RESPONDE	>480m	0	0	n.a.		77
s VITON	FYREPEL	<1m			0.23		69
g VITON	NORTH F-091	1m			0.26	degrad.	34
s VITON/NEOPRENE	MSA VAUTEX	3m			n.a.		69

methyl mercaptan (synonym: thiomethanol)
CAS Number: 74-93-1
Primary Class: 501   Sulfur Compounds, Thiols

s UNKNOWN MATERIAL	LIFE-GUARD RESPONDE	>480m	0	0	n.a.		77

methyl methacrylate (synonym: methyl 2-methylpropenoate)
CAS Number: 80-62-6
Primary Class: 223   Esters, Carboxylic, Acrylates and Methacrylates

g BUTYL	BEST 878	63m	310	3	0.75		53
g BUTYL	NORTH B-161	106m	320	3	0.45		30
g BUTYL	NORTH B-174	300m	220	3	0.68		34
g NATURAL RUBBER	ACKWELL 5-109	1m	96000	5	0.15		34
g NATURAL RUBBER	ANSELLEDMONT 36-124				0.46	degrad.	6

G A R M E Class & Number/ N test chemical/ T MATERIAL NAME	MANUFACTURER & PRODUCT IDENTIFICATION	Break- through Time in Minutes	Perm- eation Rate in mg/sq m /min.	I N D E X	Thick- ness in mm	R degrad. and comments	R e f s
g NATURAL RUBBER	ANSELLEDMONT 392				0.48	degrad.	6
g NATURAL RUBBER	MARIGOLD BLACK HEVY	2m	6424	5	0.65	degrad.	79
g NATURAL RUBBER	MARIGOLD FEATHERLIT	1m	11000	5	0.15	degrad.	79
g NATURAL RUBBER	MARIGOLD MED WT EXT	3m	15000	5	0.45	degrad.	79
g NATURAL RUBBER	MARIGOLD MEDICAL	1m	12667	5	0.28	degrad.	79
g NATURAL RUBBER	MARIGOLD ORANGE SUP	5m	3500	5	0.70		32
g NATURAL RUBBER	MARIGOLD ORANGE SUP	1m	5567	5	0.73	degrad.	79
g NATURAL RUBBER	MARIGOLD R SURGEONS	1m	12667	5	0.28	degrad.	79
g NATURAL RUBBER	MARIGOLD RED LT WT	3m	14667	5	0.43	degrad.	79
g NATURAL RUBBER	MARIGOLD SENSOTECH	1m	12667	5	0.28	degrad.	79
g NATURAL RUBBER	MARIGOLD SUREGRIP	2m	8997	5	0.58	degrad.	79
g NATURAL RUBBER	MARIGOLD YELLOW	2m	14000	5	0.38	degrad.	79
g NATURAL RUBBER	PERRX-AM TEX	2m	6700	5	0.12		32
g NATURAL RUBBER	PERRX-AM TEX	1m	11300	5	0.12	35°C	32
g NATURAL+NEOPRENE	MARIGOLD FEATHERWT	2m	13667	5	0.35	degrad.	79
g NEOPRENE	ANSELLEDMONT 29-840				0.38	degrad.	6
g NEOPRENE	ANSELLEDMONT 29-870				0.48	degrad.	34
g NEOPRENE	BEST 32	17m	1790	4	0.38	degrad.	53
g NEOPRENE	BEST 6780	29m	1350	4	n.a.		53
g NITRILE	ANSELLEDMONT 37-155				0.40	degrad.	34
g NITRILE	ANSELLEDMONT 37-165				0.60	degrad.	6
g NITRILE	MARIGOLD BLUE	31m	6451	4	0.45	degrad.	79
g NITRILE	MARIGOLD GREEN SUPA	15m	8167	4	0.30	degrad.	79
g NITRILE	MARIGOLD NITROSOLVE	34m	6107	4	0.75	degrad.	79
g PE	GYNO-DAKTARIN	2m	1740	5	0.03		32
g PE	GYNO-DAKTARIN	1m	6100	5	0.03	35°C	32
g PE/EVAL/PE	SAFETY4 4H	>240m	0	0	0.07	35°C	60
g PE/EVAL/PE	SAFETY4 4H	>1440m	0	0	0.07		60
g PVAL	ANSELLEDMONT PVA	>360m		0	n.a.		6
g PVAL	EDMONT 15-552	>360m	0	0	n.a.		6
g PVAL	EDMONT 25-545	>360m	0	0	n.a.		6
g PVAL	EDMONT 25-950	>480m	0	0	0.58		34
g PVC	ANSELLEDMONT 34-100	1m	96000	5	0.20	degrad.	34
g PVC	ANSELLEDMONT 34-500	1m	10100	5	0.11		32
g PVC	ANSELLEDMONT SNORKE				n.a.	degrad.	6
g SBR	ELASTYREN	2m	6888	5	0.16		32
g SBR	ELASTYREN	1m	15372	5	0.16	35°C	32
s TEFLON	CHEMFAB CHALL. 5100	>190m	0	1	n.a.		65
s UNKNOWN MATERIAL	DU PONT BARRICADE	>480m	0	0	n.a.		8
g VITRON	NORTH F-091				0.25	degrad.	34
s VITON/BUTYL	DRAGER 500 OR 710	>30m			n.a.		91

methyl parathion, 30-70%
CAS Number: 298-00-0
Primary Class: 462   Organophosphorus Compounds and Derivatives of Phosphorus-based Acids
Related Class: 233   Esters, Non-Carboxylic, Carbamates and Others
              840   Pesticides, Mixtures and Formulations

c PE	DU PONT TYVEK QC	15m	0.9	2	n.a.		8
s SARANEX-23	DU PONT TYVEK SARAN	120m	<0.1	1	n.a.		8

	MANUFACTURER & PRODUCT IDENTIFICATION	Break-through Time in Minutes	Perm-eation Rate in mg/sq m /min.	I N D E X	Thick-ness in mm	degrad. and comments	R e f s

methyl parathion, <30%
Primary Class: 462   Organophosphorus Compounds and Derivatives of Phosphorus-based Acids
Related Class: 233   Esters, Non-Carboxylic, Carbamates and Others
         840   Pesticides, Mixtures and Formulations

c PE	DU PONT TYVEK QC	30m	2	2	n.a.		8
s SARANEX-23	DU PONT TYVEK SARAN	240m	<0.1	0	n.a.		8

methyl pentyl ketone (synonym: 2-heptanone)
CAS Number: 110-43-0
Primary Class: 391 Ketones, Aliphatic and Alicyclic

g PE/EVAL/PE	SAFETY4 4H	>240m	0		0.07	35°C	60

methylacrylonitrile (synonym: 2-methyl-2-propenenitrile)
CAS Number: 126-98-7
Primary Class: 431   Nitriles, Aliphatic and Alicyclic

g BUTYL	NORTH B-161	410m	<0.1	1	0.31		7
g BUTYL	NORTH B-174	>480m	0	0	0.68		34
g HYPALON	NORTH Y-1532	24m	3420	4	0.40		7
g NATURAL RUBBER	ACKWELL 5-109	1m	18000	5	0.15		34
g NITRILE	ANSELLEDMONT 37-155				0.40	degrad.	34
g NITRILE	NORTH LA-142G	6m	33600	5	0.33		7
g PVAL	EDMONT 25-950	23m	5	2	0.61		34
g PVC	ANSELLEDMONT 34-100	2m	11400	5	0.20		34
g VITON	NORTH F-091	4m	27720	5	0.30		7
g VITON	NORTH F-091				0.26	degrad.	34

methylamine (synonym: monomethylamine)
Primary Class: 141   Amines, Aliphatic and Alicyclic, Primary

s UNKNOWN MATERIAL	CHEMRON CHEMREL	18m	42	3	n.a.	98%	67
s UNKNOWN MATERIAL	DU PONT BARRICADE	105m	400	3	n.a.		8

methylamine, 30-70% (synonym: monomethylamine, 30-70%)
CAS Number: 74-89-5
Primary Class: 141   Amines, Aliphatic and Alicyclic, Primary

g BUTYL	NORTH B-161	>900m	0	0	0.45		7
g NATURAL RUBBER	ANSELLEDMONT 36-124	25m	<900	3	0.46		6
g NATURAL RUBBER	ANSELLEDMONT 392	55m	<90	3	0.48		6
g NEOPRENE	ANSELLEDMONT 29-840	270m	<900	3	0.38		6
g NEOPRENE	ANSELLEDMONT NEOX	360m	<9	1	n.a.		6
g NITRILE	ANSELLEDMONT 37-165	>360m	0	0	0.60		6
g NITRILE	NORTH LA-142G	>480m	0	0	0.36		7

G A R M E Class & Number/ N test chemical/ T MATERIAL NAME	MANUFACTURER & PRODUCT IDENTIFICATION	Break- through Time in Minutes	Perm- eation Rate in mg/sq m /min.	I N D E X	Thick- ness in mm	R degrad. and comments	R e f s
g PE/EVAL/PE	SAFETY4 4H	90m			0.07		60
g PE/EVAL/PE	SAFETY4 4H	80m	13	2	0.07	35°C	60
g PVAL	ANSELLEDMONT PVA				n.a.	degrad.	6
g PVAL	EDMONT 15-552				n.a.	degrad.	6
g PVC	ANSELLEDMONT M GRIP	132m	<90	2	n.a.		6
g UNKNOWN MATERIAL	NORTH SILVERSHIELD	116m	104	3	0.10		7
g VITON	NORTH F-091	>960m	0	0	0.24		7

methylene bromide (synonym: dibromomethane)
CAS Number: 74-95-3
Primary Class: 261   Halogen Compounds, Aliphatic and Alicyclic

g NATURAL RUBBER	ANSELLEDMONT 36-124				0.46	degrad.	6
g NATURAL RUBBER	ANSELLEDMONT 392				0.48	degrad.	6
g NATURAL+NEOPRENE	ANSELL OMNI 276				0.45	degrad.	55
g NEOPRENE	ANSELLEDMONT 29-840				0.38	degrad.	6
g NEOPRENE/NATURAL	ANSELL CHEMI-PRO				n.a.	degrad.	55
g NITRILE	ANSELLEDMONT 37-165				0.60	degrad.	6
g PVAL	ANSELLEDMONT PVA	>360m	0		n.a.		6
g PVAL	EDMONT 15-552	>360m	0	0	n.a.		6
g PVAL	EDMONT 25-545	>360m	0	0	n.a.		6
g PVC	ANSELLEDMONT SNORKE				n.a.	degrad.	6

methylene chloride (synonym: dichloromethane)
CAS Number: 75-09-2
Primary Class: 261   Halogen Compounds, Aliphatic and Alicyclic

g BUTYL	ANSELL LAMPRECHT	22m	6970	4	0.70		38
g BUTYL	BEST 878	28m	3240	4	0.75		53
g BUTYL	BRUNSWICK BUTYL STD	20m	>5000	4	0.63		89
g BUTYL	BRUNSWICK BUTYL-XTR	23m	>5000	4	0.85		89
g BUTYL	COMASEC BUTYL	12m			0.55		40
s BUTYL	DRAGER 500 OR 710	>30m			n.a.		91
s BUTYL	LIFE-GUARD BUTYL	4m	58	4	n.a.		77
g BUTYL	NORTH B-131				n.a.	degrad.	65
g BUTYL	NORTH B-161	10m	6960	5	0.40		35
m BUTYL	PLYMOUTH RUBBER	1m			n.a.		46
s BUTYL	WHEELER ACID KING	333m	<9000	4	n.a.		85
g BUTYL/ECO	BRUNSWICK BUTYL-POL	13m	>4200	5	0.63		89
g BUTYL/NEOPRENE	COMASEC BUTYL PLUS	12m	0.60		40		
s BUTYL/NEOPRENE	MSA BETEX	<5m	2600	5	n.a.		48
s CPE	ILC DOVER CLOROPEL	20m			n.a.		46
s CPE	STD. SAFETY	6m	192	4	n.a.		75
g HYPALON	COMASEC DIPCO	10m	6966	5	0.59		40
g HYPALON/NEOPRENE	NORTH		20000	5	0.88		50
g NAT+NEOPR+NITRILE	MAPA-PIONEER TRIONI	<4m	>5000	5	0.49		36

G A R M E N T	Class & Number/ test chemical/ MATERIAL NAME	MANUFACTURER & PRODUCT IDENTIFICATION	Break- through Time in Minutes	Perm- eation Rate in mg/sq m /min.	I N D E X	Thick- ness in mm	degrad. and comments	R e f s
g	NAT+NEOPR+NITRILE	MAPA-PIONEER TRIONI	3m	22000	5	0.63		70
g	NATURAL RUBBER	ANSELL PVL 040				n.a.	degrad.	55
g	NATURAL RUBBER	ANSELLEDMONT 36-124				0.46	degrad.	6
g	NATURAL RUBBER	ANSELLEDMONT 392				0.48	degrad.	6
g	NATURAL RUBBER	ANSELLEDMONT 46-320	1m	82000	5	0.31		5
g	NATURAL RUBBER	COMASEC FLEXIGUM	18m	9120	4	0.95		40
u	NATURAL RUBBER	Unknown	2m	>1500	5	n.a.		10
g	NATURAL+NEOPRENE	ANSELL OMNI 276	3m	4600	5	0.45		5
g	NATURAL+NEOPRENE	ANSELL OMNI 276				0.45	degrad.	55
g	NATURAL+NEOPRENE	PLAYTEX ARGUS 123	2m	12720	5	n.a.		72
g	NEOPRENE	ANSELL NEOPRENE 520				n.a.	degrad.	55
g	NEOPRENE	ANSELLEDMONT 29-840				0.38	degrad.	6
g	NEOPRENE	ANSELLEDMONT 29-870	<1m	26820	5	0.48		35
g	NEOPRENE	ANSELLEDMONT 29-870	6m	37000	5	0.53		70
g	NEOPRENE	ANSELLEDMONT NEOX	8m	11000	5	0.77		5
g	NEOPRENE	BEST 32	3m	1520	5	n.a.	degrad.	53
g	NEOPRENE	BEST 6780	4m	588	4	n.a.		53
g	NEOPRENE	BRUNSWICK NEOPRENE	8m	>5000	5	0.90		89
g	NEOPRENE	COMASEC COMAPRENE	5m	18000	5	n.a.		40
s	NEOPRENE	LIFE-GUARD NEOPRENE	6m	163	4	n.a.		77
g	NEOPRENE	MAPA-PIONEER N-44	6m	14340	5	0.56	degrad.	36
g	NEOPRENE	NORTH		16200	5	0.38		50
g	NEOPRENE	NORTH		23000	5	0.76		50
g	NEOPRENE	PIONEER		20000	5	0.84		50
b	NEOPRENE	RAINFAIR	13m	>48	4	n.a.		90
b	NEOPRENE	RANGER	16m	>44	3	n.a.		90
b	NEOPRENE	SERVUS NO 22204	28m	>80	3	n.a.		90
u	NEOPRENE	Unknown	12m	>1500	5	0.93		10
g	NEOPRENE/NATURAL	ANSELL CHEMI-PRO				n.a.	degrad.	55
g	NITRILE	ANSELLEDMONT 37-165				0.60	degrad.	6
g	NITRILE	ANSELLEDMONT 37-165	6m	58000	5	0.55		70
g	NITRILE	ANSELLEDMONT 37-175	2m	130000	5	0.37		5
g	NITRILE	ANSELLEDMONT 37-175	<1m	56280	5	0.34		35
g	NITRILE	BEST 22R		40080	5	n.a.		53
g	NITRILE	BEST 22R	2m	28000	5	n.a.	degrad.	70
g	NITRILE	BEST 727				0.38	degrad.	53
g	NITRILE	COMASEC COMATRIL	2m	76200	5	0.55		40
g	NITRILE	COMASEC COMATRIL SU	3m	54000	5	0.60		40
g	NITRILE	COMASEC FLEXITRIL	6m	55800	5	n.a.		40
g	NITRILE	MARIGOLD NITROSOLVE	5m	3244	5	0.75	degrad.	79
g	NITRILE	NORTH LA-142G	4m	45960	5	0.36		34
b	NITRILE+PUR+PVC	BATA HAZMAX	65m	>100	3	n.a.		86
g	NITRILE+PVC	COMASEC MULTIPLUS	12m	26400	5	n.a.		40
b	NITRILE+PVC	SERVUS NO 73101	48m	>140	3	n.a.		90
b	NITRILE+PVC	TINGLEY	44m	>110	3	n.a.		90
G	PE	ANSELLEDMONT 35-125	<1m	3000	5	0.03		5
G	PE	ANSELLEDMONT 35-125	<1m	4200	5	0.07		35

G A R M E Class & Number/ N test chemical/ T MATERIAL NAME	MANUFACTURER & PRODUCT IDENTIFICATION	Break- through Time in Minutes	Perm- eation Rate in mg/sq m /min.	I N D E X	Thick- ness in mm	degrad. and comments	R e f s
c PE	DU PONT TYVEK QC	<1m	6000	5	n.a.		8
g PE/EVAL/PE	SAFETY4 4H	>240m	0	0	0.07	35°C	60
g PE/EVAL/PE	SAFETY4 4H	>1440m	0	0	0.07		60
g PVAL	ANSELLEDMONT PVA	>360m	0		n.a.		6
g PVAL	COMASEC SOLVATRIL	>480m	0	0	n.a.		40
g PVAL	EDMONT 15-552	17m	<9	2	n.a.		6
g PVAL	EDMONT 15-552	>60m			n.a.		10
g PVAL	EDMONT 15-554	>120m			n.a.		70
g PVAL	EDMONT 25-545	>360m	0	0	n.a.		6
g PVAL	EDMONT 25-545	>480m	0	0	0.45		35
g PVAL	EDMONT 25-950	>480m	0	0	0.36		34
g PVC	ANSELLEDMONT 3-318	8m	12500	5	n.a.		70
g PVC	ANSELLEDMONT 3-318	8m	130	4	n.a.		81
g PVC	ANSELLEDMONT 3-318	8m	120	4	n.a.	pre-exp 48h	81
g PVC	ANSELLEDMONT SNORKE				n.a.	degrad.	6
b PVC	BATA STANDARD	42m	368	3	n.a.		86
g PVC	BEST 812		25500	5	n.a.		53
g PVC	BEST 812	7m	24000	5	n.a.		70
g PVC	COMASEC MULTIPOST	4m	35580	5	n.a.		40
g PVC	COMASEC MULTITOP	8m	30720	5	n.a.		40
g PVC	COMASEC NORMAL	6m	34800	5	n.a.		40
g PVC	COMASEC OMNI	5m	37020	5	n.a.		40
b PVC	JORDAN DAVID	65m	>80	2	n.a.		90
g PVC	MAPA-PIONEER V-5	<1m			0.10	degrad.	35
b PVC	STD. SAFETY GL-20	4m	>4000	5	n.a.		75
s PVC	STD. SAFETY WG-20	4m	>4000	5	n.a.		75
u PVC	Unknown	9m	>1500	5	n.a.		10
s PVC	WHEELER ACID KING	3m	<9000	5	n.a.	degrad.	85
b PVC+POLYURETHANE	BATA POLYBLEND	43m	336	3	n.a.		86
b PVC+POLYURETHANE	BATA POLYBLEND	49m	>50	3	n.a.		90
b PVC+POLYURETHANE	BATA POLYMAX	53m	407	3	n.a.		86
b PVC+POLYURETHANE	BATA SUPER POLY	61m	240	3	n.a.		86
s SARANEX-23	DU PONT TYVEK SARAN	2m	3180	5	n.a.		8
s SARANEX-23 2-PLY	DU PONT TYVEK SARAN	10m	1560	5	n.a.		8
u TEFLON	CHEMFAB	2m	83	4	0.10		65
s TEFLON	CHEMFAB CHALL. 5100	46m	0.23	2	n.a.		65
s TEFLON	CHEMFAB CHALL. 5200	64m	1.2	1	n.a.		80
s TEFLON	CHEMFAB CHALL. 6000	>180m			n.a.		80
s TEFLON	LIFE-GUARD TEFGUARD	>480m	0	0	n.a.		77
s TEFLON	WHEELER ACID KING	>480m	0	0	n.a.		85
s UNKNOWN MATERIAL	CHEMRON CHEMREL	4m	30	4	n.a.		67
s UNKNOWN MATERIAL	CHEMRON CHEMREL MAX	>480m	0	0	n.a.		67
c UNKNOWN MATERIAL	CHEMRON CHEMTUFF	32m	3	2	n.a.		67
s UNKNOWN MATERIAL	DU PONT BARRICADE	423m	0.5	0	n.a.		8
s UNKNOWN MATERIAL	KAPPLER CPF III	7m	>110	4	n.a.		77
s UNKNOWN MATERIAL	LIFE-GUARD RESPONDE	>480m	0	0	n.a.		77

G A R M E Class & Number/ N test chemical/ T MATERIAL NAME	MANUFACTURER & PRODUCT IDENTIFICATION	Break- through Time in Minutes	Perm- eation Rate in mg/sq m /min.	I N D E X	Thick- ness in mm	degrad. and comments	R e f s
g UNKNOWN MATERIAL	NORTH SILVERSHIELD	114m	1	1	0.08		7
g VITRON	NORTH F-091	60m	440	3	0.24		34
g VITRON	NORTH F-091	46m	478	3	0.26		38
g VITRON	NORTH F-091	80m	194	3	0.30		62
g VITRON	NORTH F-121	60m	>100	3	0.31		10
g VITRON	NORTH F-121	83m	228	3	0.30		35
s VITON/BUTYL	DRAGER 500 OR 710	>60m			n.a.		91
g VITON/BUTYL	NORTH	38m	317	3	0.45		30
s VITON/BUTYL/UNKN.	TRELLEBORG HPS	>180m			n.a.		71
s VITON/CHLOROBUTYL	LIFE-GUARD VC-100	28m			n.a.		77
s VITON/CHLOROBUTYL	LIFE-GUARD VNC 200	16m	11	3	n.a.		77
u VITON/CHLOROBUTYL	Unknown	30m			n.a.		43
g VITON/NEOPRENE	ERISTA VITRIC	>60m			0.60		38
s VITON/NEOPRENE	MSA VAUTEX	20m	5260	4	n.a.		48

methylene chloride, 30-70% & phenol, <30% & formic acid, <30%
Primary Class: 800   Multicomponent Mixtures With >2 Components

c PE	MOLNLYCKE H D	1m	7724	5	n.a.		30
g PE/EVAL/PE	SAFETY4 4H	69m			0.07		60
g PE/EVAL/PE	SAFETY4 4H	8m			0.07	35°C	60
g PVAL	EDMONT 15-552	3m	4433	5	n.a.		30
s PVC	TRELLEBORG TRELLCHE	2m	12600	5	n.a.		30
s VITON/BUTYL	TRELLEBORG TRELLCHE	55m	596	3	0.43		30
g VITON/NEOPRENE	ERISTA VITRIC	34m	1191	4	0.47		30

methylene chloride, 50% & hexane, 50%
Primary Class: 261   Halogen Compounds, Aliphatic and Alicyclic

methylene chloride, 50% & toluene, 50%
Primary Class: 261   Halogen Compounds, Aliphatic and Alicyclic

u VITON/CHLOROBUTYL	Unknown	50/62m			n.a.		43

methylene chloride, 70% & paint stripper D23, 30%
Primary Class: 261   Halogen Compounds, Aliphatic and Alicyclic

g PE/EVAL/PE	SAFETY4 4H	>240m		0	0.07		60

methylene chloride, 90% & isopropyl alcohol, 10%
Primary Class: 261   Halogen Compounds, Aliphatic and Alicyclic

g PE/EVAL/PE	SAFETY4 4H	>240m	0	0	0.07	35°C	60

G A R M E Class & Number/ N test chemical/ T MATERIAL NAME	MANUFACTURER & PRODUCT IDENTIFICATION	Break- through Time in Minutes	Perm- eation Rate in mg/sq m /min.	I N D E X	Thick- ness in mm	degrad. and comments	R e f s

methylene chloride, >70% & ethanol, <30%
Primary Class: 261   Halogen Compounds, Aliphatic and Alicyclic
Related Class: 311   Hydroxy Compounds, Aliphatic and Alicyclic, Primary

g PE/EVAL/PE	SAFETY4 4H	>240m	0	0	0.07	35°C	60

methylene chloride, >70% & phenol, <30%
Primary Class: 261   Halogen Compounds, Aliphatic and Alicyclic

g NEOPRENE	NOLATO 1505	18m	13415	5	1.10		33
g PE/EVAL/PE	SAFETY4 4H	>240m	0	0	0.07		60
g PE/EVAL/PE	SAFETY4 4H	200m			0.07	35°C	60
g PVAL	EDMONT 15-552	>240m	0	0	n.a.		33

methylenebisphenyl-4,4'-diisocyanate (synonym: MDI)
CAS Number: 101-68-8
Primary Class: 212   Isocyanates, Aromatic

g PE/EVAL/PE	SAFETY4 4H	>240m	0	0	0.07	35°C	60
g PE/EVAL/PE	SAFETY4 4H	>480m	0	0	0.07		60
s UNKNOWN MATERIAL	DU PONT BARRICADE	>480m	0	0	n.a.		8
s VITON/BUTYL	DRAGER 500 OR 710	>30m			n.a.	50°C	91

methylhydrazine
CAS Number: 60-34-4
Primary Class: 280   Hydrazines

b BUTYL	BATA	>120m			1.30	preexp. 24h	20
g BUTYL	NORTH B-161	>120m			0.50	preexp. 24h	20
s CHLOROBUTYL	FAIRPRENE	>120m			0.40	preexp. 24h	20
v CR39	Unknown	>120m			1.70	preexp. 24h	20
g NITRILE	ANSELLEDMONT 37-155				0.40	degrad.	34
g PVAL	EDMONT 25-950				0.50	degrad.	34
b PVC	BATA STANDARD	>120m			2.00	preexp. 24h	20
g PVC	BEST 814	114m			n.a.	preexp. 24h	20
u TEFLON	Unknown	<6m			n.a.	preexp. 24h	20
g VITRON	NORTH F-091	89m			0.36	preexp. 24h	20
g VITRON	NORTH F-091				0.25	degrad.	34

methylnadic anhydride (synonym: methyl-5-norbornene-2,3-dicarboxyl anhydride)
CAS Number: 25135-21-8
Primary Class: 161   Anhydrides, Aliphatic and Alicyclic

g PE/EVAL/PE	SAFETY4 4H	>240m	0	0	0.07		60

G A R M E Class & Number/ N test chemical/ T MATERIAL NAME	MANUFACTURER & PRODUCT IDENTIFICATION	Break- through Time in Minutes	Perm- eation Rate in mg/sq m /min.	I N D E X	Thick- ness in mm	R degrad. and comments	R e f s
methyltrichlorosilane (synonym: trichloromethylsilane) CAS Number: 75-79-6 Primary Class: 480 Organosilicon Compounds							
g PE/EVAL/PE	SAFETY4 4H	>240m		0	0.07		60
methyltriglycol (synonym: triethylene glycol monomethyl ether) CAS Number: 112-35-6 Primary Class: 245 Ethers, Glycols							
g BUTYL	NORTH B-161	>480m	0	0	0.40		68
g NEOPRENE	ANSELLEDMONT NEOX	>480m	0	0	n.a.		68
g NITRILE	MAPA-PIONEER AF-18	285m	108	3	0.56		68
g PVC	ANSELLEDMONT SNORKE	210m	56	2	n.a.		68
misityl oxide (synonym: isopropylidene acetone) CAS Number: 141-79-7 Primary Class: 391 Ketones, Aliphatic and Alicyclic							
s UNKNOWN MATERIAL	LIFE-GUARD RESPONDE	>480m	0	0	n.a.		77
Monsanto santicizer 2037 & nonyltoluene & dinonyltoluene Primary Class: 900 Miscellaneous Unclassified Chemicals							
s NEOPRENE	COMASEC ASTRO-PRENE	260m	24	2	n.a.		76
g NITRILE	COMASEC 7414	360m	6	1	n.a.		76
g PVC	BEST 7712R	330m	42	2	n.a.		76
Monsanto skydrol R Primary Class: 900 Miscellaneous Unclassified Chemicals							
g NATURAL RUBBER	ANSELLEDMONT 392				0.48	degrad.	6
g NEOPRENE	ANSELLEDMONT 29-840				0.38	degrad.	6
g NITRILE	ANSELLEDMONT 37-145				0.54	degrad.	6
g PE/EVAL/PE	SAFETY4 4H	>240m	0	0	n.a.	35°C	60
g PVAL	ANSELLEDMONT PVA				n.a.	degrad.	6
g PVC	ANSELLEDMONT SNORKE				n.a.	degrad.	6
Monsanto therminol VP-1 heat transfer fluid Primary Class: 900 Miscellaneous Unclassified Chemicals							
s NEOPRENE	COMASEC ASTRO-PRENE	45m	3300	4	n.a.		76
g PVC	ANSELLEDMONT SNORKE	85m	420	3	n.a.		76

G A R M E Class & Number/ N test chemical/ T MATERIAL NAME	MANUFACTURER & PRODUCT IDENTIFICATION	Break-through Time in Minutes	Perm-eation Rate in mg/sq m /min.	I N D E X	Thick-ness in mm	R degrad. e and f comments s

morpholine (synonym: tetrahydro-1,4-oxazine)
CAS Number: 110-91-8
Primary Class: 142   Amines, Aliphatic and Alicyclic, Secondary

g BUTYL	NORTH B-161	>960m	0	0	0.43		34
g NATURAL RUBBER	ANSELLEDMONT 36-124	30m	<90	3	0.46		6
g NATURAL RUBBER	ANSELLEDMONT 392	20m	<9	2	0.48		6
g NATURAL RUBBER	MARIGOLD BLACK HEVY	29m	1370	4	0.65	degrad.	79
g NATURAL RUBBER	MARIGOLD FEATHERLIT	15m	3467	4	0.15		79
g NATURAL RUBBER	MARIGOLD MED WT EXT	15m	3067	4	0.45	degrad.	79
g NATURAL RUBBER	MARIGOLD MEDICAL	15m	3300	4	0.28	degrad.	79
g NATURAL RUBBER	MARIGOLD ORANGE SUP	30m	1200	4	0.73	degrad.	79
g NATURAL RUBBER	MARIGOLD R SURGEONS	15m	3300	4	0.28	degrad.	79
g NATURAL RUBBER	MARIGOLD RED LT WT	15m	3100	4	0.43	degrad.	79
g NATURAL RUBBER	MARIGOLD SENSOTECH	15m	3300	4	0.28	degrad.	79
g NATURAL RUBBER	MARIGOLD SUREGRIP	25m	1879	4	0.58	degrad.	79
g NATURAL RUBBER	MARIGOLD YELLOW	15m	3167	4	0.38	degrad.	79
g NATURAL+NEOPRENE	MARIGOLD FEATHERWT	15m	3200	4	0.35	degrad.	79
g NEOPRENE	ANSELLEDMONT 29-840				0.38	degrad.	6
g NEOPRENE	ANSELLEDMONT 29-870				0.48	degrad.	34
g NITRILE	ANSELLEDMONT 37-165				0.60	degrad.	6
g NITRILE	MARIGOLD NITROSOLVE				0.75	degrad.	79
g NITRILE	NORTH LA-142G	44m	12380	5	0.33	degrad.	34
g PE/EVAL/PE	SAFETY4 4H	>240m	0	0	0.07		60
g PE/EVAL/PE	SAFETY4 4H	35m	25	3	0.07	35°C	60
g PVAL	ANSELLEDMONT PVA	90m	<9	1	n.a.		6
g PVAL	EDMONT 15-552	180m	<9	1	n.a.		6
g PVAL	EDMONT 25-545	>360m	0	0	n.a.		6
g PVAL	EDMONT 25-950	372m	252	3	0.33		34
g PVC	ANSELLEDMONT 34-100				0.20	degrad.	34
g PVC	ANSELLEDMONT SNORKE				n.a.	degrad.	6
g UNKNOWN MATERIAL	NORTH SILVERSHIELD	>480m	0	0	0.10		7
g VITON	NORTH F-091	108m	5800	4	0.24	degrad.	34

mustard gas (synonym: bis(2-chloroethyl)sulfide)
CAS Number: 505-60-2
Primary Class: 502   Sulfur Compounds, Sulfides and Disulfides

g PE/EVAL/PE	SAFETY4 4H	>240m	0	0	0.07	35°C	60

G A R M E Class & Number/ N test chemical/ T MATERIAL NAME	MANUFACTURER & PRODUCT IDENTIFICATION	Break- through Time in Minutes	Perm- eation Rate in mg/sq m /min.	I N D E X	Thick- ness in mm	degrad. and comments	R e f s
N,N,N',N'-tetramethylethylenediamine (synonym: TEMEDA) CAS Number: 110-18-9 Primary Class: 148 Amines, Poly, Aliphatic and Alicyclic							
g BUTYL	NORTH B-174	65m	460	3	0.58	degrad.	34
g NATURAL RUBBER	ACKWELL 5-109				0.15	degrad.	34
g NEOPRENE	ANSELLEDMONT 29-870				0.48	degrad.	34
g NITRILE	ANSELLEDMONT 37-155	108m	895	3	0.36	degrad.	34
g PVC	ANSELLEDMONT 34-100	2m	19260	5	0.20		34
g VITON	NORTH F-091	26m	17230	5	0.33	degrad.	34
N,N-dimethylaniline (synonym: DMA) CAS Number: 121-69-7 Primary Class: 146 Amines, Aromatic, Secondary and Tertiary							
g PE/EVAL/PE	SAFETY4 4H	>240m		0	0.07	35°C	60
N,N-dimethylethylamine (synonym: N-ethyldimethylamine) CAS Number: 598-56-1 Primary Class: 143 Amines, Aliphatic and Alicyclic, Tertiary							
g PE/EVAL/PE	SAFETY4 4H	2m			0.07	35°C	60
g PE/EVAL/PE	SAFETY4 4H	9m			0.07		60
n-amylamine (synonyms: 1-aminopentane; n-pentylamine) CAS Number: 110-58-7 Primary Class: 141 Amines, Aliphatic and Alicyclic, Primary							
g BUTYL	NORTH B-174	150m	3603		0.64		34
g NATURAL RUBBER	ACKWELL 5-109	1m	15720	5	0.15	degrad.	34
g NEOPRENE	ANSELLEDMONT 29-870	28m	5040	4	0.55	degrad.	34
g NITRILE	ANSELLEDMONT 37-155				0.40	degrad.	34
g PVC	ANSELLEDMONT 34-100	2m	24960	5	0.20	degrad.	34
g VITON	NORTH F-091				0.26	degrad.	34
n-butyl chloride (synonym: 1-chlorobutane) CAS Number: 109-69-3 Primary Class: 261 Halogen Compounds, Aliphatic and Alicyclic							
g BUTYL	NORTH B-174				0.66	degrad.	34
g NATURAL RUBBER	ACKWELL 5-109				0.15	degrad.	34
g NEOPRENE	ANSELLEDMONT 29-870				0.48	degrad.	34
g NITRILE	ANSELLEDMONT 37-155	12m	6600	5	0.40	degrad.	34
g PVAL	EDMONT 25-950	>480m	0	0	0.61		34
g PVC	ANSELLEDMONT 34-100	1m	9627	5	0.20		34
g VITON	NORTH F-091	265m	34	2	0.38		34

G A R M E Class & Number/ N test chemical/ T MATERIAL NAME	MANUFACTURER & PRODUCT IDENTIFICATION	Break- through Time in Minutes	Perm- eation Rate in mg/sq m /min.	I N D E X	Thick- ness in mm	degrad. and comments	R e f s

n-butylamine (synonym: 1-aminobutane)
CAS Number: 109-73-9
Primary Class: 141   Amines, Aliphatic and Alicyclic, Primary

g BUTYL	NORTH B-131	16m			0.35		83
g BUTYL	NORTH B-174	103m	3000	4	0.96	degrad.	34
u CPE	ILC DOVER	20m			n.a.		29
g NATURAL RUBBER	ACKWELL 5-109	1m	46380	5	0.15	degrad.	34
g NEOPRENE	ANSELLEDMONT 29-870	12m	14760	5	0.48	degrad.	34
g NITRILE	ANSELLEDMONT 37-155				0.40	degrad.	34
g PVAL	EDMONT 25-950				0.40	degrad.	34
g PVC	ANSELLEDMONT 34-100	1m	33060	5	0.20	degrad.	34
s TEFLON	CHEMFAB CHALL. 5100	>180m	0	1	n.a.		65
g UNKNOWN MATERIAL	NORTH SILVERSHIELD	48m			0.08		83
g VITON	NORTH F-091				0.26	degrad.	34
g VITON	NORTH F-091	<4m			0.25		83
u VITON/CHLOROBUTYL	Unknown	21m			n.a.		29

N-methyl-2-pyrrolidone (synonym: 1-methyl-2-pyrrolidone)
CAS Number: 872-50-4
Primary Class: 132   Amides, Aliphatic and Alicyclic
Related Class: 391   Ketones, Aliphatic and Alicyclic

g BUTYL	NORTH B-174	>480m	0	0	0.74		34
g NATURAL RUBBER	ACKWELL 5-109	15m	<1	2	0.15		34
g NATURAL RUBBER	ANSELLEDMONT 36-124	45m	<900	3	0.46		6
g NATURAL RUBBER	ANSELLEDMONT 392	80m	<90	2	0.48		6
g NATURAL RUBBER	BEST 65NFW	>480m	0	0	n.a.		53
g NATURAL RUBBER	BEST 67NFW		36	5	n.a.		53
g NEOPRENE	ANSELLEDMONT 29-840				0.38	degrad.	6
g NEOPRENE	ANSELLEDMONT 29-870				0.48	degrad.	34
g NEOPRENE	ANSELLEDMONT NEOX				n.a.	degrad.	6
g NEOPRENE	BEST 6780		60	5	n.a.		53
g NITRILE	ANSELLEDMONT 37-155				0.40	degrad.	34
g NITRILE	ANSELLEDMONT 37-165				0.60	degrad.	6
g NITRILE	BEST 22R		240	5	n.a.		53
g PE/EVAL/PE	SAFETY4 4H	>240m	0	0	0.07	35°C	60
g PVAL	ANSELLEDMONT PVA				n.a.	degrad.	6
g PVAL	EDMONT 15-552				n.a.	degrad.	6
g PVAL	EDMONT 25-950	48m	<1	2	0.28		34
g PVC	ANSELLEDMONT 34-100				0.20	degrad.	34
g PVC	ANSELLEDMONT SNORKE				n.a.	degrad.	6
g PVC	BEST 812		240	5	n.a.		53
g VITON	NORTH F-091	37m	30	3	0.28	degrad.	34

G A R M E Class & Number/ N test chemical/ T MATERIAL NAME	MANUFACTURER & PRODUCT IDENTIFICATION	Break- through Time in Minutes	Perm- eation Rate in mg/sq m /min.	I N D E X	Thick- ness in mm	degrad. and comments	R e f s

N-methylethanolamine (synonym: 2-methylaminoethanol)
CAS Number: 109-83-1
Primary Class: 142   Amines, Aliphatic and Alicyclic, Secondary
Related Class: 311   Hydroxy Compounds, Aliphatic and Alicyclic, Primary

g BUTYL	NORTH B-174	>480m	0		0	0.50		34
g NATURAL RUBBER	ACKWELL 5-109	5m	908	4	0.15			34
g NEOPRENE	ANSELLEDMONT 29-870	>480m	0		0	0.38		34
g VITRON	NORTH F-091	>480m	0		0	0.25		34

N-methylmethacrylamide
CAS Number: 3887-02-3
Primary Class: 133   Acrylamides

s UNKNOWN MATERIAL	LIFE-GUARD RESPONDE	>480m	0		0	n.a.		77

N-nitrosodiethylamine (synonym: N,N-diethylnitrosamine)
CAS Number: 55-18-5
Primary Class: 450   Nitroso Compounds

g BUTYL	NORTH B-174	>480m	0		0	0.43		34
g NITRILE	ANSELLEDMONT 37-175	15m	1233	4	0.38			34
g PE/EVAL/PE	SAFETY4 4H	>240m			0	0.07		60

N-nitrosodimethylamine (synonym: N,N-dimethylnitrosamine)
CAS Number: 62-75-9
Primary Class: 450   Nitroso Compounds

g NEOPRENE	ANSELLEDMONT 29-870					0.48	degrad.	34
g NITRILE	ANSELLEDMONT 37-155					0.40	degrad.	34
g PVAL	EDMONT 25-950					0.60	degrad.	34
g PVC	ANSELLEDMONT 34-100					0.20	degrad.	34
g VITON	NORTH F-091					0.25	degrad.	34

N-vinylpyrrolidone (synonym: 1-vinyl-2-pyrrolidone)
CAS Number: 88-12-0
Primary Class: 132   Amides, Aliphatic and Alicyclic

g PE/EVAL/PE	SAFETY4 4H	>240m	0		0	0.07	35°C	60

G A R M E Class & Number/ N test chemical/ T MATERIAL NAME	MANUFACTURER & PRODUCT IDENTIFICATION	Break- through Time in Minutes	Perm- eation Rate in mg/sq m /min.	I N D E X	Thick- ness in mm	R degrad. e and f comments s

naled
CAS Number: 300-76-5
Primary Class: 462   Organophosphorus Compounds and Derivatives of Phosphorus-based Acids
Related Class: 233   Esters, Non-Carboxylic, Carbamates and Others

s TEFLON	CHEMFAB CHALL. 5100	>204m	0	1	n.a.	65

naphtha with 10-15% aromatics (b.p. 50-120°C) (synonym: rubber solvent)
Primary Class: 291   Hydrocarbons, Aliphatic and Alicyclic, Saturated

g NATURAL RUBBER	ANSELL ORANGE 208	12m	9	3	0.76	55
g NATURAL RUBBER	ANSELLEDMONT 36-124				0.46	degrad. 6
g NATURAL RUBBER	ANSELLEDMONT 392				0.48	degrad. 6
g NATURAL+NEOPRENE	ANSELL OMNI 276	8m	15	4	0.56	55
g NEOPRENE	ANSELL NEOPRENE 530	>360m		0	0.46	55
g NEOPRENE	ANSELLEDMONT 29-840	30m	<900	3	0.38	6
g NEOPRENE	ANSELLEDMONT NEOX	60m	<900	3	n.a.	6
g NEOPRENE/NATURAL	ANSELL CHEMI-PRO	>360m	0	0	0.72	55
g NITRILE	ANSELL CHALLENGER	>360m	0	0	0.38	55
g NITRILE	ANSELLEDMONT 37-165	>360m	0	0	0.60	6
g PVAL	ANSELLEDMONT PVA	>360m		0	n.a.	degrad. 6
g PVAL	EDMONT 15-552	90m	<9	1	n.a.	6
g PVAL	EDMONT 25-545	>360m	0	0	n.a.	6
g PVC	ANSELLEDMONT SNORKE				n.a.	degrad. 6

naphtha with 10-15% aromatics (b.p. 120-140°C) (synonym: Naphtha VMP)
Primary Class: 291   Hydrocarbons, Aliphatic and Alicyclic, Saturated

s BUTYL	MSA CHEMPRUF	<60m			n.a.	48
g NATURAL RUBBER	ANSELLEDMONT 36-124				0.46	degrad. 6
g NATURAL RUBBER	ANSELLEDMONT 392				0.48	degrad. 6
g NATURAL+NEOPRENE	PLAYTEX ARGUS 123	4m	960	4	n.a.	72
g NEOPRENE	ANSELL NEOPRENE 530	>360m	0	0	0.46	55
g NEOPRENE	ANSELLEDMONT 29-840	15m	<9000	4	0.38	6
g NEOPRENE	ANSELLEDMONT NEOX	>360m	0	0	n.a.	6
g NEOPRENE/NATURAL	ANSELL CHEMI-PRO	>360m		0	0.72	55
g NITRILE	ANSELL CHALLENGER	>360m		0	0.38	55
g NITRILE	ANSELLEDMONT 37-165	>360m	0	0	0.60	6
g NITRILE	MARIGOLD BLUE	128m	39	2	0.45	degrad. 79
g NITRILE	MARIGOLD GREEN SUPA	195m	59	2	0.30	degrad. 79
g PVAL	ANSELLEDMONT PVA	>420m		0	n.a.	6
g PVAL	EDMONT 15-552	>420m	0	0	n.a.	6
g PVAL	EDMONT 25-545	>360m	0	0	n.a.	6
g PVC	ANSELLEDMONT M GRIP	120m	<90	2	n.a.	6
s PVC	MSA UPC	<120m	0.20			48
g UNKNOWN MATERIAL	MARIGOLD R MEDIGLOV	15m	11200	5	0.03	degrad. 79

G A R M E	Class & Number/ test chemical/	MANUFACTURER	Break-	Perm- eation	I N D	Thick-		R e
N T	MATERIAL NAME	& PRODUCT IDENTIFICATION	through Time in Minutes	Rate in mg/sq m /min.	E X	ness in mm	degrad. and comments	f s

naphtha with 10-15% aromatics (b.p. 150-200°C) (synonym: Stoddard solvent)
Primary Class: 291   Hydrocarbons, Aliphatic and Alicyclic, Saturated

g BUTYL	BEST 878	77m	1560	4	0.75		53
g NATURAL RUBBER	ANSELL ORANGE 208	14m	9	3	0.76		55
g NATURAL RUBBER	ANSELLEDMONT 36-124				0.46	degrad.	6
g NATURAL RUBBER	ANSELLEDMONT 392				0.48	degrad.	6
g NEOPRENE	ANSELL NEOPRENE 530	>360m	0	0	0.46		55
g NEOPRENE	ANSELLEDMONT 29-840	180m	<90	2	0.38		6
g NEOPRENE	ANSELLEDMONT NEOX	>360m	0	0	n.a.		6
g NEOPRENE	BEST 6780	>480m	0	0	n.a.		53
g NEOPRENE/NATURAL	ANSELL CHEMI-PRO	>360m	0	0	0.72		55
g NITRILE	ANSELL CHALLENGER	>360m	0	0	0.38		55
g NITRILE	ANSELLEDMONT 37-165	>360m	0	0	0.60		6
g NITRILE	BEST 727	>480m	0	0	0.38		53
g PVAL	ANSELLEDMONT PVA	>360m		0	n.a.		6
g PVAL	EDMONT 15-552	>360m	0	0	n.a.		6
g PVAL	EDMONT 25-545	>360m	0	0	n.a.		6
g PVC	ANSELLEDMONT M GRIP	360m	<9	1	n.a.		6

naphtha with 15-20% aromatics (b.p. 150-200°C) (synonym: mineral spirits)
CAS Number: 67742-88-7
Primary Class: 291   Hydrocarbons, Aliphatic and Alicyclic, Saturated

g NATURAL RUBBER	BEST 65NFW				n.a.	degrad.	53
g NATURAL RUBBER	MARIGOLD ORANGE SUP	9m	580	4	0.70		37
u NATURAL RUBBER	Unknown	<1m	760	4	0.12		37
g NATURAL+NEOPRENE	PLAYTEX ARGUS 123	13m	1380	5	n.a.		72
g NEOPRENE	MAPA-PIONEER N-44	126m	120	3	0.56		36
u NEOPRENE	Unknown	66m	360	3	0.61		37
g NITRILE	MAPA-PIONEER A-14	>480m	0	0	0.56		36
g NITRILE	MAPA-PIONEER A-15	>480m	0	0	0.36		36
g NITRILE	MARIGOLD NITROSOLVE	114m	35	2	0.75	degrad.	79
c PE	DU PONT TYVEK QC	<2m	70	4	n.a.		8
g PE/EVAL/PE	SAFETY4 4H	>240m	0	0	0.07	35°C	60
s SARANEX-23	DU PONT TYVEK SARAN	>480m	0	0	n.a.		8
s UNKNOWN MATERIAL	DU PONT BARRICADE	>480m	0	0	n.a.		8

naphtha with 15-20% aromatics (b.p. 180-260°C) (synonym: kerosene)
CAS Number: 8008-20-6
Primary Class: 291   Hydrocarbons, Aliphatic and Alicyclic, Saturated

g BUTYL	BEST 878				0.75	degrad.	53
s BUTYL	MSA CHEMPRUF	<60m			n.a.		48
g NATURAL RUBBER	ANSELL FL 200 254	8m	15	4	0.51		55
g NATURAL RUBBER	ANSELL ORANGE 208	19m	6	2	0.76		55
g NATURAL RUBBER	ANSELL PVL 040	16m	10	3	0.46		55

G								

G A R M E Class & Number/ N test chemical/ T MATERIAL NAME	MANUFACTURER & PRODUCT IDENTIFICATION	Break-through Time in Minutes	Perm-eation Rate in mg/sq m /min.	I N D E X	Thick-ness in mm	degrad. and comments	R e f s
g NATURAL RUBBER	ANSELL STERILE 832	1m	37	4	0.23		55
g NATURAL RUBBER	ANSELLEDMONT 36-124	30m	<90	3	0.46		6
g NATURAL RUBBER	ANSELLEDMONT 392				0.48	degrad.	6
g NATURAL RUBBER	BEST 65NFW				n.a.	degrad.	53
g NATURAL RUBBER	MARIGOLD BLACK HEVY	15m	3152	4	0.65	degrad.	79
g NATURAL RUBBER	MARIGOLD FEATHERLIT	15m	10512	5	0.15	degrad.	79
g NATURAL RUBBER	MARIGOLD MED WT EXT	15m	4250	4	0.45	degrad.	79
g NATURAL RUBBER	MARIGOLD MEDICAL	15m	7903	4	0.28	degrad.	79
g NATURAL RUBBER	MARIGOLD ORANGE SUP	35m	4200	4	0.74		12
g NATURAL RUBBER	MARIGOLD ORANGE SUP	15m	3042	4	0.73	degrad.	79
g NATURAL RUBBER	MARIGOLD R SURGEONS	15m	7903	4	0.28	degrad.	79
g NATURAL RUBBER	MARIGOLD RED LT WT	15m	4772	4	0.43	degrad.	79
g NATURAL RUBBER	MARIGOLD SENSOTECH	15m	7903	4	0.28	degrad.	79
g NATURAL RUBBER	MARIGOLD SUREGRIP	15m	3481	4	0.58	degrad.	79
g NATURAL RUBBER	MARIGOLD YELLOW	15m	5815	4	0.38	degrad.	79
g NATURAL+NEOPRENE	ANSELL OMNI 276	13m	8	3	0.56		55
g NATURAL+NEOPRENE	ANSELL TECHNICIANS	6m	17	4	0.43		55
g NATURAL+NEOPRENE	MARIGOLD FEATHERWT	15m	6337	4	0.35	degrad.	79
g NATURAL+NEOPRENE	PLAYTEX ARGUS 123	36m	120	3	n.a.		72
g NEOPRENE	ANSELL NEOPRENE 530	81m	1.2	1	0.46		55
g NEOPRENE	ANSELLEDMONT 29-840	>360m	0	0	0.38		6
g NEOPRENE	ANSELLEDMONT NEOX	>360m	0	0	n.a.		6
g NEOPRENE	MAPA-PIONEER N-44	>480m	0	0	0.56		36
g NEOPRENE/NATURAL	ANSELL CHEMI-PRO	20m	3	2	0.72		55
g NITRILE	ANSELL CHALLENGER	>360m	0	0	0.38		55
g NITRILE	ANSELLEDMONT 37-165	>360m	0	0	0.60		6
g NITRILE	ANSELLEDMONT 37-175	>120m			0.34		12
g NITRILE	MAPA-PIONEER A-14	>480m	0	0	0.56		36
g NITRILE	MARIGOLD BLUE	>480m	0	0	0.45		79
g NITRILE	MARIGOLD GREEN SUPA	>480m	0	0	0.30		79
g NITRILE	MARIGOLD NITROSOLVE	>480m	0	0	0.75		79
g PVAL	ANSELLEDMONT PVA	>360m		0	n.a.		6
g PVAL	EDMONT 15-552	>360m	0	0	n.a.		6
g PVAL	EDMONT 25-545	>360m	0	0	n.a.		6
g PVC	ANSELLEDMONT M GRIP	>360m	0	0	n.a.		6
s PVC	MSA UPC	<120m			0.20		48
g UNKNOWN MATERIAL	MARIGOLD R MEDIGLOV	15m	1945	4	0.03	degrad.	79

naphtha with <3% aromatics (b.p. 150-200°C) (synonym: Rule 66)
Primary Class: 291   Hydrocarbons, Aliphatic and Alicyclic, Saturated

g BUTYL	ERISTA BX	65m	814	3	0.66		33
g BUTYL	ERISTA BX	64m	478	3	0.66	preexp. 20h	33
g BUTYL	ERISTA BX	80m	889	3	0.70		33
g BUTYL	ERISTA BX	63m	694	3	0.70	preexp. 20h	33
g NATURAL RUBBER	ANSELLEDMONT 36-124				n.a.	degrad.	6
g NATURAL RUBBER	ANSELLEDMONT 392				0.48	degrad.	6
g NATURAL RUBBER	MAPA-PIONEER L-118	9m	1260	5	0.46		36
g NATURAL RUBBER	MARIGOLD ORANGE SUP	12m	780	4	0.70		37

G A R M E Class & Number/ N test chemical/ T MATERIAL NAME	MANUFACTURER & PRODUCT IDENTIFICATION	Break- through Time in Minutes	Perm- eation Rate in mg/sq m /min.	I N D E X	Thick- ness in mm	degrad. and comments	R e f s
u NATURAL RUBBER	Unknown	<1m	740	4	0.12		37
g NEOPRENE	ANSELL NEOPRENE 530	>360m	0	0	0.46		55
g NEOPRENE	ANSELLEDMONT 29-840	90m	<90	2	0.38		6
g NEOPRENE	ANSELLEDMONT NEOX	>360m	0	0	n.a.		6
g NEOPRENE	MAPA-PIONEER N-44	330m	24	2	0.56		36
g NEOPRENE	NOLATO 1505	>180m			0.85		11
u NEOPRENE	Unknown	126m	250	3	0.61		37
g NEOPRENE/NATURAL	ANSELL CHEMI-PRO	>360m		0	0.72		55
g NITRILE	ANSELL CHALLENGER	>360m	0	0	0.38		55
g NITRILE	ANSELLEDMONT 37-165	>360m	0	0	0.60		6
g NITRILE	MAPA-PIONEER A-14	>480m	0	0	0.56		36
g NITRILE	NORTH LA-102G	>120m			0.28		11
u PE	Unknown	3m	504	4	0.04		30
g PVAL	ANSELLEDMONT PVA	>360m		0	n.a.		6
g PVAL	EDMONT 25-545	>360m	0	0	n.a.		6
g PVAL	EDMONT 25-545	>360m	0	0	n.a.		6
g PVC	ANSELLEDMONT M GRIP	150m	<90	2	n.a.		6
g PVC	KID 490	>210m			0.55		11
g PVC	MAPA-PIONEER V-20	42m	588	3	0.51		36
u PVDC/PE/PVDC	MOLNLYCKE	>240m	0	0	0.07		54

naphtha with <3% aromatics (b.p. 150-200°C), 50% & 2-butoxyethanol, 50%
Primary Class: 291   Hydrocarbons, Aliphatic and Alicyclic, Saturated

g NITRILE	ERISTA SPECIAL	252m		0	0.32		11

naphtha with <3% aromatics (b.p. 150-200°C), 95% & 2-butoxyethanol, 5%
Primary Class: 291   Hydrocarbons, Aliphatic and Alicyclic, Saturated

g BUTYL	ERISTA BX	90m			0.59	37°C TL	11
g NITRILE	ANSELLEDMONT 37-175	888m	0	0	0.37	37°C TL	11
g NITRILE	ERISTA SPECIAL	137m			0.40	37°C TL	11
g PVC	KID 490	80m			n.a.	37°C TL	11

naphtha with <3% aromatics (b.p. 180-260°C)
Primary Class: 291   Hydrocarbons, Aliphatic and Alicyclic, Saturated

g NATURAL RUBBER	MARIGOLD ORANGE SUP	24m	54	3	0.70		37
u NATURAL RUBBER	Unknown	<1m	350	4	0.12		37
u NEOPRENE	Unknown	222m	48	2	0.61		37

naphthalene
CAS Number: 91-20-3
Primary Class: 293   Hydrocarbons, Polynuclear Aromatic

s PVC	TRELLEBORG TRELLCHE				n.a.	degrad.	71
s TEFLON	CHEMFAB CHALL. 5100	>790m	0	0	n.a.		65

G A R M E Class & Number/ N test chemical/ T MATERIAL NAME	MANUFACTURER & PRODUCT IDENTIFICATION	Break- through Time in Minutes	Perm- eation Rate in mg/sq m /min.	I N D E X	Thick- ness in mm	degrad. and comments	R e f s
nickel subsulfide CAS Number: 12035-72-2 Primary Class: 380 Inorganic Bases							
g NATURAL RUBBER	ANSELL CANNERS 334	>480m	0	0	n.a.		34
g PVC	ANSELLEDMONT 34-100	>480m	0	0	0.20		34
nicotine CAS Number: 54-11-5 Primary Class: 271 Heterocyclic Compounds, Nitrogen, Pyridines							
u CPE	ILC DOVER	>180m			n.a.		29
g PE/EVAL/PE	SAFETY4 4H	>240m	0	0	0.07	35°C	60
u VITON/CHLOROBUTYL	Unknown	>180m			n.a.		29
nitric acid, 30-70% CAS Number: 7697-37-2 Primary Class: 370 Inorganic Acids							
g BUTYL	BEST 878	>480m	0	0	0.75		53
s BUTYL	MSA CHEMPRUF	>480m	0	0	n.a.		48
g HYPALON	COMASEC DIPCO	>480m	0	0.51			40
g NATURAL RUBBER	ANSELL FL 200 254				0.51	degrad.	55
g NATURAL RUBBER	ANSELL PVL 040				0.46	degrad.	55
g NATURAL RUBBER	ANSELLEDMONT 36-124				0.46	degrad.	6
g NATURAL RUBBER	ANSELLEDMONT 392				n.a.	degrad.	6
g NATURAL RUBBER	COMASEC FLEXIGUM	246m	2280	4	0.95		40
g NATURAL RUBBER	MAPA-PIONEER L-118	233m	>10000	5	0.46		36
g NATURAL RUBBER	MARIGOLD BLACK HEVY	>480m	0	0	0.65		79
g NATURAL RUBBER	MARIGOLD FEATHERLIT	1m	221	4	0.15		79
g NATURAL RUBBER	MARIGOLD MED WT EXT	>480m	0	0	0.45		79
g NATURAL RUBBER	MARIGOLD MEDICAL	165m	129	3	0.28	degrad.	79
g NATURAL RUBBER	MARIGOLD ORANGE SUP	>480m	0	0	0.73		79
g NATURAL RUBBER	MARIGOLD R SURGEONS	165m	129	3	0.28	degrad.	79
g NATURAL RUBBER	MARIGOLD RED LT WT	435m	18	2	0.43		79
g NATURAL RUBBER	MARIGOLD SENSOTECH	165m	129	3	0.28	degrad.	79
g NATURAL RUBBER	MARIGOLD SUREGRIP	>480m	0	0	0.58		79
g NATURAL RUBBER	MARIGOLD YELLOW	345m	55	2	0.38		79
u NATURAL RUBBER	Unknown	>60m			n.a.	degrad.	9
u NATURAL RUBBER	Unknown	>60m			0.43	degrad.	9
g NATURAL+NEOPRENE	ANSELL OMNI 276	>			0.56	degrad.	55
g NATURAL+NEOPRENE	ANSELL TECHNICIANS				0.43	degrad.	55
g NATURAL+NEOPRENE	MARIGOLD FEATHERWT	300m	74	2	0.35		79
u NATURAL+NEOPRENE	Unknown	>60m			n.a.		9
g NEOPRENE	ANSELL NEOPRENE 530	>360m	0	0	0.46		55
g NEOPRENE	ANSELLEDMONT 29-840	150m			0.38	70%	6
g NEOPRENE	ANSELLEDMONT NEOX	>360m	0	0	n.a.	70%	6
g NEOPRENE	BEST 32	>480m	0	0	n.a.		53

G A R M E N T	Class & Number/ test chemical/ MATERIAL NAME	MANUFACTURER & PRODUCT IDENTIFICATION	Break- through Time in Minutes	Perm- eation Rate in mg/sq m /min.	I N D E X	Thick- ness in mm	degrad. and comments	R e f s
g	NEOPRENE	BEST 6780	>480m	0	0	n.a.		53
g	NEOPRENE	COMASEC COMAPRENE	120m			n.a.		40
g	NEOPRENE	MAPA-PIONEER N-44	>480m	0	0	0.56		36
u	NEOPRENE	Unknown	>60m			0.61		9
g	NEOPRENE/NATURAL	ANSELL CHEMI-PRO	>360m	0	0	0.72		55
g	NITRILE	ANSELL CHALLENGER	>360m	0	0	0.38	degrad.	55
g	NITRILE	ANSELLEDMONT 37-165				0.60	degrad.	6
g	NITRILE	BEST 22R	60m	27420	5	n.a.		53
g	NITRILE	BEST 727				0.38	degrad.	53
g	NITRILE	COMASEC COMATRIL	180m	69060	5	0.55		40
g	NITRILE	COMASEC COMATRIL SU	240m	109440	5	0.60		40
g	NITRILE	COMASEC FLEXITRIL	25m	96600	5	n.a.		40
g	NITRILE	MAPA-PIONEER A-14	72m	12060	5	0.56	degrad.	36
g	NITRILE	MARIGOLD NITROSOLVE	288m	28	2	0.75		79
u	NITRILE	Unknown	>60m			0.45	degrad.	9
g	NITRILE+PVC	COMASEC MULTIPLUS	270m	1380	4	n.a.		40
c	PE	DU PONT TYVEK QC	404m	16	2	n.a.		8
g	PE/EVAL/PE	SAFETY4 4H	>240m	0	0	0.07	35°C	60
g	PVAL	ANSELLEDMONT PVA				n.a.	degrad.	6
g	PVAL	EDMONT 15-552				n.a.	degrad.	6
g	PVC	ANSELLEDMONT M GRIP	345m	0	0	n.a.	70%	6
g	PVC	BEST 725R	240m	24360	5	n.a.		53
g	PVC	COMASEC MULTIPOST	240m	2220	4	n.a.		40
g	PVC	COMASEC MULTITOP	252m	1920	4	n.a.		40
g	PVC	COMASEC NORMAL	220m	2760	4	n.a.		40
g	PVC	COMASEC OMNI	80m	274800	5	n.a.		40
g	PVC	MAPA-PIONEER V-20	114m	13	2	0.51		36
s	PVC	MSA UPC	<240m		0	n.a.		48
u	PVC	Unknown	6m			0.22	degrad.	9
u	PVC	Unknown	>60m			n.a.		9
s	SARANEX-23	DU PONT TYVEK SARAN	>480m	0	0	n.a.	70%	8
s	UNKNOWN MATERIAL	CHEMRON CHEMREL	>1440m	0	0	n.a.	70%	67
s	UNKNOWN MATERIAL	CHEMRON CHEMREL MAX	>1440m	0	0	n.a.		67
c	UNKNOWN MATERIAL	CHEMRON CHEMTUFF	>480m	0	0	n.a.		67
s	UNKNOWN MATERIAL	DU PONT BARRICADE	>480m	0	0	n.a.	70%	8
g	VITON	NORTH F-121	>60m			0.31		9

nitric acid, <30%
Primary Class: 370   Inorganic Acids

s	BUTYL	MSA CHEMPRUF	>480m	0	0	n.a.		48
g	NAT+NEOPR+NITRILE	MAPA-PIONEER TRIONI	>840m	0	0	0.48		36
g	NATURAL RUBBER	ANSELL CONFORM 4205	>360m	0	0	0.13		55
g	NATURAL RUBBER	ANSELL CONT ENVIRON	>480m	0	0	0.55		55
g	NATURAL RUBBER	ANSELL FL 200 254	>360m	0	0	0.51		55
g	NATURAL RUBBER	ANSELL ORANGE 208	>360m	0	0	0.76		55
g	NATURAL RUBBER	ANSELL PVL 040	>360m	0	0	0.46		55
g	NATURAL RUBBER	ANSELL STERILE 832	>360m	0	0	0.23		55
g	NATURAL RUBBER	ANSELLEDMONT 36-124	>360m	0	0	0.46		6
g	NATURAL RUBBER	ANSELLEDMONT 392	>360m	0	0	0.48		6

G A R M E Class & Number/ N test chemical/ T MATERIAL NAME	MANUFACTURER & PRODUCT IDENTIFICATION	Break-through Time in Minutes	Perm-eation Rate in mg/sq m /min.	I N D E X	Thick-ness in mm	degrad. and comments	R e f s
g NATURAL+NEOPRENE	ANSELL OMNI 276	>360m	0	0	0.56		55
g NATURAL+NEOPRENE	ANSELL TECHNICIANS	>360m	0	0	0.43		55
g NATURAL+NEOPRENE	PLAYTEX ARGUS 123	>480m	0	0	n.a.		72
g NEOPRENE	ANSELL NEOPRENE 530	>360m	0	0	0.46		55
g NEOPRENE	ANSELLEDMONT 29-840	>360m	0	0	0.31		6
g NEOPRENE	ANSELLEDMONT NEOX	>360m	0	0	n.a.		6
g NEOPRENE/NATURAL	ANSELL CHEMI-PRO	>360m	0	0	0.72		55
g NITRILE	ANSELL CHALLENGER	>360m	0	0	0.38		55
g NITRILE	ANSELLEDMONT 37-165	>360m	0	0	0.60		6
g PE/EVAL/PE	SAFETY4 4H	>240m	0	0	0.07	35°C	60
g PVAL	ANSELLEDMONT PVA				n.a.	degrad.	6
g PVAL	EDMONT 15-552				n.a.	degrad.	6
g PVC	ANSELLEDMONT M GRIP	>360m	0	0	n.a.		6
s PVC	MSA UPC	>480m	0	0	0.20		48
g UNKNOWN MATERIAL	NORTH SILVERSHIELD	>360m	0	0	0.08		7

nitric acid, >70%
Primary Class: 370   Inorganic Acids

g BUTYL	BRUNSWICK BUTYL STD	>480m	0	0	0.63	concentr.	89
g BUTYL	BRUNSWICK BUTYL-XTR	>480m	0	0	0.63	concentr.	89
g BUTYL/ECO	BRUNSWICK BUTYL-POL	>480m	0	0	0.63	concentr.	89
s BUTYL/NEOPRENE	MSA BETEX	>480m	0	0	n.a.		48
u CPE	ILC DOVER	>180m			n.a.		29
g NAT+NEOPR+NITRILE	MAPA-PIONEER TRIONI	>210m			0.48		36
g NITRILE	MARIGOLD NITROSOLVE				0.75	degrad.	79
g PE/EVAL/PE	SAFETY4 4H	180m			0.07		60
g PE/EVAL/PE	SAFETY4 4H	60m			0.07	35°C	60
s SARANEX-23	DU PONT TYVEK SARAN	107m			n.a.		8
s UNKNOWN MATERIAL	LIFE-GUARD RESPONDE	>180m			n.a.		77
u VITON/CHLOROBUTYL	Unknown	>180m			n.a.		29
s VITON/NEOPRENE	MSA VAUTEX	>480m	0	0	n.a.		48

nitric acid, red fuming
Primary Class: 370   Inorganic Acids

g BUTYL	BRUNSWICK BUTYL-XTR	>480m	0	0	0.63		89
g BUTYL	NORTH B-161				0.43	degrad.	7
g BUTYL	NORTH B-174	>90m			0.43		21
m CHLOROBUTYL	ARROWHEAD PRODUCTS	>90m			0.51		21
s CPE	ILC DOVER CLOROPEL	27m			n.a.		21
g NATURAL RUBBER	ANSELLEDMONT 36-124				0.46	degrad.	6
g NATURAL RUBBER	ANSELLEDMONT 392				0.48	degrad.	6
g NATURAL RUBBER	ANSELLEDMONT 46-322	>90m			0.43		21
g NEOPRENE	ANSELLEDMONT 29-840				0.38	degrad.	6
g NEOPRENE	ANSELLEDMONT 29-870	>90m			0.46		21
g NEOPRENE	ANSELLEDMONT NEOX				n.a.	degrad.	6
g NEOPRENE	ANSELLEDMONT NEOX	>90m			n.a.		21
g NITRILE	ANSELLEDMONT 37-155	>90m			0.38		21

G A R M E Class & Number/ N test chemical/ T MATERIAL NAME	MANUFACTURER & PRODUCT IDENTIFICATION	Break- through Time in Minutes	Perm- eation Rate in mg/sq m /min.	I N D E X	Thick- ness in mm	degrad. and comments	R e f s
g NITRILE	ANSELLEDMONT 37-165				0.60	degrad.	6
g NITRILE	NORTH LA-111EB	>90m			0.28		21
g NITRILE	NORTH LA-142G				0.38	degrad.	7
g PE/EVAL/PE	SAFETY4 4H	60m			0.07	35°C	60
g PVAL	ANSELLEDMONT PVA				n.a.	degrad.	6
g PVAL	EDMONT 15-552				n.a.	degrad.	6
g PVAL	EDMONT 15-552	<1m			n.a.		21
g PVC	ANSELLEDMONT SNORKE				n.a.	degrad.	6
g PVC	ANSELLEDMONT SNORKE	50m			n.a.		21
m PVC	RICH INDUSTRIES	7m			0.38		21
s UNKNOWN MATERIAL	CHEMRON CHEMREL	89m	30	2	n.a.	86%	67
s UNKNOWN MATERIAL	CHEMRON CHEMREL MAX	>1440m	0	0	n.a.		67
s UNKNOWN MATERIAL	LIFE-GUARD RESPONDE	>180m			n.a.		77
g VITON	NORTH F-091				0.25	degrad.	7
g VITON	NORTH F-121	>90m			0.30		21

nitrobenzene
CAS Number: 98-95-3
Primary Class: 441   Nitro Compounds, Unsubstituted

g BUTYL	BEST 878	>240m		0	0.75		53
g BUTYL	BRUNSWICK BUTYL STD	>480m	0	0	0.69		89
g BUTYL	BRUNSWICK BUTYL-XTR	>480m	0	0	0.72		89
s BUTYL	LIFE-GUARD BUTYL	>480m	0	0	n.a.		77
g BUTYL	NORTH B-174	>1380m	0	0	0.63		34
s BUTYL	WHEELER ACID KING	300m	<90	2	n.a.		85
g BUTYL/ECO	BRUNSWICK BUTYL-POL	277m	300	3	0.63		89
s BUTYL/NEOPRENE	MSA BETEX	>240m	0	0	n.a.		48
u CPE	ILC DOVER	62m			n.a.		29
s CPE	STD. SAFETY	140m	27	2	n.a.		75
g NATURAL RUBBER	ACKWELL 5-109				0.15	degrad.	34
g NATURAL RUBBER	ANSELL PVL 040				0.46	degrad.	55
g NATURAL RUBBER	ANSELLEDMONT 36-124	5m	<900	4	0.46		6
g NATURAL RUBBER	ANSELLEDMONT 392	15m	<900	3	0.48		6
g NATURAL RUBBER	MARIGOLD BLACK HEVY	15m	1427	4	0.65	degrad.	79
g NATURAL RUBBER	MARIGOLD FEATHERLIT	15m	1767	4	0.15	degrad.	79
g NATURAL RUBBER	MARIGOLD MED WT EXT	15m	1367	4	0.45	degrad.	79
g NATURAL RUBBER	MARIGOLD MEDICAL	15m	1600	4	0.28	degrad.	79
g NATURAL RUBBER	MARIGOLD ORANGE SUP	15m	1433	4	0.73	degrad.	79
g NATURAL RUBBER	MARIGOLD R SURGEONS	15m	1600	4	0.28	degrad.	79
g NATURAL RUBBER	MARIGOLD RED LT WT	15m	1400	4	0.43	degrad.	79
g NATURAL RUBBER	MARIGOLD SENSOTECH	15m	1600	4	0.28	degrad.	79
g NATURAL RUBBER	MARIGOLD SUREGRIP	15m	1409	4	0.58	degrad.	79
g NATURAL RUBBER	MARIGOLD YELLOW	15m	1467	4	0.38	degrad.	79
g NATURAL+NEOPRENE	ANSELL OMNI 276				0.56	degrad.	55
g NATURAL+NEOPRENE	MARIGOLD FEATHERWT	15m	1500	4	0.35	degrad.	79
g NEOPRENE	ANSELL NEOPRENE 530				0.46	degrad.	55
g NEOPRENE	ANSELLEDMONT 29-840				0.38	degrad.	6
g NEOPRENE	ANSELLEDMONT 29-870	45m	11	3	0.53	degrad.	34

G A R M E N T	Class & Number/ test chemical/ MATERIAL NAME	MANUFACTURER & PRODUCT IDENTIFICATION	Break-through Time in Minutes	Perm-eation Rate in mg/sq m /min.	I N D E X	Thick-ness in mm	degrad. and comments	R e f s
g	NEOPRENE	BEST 32	321m	70	2	n.a.	degrad.	53
g	NEOPRENE	BEST 6780	>480m	0	0	n.a.		53
g	NEOPRENE	BRUNSWICK NEOPRENE	84m	>5000	4	0.92		89
s	NEOPRENE	LIFE-GUARD NEOPRENE	45m	5	2	n.a.		77
g	NEOPRENE	MAPA-PIONEER N-44	60m	1200	4	0.74	degrad.	36
b	NEOPRENE	RAINFAIR	120m	6	1	n.a.		90
b	NEOPRENE	RANGER	71m	350	3	n.a.		90
b	NEOPRENE	SERVUS NO 22204	108m	45	2	n.a.		90
g	NEOPRENE/NATURAL	ANSELL CHEMI-PRO				0.72	degrad.	55
g	NITRILE	ANSELL CHALLENGER				0.38	degrad.	55
g	NITRILE	ANSELLEDMONT 37-165				0.60	degrad.	6
g	NITRILE	BEST 22R				0.38	degrad.	53
g	NITRILE	BEST 727	52m	1250	4	n.a.	degrad.	53
g	NITRILE	MAPA-PIONEER A-14	60m	900	3	0.54	degrad.	36
g	NITRILE	MARIGOLD NITROSOLVE				0.75	degrad.	79
g	NITRILE	NORTH LA-142G	29m	102	3	0.38	degrad.	34
b	NITRILE+PUR+PVC	BATA HAZMAX	>180m			n.a.		86
b	NITRILE+PVC	SERVUS NO 73101	>180m			n.a.		90
b	NITRILE+PVC	TINGLEY	>180m			n.a.		90
c	PE	DU PONT TYVEK QC	<1m	24	4	n.a.		8
g	PE/EVAL/PE	SAFETY4 4H	>240m	0	0	0.07	35°C	60
g	PE/EVAL/PE	SAFETY4 4H	>1440m	0	0	0.07		60
g	PVAL	EDMONT 15-552	>360m	0	0	n.a.		6
g	PVAL	EDMONT 25-545	>360m	0	0	n.a.		6
g	PVAL	EDMONT 25-950	>960m	0	0	0.33		34
g	PVC	ANSELLEDMONT 34-100				0.20	degrad.	34
g	PVC	ANSELLEDMONT SNORKE				n.a.	degrad.	6
b	PVC	BATA STANDARD	360m	2	1	n.a.		86
g	PVC	BEST 725R				n.a.	degrad.	53
b	PVC	JORDAN DAVID	>180m			n.a.		90
b	PVC	STD. SAFETY GL-20	31m	1861	4	n.a.		75
s	PVC	STD. SAFETY WG-20	21m	1017	4	n.a.		75
s	PVC	WHEELER ACID KING	25m	<900	3	n.a.	degrad.	85
b	PVC+POLYURETHANE	BATA POLYBLEND	360m	2	1	n.a.		86
b	PVC+POLYURETHANE	BATA POLYBLEND	>180m			n.a.		90
b	PVC+POLYURETHANE	BATA POLYMAX	360m	0.03	0	n.a.		86
b	PVC+POLYURETHANE	BATA SUPER POLY	>480m	0	0	n.a.		86
s	SARANEX-23	DU PONT TYVEK SARAN	205m	6	1	n.a.		8
s	SARANEX-23 2-PLY	DU PONT TYVEK SARA	>480m	0	0	n.a.		8
u	TEFLON	CHEMFAB	2m	9	3	0.10		65
s	TEFLON	CHEMFAB CHALL. 5100	>480m	0	0	n.a.		56
s	TEFLON	CHEMFAB CHALL. 5100	>180m	0	1	n.a.		65
s	TEFLON	CHEMFAB CHALL. 5200	>300m	0	0	n.a.		80
s	TEFLON	CHEMFAB CHALL. 6000	>180m			n.a.		80
s	TEFLON	LIFE-GUARD TEFGUARD	>480m	0	0	n.a.		77
s	TEFLON	WHEELER ACID KING	>480m	0	0	n.a.		85
v	TEFLON-FEP	CHEMFAB CHALL.	>180m			0.25		65
s	UNKNOWN MATERIAL	CHEMRON CHEMREL MAX	>1440m	0	0	n.a.		67
c	UNKNOWN MATERIAL	CHEMRON CHEMTUFF	309m	0.4	0	n.a.		67

G A R M E Class & Number/ N test chemical/ T MATERIAL NAME	MANUFACTURER & PRODUCT IDENTIFICATION	Break- through Time in Minutes	Perm- eation Rate in mg/sq m /min.	I N D E X	Thick- ness degrad. in and mm comments	R e f s
s UNKNOWN MATERIAL	DU PONT BARRICADE	>480m	0	0	n.a.	8
s UNKNOWN MATERIAL	KAPPLER CPF III	>480m		0	n.a.	77
s UNKNOWN MATERIAL	LIFE-GUARD RESPONDE	>480m	0	0	n.a.	77
g UNKNOWN MATERIAL	NORTH SILVERSHIELD	>480m	0	0	0.10	7
g VITON	NORTH F-091	>480m	0	0	0.25	34
g VITON	NORTH F-091	1335m	8	1	0.30	62
s VITON/BUTYL/UNKN.	TRELLEBORG HPS	>180m			n.a.	71
s VITON/CHLOROBUTYL	LIFE-GUARD VC-100	289m		0	n.a.	77
s VITON/CHLOROBUTYL	LIFE-GUARD VNC 200	>480m	0	0	n.a.	77
u VITON/CHLOROBUTYL	Unknown	175m			n.a.	46
s VITON/NEOPRENE	MSA VAUTEX	>240m	0	0	n.a.	48

nitroethane
CAS Number: 79-24-3
Primary Class: 441   Nitro Compounds, Unsubstituted

g BUTYL	NORTH B-174	>480m	0	0	0.68	34
g NATURAL RUBBER	ACKWELL 5-109	2m	1888	5	0.15	34
g NATURAL+NEOPRENE	PLAYTEX ARGUS 123	11m	480	4	n.a.	72
g NEOPRENE	ANSELLEDMONT 29-870	49m	1434	4	0.43	34
g NITRILE	ANSELLEDMONT 37-155				0.40	34
g PE/EVAL/PE	SAFETY4 4H	>240m	0	0	0.07 35°C	60
g PVAL	EDMONT 25-950	210m	25	2	0.74	34
g PVC	ANSELLEDMONT 34-100				0.20 degrad.	34
s UNKNOWN MATERIAL	CHEMRON CHEMREL	>1440m	0	0	n.a.	67
g VITON	NORTH F-091				0.26 degrad.	34
u VITON/CHLOROBUTYL	Unknown	>180m			0.36	29

nitrogen dioxide
CAS Number: 10102-44-0
Primary Class: 350   Inorganic Gases and Vapors
Related Class: 365   Inorganic Acid Oxides

s PVC	TRELLEBORG TRELLCHE				n.a. degrad.	71
s SARANEX-23	DU PONT TYVEK SARAN	>480m	0	0	n.a.	8

nitrogen tetroxide
CAS Number: 10544-72-6
Primary Class: 350   Inorganic Gases and Vapors
Related Class: 365   Inorganic Acid Oxides

b BUTYL	BATA	>120m			1.10 preexp. 24h	20
g BUTYL	NORTH B-161	>120m			0.41 preexp. 24h	20
s BUTYL/NEOPRENE	MSA BETEX	15m			n.a.	48
s CHLOROBUTYL	FAIRPRENE	>120m			0.05 preexp. 24h	20
v CR39	Unknown	>120m			1.70 preexp. 24h	20
b PVC	BATA STANDARD	>120m			2.00 preexp. 24h	20
g PVC	BEST 814	12m			n.a. preexp. 24h	20

G A R M E Class & Number/ N test chemical/ T MATERIAL NAME	MANUFACTURER & PRODUCT IDENTIFICATION	Break- through Time in Minutes	Perm- eation Rate in mg/sq m /min.	I N D E X	Thick- ness in mm	degrad. and comments	R e f s
u TEFLON	Unknown	<1m			n.a.	preexp. 24h	20
s UNKNOWN MATERIAL	DU PONT BARRICADE	24m	657	3	n.a.		8
s UNKNOWN MATERIAL	LIFE-GUARD RESPONDE	220m	70	2	n.a.		77
g VITON	NORTH F-091	46m			0.35	preexp. 24h	20
s VITON/NEOPRENE	MSA VAUTEX	15m			n.a.		48

nitroglycerol (synonym: nitroglycerin)
CAS Number: 55-63-0
Primary Class: 442   Nitro Compounds, Substituted
Related Class: 314   Hydroxy Compounds, Aliphatic and Alicyclic, Polyols

g PE/EVAL/PE	SAFETY4 4H	>240m		0	0.07		60

nitroglycol
Primary Class: 442   Nitro Compounds, Substituted

g PE/EVAL/PE	SAFETY4 4H	>240m	0		0.07		60

nitromethane
CAS Number: 75-52-5
Primary Class: 441   Nitro Compounds, Unsubstituted

g BUTYL	NORTH B-174	>480m	0	0	0.65		34
g NATURAL RUBBER	ACKWELL 5-109	1m	960	4	0.15		34
g NATURAL RUBBER	ANSELLEDMONT 36-124	4m	<9	3	0.46		6
g NATURAL RUBBER	ANSELLEDMONT 392	10m	<900	4	0.48		6
g NEOPRENE	ANSELLEDMONT 29-840	60m	<90	2	0.31		6
g NEOPRENE	ANSELLEDMONT 29-870	63m	30	2	0.50		34
g NEOPRENE	ANSELLEDMONT NEOX	90m	<9	1	n.a.		6
g NITRILE	ANSELLEDMONT 37-155				0.40	degrad.	34
g NITRILE	ANSELLEDMONT 37-165	30m	<9000	4	0.60		6
g PE/EVAL/PE	SAFETY4 4H	>240m	0	0	0.07	35°C	60
g PVAL	ANSELLEDMONT PVA	>360m		0	n.a.		6
g PVAL	EDMONT 15-552	>360m	0	0	n.a.		6
g PVAL	EDMONT 25-545	>360m	0	0	n.a.		6
g PVAL	EDMONT 25-950	10m	288	4	0.66		34
g PVC	ANSELLEDMONT 34-100				0.20	degrad.	34
g PVC	ANSELLEDMONT SNORKE				n.a.	degrad.	6
s UNKNOWN MATERIAL	CHEMRON CHEMREL	>1440m	0	0	n.a.		67
s UNKNOWN MATERIAL	CHEMRON CHEMREL MAX	>1440m	0	0	n.a.		67
s UNKNOWN MATERIAL	DU PONT BARRICADE	>480m	0	0	n.a.		8

nonylamine
CAS Number: 112-20-9
Primary Class: 141   Amines, Aliphatic and Alicyclic, Primary

s UNKNOWN MATERIAL	LIFE-GUARD RESPONDE	>480m	0		n.a.		77

```
G
A
R                                       Perm-   I
M                               Break-   eation  N  Thick-           R
E  Class & Number/      MANUFACTURER    through  Rate in  D  ness degrad.  e
N  test chemical/        & PRODUCT      Time in  mg/sq m  E   in    and    f
T  MATERIAL NAME        IDENTIFICATION  Minutes  /min.    X  mm comments   s
```

nonylphenol
CAS Number: 25154-52-3
Primary Class: 316   Hydroxy Compounds, Aromatic (Phenols)

g NEOPRENE	ANSELLEDMONT 29-840	>1200m	0		0	0.45	33
g NITRILE	ANSELLEDMONT 37-175	>240m	0		0	0.40	33

o-chlorotoluene (synonyms: 2-chlorotoluene; 2-methylchlorobenzene)
CAS Number: 95-49-8
Primary Class: 264   Halogen Compounds, Aromatic

g NITRILE	GRANET N11F	17m	11617	5	n.a.		39
g NITRILE	MAPA-PIONEER A-15	52m	9867	4	0.35		39
s SARANEX-23	DU PONT TYVEK SARAN	26m	300	3	n.a.		8
s UNKNOWN MATERIAL	DU PONT BARRICADE	>480m	0	0	n.a.		8
g VITON	NORTH F-121	>240m	0	0	0.31		39

o-toluidine
CAS Number: 95-53-4
Primary Class: 145   Amines, Aromatic, Primary

c PE	DU PONT TYVEK QC	<1m	10	4	n.a.		8
s SARANEX-23	DU PONT TYVEK SARAN	255m	4	1	n.a.		8
s TEFLON	CHEMFAB CHALL. 5100	>200m	0	1	n.a.		65

o-xylene (synonym: 1,2-dimethylbenzene)
CAS Number: 95-47-6
Primary Class: 292   Hydrocarbons, Aromatic

g BUTYL	NORTH B-174	52m	1164	4	0.63	degrad.	34
g NATURAL RUBBER	ACKWELL 5-109				0.15	degrad.	34
g NEOPRENE	ANSELLEDMONT 29-870				0.48	degrad.	34
g NITRILE	ANSELLEDMONT 37-155	12m	7600	5	0.36	degrad.	34
g PVAL	EDMONT 25-950	>761m	0	0	0.25		34
g VITON	NORTH F-091	>480m	0	0	0.27		34

octane (synonym: n-octane)
CAS Number: 111-65-9
Primary Class: 291   Hydrocarbons, Aliphatic and Alicyclic, Saturated

s BUTYL	TRELLEBORG TRELLCHE				n.a.	degrad.	71
g BUTYL/NEOPRENE	COMASEC BUTYL PLUS	8m	2520	5	0.50		40
g HYPALON	COMASEC DIPCO	>480m	0	0	0.60		40
g NATURAL RUBBER	COMASEC FLEXIGUM	24m	960	3	0.95		40
g NEOPRENE	COMASEC COMAPRENE	420m	2160	4	n.a.		40
g NITRILE	COMASEC COMATRIL	200m	900	3	0.55		40
g NITRILE	COMASEC COMATRIL SU	>480m	0	0	0.60		40

G A R M E Class & Number/ N test chemical/ T MATERIAL NAME	MANUFACTURER & PRODUCT IDENTIFICATION	Break-through Time in Minutes	Perm-eation Rate in mg/sq m /min.	I N D E X	Thick-ness in mm	degrad. and comments	R e f s
g NITRILE	COMASEC FLEXITRIL	22m	1260	4	n.a.		40
g NITRILE+PVC	COMASEC MULTIMAX	>480m	0	0	n.a.		40
g NITRILE+PVC	COMASEC MULTIPLUS	108m	720	3	n.a.		40
g PVC	COMASEC MULTIPOST	80m	960	3	n.a.		40
g PVC	COMASEC MULTITOP	90m	780	3	n.a.		40
g PVC	COMASEC NORMAL	75m	1080	4	n.a.		40
g PVC	COMASEC OMNI	55m	1080	4	n.a.		40
s UNKNOWN MATERIAL	LIFE-GUARD RESPONDE	>480m	0	0	n.a.		77

octyl alcohol   (synonyms: 1-octanol; n-octanol)
CAS Number: 111-87-5
Primary Class: 311   Hydroxy Compounds, Aliphatic and Alicyclic, Primary

g BUTYL	BEST 878	>480m	0	0	0.75		53
g NATURAL RUBBER	ANSELL CONFORM 4205	>360m		0	0.13		55
g NATURAL RUBBER	ANSELL FL 200 254	>360m		0	0.51		55
g NATURAL RUBBER	ANSELL ORANGE 208	>360m		0	0.76		55
g NATURAL RUBBER	ANSELL PVL 040	>360m		0	0.46		55
g NATURAL RUBBER	ANSELL STERILE 832	>360m		0	0.23		55
g NATURAL RUBBER	ANSELLEDMONT 36-124	60m	<9	1	0.46		6
g NATURAL RUBBER	ANSELLEDMONT 392	30m	<90	3	0.48		6
g NATURAL RUBBER	COMASEC FLEXIGUM	280m	42	2	0.95		40
g NATURAL+NEOPRENE	ANSELL OMNI 276	>360m	0	0	0.56		55
g NATURAL+NEOPRENE	ANSELL TECHNICIANS	>360m	0	0	0.43		55
g NEOPRENE	ANSELL NEOPRENE 530	>360m	0	0	0.46		55
g NEOPRENE	ANSELLEDMONT 29-840	420m	<9	1	0.31		6
g NEOPRENE	ANSELLEDMONT NEOX	>420m	0	0	n.a.		6
g NEOPRENE	BEST 32	>480m	0	0	n.a.		53
g NEOPRENE	BEST 6780	>480m	0	0	n.a.		53
g NEOPRENE	COMASEC COMAPRENE	>360m	0	0	n.a.		40
g NEOPRENE	MAPA-PIONEER NS-420	195m	4	1	0.89		36
g NEOPRENE/NATURAL	ANSELL CHEMI-PRO	>360m		0	0.72		55
g NITRILE	ANSELL CHALLENGER	>360m	0	0	0.38		55
g NITRILE	ANSELLEDMONT 37-165	>360m	0	0	0.60		6
g NITRILE	BEST 727	>480m	0	0	0.38		53
g NITRILE	COMASEC COMATRIL	>480m	0	0	0.55		40
g NITRILE	COMASEC COMATRIL SU	>480m	0	0	0.60		40
g NITRILE	COMASEC FLEXITRIL	55m	114	3	n.a.		40
g NITRILE+PVC	COMASEC MULTIPLUS	>480m	0	0	n.a.		40
g PVAL	ANSELLEDMONT PVA	>360m		0	n.a.		6
g PVAL	EDMONT 15-552	240m	<9	1	n.a.		6
g PVAL	EDMONT 25-545	>360m	0	0	n.a.		6
g PVC	ANSELLEDMONT M GRIP	>360m	0	0	n.a.		6
g PVC	COMASEC MULTIPOST	>480m	0	0	n.a.		40
g PVC	COMASEC MULTITOP	>480m	0	0	n.a.		40
g PVC	COMASEC NORMAL	>480m	0	0	n.a.		40
g PVC	COMASEC OMNI	330m	120	3	n.a.		40

G A R M E Class & Number/ N test chemical/ T MATERIAL NAME	MANUFACTURER & PRODUCT IDENTIFICATION	Break- through Time in Minutes	Perm- eation Rate in mg/sq m /min.	I N D E X	Thick- ness in mm	degrad. and comments	R e f s
oleic acid							
CAS Number: 112-80-1							
Primary Class: 102 Acids, Carboxylic, Aliphatic and Alicyclic, Unsubstituted							
g BUTYL	BEST 878	>480m	0	0	0.75		53
g NATURAL RUBBER	ACKWELL 5-109				0.15	degrad.	34
g NATURAL RUBBER	ANSELL CONFORM 4205	240m		0	0.13		55
g NATURAL RUBBER	ANSELL FL 200 254	>360m		0	0.51		55
g NATURAL RUBBER	ANSELL ORANGE 208	>360m	0	0	0.76		55
g NATURAL RUBBER	ANSELL PVL 040	>360m		0	0.46		55
g NATURAL RUBBER	ANSELL STERILE 832	>360m		0	0.23		55
g NATURAL RUBBER	ANSELLEDMONT 36-124	30m	<90	3	0.46		6
g NATURAL RUBBER	ANSELLEDMONT 392	>360m		0	0.48		6
g NATURAL RUBBER	BEST 65NFW	>480m	0	0	n.a.		53
g NATURAL+NEOPRENE	ANSELL OMNI 276	>360m	0	0	0.56		55
g NATURAL+NEOPRENE	ANSELL TECHNICIANS	>360m	0	0	0.43		55
g NEOPRENE	ANSELL NEOPRENE 530	>360m	0	0	0.46		55
g NEOPRENE	ANSELLEDMONT 29-840	60m	<90	2	0.38		6
g NEOPRENE	ANSELLEDMONT NEOX	150m	<9	1	n.a.		6
g NEOPRENE	BEST 32	>480m	0	0	n.a.		53
g NEOPRENE	BEST 6780	>480m	0	0	n.a.		53
g NEOPRENE/NATURAL	ANSELL CHEMI-PRO	>360m	0	0	0.72		55
g NITRILE	ANSELL CHALLENGER	>360m	0	0	0.38		55
g NITRILE	ANSELLEDMONT 37-165	>360m	0	0	0.60		6
g NITRILE	BEST 22R	>480m	0	0	n.a.		53
g NITRILE	BEST 727	>480m	0	0	0.38		53
g PVAL	ANSELLEDMONT PVA	60m	<9	1	n.a.		6
g PVAL	EDMONT 15-552	60m	<9	1	n.a.		6
g PVAL	EDMONT 25-545	>360m	0	0	n.a.		6
g PVC	ANSELLEDMONT SNORKE	90m	<90	2	n.a.		6
g PVC	BEST 725R	>480m	0	0	n.a.		53
organo-tin paint							
Primary Class: 900 Miscellaneous Unclassified Chemicals							
s UNKNOWN MATERIAL	LIFE-GUARD RESPONDE	>480m	0		n.a.		77
s VITON/CHLOROBUTYL	LIFE-GUARD VC-100	>240m	0		n.a.		77
Orthocid 83							
CAS Number: 135-06-2							
Primary Class: 840 Pesticides, Mixtures and Formulations							
g NITRILE	ANSELLEDMONT 37-145	>450m		0	0.30		57
c PE	MOLNLYCKE H D	15m			n.a.		57
g PE/EVAL/PE	SAFETY4 4H	>240m	0	0	0.07		60

G A R M E Class & Number/ N test chemical/ T MATERIAL NAME	MANUFACTURER & PRODUCT IDENTIFICATION	Break- through Time in Minutes	Perm- eation Rate in mg/sq m /min.	I N D E X	Thick- ness in mm	degrad. and comments	R e f s

oxalic acid
CAS Number: 144-62-7
Primary Class: 104   Acids, Carboxylic, Aliphatic and Alicyclic, Polybasic

g BUTYL	NORTH B-174	>480m	0	0	0.66		34
g NATURAL RUBBER	ANSELL CONFORM 4205	>360m	0	0	0.13		55
g NATURAL RUBBER	ANSELL FL 200 254	>360m	0	0	0.51		55
g NATURAL RUBBER	ANSELL ORANGE 208	>360m	0	0	0.76		55
g NATURAL RUBBER	ANSELL PVL 040	>360m	0	0	0.46		55
g NATURAL RUBBER	ANSELL STERILE 832	>360m	0	0	0.23		55
g NATURAL RUBBER	ANSELLEDMONT 36-124	>360m	0	0	0.46		6
g NATURAL RUBBER	ANSELLEDMONT 392	>360m	0	0	0.48		6
g NATURAL RUBBER	COMASEC FLEXIGUM	>480m	0	0	0.95		40
g NATURAL RUBBER	MARIGOLD BLACK HEVY	>480m	0	0	0.65		79
g NATURAL RUBBER	MARIGOLD FEATHERLIT	>480m	0	0	0.15	degrad.	79
g NATURAL RUBBER	MARIGOLD MED WT EXT	>480m	0	0	0.45		79
g NATURAL RUBBER	MARIGOLD MEDICAL	>480m	0	0	0.28		79
g NATURAL RUBBER	MARIGOLD ORANGE SUP	>480m	0	0	0.73		79
g NATURAL RUBBER	MARIGOLD R SURGEONS	>480m	0	0	0.28		79
g NATURAL RUBBER	MARIGOLD RED LT WT	>480m	0	0	0.43		79
g NATURAL RUBBER	MARIGOLD SENSOTECH	>480m	0	0	0.28		79
g NATURAL RUBBER	MARIGOLD SUREGRIP	>480m	0	0	0.58		79
g NATURAL RUBBER	MARIGOLD YELLOW	>480m	0	0	0.38		79
g NATURAL+NEOPRENE	ANSELL OMNI 276	>360m	0	0	0.56		55
g NATURAL+NEOPRENE	ANSELL TECHNICIANS	>360m	0	0	0.43		55
g NATURAL+NEOPRENE	MARIGOLD FEATHERWT	>480m	0	0	0.35		79
g NEOPRENE	ANSELL NEOPRENE 530	>360m	0	0	0.46		55
g NEOPRENE	ANSELLEDMONT 29-840	>360m	0	0	0.38		6
g NEOPRENE	ANSELLEDMONT 29-870	>480m	0	0	0.46		34
g NEOPRENE	ANSELLEDMONT NEOX	>360m	0	0	n.a.		6
g NEOPRENE	COMASEC COMAPRENE	>360m	0	0	n.a.		40
g NEOPRENE/NATURAL	ANSELL CHEMI-PRO	>360m	0	0	0.72		55
g NITRILE	ANSELL CHALLENGER	>360m	0	0	0.38		55
g NITRILE	ANSELLEDMONT 37-155	>480m	0	0	0.40		34
g NITRILE	ANSELLEDMONT 37-165	>360m	0	0	0.60		6
g NITRILE	COMASEC COMATRIL	>480m	0	0	0.55		40
g NITRILE	COMASEC COMATRIL SU	>480m	0	0	0.60		40
g NITRILE	COMASEC FLEXITRIL	130m			n.a.		40
g NITRILE	MARIGOLD BLUE	>480m	0	0	0.45		79
g NITRILE	MARIGOLD GREEN SUPA	>480m	0	0	0.30		79
g NITRILE	MARIGOLD NITROSOLVE	>480m	0	0	0.75		79
g NITRILE+PVC	COMASEC MULTIPLUS	>480m	0	0	n.a.		40
g PVAL	ANSELLEDMONT PVA				n.a.	degrad.	6
g PVAL	EDMONT 15-552				n.a.	degrad.	6
g PVC	ANSELLEDMONT SNORKE	>360m	0	0	n.a.		6
g PVC	COMASEC MULTIPOST	>480m	0	0	n.a.		40
g PVC	COMASEC MULTITOP	>480m	0	0	n.a.		40
g PVC	COMASEC NORMAL	>480m	0	0	n.a.		40

G A R M E Class & Number/ N test chemical/ T MATERIAL NAME	MANUFACTURER & PRODUCT IDENTIFICATION	Break- through Time in Minutes	Perm- eation Rate in mg/sq m /min.	I N D E X	Thick- ness in mm	degrad. and comments	R e f s
g PVC	COMASEC OMNI	>480m	0	0	n.a.		40
g VITON	NORTH F-091	>480m	0	0	0.33		34

p-benzoquinone (synonym: benzoquinone)
CAS Number: 106-51-4
Primary Class: 410   Quinones

s SARANEX-23	DU PONT TYVEK SARAN	>480m	0	0	n.a.		8

p-chlorotoluene (synonyms: 4-chlorotoluene; 4-methylchlorobenzene)
CAS Number: 106-43-4
Primary Class: 264   Halogen Compounds, Aromatic

g NITRILE	GRANET N11F	15m	12217	5	n.a.		39
g NITRILE	MAPA-PIONEER A-15	25m	8883	4	0.35		39
g VITON	NORTH F-121	>240m	0	0	0.31		39

p-cresol (synonym: 4-methylphenol)
CAS Number: 106-44-5
Primary Class: 316   Hydroxy Compounds, Aromatic (Phenols)
Related Class: 292   Hydrocarbons, Aromatic

g PE/EVAL/PE	SAFETY4 4H	>240m	0	0.07	35°C		60

p-tert-butyltoluene (synonym: 4-methyl-t-butylbenzene)
CAS Number: 98-51-1
Primary Class: 292   Hydrocarbons, Aromatic

g BUTYL	BRUNSWICK BUTYL STD	91m	>320	3	0.63		89
g BUTYL	BRUNSWICK BUTYL-XTR	59m	>330	3	0.63		89
g BUTYL	NORTH B-174	107m	480	3	0.64		34
g BUTYL/ECO	BRUNSWICK BUTYL-POL	457m	0.03	0	0.63		89
g NEOPRENE	ANSELLEDMONT 29-870	69m	4200	4	0.51		34
g NITRILE	ANSELLEDMONT 37-155	>360m	0	0	0.38		34
g NITRILE	NORTH LA-142G				0.38	degrad.	7
g PVAL	EDMONT 25-950	>420m	0	0	0.25		34
g UNKNOWN MATERIAL	NORTH SILVERSHIELD	>480m	0	0	0.10		7
g VITON	NORTH F-091	>480m	0	0	0.23		34

p-toluenesulfonic acid (synonym: 4-methylbenzenesulfonic acid)
CAS Number: 536-57-2
Primary Class: 504   Sulfur Compounds, Sulfonic Acids

u CPE	ILC DOVER	>180m			n.a.		29
g NEOPRENE	ANSELLEDMONT 29-840	>240m		0	0.45		33
g PVC	KID VINYLPRODUKTER	>240m		0	0.45		33
u VITON/CHLOROBUTYL	Unknown	>180m			n.a.		29

G A R M E Class & Number/ N test chemical/ T MATERIAL NAME	MANUFACTURER & PRODUCT IDENTIFICATION	Break-through Time in Minutes	Perm-eation Rate in mg/sq m /min.	I N D E X	Thick-ness in mm	degrad. and comments	R e f s
p-xylene (synonym: 1,4-dimethylbenzene)							
CAS Number: 106-42-3							
Primary Class: 292 Hydrocarbons, Aromatic							
g BUTYL	NORTH B-174	27m	906	3	0.64	degrad.	34
g NATURAL RUBBER	ACKWELL 5-109				0.15	degrad.	34
g NEOPRENE	ANSELLEDMONT 29-870				0.48	degrad.	34
g NITRILE	ANSELL 632	11m			0.36		58
g NITRILE	ANSELLEDMONT 37-155	52m	858	3	0.38	degrad.	34
g NITRILE	ANSELLEDMONT 37-155	54m	4260	4	0.36		41
g NITRILE	ANSELLEDMONT 37-155	33m			0.32		58
g NITRILE	BEST 727	27m			0.28		58
g NITRILE	GRANET 490	24m			0.32		58
g NITRILE	MAPA-PIONEER A-15	60m			0.33		58
g NITRILE	NORTH LA-142G	33m			0.32		58
g NITRILE	SURETY 315R	54m			0.32		58
g NITRILE	UNIROYAL 4-15	28m			0.32		58
g PVAL	EDMONT 25-950	>840m	0	0	0.26		34
g PVC	ANSELLEDMONT 34-100	<1m	1848	5	0.18		34
g VITON	NORTH F-091	>960m	0	0	0.25		34

p-xylene, 50% & toluene, 50%							
Primary Class: 292 Hydrocarbons, Aromatic							
g NITRILE	ANSELLEDMONT 37-155	35/33m	3120/	4	0.37		41

p-xylene, 75% & toluene, 25%							
Primary Class: 292 Hydrocarbons, Aromatic							
g NITRILE	ANSELLEDMONT 37-155	55/55m	2940/	4	0.36		41

paint and varnish remover							
Primary Class: 900 Miscellaneous Unclassified Chemicals							
g NATURAL+NEOPRENE	PLAYTEX ARGUS 123	4m	6720	5	0.91		72

palmitic acid							
CAS Number: 57-10-3							
Primary Class: 102 Acids, Carboxylic, Aliphatic and Alicyclic, Unsubstituted							
g NATURAL RUBBER	ANSELL CONFORM 4205	>360m	0	0	0.13		55
g NATURAL RUBBER	ANSELL FL 200 254	>360m	0	0	0.51		55
g NATURAL RUBBER	ANSELL ORANGE 208	>360m	0	0	0.76		55
g NATURAL RUBBER	ANSELL PVL 040	>360m	0	0	0.46		55
g NATURAL RUBBER	ANSELL STERILE 832	>360m	0	0	0.23		55
g NATURAL RUBBER	ANSELLEDMONT 36-124	5m			0.46		6

G A R M E Class & Number/ N test chemical/ T MATERIAL NAME	MANUFACTURER & PRODUCT IDENTIFICATION	Break- through Time in Minutes	Perm- eation Rate in mg/sq m /min.	I N D E X	Thick- ness in mm	R degrad. and comments	R e f s
g NATURAL RUBBER	ANSELLEDMONT 392	5m			0.48		6
g NATURAL+NEOPRENE	ANSELL OMNI 276	>360m	0	0	0.56		55
g NATURAL+NEOPRENE	ANSELL TECHNICIANS	>360m	0	0	0.43		55
g NEOPRENE	ANSELL NEOPRENE 530	>360m	0	0	0.46		55
g NEOPRENE	ANSELLEDMONT 29-840	>360m	0	0	0.38		6
g NEOPRENE	ANSELLEDMONT NEOX	>360m	0	0	n.a.		6
g NEOPRENE/NATURAL	ANSELL CHEMI-PRO	>360m	0	0	0.72		55
g NITRILE	ANSELL CHALLENGER	>360m	0	0	0.38		55
g NITRILE	ANSELLEDMONT 37-165	30m			0.60		6
g PVAL	ANSELLEDMONT PVA				n.a.	degrad.	6
g PVAL	EDMONT 15-552				n.a.	degrad.	6
g PVC	ANSELLEDMONT M GRIP	75m			n.a.		6

PCB & transformer oil
Primary Class: 264   Halogen Compounds, Aromatic

s UNKNOWN MATERIAL	LIFE-GUARD RESPONDE	>240m	0	0	n.a.		77
s VITON/CHLOROBUTYL	LIFE-GUARD VC-100	>240m		0	n.a.		77

PCB, 1% & naphtha, 99%
Primary Class: 264   Halogen Compounds, Aromatic
Related Class: 292   Hydrocarbons, Aromatic

s SARANEX-23	DU PONT TYVEK SARAN	>480m	0	0	n.a.		8

PCB, 4% & 1,2,4-trichlorobenzene, 6% & naphtha, 90%
Primary Class: 264   Halogen Compounds, Aromatic

s SARANEX-23	DU PONT TYVEK SARAN	>60m	0.1	1	n.a.		8

PCB, 50% & naphtha, 50% (synonym: PCB, 50% & mineral spirits, 50%)
Primary Class: 264   Halogen Compounds, Aromatic

s SARANEX-23	DU PONT TYVEK SARAN	>480m	0	0	n.a.		8
s UNKNOWN MATERIAL	CHEMRON CHEMREL	>480m	0	0	n.a.		67

pentachlorophenol
CAS Number: 87-86-5
Primary Class: 316   Hydroxy Compounds, Aromatic (Phenols)

g NATURAL RUBBER	ANSELL CONFORM 4205	150m			0.13		55
g NATURAL RUBBER	ANSELLEDMONT 36-124				0.46	degrad.	6
g NATURAL RUBBER	ANSELLEDMONT 392				0.48	degrad.	6
g NEOPRENE	ANSELLEDMONT 29-840	6m<	9	3	0.38		6
g NEOPRENE	ANSELLEDMONT NEOX	6m<	9	3	n.a.		6
g NEOPRENE	BRUNSWICK NEOPRENE	>480m	0	0	0.88		89
g NEOPRENE/NATURAL	ANSELL CHEMI-PRO	>360m	0	0	0.72		55

G A R M E Class & Number/ N test chemical/ T MATERIAL NAME	MANUFACTURER & PRODUCT IDENTIFICATION	Break- through Time in Minutes	Perm- eation Rate in mg/sq m /min.	I N D E X	Thick- ness degrad. in and mm comments	R e f s
g NITRILE	ANSELL CHALLENGER	>360m	0	0	0.38	55
g NITRILE	ANSELLEDMONT 37-165	>360m	0	0	0.60	6
g NITRILE	NORTH LA-142G	>960m	0	0	0.36 1%, naphtha	7
g PVAL	ANSELLEDMONT PVA	5m	<9000	5	n.a.	6
g PVAL	EDMONT 15-552	7m	<9000	5	n.a.	6
g PVAL	EDMONT 25-545	15m	<900	3	n.a.	6
g PVC	ANSELLEDMONT SNORKE	180m	<9	1	n.a.	6
g UNKNOWN MATERIAL	NORTH SILVERSHIELD	>480m	0	0	0.10	7
g VITON	NORTH F-091	>780m	0	0	0.24 1%, naphtha	7

pentachlorophenol, 4.3% in diesel oil
Primary Class: 860   Coals, Charcoals, Oils

g NATURAL RUBBER	BEST 65NFW	<1m	0.2	1	n.a.	64
g NATURAL+NEOPRENE	PLAYTEX 835	60m	13	2	0.41	64
g NITRILE	ANSELLEDMONT 37-165	>480m	0	0	0.64	64
g PVC	DAYTON TRIFLEX	<1m	3	3	0.19	64
g PVC	GRANET SAFETY 1012	>960m	0	0	n.a.	64

pentane (synonym: n-pentane)
CAS Number: 109-66-0
Primary Class: 291   Hydrocarbons, Aliphatic and Alicyclic, Saturated

g BUTYL	BEST 878	12m	2350	5	0.75 degrad.	53
g BUTYL	NORTH B-161				0.45 degrad.	7
g BUTYL/NEOPRENE	COMASEC BUTYL PLUS	10m			0.50	40
g NATURAL RUBBER	ANSELLEDMONT 36-124				0.46 degrad.	6
g NATURAL RUBBER	ANSELLEDMONT 392				0.48 degrad.	6
g NATURAL RUBBER	ANSELLEDMONT 46-320	2m	27000	5	0.31	5
g NATURAL RUBBER	COMASEC FLEXIGUM	12m	8400	5	0.95	40
g NATURAL+NEOPRENE	ANSELL OMNI 276	4m	18000	5	0.45	5
g NEOPRENE	ANSELL NEOPRENE 530	33m	8	2	0.46	55
g NEOPRENE	ANSELLEDMONT 29-840	30m	<9000	4	0.38	6
g NEOPRENE	ANSELLEDMONT 29-870	87m	360	3	0.51	34
g NEOPRENE	ANSELLEDMONT NEOX	45m	<90	3	n.a.	6
g NEOPRENE	BEST 32	27m	100	3	n.a.	53
g NEOPRENE	BEST 6780	79m	200	3	n.a.	53
g NEOPRENE	BRUNSWICK NEOPRENE	36m	10	3	0.63	89
g NEOPRENE	COMASEC COMAPRENE	30m	6660	4	n.a.	40
g NEOPRENE	MAPA-PIONEER N-73	38m	160	3	0.46	5
g NEOPRENE/NATURAL	ANSELL CHEMI-PRO	3m	206	4	0.72	55
g NITRILE	ANSELL CHALLENGER	>360m		0	0.38	55
g NITRILE	ANSELLEDMONT 37-155	2m	<1	3	0.38	34
g NITRILE	ANSELLEDMONT 37-165	> 60m			0.37	5
g NITRILE	ANSELLEDMONT 37-165	>360m	0	0	0.60	6
g NITRILE	BEST 22R	90m	40	2	n.a.	53
g NITRILE	BEST 727	>480m	0	0	0.38	53

GARMENT Class & Number/ test chemical/ MATERIAL NAME	MANUFACTURER & PRODUCT IDENTIFICATION	Break-through Time in Minutes	Perm-eation Rate in mg/sq m /min.	INDEX	Thick-ness in mm	degrad. and comments	Refs
g NITRILE	COMASEC COMATRIL	210m	1320	4	0.55		40
g NITRILE	COMASEC COMATRIL SU	>480m	0	0	0.60		40
g NITRILE	COMASEC FLEXITRIL	35m	9660	4	n.a.		40
g NITRILE	MARIGOLD BLUE	>480m	0	0	0.45		79
g NITRILE	MARIGOLD GREEN SUPA	>480m	0	0	0.30		79
g NITRILE	MARIGOLD NITROSOLVE	480m	0	0	0.75		79
g NITRILE+PVC	COMASEC MULTIPLUS	75m	900	3	n.a.		40
g PE	ANSELLEDMONT 35-125	<1m	4000	5	0.03		5
g PE/EVAL/PE	SAFETY4 4H	>480m	0	0	0.07		60
g PVAL	ANSELLEDMONT PVA	>360m		0	n.a.		6
g PVAL	EDMONT 15-552	>360m	0	0	n.a.		6
g PVAL	EDMONT 25-545	>360m	0	0	n.a.		6
g PVAL	EDMONT 25-950	15m	<1	2	0.25		34
g PVC	ANSELLEDMONT SNORKE				n.a.	degrad.	6
g PVC	COMASEC MULTIPOST	45m	1320	4	n.a.		40
g PVC	COMASEC MULTITOP	60m	1200	4	n.a.		40
g PVC	COMASEC NORMAL	45m	1500	4	n.a.		40
g PVC	COMASEC OMNI	20m	2100	4	n.a.		40
g PVC	MAPA-PIONEER V-20	9m	1000	5	0.31		5
g UNKNOWN MATERIAL	MARIGOLD R MEDIGLOV	1m	29333	5	0.03	degrad.	79
g VITON	NORTH F-091	>480m	0	0	0.23		34

pentane, 50% & trichloroethylene, 50%
Primary Class: 291   Hydrocarbons, Aliphatic and Alicyclic, Saturated

perchloric acid, 30-70%
CAS Number: 7601-90-3
Primary Class: 370   Inorganic Acids

g NATURAL RUBBER	ANSELL FL 200 254	>360m		0	0.51		55
g NATURAL RUBBER	ANSELL ORANGE 208	>360m		0	0.76		55
g NATURAL RUBBER	ANSELL PVL 040	>360m		0	0.46		55
g NATURAL RUBBER	ANSELLEDMONT 36-124	>360m	0	0	0.46		6
g NATURAL RUBBER	ANSELLEDMONT 392	>360m		0	0.48		6
g NATURAL RUBBER	COMASEC FLEXIGUM	>480m	0	0	0.95		40
g NATURAL+NEOPRENE	ANSELL OMNI 276	>360m		0	0.56		55
g NATURAL+NEOPRENE	ANSELL TECHNICIANS	>360m		0	0.43		55
g NEOPRENE	ANSELL NEOPRENE 530	>360m		0	0.46		55
g NEOPRENE	ANSELLEDMONT 29-840	>360m	0	0	0.31		6
g NEOPRENE	ANSELLEDMONT NEOX	>360m	0	0	n.a.		6
g NEOPRENE	COMASEC COMAPRENE	>360m	0	0	n.a.		40
g NEOPRENE/NATURAL	ANSELL CHEMI-PRO	>360m		0	0.72		55
g NITRILE	ANSELL CHALLENGER	>360m		0	0.38		55
g NITRILE	ANSELLEDMONT 37-165	>360m	0	0	0.60		6
g NITRILE	COMASEC COMATRIL	>480m	0	0	0.55		40
g NITRILE	COMASEC COMATRIL SU	>480m	0	0	0.60		40

G A R M E Class & Number/ N test chemical/ T MATERIAL NAME	MANUFACTURER & PRODUCT IDENTIFICATION	Break-through Time in Minutes	Perm-eation Rate in mg/sq m /min.	I N D E X	Thick-ness in mm	degrad. and comments	R e f s
g NITRILE	COMASEC FLEXITRIL	>480m	0	0	n.a.		40
g NITRILE+PVC	COMASEC MULTIPLUS	>480m	0	0	n.a.		40
g PE/EVAL/PE	SAFETY4 4H	>240m	0	0	0.07	70% 35°C	60
g PVAL	ANSELLEDMONT PVA				n.a.	degrad.	6
g PVAL	EDMONT 15-552				n.a.	degrad.	6
g PVC	ANSELLEDMONT M GRIP	>360m	0	0	n.a.		6
g PVC	COMASEC MULTIPOST	>480m	0	0	n.a.		40
g PVC	COMASEC MULTITOP	>480m	0	0	n.a.		40
g PVC	COMASEC NORMAL	>480m	0	0	n.a.		40
g PVC	COMASEC OMNI	>480m	0	0	n.a.		40

Perma Fluid (synonym: ammonium thioglycolate, 8% in water)
Primary Class: 900   Miscellaneous Unclassified Chemicals

g PE/EVAL/PE	SAFETY4 4H	>240m	0	0	0.07	35°C	60

peroxyacetic acid
CAS Number: 79-21-0
Primary Class: 300   Peroxides

g BUTYL	NORTH B-174	>480m	0	0	0.62		34
g NATURAL RUBBER	ACKWELL 5-109				0.15	degrad.	34
g NITRILE	ANSELLEDMONT 37-155	81m	1140	4	0.36		34
g PVC	ANSELLEDMONT 34-100				0.20	degrad.	34
g VITON	NORTH F-091	>444m	4	1	0.27		34

petroleum ether with <0.5% aromatics (b.p. 80-110°C)
CAS Number: 8032-32-4
Primary Class: 291   Hydrocarbons, Aliphatic and Alicyclic, Saturated

g NEOPRENE	ANSELLEDMONT 29-875	65m	780	3	0.52		30
g NEOPRENE	BEST 32	47m	1110	4	n.a.		53
g NEOPRENE	BEST 6780	94m	840	3	n.a.		53
g NITRILE	ANSELLEDMONT 37-145	>180m			0.32		30
g NITRILE	ANSELLEDMONT 37-175	>180m			0.43		30
g NITRILE	BEST 22R	112m	560	3	n.a.		53
g NITRILE	BEST 727	>480m	0	0	0.38		53
g NITRILE	MARIGOLD BLUE	>480m	0	0	0.45		79
g NITRILE	MARIGOLD GREEN SUPA	>480m	0	0	0.30		79
g NITRILE	MARIGOLD NITROSOLVE	>480m	0	0	0.75		79
g PE/EVAL/PE	SAFETY4 4H	>240m	0	0	0.07	35°C	60
g PE/EVAL/PE	SAFETY4 4H	>480m	0	0	0.07		60
g PVC	BEST 725R	19m	1350	4	n.a.		53
g PVC	KID 490	>210m			n.a.		30
s TEFLON	CHEMFAB CHALL. 5100	>204m	0	1	n.a.		65
g UNKNOWN MATERIAL	MARIGOLD R MEDIGLOV	1m	2600	5	0.03	degrad.	79

			Perm-	I			
G							
A							
R			Break-	eation	N	Thick-	R
M			through	Rate in	D	ness degrad.	e
E Class & Number/	MANUFACTURER		Time in	mg/sq m	E	in and	f
N test chemical/	& PRODUCT		Minutes	/min.	X	mm comments	s
T MATERIAL NAME	IDENTIFICATION						

petroleum, Shell speciality
Primary Class: 860   Coals, Charcoals, Oils
Related Class: 900   Miscellaneous Unclassified Chemicals

g PE/EVAL/PE	SAFETY4 4H	>240m	0		0	0.07	35°C	60

phenol (synonym: carbolic acid)
CAS Number: 108-95-2
Primary Class: 316   Hydroxy Compounds, Aromatic (Phenols)ww

		Break-through Time	Perm Rate	IND EX	Thickness mm	degrad. and comments	Refs
g BUTYL	BRUNSWICK BUTYL STD	456m	3	1	0.63		89
s BUTYL	MSA CHEMPRUF	>480m	0	0	n.a.		48
g BUTYL	NORTH B-131	31m			0.35	85% in H_2O	83
g BUTYL	NORTH B-174	>1200m	0	0	0.61	85% in H_2O	34
g BUTYL/NEOPRENE	COMASEC BUTYL PLUS	>480m	0	0	0.50		40
s BUTYL/NEOPRENE	MSA BETEX	>480m	0	0	n.a.		48
g NAT+NEOPR+NITRILE	MAPA-PIONEER TRIONI	102m	108	3	0.46		36
g NATURAL RUBBER	ANSELL CONFORM 4205	>360m		0	0.13		55
g NATURAL RUBBER	ANSELL FL 200 254	>360m		0	0.51		55
g NATURAL RUBBER	ANSELL ORANGE 208	>360m		0	0.76		55
g NATURAL RUBBER	ANSELL PVL 040	>360m		0	0.46		55
g NATURAL RUBBER	ANSELL STERILE 832	>360m		0	0.23		55
g NATURAL RUBBER	ANSELLEDMONT 30-139	60m	<90	2	0.25		42
g NATURAL RUBBER	ANSELLEDMONT 36-124	60m	<900	3	0.46		6
g NATURAL RUBBER	ANSELLEDMONT 392	150m			0.48		6
g NATURAL RUBBER	ANSELLEDMONT 46-320	>60m			0.31		5
g NATURAL RUBBER	COMASEC FLEXIGUM	54m	2460	4	0.95		40
g NATURAL RUBBER	MAPA-PIONEER L-118	>480m	0	0	0.46		36
g NATURAL RUBBER	MARIGOLD BLACK HEVY	39m	34	3	0.65		79
g NATURAL RUBBER	MARIGOLD FEATHERLIT	4m	179	4	0.15	degrad.	79
g NATURAL RUBBER	MARIGOLD MED WT EXT	30m	154	3	0.45	degrad.	79
g NATURAL RUBBER	MARIGOLD MEDICAL	15m	169	3	0.28	degrad.	79
g NATURAL RUBBER	MARIGOLD ORANGE SUP	40m	22	3	0.73		79
g NATURAL RUBBER	MARIGOLD R SURGEONS	15m	169	3	0.28	degrad.	79
g NATURAL RUBBER	MARIGOLD RED LT WT	28m	156	3	0.43	degrad.	79
g NATURAL RUBBER	MARIGOLD SENSOTECH	15m	169	3	0.28	degrad.	79
g NATURAL RUBBER	MARIGOLD SUREGRIP	36m	70	3	0.58		79
g NATURAL RUBBER	MARIGOLD YELLOW	24m	160	3	0.38	degrad.	79
g NATURAL+NEOPRENE	ANSELL OMNI 276	>60m			0.45		5
g NATURAL+NEOPRENE	ANSELL OMNI 276	>360m		0	0.56		55
g NATURAL+NEOPRENE	ANSELL TECHNICIANS	>360m		0	0.43		55
g NATURAL+NEOPRENE	MARIGOLD FEATHERWT	21m	163	3	0.35	degrad.	79
g NEOPRENE	ANSELL NEOPRENE 530	>360m	0	0	0.46		55
g NEOPRENE	ANSELLEDMONT 29-840	180m	<900	3	0.38		6
g NEOPRENE	ANSELLEDMONT 29-870	>641m	0	0	0.46	85% in H_2O	34
g NEOPRENE	ANSELLEDMONT NEOX	390m	0	0	n.a.		6
g NEOPRENE	BEST 6780	>480m	0	0	n.a.	85% in H_2O	53
g NEOPRENE	MAPA-PIONEER N-44	>480m	0	0	0.56		36
g NEOPRENE	MAPA-PIONEER N-73	>60m			0.46		5

G A R M E N T	Class & Number/ test chemical/ MATERIAL NAME	MANUFACTURER & PRODUCT IDENTIFICATION	Break- through Time in Minutes	Perm- eation Rate in mg/sq m /min.	I N D E X	Thick- ness in mm	degrad. and comments	R e f s
g	NEOPRENE/NATURAL	ANSELL CHEMI-PRO	>360m	0	0	0.72		55
g	NITRILE	ANSELL CHALLENGER	>360m	0	0	0.38		55
g	NITRILE	ANSELLEDMONT 37-165	>60m			0.55		5
g	NITRILE	ANSELLEDMONT 37-165				0.60	degrad.	6
g	NITRILE	ANSELLEDMONT 37-175	32m	3000	4	0.37		5
g	NITRILE	BEST 22R		180	5	n.a.	85% in H_2O	53
g	NITRILE	COMASEC COMATRIL	30m			0.55		40
g	NITRILE	COMASEC COMATRIL SU	40m			0.60		40
g	NITRILE	COMASEC FLEXITRIL	22m	2460	4	n.a.		40
g	NITRILE	MAPA-PIONEER A-14	>480m	0	0	0.54		36
g	NITRILE	MARIGOLD NITROSOLVE				0.75	degrad.	79
g	NITRILE	NORTH LA-142G	35m	12720	5	0.33	degr., 85%	34
g	NITRILE+PVC	COMASEC MULTIPLUS	120m			n.a.		40
g	PE	ANSELLEDMONT 35-125	> 60m			0.03		5
c	PE	DU PONT TYVEK QC	<1m	4	3	n.a.		8
g	PE/EVAL/PE	SAFETY4 4H	>240m		0	0.07	35°C	60
g	PVAL	EDMONT 15-552	30m	<900	3	n.a.		6
g	PVAL	EDMONT 25-950				0.40	degrad.	34
g	PVC	ANSELLEDMONT M GRIP	75m	<90	2	1.40		6
g	PVC	BEST 812		180	5	n.a.	85% in H_2O	3
g	PVC	COMASEC MULTIPOST	90m			n.a.		40
g	PVC	COMASEC MULTITOP	105m			n.a.		40
g	PVC	COMASEC NORMAL	80m			n.a.		40
g	PVC	COMASEC OMNI	50m			n.a.		40
g	PVC	MAPA-PIONEER V-20	32m	770	3	0.31		5
s	PVC	MSA UPC	<30m			0.20		48
s	SARANEX-23	DU PONT TYVEK SARAN	>480m	0	0	n.a.		8
s	TEFLON	CHEMFAB CHALL. 5100	>180m	0	1	n.a.		65
s	UNKNOWN MATERIAL	CHEMRON CHEMREL	>480m	0	0	n.a.	85% in H_2O	67
s	UNKNOWN MATERIAL	CHEMRON CHEMREL MAX	>1440m	0	0	n.a.		67
s	UNKNOWN MATERIAL	DU PONT BARRICADE	>480m	0	0	n.a.		8
s	UNKNOWN MATERIAL	KAPPLER CPF III	85m	15	2	n.a.		77
s	UNKNOWN MATERIAL	LIFE-GUARD RESPONDE	>480m	0	0	n.a.		77
g	UNKNOWN MATERIAL	NORTH SILVERSHIELD	>180m	0	1	0.08	85% in H_2O	83
g	VITON	NORTH F-091	>900m	0	0	0.25	85% in H_2O	34
g	VITON	NORTH F-091	>900m	0	0	0.30	85% in H_2O	83
s	VITON/CHLOROBUTYL	LIFE-GUARD VC-100	>240m	0	n.a.			77
s	VITON/NEOPRENE	MSA VAUTEX	>480m	0	0	n.a.		48

phenol & methyl ethyl ketone, 30-70% & methyl isobutyl ketone, <30%
Primary Class: 830   Lacquer Products

g	BUTYL	ERISTA BX	>240m	0	0	0.69		33
g	BUTYL	ERISTA BX	>240m	0	0	0.69	preexp. 20h	33
g	NITRILE	ANSELLEDMONT 37-175	19m	2600	4	0.39		33
g	PVC	KID VINYLPRODUKTER	9m	3140	5	0.55		33

G							
A							
R			Perm-	I			
M		Break-	eation	N	Thick-		R
E Class & Number/	MANUFACTURER	through	Rate in	D	ness	degrad.	e
N test chemical/	& PRODUCT	Time in	mg/sq m	E	in	and	f
T MATERIAL NAME	IDENTIFICATION	Minutes	/min.	X	mm	comments	s

phenol, 50% & chloroform, 48% & isopentyl alcohol, 2%
Primary Class: 800   Multicomponent Mixtures With >2 Components

g PE/EVAL/PE	SAFETY4 4H	70m			0.07	35°C	60
g PE/EVAL/PE	SAFETY4 4H	130m			0.07		60

phenol, 50% & methyl ethyl ketone, 50%
Primary Class: 316   Hydroxy Compounds, Aromatic (Phenols)

phenol-formaldehyde resin in solution (methyl alcohol, 5% & water)
Primary Class: 830   Lacquer Products

g BUTYL	ERISTA BX	>240m	0	0	0.68		33
g BUTYL	ERISTA BX	>240m	0	0	0.68	preexp. 20h	33
g PVAL	EDMONT 15-552	>240m	0	0	n.a.		33
g PVAL	EDMONT 15-552	>240m	0	0	n.a.	preexp. 20h	33

phenolphthalein
CAS Number: 77-09-8
Primary Class: 316   Hydroxy Compounds, Aromatic (Phenols)

g NATURAL RUBBER	ACKWELL 5-109	>480m	0	0	0.15		34
g NEOPRENE	ANSELLEDMONT 29-870	>480m	0	0	0.41		34
g NITRILE	ANSELLEDMONT 37-155	>480m	0	0	0.39		34
g PVC	ANSELLEDMONT 34-100	>480m	0	0	0.18		4

phenolsulfonic acid (synonym: hydroxybenzenesulfonic acid)
CAS Number: 1333-39-7
Primary Class: 504   Sulfur Compounds, Sulfonic Acids

g NEOPRENE	ANSELLEDMONT 29-840	>1200m	0	0	0.45		33
g NITRILE	ERISTA SPECIAL	>240m	0	0	0.40	degrad.	33

phosgene (synonym: carbonyl chloride)
CAS Number: 75-44-5
Primary Class: 350   Inorganic Gases and Vapors
Related Class: 360   Inorganic Acid Halides

s TEFLON	CHEMFAB CHALL. 5200	>480m	0	0	n.a.		80
s UNKNOWN MATERIAL	LIFE-GUARD RESPONDE	>480m	0	0	n.a.		77

G A R M E Class & Number/ N test chemical/ T MATERIAL NAME	MANUFACTURER & PRODUCT IDENTIFICATION	Break- through Time in Minutes	Perm- eation Rate in mg/sq m /min.	I N D E X	Thick- ness degrad. in and mm comments	R e f s
phosphine CAS Number: 7803-51-2 Primary Class: 350 Inorganic Gases and Vapors						
u NEOPRENE	Unknown	25m	4783	4	0.53	44
u NEOPRENE	Unknown	10m	22167	5	n.a.	44
u PE	Unknown	7m	12717	5	0.07	44
u PE	Unknown	15m	8950	4	0.10	44
u PE	Unknown	20m	4317	4	0.30	44
u PE	Unknown	55m	8083	4	0.32	44
u PE	Unknown	4m	44667	5	0.02	44
u PE	Unknown	5m	39667	5	0.05	44
u PE	Unknown	10m	19000	5	0.07	44
u PE	Unknown	12m	12200	5	0.15	44
u PE	Unknown	25m	9083	4	n.a.	44
u PVC	Unknown	40m	5233	4	0.16	44
u PVC	Unknown	100m	4617	4	0.23	44
phosphoric acid & nitric acid (synonym: aluminum etch) Primary Class: 820 Etching Products						
g NAT+NEOPR+NITRILE	MAPA-PIONEER TRIONI	>960m	0	0	0.46	36
g NATURAL RUBBER	ANSELL CONT ENVIRON	120m	>30000	5	0.55	55
g NATURAL RUBBER	ANSELL FL 200 254	120m	>30000	5	0.51	55
g NATURAL RUBBER	ANSELL PVL 040	120m	>30000	5	0.46	55
g NATURAL+NEOPRENE	ANSELL OMNI 276	120m	>30000	5	0.56	55
g NEOPRENE	ANSELL NEOPRENE 530	120m	>30000	5	0.46	55
g NEOPRENE/NATURAL	ANSELL CHEMI-PRO	120m	>30000	5	0.72	55
phosphoric acid & nitric acid & acetic acid (synonym: slope etch) Primary Class: 820 Etching Products						
g NAT+NEOPR+NITRILE	MAPA-PIONEER TRIONI	260m	<1	1	0.46	36
phosphoric acid, >70% CAS Number: 7664-38-2 Primary Class: 370 Inorganic Acids						
g BUTYL	BEST 878	>480m	0	0	0.75	53
g BUTYL	BRUNSWICK BUTYL STD	>480m	0	0	0.63	89
g BUTYL	BRUNSWICK BUTYL-XTR	>480m		0	0.63	89
g BUTYL/ECO	BRUNSWICK BUTYL-POL	>480m	0	0	0.63	89
g NAT+NEOPR+NITRILE	MAPA-PIONEER TRIONI	>1080m	0	0	0.48	36
g NATURAL RUBBER	ANSELL CONFORM 4205	>360m	0	0	0.13	55
g NATURAL RUBBER	ANSELL FL 200 254	>360m	0	0	0.51	55
g NATURAL RUBBER	ANSELL ORANGE 208	>360m	0	0	0.76	55

G A R M E Class & Number/ N test chemical/ T MATERIAL NAME	MANUFACTURER & PRODUCT IDENTIFICATION	Break- through Time in Minutes	Perm- eation Rate in mg/sq m /min.	I N D E X	Thick- ness in mm	degrad. and comments	R e f s
g NATURAL RUBBER	ANSELL PVL 040	>360m	0	0	0.46		55
g NATURAL RUBBER	ANSELL STERILE 832	>360m	0	0	0.23		55
g NATURAL RUBBER	ANSELLEDMONT 36-124	>360m	0	0	0.46		6
g NATURAL RUBBER	ANSELLEDMONT 392	>360m		0	0.48		6
g NATURAL RUBBER	BEST 65NFW	>480m	0	0	n.a.		53
g NATURAL RUBBER	COMASEC FLEXIGUM	>480m	0	0	0.95		40
g NATURAL RUBBER	MAPA-PIONEER L-118	>480m	0	0	0.46		36
g NATURAL RUBBER	MARIGOLD BLACK HEVY	>480m	0	0	0.65		79
g NATURAL RUBBER	MARIGOLD FEATHERLIT	>480m	0	0	0.15		79
g NATURAL RUBBER	MARIGOLD MED WT EXT	>480m	0	0	0.45		79
g NATURAL RUBBER	MARIGOLD MEDICAL	>480m	0	0	0.28		79
g NATURAL RUBBER	MARIGOLD ORANGE SUP	>480m	0	0	0.73		79
g NATURAL RUBBER	MARIGOLD R SURGEONS	>480m	0	0	0.28		79
g NATURAL RUBBER	MARIGOLD RED LT WT	>480m	0	0	0.43		79
g NATURAL RUBBER	MARIGOLD SENSOTECH	>480m	0	0	0.28		79
g NATURAL RUBBER	MARIGOLD SUREGRIP	>480m	0	0	0.58		79
g NATURAL RUBBER	MARIGOLD YELLOW	>480m	0	0	0.38		79
g NATURAL+NEOPRENE	ANSELL OMNI 276	>360m	0	0	0.56		55
g NATURAL+NEOPRENE	ANSELL TECHNICIANS	>360m	0	0	0.43		55
g NATURAL+NEOPRENE	MARIGOLD FEATHERWT	>480m	0	0	0.35		79
g NATURAL+NEOPRENE	PLAYTEX ARGUS 123	>480m	0	0	n.a.		72
g NEOPRENE	ANSELL NEOPRENE 530	>120m	0	1	0.46		55
g NEOPRENE	ANSELLEDMONT 29-840	>360m	0	0	0.31		6
g NEOPRENE	ANSELLEDMONT NEOX	>360m	0	0	n.a.		6
g NEOPRENE	BEST 32	>480m	0	0	n.a.		53
g NEOPRENE	BEST 6780	>480m	0	0	n.a.		53
g NEOPRENE	COMASEC COMAPRENE	>360m	0	0	n.a.		40
g NEOPRENE	MAPA-PIONEER N-44	>480m	0	0	0.56		36
g NEOPRENE/NATURAL	ANSELL CHEMI-PRO	>360m	0	0	0.72		55
g NITRILE	ANSELL CHALLENGER	>360m	0	0	0.38		55
g NITRILE	ANSELLEDMONT 37-165	>360m	0	0	0.60		6
g NITRILE	BEST 22R	>480m	0	0	n.a.		53
g NITRILE	BEST 727	>480m	0	0	0.38		53
g NITRILE	COMASEC COMATRIL	>480m	0	0	0.55		40
g NITRILE	COMASEC COMATRIL SU	>480m	0	0	0.60		40
g NITRILE	COMASEC FLEXITRIL	>480m	0	0	n.a.		40
g NITRILE	MAPA-PIONEER A-14	>480m	0	0	0.56		36
g NITRILE	MARIGOLD BLUE	>480m	0	0	0.45		79
g NITRILE	MARIGOLD GREEN SUPA	>480m	0	0	0.30		79
g NITRILE	MARIGOLD NITROSOLVE	480m	0	0	0.75		79
g NITRILE+PVC	COMASEC MULTIPLUS	>480m	0	0	n.a.		40
c PE	DU PONT TYVEK QC	>840m	0	0	n.a.		8
g PE/EVAL/PE	SAFETY4 4H	>240m		0	0.07	35°C	60
g PVAL	ANSELLEDMONT PVA				n.a.	degrad.	6
g PVAL	EDMONT 15-552				n.a.	degrad.	6
g PVC	ANSELLEDMONT M GRIP	>360m	0	0	1.40		6
g PVC	BEST 725R	>480m	0	0	n.a.		53
g PVC	COMASEC MULTIPOST	>480m	0	0	n.a.		40

G A R M E Class & Number/ N test chemical/ T MATERIAL NAME	MANUFACTURER & PRODUCT IDENTIFICATION	Break- through Time in Minutes	Perm- eation Rate in mg/sq m /min.	I N D E X	Thick- ness in mm	degrad. and comments	R e f s
g PVC	COMASEC MULTITOP	>480m	0	0	n.a.		40
g PVC	COMASEC NORMAL	>480m	0	0	n.a.		40
g PVC	COMASEC OMNI	>480m	0	0	n.a.		40
g PVC	MAPA-PIONEER V-20	>480m	0	0	0.51		36
s SARANEX-23	DU PONT TYVEK SARAN	>840m	0	0	n.a.		8
c UNKNOWN MATERIAL	CHEMRON CHEMTUFF	>480m	0	0	n.a.		67
s UNKNOWN MATERIAL	DU PONT BARRICADE	>480m		0	n.a.		8

phosphorus oxychloride (synonym: phosphoryl chloride)
CAS Number: 10025-87-3
Primary Class: 370   Inorganic Acids

u CPE	ILC DOVER	50m			n.a.	degrad.	9
u NEOPRENE	Unknown	<1m			n.a.	degrad.	9
u NEOPRENE	Unknown	33m			0.61	degrad.	9
u NEOPRENE	Unknown	>60m			0.94		9
u NITRILE+PVC	Unknown	29m			n.a.	degrad.	9
c PE	DU PONT TYVEK QC	6m			0.15		9
g PE/EVAL/PE	SAFETY4 4H	>240m		0	0.07	35°C	60
u PVC	Unknown	<1m			n.a.	degrad.	9
s SARANEX-23	DU PONT TYVEK SARAN	50m			0.15		9
s TEFLON	CHEMFAB CHALL. 5100	>180m	0	1	n.a.		65
g VITON	NORTH F-121	16m			0.31		9

phosphorus tribromide
CAS Number: 7789-60-8
Primary Class: 360   Inorganic Acid Halides

s PVC	TRELLEBORG TRELLCHE				n.a.	degrad.	71

phosphorus trichloride
CAS Number: 7719-12-2
Primary Class: 360   Inorganic Acid Halides

s PVC	TRELLEBORG TRELLCHE				n.a.	degrad.	71
s TEFLON	CHEMFAB CHALL. 5100	>180m	0	1	n.a.		65

Photo Resist 1450 (synonym: 2-ethoxyethanol & xylene & butyl acetate)
Primary Class: 900   Miscellaneous Unclassified Chemicals
Related Class: 222   Esters, Carboxylic, Acetates
                245   Ethers, Glycols
                292   Hydrocarbons, Aromatic

g PE/EVAL/PE	SAFETY4 4H	>240m	0	0	0.07	35°C	60

G A R M E Class & Number/ N test chemical/ T MATERIAL NAME	MANUFACTURER & PRODUCT IDENTIFICATION	Break- through Time in Minutes	Perm- eation Rate in mg/sq m /min.	I N D E X	Thick- ness in mm	degrad. and comments	R e f s
phthalic acid anhydride							
CAS Number: 85-44-9							
Primary Class: 161 Anhydrides, Aliphatic and Alicyclic							
g PE/EVAL/PE	SAFETY4 4H	>240m	0		0.07		60
picric acid							
CAS Number: 88-89-1							
Primary Class: 442 Nitro Compounds, Substituted							
Related Class: 316 Hydroxy Compounds, Aromatic (Phenols)							
g NATURAL RUBBER	ANSELLEDMONT 36-124	3m	<90	4	0.46		6
g NEOPRENE	ANSELLEDMONT 29-840	150m	<90	2	0.31		6
g NEOPRENE	ANSELLEDMONT NEOX	180m	<90	2	n.a.		6
g NITRILE	ANSELLEDMONT 37-165	162m	<90	2	0.60		6
g PVAL	ANSELLEDMONT PVA				n.a.	degrad.	6
g PVAL	EDMONT 15-552				n.a.	degrad.	6
g PVC	ANSELLEDMONT M GRIP	40m	<90	3	n.a.		6
pigment yellow 74 (synonym: C.I. Pigment Yellow 74)							
CAS Number: 6358-31-2							
Primary Class: 170 Azo and Azoxy Compounds							
g NATURAL RUBBER	ANSELL CANNERS 334	>480m	0	0	n.a.		34
g NEOPRENE	ANSELLEDMONT 29-870	>480m	0	0	0.46		34
pigment yellow 74 (synonym: C.I. Pigment Yellow 74)							
CAS Number: 110-89-4							
Primary Class: 170 Azo and Azoxy Compounds							
Related Class: 274 Heterocyclic Compounds, Nitrogen, Others							
u BUTYL	Unknown	18m			0.29	degrad.	49
u NEOPRENE	Unknown	40m			0.52	degrad.	49
u PVC	Unknown	42m			0.48	degrad.	49
piperazine (synonym: diethylenediamine)							
CAS Number: 110-85-0							
Primary Class: 148 Amines, Poly, Aliphatic and Alicyclic							
Related Class: 142 Amines, Aliphatic and Alicyclic, Secondary							
274 Heterocyclic Compounds, Nitrogen, Others							
g BUTYL	NORTH B-174	65m	460	3	0.58	degrad.	34
g NATURAL RUBBER	ACKWELL 5-109				0.15	degrad.	34
g NEOPRENE	ANSELLEDMONT 29-870				0.48	degrad.	34
g NITRILE	ANSELLEDMONT 37-155	108m	895	3	0.36	degrad.	34
g PVC	ANSELLEDMONT 34-100	2m	19260	5	0.20		34
g VITON	NORTH F-091	26m	17230	5	0.33	degrad.	34

G A R M E Class & Number/ N test chemical/ T MATERIAL NAME	MANUFACTURER & PRODUCT IDENTIFICATION	Break- through Time in Minutes	Perm- eation Rate in mg/sq m /min.	I N D E X	Thick- ness in mm	degrad. and comments	R e f s

piperidine (synonym: hexahydropyridine)
CAS Number: 110-89-4
Primary Class: 142   Amines, Aliphatic and Alicyclic, Secondary

u BUTYL	Unknown	18m			0.29	degrad.	49
u NEOPRENE	Unknown	40m			0.52	degrad.	49
u PVC	Unknown	42m			0.48	degrad.	49

polyamide (accelerator), 30-70% & isobutyl alcohol, isopropyl alcohol, MIBK
Primary Class: 810   Epoxy Products

g PVAL	EDMONT 15-552	>240m	0	0	n.a.		33

polychlorinated biphenyls (synonym: PCBs)
CAS Number: 40817-08-1
Primary Class: 264   Halogen Compounds, Aromatic

g BUTYL	NORTH B-161	1440m		0	0.38		24
u CPE	ILC DOVER	>180m			n.a.		29
u NATURAL RUBBER	Unknown	30m			0.25		24
g NEOPRENE	ANSELLEDMONT 29-840	1440m		0	0.43		24
g NEOPRENE	MAPA-PIONEER N-44	>480m	0	0	0.56		36
g NITRILE	ANSELLEDMONT 37-145	>60m			0.32		30
g NITRILE	MAPA-PIONEER A-14	342m	2160	4	0.56		36
g NITRILE	SURETY	>480m	0	0	0.38		24
g PE	ANSELLEDMONT 35-125	60m			0.03		24
c PE	DU PONT TYVEK QC	41m	36	3	n.a.		8
s SARANEX-23	DU PONT TYVEK SARAN	360m	0	0	n.a.		24
g TEFLON	CLEAN ROOM PRODUCTS	>1440m	0	0	0.05		24
s UNKNOWN MATERIAL	LIFE-GUARD RESPONDE	>480m	0	0	n.a.		77
g UNKNOWN MATERIAL	NORTH SILVERSHIELD	>480m	0	0	0.10		7
g VITON	NORTH F-091	>1440m	0	0	0.24		24
g VITON	NORTH SF	>1440m	0	0	0.24		24
s VITON/CHLOROBUTYL	LIFE-GUARD VC-100	>240m	0	0	n.a.		77
u VITON/CHLOROBUTYL	Unknown	>180m			n.a.		29
g VITON/NITRILE	NORTH VITRILE	>1440m	0	0	0.20		24

polychlorinated biphenyls, 58% & 1,2,4-trichlorobenzene, 42%
Primary Class: 264   Halogen Compounds, Aromatic

g BUTYL	NORTH B-161	180m			0.38		24
u NATURAL RUBBER	Unknown	5m			0.23		24
g NEOPRENE	ANSELLEDMONT 29-840	120m			0.43		24
g PE	ANSELLEDMONT 35-125	5m			0.03		24
s SARANEX-23	DU PONT TYVEK SARAN	60m	<1	1	n.a.		8
g TEFLON	CLEAN ROOM PRODUCTS	>1440m	0	0	0.05		24

Class & Number/ test chemical/ MATERIAL NAME	MANUFACTURER & PRODUCT IDENTIFICATION	Break- through Time in Minutes	Perm- eation Rate in mg/sq m /min.	I N D E X	Thick- ness in mm	degrad. and comments	R e f s
s UNKNOWN MATERIAL	CHEMRON CHEMREL MAX	>1440m	0	0	n.a.		67
c UNKNOWN MATERIAL	CHEMRON CHEMTUFF	>480m	0	0	n.a.		67
s UNKNOWN MATERIAL	DU PONT BARRICADE	>480m	0	0	n.a.		8
g VITON	NORTH F-091	240m	0	0	0.23		24

polymer 14435W8
Primary Class: 900   Miscellaneous Unclassified Chemicals

g PE/EVAL/PE	SAFETY4 4H	>240m		0	0.07		60

Posistrip LE (synonym: morpholine & butyrolactone & N-methyl-2-pyrrolidone)
Primary Class: 900   Miscellaneous Unclassified Chemicals
Related Class: 132   Amides, Aliphatic and Alicyclic
              142   Amines, Aliphatic and Alicyclic, Secondary

g PE/EVAL/PE	SAFETY4 4H	>480m	0	0	0.07	35°C	60

potassium acetate (saturated solution)
CAS Number: 127-08-2
Primary Class: 340 Inorganic Salts

s SARANEX-23	DU PONT TYVEK SARAN	>480m	0		n.a.		8
s UNKNOWN MATERIAL	DU PONT BARRICADE	>480m	0		n.a.		8

potassium chromate (saturated solution)
CAS Number: 7789-00-6
Primary Class: 340   Inorganic Salts

s SARANEX-23	DU PONT TYVEK SARAN	>480m		0	n.a.		8
s UNKNOWN MATERIAL	DU PONT BARRICADE	>480m		0	n.a.		8

potassium hydroxide & butyl alcohol & propyl alcohol (synonym: KOH etch)
Primary Class: 820   Etching Products

g NAT+NEOPR+NITRILE	MAPA-PIONEER TRIONI	278m	<1	1	0.46		36

potassium hydroxide, 30-70%
CAS Number: 1310-58-3
Primary Class: 380   Inorganic Bases

g BUTYL	BEST 878	>480m	0	0	0.75		53
g BUTYL	BRUNSWICK BUTYL STD	>480m	0	0	0.63		89
g BUTYL	BRUNSWICK BUTYL-XTR	>480m	0	0	0.63		89
s BUTYL	MSA CHEMPRUF	>480m	0	0	n.a.		48
g BUTYL/ECO	BRUNSWICK BUTYL-POL	>480m	0	0	0.63		89

G A R M E Class & Number/ N test chemical/ T MATERIAL NAME	MANUFACTURER & PRODUCT IDENTIFICATION	Break-through Time in Minutes	Perm-eation Rate in mg/sq m /min.	I N D E X	Thick-ness in mm	degrad. and comments	R e f s
g NATURAL RUBBER	ANSELL CONFORM 4205	>360m	0		0.13		55
g NATURAL RUBBER	ANSELL FL 200 254	>360m	0		0.51		55
g NATURAL RUBBER	ANSELL ORANGE 208	>360m	0		0.76		55
g NATURAL RUBBER	ANSELL PVL 040	>360m	0		0.46		55
g NATURAL RUBBER	ANSELL STERILE 832	>360m	0		0.23		55
g NATURAL RUBBER	ANSELLEDMONT 36-124	>360m	0	0	0.46		6
g NATURAL RUBBER	ANSELLEDMONT 392	>360m		0	0.48		6
g NATURAL RUBBER	COMASEC FLEXIGUM	126m	1080	4	0.95		40
g NATURAL RUBBER	MAPA-PIONEER L-118	>480m	0	0	0.46		36
g NATURAL RUBBER	MARIGOLD BLACK HEVY	>480m	0	0	0.65		79
g NATURAL RUBBER	MARIGOLD FEATHERLIT	429m	34	2	0.15		79
g NATURAL RUBBER	MARIGOLD MED WT EXT	>480m	0	0	0.45		79
g NATURAL RUBBER	MARIGOLD MEDICAL	450m	20	2	0.28		79
g NATURAL RUBBER	MARIGOLD ORANGE SUP	>480m	0	0	0.73		79
g NATURAL RUBBER	MARIGOLD R SURGEONS	450m	20	2	0.28		79
g NATURAL RUBBER	MARIGOLD RED LT WT	476m	3	1	0.43		79
g NATURAL RUBBER	MARIGOLD SENSOTECH	450m	20	2	0.28		79
g NATURAL RUBBER	MARIGOLD SUREGRIP	>480m	0	0	0.58		79
g NATURAL RUBBER	MARIGOLD YELLOW	467m	8	1	0.38		79
g NATURAL+NEOPRENE	ANSELL OMNI 276	>360m		0	0.56		55
g NATURAL+NEOPRENE	ANSELL TECHNICIANS	>360m		0	0.43		55
g NATURAL+NEOPRENE	MARIGOLD FEATHERWT	463m	11	2	0.35		79
g NATURAL+NEOPRENE	PLAYTEX ARGUS 123	>480m	0	0	n.a.		72
g NEOPRENE	ANSELL NEOPRENE 530	>360m		0	0.46		55
g NEOPRENE	ANSELLEDMONT 29-840	>360m	0	0	0.31		6
g NEOPRENE	ANSELLEDMONT NEOX	>360m	0	0	n.a.		6
g NEOPRENE	BEST 32	>480m	0	0	n.a.		53
g NEOPRENE	BEST 6780	>480m	0	0	n.a.		53
g NEOPRENE	BRUNSWICK NEOPRENE	>480m	0	0	0.88		89
g NEOPRENE	MAPA-PIONEER N-44	>480m	0	0	0.56		36
g NEOPRENE/NATURAL	ANSELL CHEMI-PRO	>360m		0	0.72		55
g NITRILE	ANSELL CHALLENGER	>360m		0	0.38		55
g NITRILE	ANSELLEDMONT 37-165	>360m	0	0	0.60		6
g NITRILE	COMASEC COMATRIL	>480m	0	0	0.55		40
g NITRILE	COMASEC COMATRIL SU	>480m	0	0	0.60		40
g NITRILE	COMASEC FLEXITRIL	70m	24540	5	n.a.		40
g NITRILE	MAPA-PIONEER A-14	>480m	0	0	0.56		36
g NITRILE	MARIGOLD BLUE	>480m	0	0	0.45		79
g NITRILE	MARIGOLD GREEN SUPA	>480m	0	0	0.30		79
g NITRILE	MARIGOLD NITROSOLVE	>480m	0	0	0.75		79
g NITRILE+PVC	COMASEC MULTIPLUS	>480m	0	0	n.a.		40
g PE/EVAL/PE	SAFETY4 4H	>240m	0	0	0.07	35°C	60
g PVAL	ANSELLEDMONT PVA				n.a.	degrad.	6
g PVAL	EDMONT 15-552				n.a.	degrad.	6
g PVC	ANSELLEDMONT SNORKE	>360m	0	0	1.40		6
g PVC	COMASEC MULTIPOST	>480m	0	0	n.a.		40
g PVC	COMASEC MULTITOP	>480m	0	0	n.a.		40
g PVC	COMASEC NORMAL	>480m	0	0	n.a.		40

G A R M E Class & Number/ N test chemical/ T MATERIAL NAME	MANUFACTURER & PRODUCT IDENTIFICATION	Break- through Time in Minutes	Perm- eation Rate in mg/sq m /min.	I N D E X	Thick- ness in mm	degrad. and comments	R e f s
g PVC	COMASEC OMNI	>480m	0	0	n.a.		40
g PVC	MAPA-PIONEER V-20	>480m	0	0	0.51		36
s PVC	MSA UPC	>480m	0	0	0.20		48

potassium iodide (saturated solution)
CAS Number: 7681-11-0
Primary Class: 340   Inorganic Salts

g NATURAL RUBBER	MARIGOLD BLACK HEVY	>480m	0	0	0.65		79
g NATURAL RUBBER	MARIGOLD FEATHERLIT	>480m	0	0	0.15		79
g NATURAL RUBBER	MARIGOLD MED WT EXT	>480m	0	0	0.45		79
g NATURAL RUBBER	MARIGOLD MEDICAL	>480m	0	0	0.28		79
g NATURAL RUBBER	MARIGOLD ORANGE SUP	>480m	0	0	0.73		79
g NATURAL RUBBER	MARIGOLD R SURGEONS	>480m	0	0	0.28		79
g NATURAL RUBBER	MARIGOLD RED LT WT	>480m	0	0	0.43		79
g NATURAL RUBBER	MARIGOLD SENSOTECH	>480m	0	0	0.28		79
g NATURAL RUBBER	MARIGOLD SUREGRIP	>480m	0	0	0.58		79
g NATURAL RUBBER	MARIGOLD YELLOW	>480m	0	0	0.38		79
g NATURAL+NEOPRENE	MARIGOLD FEATHERWT	>480m	0	0	0.35		79
g NITRILE	MARIGOLD BLUE	>480m	0	0	0.45		79
g NITRILE	MARIGOLD GREEN SUPA	>480m	0	0	0.30		79
g NITRILE	MARIGOLD NITROSOLVE	>480m	0	0	0.75		79
g UNKNOWN MATERIAL	MARIGOLD R MEDIGLOV	>480m	0	0	0.03		79

potassium permanganate
CAS Number: 7722-64-7
Primary Class: 340   Inorganic Salts

g PE/EVAL/PE	SAFETY4 4H	>240m	0	0	0.07	sat. 35°	60

Pramitol
CAS Number: 1610-18-0
Primary Class: 840   Pesticides, Mixtures and Formulations

g PE/EVAL/PE	SAFETY4 4H	>240m	0	0	0.07	35°C	60

Pro Strip (mixture)
Primary Class: 900   Miscellaneous Unclassified Chemicals

g PE/EVAL/PE	SAFETY4 4H	>240m		0	0.07		60

promethazine hydrochloride
CAS Number: 58-33-3
Primary Class: 550   Organic Salts (Solutions)
Related Class: 143   Amines, Aliphatic and Alicyclic, Tertiary
                274   Heterocyclic Compounds, Nitrogen, Others

G A R M E Class & Number/ N test chemical/ T MATERIAL NAME	MANUFACTURER & PRODUCT IDENTIFICATION	Break- through Time in Minutes	Perm- eation Rate in mg/sq m /min.	I N D E X	Thick- ness in mm	degrad. and comments	R e f s
g BUTYL	NORTH B-174	>480m	0	0	0.62		34
g NATURAL RUBBER	ACKWELL 5-109	>480m	0	0	0.18		34
g NEOPRENE	ANSELLEDMONT 29-870	>480m	0	0	0.50		34
g PVC	ANSELLEDMONT 34-100	>480m	0	0	0.20		34

propane
CAS Number: 74-98-6
Primary Class: 291   Hydrocarbons, Aliphatic and Alicyclic, Saturated

s UNKNOWN MATERIAL	LIFE-GUARD RESPONDE	>480m		0	n.a.		77

propionaldehyde (synonym: propanal)
CAS Number: 123-38-6
Primary Class: 121   Aldehydes, Aliphatic and Alicyclic

g BUTYL	NORTH B-174	>780m	0	0	0.64		34
g NEOPRENE	ANSELLEDMONT 29-870	12m	678	4	0.48		34
c PE	DU PONT TYVEK QC	5m	762	4	n.a.		8
g PVAL	EDMONT 25-950	<1m	270	4	0.36		34
s UNKNOWN MATERIAL	LIFE-GUARD RESPONDE	>480m	0	0	n.a.		77
g VITON	NORTH F-091	<1m	852	4	0.28	degrad.	34

propionic acid
CAS Number: 79-09-4
Primary Class: 102   Acids, Carboxylic, Aliphatic and Alicyclic, Unsubstituted

c PE	DU PONT TYVEK QC	3m	16	4	n.a.		8
s TEFLON	CHEMFAB CHALL. 5100	>180m			n.a.		65

propionitrile (synonym: ethyl cyanide)
CAS Number: 107-12-0
Primary Class: 431   Nitriles, Aliphatic and Alicyclic

g BUTYL	NORTH B-174	24m	1674	4	0.62		34
g NATURAL RUBBER	ACKWELL 5-109	<1m	792	4	0.10		34
g NITRILE	ANSELLEDMONT 37-155				0.40	degrad.	34
g PVAL	EDMONT 15-552	>480m	0	0	0.25		34
g PVC	ANSELLEDMONT 34-100 <	1m	180	4	0.15		34
g VITON	NORTH F-091				0.26	degrad.	34

propiophenone (synonyms: 1-phenyl-1-propanone; ethyl phenyl ketone)
CAS Number: 93-55-0
Primary Class: 392   Ketones, Aromatic

g PE/EVAL/PE	SAFETY4 4H	>240m	0	0	0.07	35°C	60

G A R M E Class & Number/ N test chemical/ T MATERIAL NAME	MANUFACTURER & PRODUCT IDENTIFICATION	Break- through Time in Minutes	Perm- eation Rate in mg/sq m /min.	I N D E X	Thick- ness in mm	degrad. and comments	R e f s
propyl acetate							
CAS Number: 109-60-4							
Primary Class: 222 Esters, Carboxylic, Acetates							
g BUTYL	BRUNSWICK BUTYL STD	109m	190	3	0.63		89
g BUTYL	BRUNSWICK BUTYL-XTR	123m	340	3	0.63		89
g BUTYL	NORTH B-161	162m	172	3	0.45		7
g BUTYL/ECO	BRUNSWICK BUTYL-POL	88m	370	3	0.63		89
g NATURAL RUBBER	ANSELL CONFORM 4205	1m	155	4	0.13		55
g NATURAL RUBBER	ANSELL FL 200 254	5m	5	3	0.51		55
g NATURAL RUBBER	ANSELL ORANGE 208	9m	4	3	0.76		55
g NATURAL RUBBER	ANSELL PVL 040	10m	4	3	0.46		55
g NATURAL RUBBER	ANSELL STERILE 832	1m	4	3	0.23		55
g NATURAL RUBBER	ANSELLEDMONT 36-124	5m	<9000	5	0.46		6
g NATURAL RUBBER	ANSELLEDMONT 392				0.48	degrad.	6
g NATURAL+NEOPRENE	ANSELL OMNI 276	6m	5	3	0.56		55
g NATURAL+NEOPRENE	ANSELL TECHNICIANS	4m	6	3	0.43		55
g NEOPRENE	ANSELL NEOPRENE 530	8m	4	3	0.46		55
g NEOPRENE	ANSELLEDMONT 29-840				0.38	degrad.	6
g NEOPRENE/NATURAL	ANSELL CHEMI-PRO	7m	5	3	0.72		55
g NITRILE	ANSELL CHALLENGER	27m	4	2	0.38		55
g NITRILE	ANSELLEDMONT 37-165	20m	<900	3	0.60		6
g NITRILE	NORTH LA-142G	17m	4350	4	0.36		7
g PE/EVAL/PE	SAFETY4 4H	>240m	0		0.07		60
g PVAL	ANSELLEDMONT PVA	120m	<90	2	n.a.		6
g PVAL	EDMONT 15-552	120m	<90	2	n.a.		6
g PVAL	EDMONT 25-545	>360m	0	0	n.a.		6
g PVC	ANSELLEDMONT SNORKE				n.a.	degrad.	6
g UNKNOWN MATERIAL	NORTH SILVERSHIELD	>360m	0	0	0.08		7
g VITON	NORTH F-091				0.25	degrad.	7
propyl alcohol (synonyms: 1-propanol; n-propanol)							
CAS Number: 71-23-8							
Primary Class: 311 Hydroxy Compounds, Aliphatic and Alicyclic, Primary							
g BUTYL	BEST 878	>480m	0	0	0.75		53
g NATURAL RUBBER	ANSELL CONFORM 4205	3m	4	3	0.13		55
g NATURAL RUBBER	ANSELL FL 200 254	32m	1	2	0.51		55
g NATURAL RUBBER	ANSELL ORANGE 208	>360m	0	0	0.76		55
g NATURAL RUBBER	ANSELL PVL 040	70m	0.1		0.46		55
g NATURAL RUBBER	ANSELL STERILE 832	8m	0.2		0.23		55
g NATURAL RUBBER	ANSELLEDMONT 36-124	20m	<90	3	0.46		6
g NATURAL RUBBER	ANSELLEDMONT 392	20m	<90	3	0.48		6
g NATURAL RUBBER	COMASEC FLEXIGUM	186m	48	2	0.95		40
g NATURAL RUBBER	MARIGOLD BLACK HEVY	20m	42	3	0.65		79
g NATURAL RUBBER	MARIGOLD FEATHERLIT	1m	162	4	0.15		79
g NATURAL RUBBER	MARIGOLD MED WT EXT	10m	53	4	0.45		79

G A R M E Class & Number/ N test chemical/ T MATERIAL NAME	MANUFACTURER & PRODUCT IDENTIFICATION	Break-through Time in Minutes	Perm-eation Rate in mg/sq m /min.	I N D E X	Thick-ness in mm	degrad. and comments	R e f s
g NATURAL RUBBER	MARIGOLD MEDICAL	5m	117	4	0.28		79
g NATURAL RUBBER	MARIGOLD ORANGE SUP	21m	41	3	0.73		79
g NATURAL RUBBER	MARIGOLD R SURGEONS	5m	117	4	0.28		79
g NATURAL RUBBER	MARIGOLD RED LT WT	10m	62	4	0.43		79
g NATURAL RUBBER	MARIGOLD SENSOTECH	5m	117	4	0.28		79
g NATURAL RUBBER	MARIGOLD SUREGRIP	17m	45	3	0.58		79
g NATURAL RUBBER	MARIGOLD YELLOW	8m	80	4	0.38		79
g NATURAL+NEOPRENE	ANSELL OMNI 276	70m	0.1		0.56		55
g NATURAL+NEOPRENE	ANSELL TECHNICIANS	26m	0.1		0.43		55
g NATURAL+NEOPRENE	MARIGOLD FEATHERWT	7m	89	4	0.35		79
g NEOPRENE	ANSELL NEOPRENE 530	>360m	0	0	0.46		55
g NEOPRENE	ANSELLEDMONT 29-840	150m	<9	1	0.31		6
g NEOPRENE	ANSELLEDMONT NEOX	>360m	0	0	n.a.		6
g NEOPRENE	BEST 32	94m	10	2	n.a.		53
g NEOPRENE	BEST 6780	>480m	0	0	n.a.		53
g NEOPRENE	COMASEC COMAPRENE	90m	60	2	n.a.		40
g NEOPRENE/NATURAL	ANSELL CHEMI-PRO	>360m	0	0	0.72		55
g NITRILE	ANSELL CHALLENGER	>360m	0	0	0.38		55
g NITRILE	ANSELLEDMONT 37-165	>360m	0	0	0.60		6
g NITRILE	BEST 22R	362m	0.30	0	n.a.		53
g NITRILE	BEST 727	>480m	0	0	0.38		53
g NITRILE	COMASEC COMATRIL	330m	150	3	0.55		40
g NITRILE	COMASEC COMATRIL SU	>480m	0	0	0.60		40
g NITRILE	COMASEC FLEXITRIL	90m	90	2	n.a.		40
g NITRILE	MARIGOLD BLUE	>480m	0	0	0.45		79
g NITRILE	MARIGOLD GREEN SUPA	>480m	0	0	0.30		79
g NITRILE	MARIGOLD NITROSOLVE	>480m	0	0	0.75		79
g NITRILE+PVC	COMASEC MULTIPLUS	>360m	0	0	n.a.		40
g PE/EVAL/PE	SAFETY4 4H	>240m	0	0	0.07	35°C	60
g PVAL	ANSELLEDMONT PVA				n.a.	degrad.	6
g PVAL	EDMONT 15-552				n.a.	degrad.	6
g PVC	ANSELLEDMONT M GRIP	90m	<90	2	n.a.		6
g PVC	COMASEC MULTIPOST	126m	132	3	n.a.		40
g PVC	COMASEC MULTITOP	270m	114	3	n.a.		40
g PVC	COMASEC NORMAL	126m	120	3	n.a.		40
g PVC	COMASEC OMNI	90m	150	3	n.a.		40
s TEFLON	CHEMFAB CHALL. 5100	>180m	0	1	n.a.		65
s UNKNOWN MATERIAL	CHEMRON CHEMREL	>1440m	0	0	n.a.		67
s UNKNOWN MATERIAL	CHEMRON CHEMREL	>1440m	0	0	n.a.		67
g UNKNOWN MATERIAL	MARIGOLD R MEDIGLOV	1m	190	4	0.03	degrad.	79
g VITON	NORTH F-091	1020m	49	2	n.a.		62

propyl methacrylate (synonym: n-propyl 2-methylpropenoate)
CAS Number: 2210-28-8
Primary Class: 223   Esters, Carboxylic, Acrylates and Methacrylates

g BUTYL	NORTH B-174	410m	480	3	0.50	degrad.	34
g NATURAL RUBBER	ACKWELL 5-109				0.15	degrad.	34
g NEOPRENE	ANSELLEDMONT 29-870				0.48	degrad.	34

G A R M E Class & Number/ N test chemical/ T MATERIAL NAME	MANUFACTURER & PRODUCT IDENTIFICATION	Break- through Time in Minutes	Perm- eation Rate in mg/sq m /min.	I N D E X	Thick- ness in mm	degrad. and comments	R e f s
g NITRILE	ANSELLEDMONT 37-155	60m	1500	4	0.36	degrad.	34
g PVAL	EDMONT 25-950	>480m	0	0	0.56		34
g PVC	ANSELLEDMONT 34-100	2m	4620	5	0.20	degrad.	34
g VITON	NORTH F-091				0.26	degrad.	34

propylamine (synonym: monopropylamine)
CAS Number: 107-10-8
Primary Class: 141   Amines, Aliphatic and Alicyclic, Primary

u CPE	ILC DOVER	9m			n.a.		29
s TEFLON	CHEMFAB CHALL. 5100	>610m	0	0	n.a.		65
u VITON/CHLOROBUTYL	Unknown	18m			n.a.		29

propylene glycol (synonym: 1,2-propanediol)
CAS Number: 57-55-6
Primary Class: 314   Hydroxy Compounds, Aliphatic and Alicyclic, Polyols

g PE/EVAL/PE	SAFETY4 4H	>240m	0	0	0.07	35°C	60

propylene glycol monoethylether acetate
CAS Number: 19234-20-9
Primary Class: 245   Ethers, Glycols

g PE/EVAL/PE	SAFETY4 4H	>240m	0	0	0.07	35°C	60

propylenediamine (synonym: 1,2-diaminopropane)
CAS Number: 78-90-0
Primary Class: 148   Amines, Poly, Aliphatic and Alicyclic
Related Class: 141   Amines, Aliphatic and Alicyclic, Primary

g BUTYL	NORTH B-174	>480m	0	0	0.53		34
g NEOPRENE	ANSELLEDMONT 29-870	>480m	0	0	0.38		34
g PVAL	EDMONT 25-950				0.40	degrad.	34
g PVC	ANSELLEDMONT 34-100	18m	56	3	0.20		34
g VITON	NORTH F-091	>480m	0	0	0.24		34

propyzamide, <30% (synonym: KERB 5O R)
CAS Number: 23950-58-5
Primary Class: 135 Amides, Aromatic, Others
Related Class: 840 Pesticides, Mixtures and Formulations

g PE/EVAL/PE	SAFETY4 4H	>240m	0	0	0.07	35°C	60

pure cutting fluid (no emulsions) with metals
Primary Class: 850   Cutting Fluids

g NEOPRENE	NOLATO 1505	112m	26000	5	0.40		12

G A R M E Class & Number/ N test chemical/ T MATERIAL NAME	MANUFACTURER & PRODUCT IDENTIFICATION	Break- through Time in Minutes	Perm- eation Rate in mg/sq m /min.	I N D E X	Thick- ness in mm	degrad. and comments	R e f s
u NITRILE	NOLATO	>300m	0	0	0.37		12
g PVC	KID VINYLPRODUKTER	101m	17000	5	0.50		12

pyridine (synonym: azine)
CAS Number: 110-86-1
Primary Class: 271   Heterocyclic Compounds, Nitrogen, Pyridines

g BUTYL	NORTH B-174	>480m	0	0	0.63		34
g NATURAL RUBBER	ACKWELL 5-109				n.a.	degrad.	34
g NATURAL RUBBER	ANSELLEDMONT 36-124	5m	<9000	5	0.46		6
g NATURAL RUBBER	ANSELLEDMONT 392	10m	<9000	5	0.48		6
g NATURAL RUBBER	ANSELLEDMONT 46-320	2m	7000	5	0.31		5
g NATURAL+NEOPRENE	ANSELL OMNI 276	8m	4000	5	0.45		5
g NEOPRENE	ANSELLEDMONT 29-840				0.38	degrad.	6
g NEOPRENE	ANSELLEDMONT 29-870	15m	6780	4	0.48	degrad.	34
g NEOPRENE	ANSELLEDMONT NEOX	39m	2000	4	n.a.		5
g NEOPRENE	ANSELLEDMONT NEOX				n.a.	degrad.	6
g NEOPRENE	MAPA-PIONEER N-73	26m	7000	4	0.46		5
g NITRILE	ANSELLEDMONT 37-155				0.40	degrad.	34
g NITRILE	ANSELLEDMONT 37-165	15m	30000	5	0.55		5
g NITRILE	ANSELLEDMONT 37-165				0.60	degrad.	6
g NITRILE	ANSELLEDMONT 37-175	9m	35000	5	0.37		5
g PE	ANSELLEDMONT 35-125	>60m			0.03		5
g PE/EVAL/PE	SAFETY4 4H	>240m	0	0	0.07	35°C	60
g PE/EVAL/PE	SAFETY4 4H	>480m	0	0	0.07		60
g PVAL	ANSELLEDMONT PVA	10m	<9000	5	n.a.		6
g PVAL	EDMONT 15-552	50m	<900	3	n.a.		6
g PVAL	EDMONT 25-950				0.62	degrad.	34
g PVC	ANSELLEDMONT 34-100				0.20	degrad.	34
g PVC	ANSELLEDMONT SNORKE				n.a.	degrad.	6
g PVC	MAPA-PIONEER V-20	<1m	>90000	5	0.31		5
s UNKNOWN MATERIAL	LIFE-GUARD RESPONDE	>480m	0	0	n.a.		77
g VITON	NORTH F-091	38m	4416	4	0.25	degrad.	34

Pyrotec HFD46
Primary Class: 900   Miscellaneous Unclassified Chemicals

g PE/EVAL/PE	SAFETY4 4H	>240m		0	0.07		60

quinoline (synonym: 1-azanaphthalene)
CAS Number: 91-22-5
Primary Class: 274   Heterocyclic Compounds, Nitrogen, Others

g PE/EVAL/PE	SAFETY4 4H	>240m	0	0	0.07	35°C	60

G A R M E Class & Number/ N test chemical/ T MATERIAL NAME	MANUFACTURER & PRODUCT IDENTIFICATION	Break- through Time in Minutes	Perm- eation Rate in mg/sq m /min.	I N D E X	Thick- ness in mm	degrad. and comments	R e f s
Round Up R solution (synonym: N,N-bis(phosphonemethyl)glycine isopropylamine) CAS Number: 1071-83-6 Primary Class: 462 Organophosphorus Compounds and Derivatives of Phosphorus-based Acids							
g PE/EVAL/PE	SAFETY4 4H	>240m	0	0	0.07		60
Sadofoss primer 17 Primary Class: 900 Miscellaneous Unclassified Chemicals							
g PE/EVAL/PE	SAFETY4 4H	>240m	0	0	0.07	35°C	60
Sadofoss primer 513 Primary Class: 900 Miscellaneous Unclassified Chemicals							
g PE/EVAL/PE	SAFETY4 4H	64m			0.07	35°C	60
sec-butyl alcohol (synonyms: 2-butanol; sec-butanol) CAS Number: 78-92-2 Primary Class: 312 Hydroxy Compounds, Aliphatic and Alicyclic, Secondary							
g NITRILE+PVC	COMASEC MULTIMAX	>480m	0	0	n.a.		40
g PE/EVAL/PE	SAFETY4 4H	>240m	0	0	0.07	35°C	60
g PE/EVAL/PE	SAFETY4 4H	>480m	0	0	0.07		60
sec-butylamine (synonym: 2-methylpropylamine) CAS Number: 13952-84-6 Primary Class: 141 Amines, Aliphatic and Alicyclic, Primary							
g BUTYL	NORTH B-174	161m	1105	4	0.68	degrad.	34
g NATURAL RUBBER	ACKWELL 5-109				0.15	degrad.	34
g NEOPRENE	ANSELLEDMONT 29-870	15m	8400	4	0.50	degrad.	34
g NITRILE	ANSELLEDMONT 37-155	20m	8900	4	0.40	degrad.	34
g PVC	ANSELLEDMONT 34-100	<1m	27120	5	0.20		34
g VITON	NORTH F-091				0.26	degrad.	34
sevin 50W (synonyms: 1-naphthol methyl carbamate; carbaryl) CAS Number: 63-25-2 Primary Class: 233 Esters, Non-Carboxylic, Carbamates and Others Related Class: 840 Pesticides, Mixtures and Formulations							
g NATURAL RUBBER	ANSELLEDMONT 36-124	>510m	0	0	0.51		74
g NEOPRENE	ANSELLEDMONT 29-865	>510m	0	0	0.51		74
g NITRILE	ANSELLEDMONT 37-175	>510m	0	0	0.46		74
g PVC	EDMONT CANADA 14112	>510m	0	0	n.a.		74

| G |
| A |
| R |
| M |

E Class & Number/	MANUFACTURER	Break-	Perm-eation	I N D E X	Thick-	R e f s
N test chemical/	& PRODUCT	through	Rate in		ness	degrad.
T MATERIAL NAME	IDENTIFICATION	Time in	mg/sq m		in	and
		Minutes	/min.		mm	comments

---

shale oil
CAS Number: 68308-34-9
Primary Class: 860   Coals, Charcoals, Oils

u NATURAL RUBBER	Unknown	20m			0.34	17
u NEOPRENE	Unknown	>60m			0.31	17
u NITRILE	Unknown	>60m			0.37	17
u PE	Unknown	16m			0.09	17
u PVC	Unknown	>60m			0.31	17

---

silicon tetrachloride
CAS Number: 10026-04-2
Primary Class: 360   Inorganic Acid Halides

s PVC	TRELLEBORG TRELLCHE				n.a.	degrad. 71
s TEFLON	CHEMFAB CHALL. 5100	>180m	0	1	n.a.	65

---

silver cyanide, < 30%
CAS Number: 506-64-9
Primary Class: 345 Inorganic Cyano Compounds

g PE/EVAL/PE	SAFETY4 4H	>240m		0	0.07	4% 60

---

sodium carbonate (saturated solution)
CAS Number: 497-19-8
Primary Class: 340 Inorganic Salts

g NATURAL RUBBER	MARIGOLD BLACK HEVY	>480m	0	0	0.65	79
g NATURAL RUBBER	MARIGOLD MED WT EXT	>480m	0	0	0.45	79
g NATURAL RUBBER	MARIGOLD ORANGE SUP	>480m	0	0	0.73	79
g NATURAL RUBBER	MARIGOLD SUREGRIP	>480m	0	0	0.58	79
g NITRILE	MARIGOLD BLUE	>480m	0	0	0.45	79
g NITRILE	MARIGOLD GREEN SUPA	>480m	0	0	0.30	79
g NITRILE	MARIGOLD NITROSOLVE	>480m	0	0	0.75	79
g UNKNOWN MATERIAL	MARIGOLD R MEDIGLOV	>480m	0	0	0.03	79

---

sodium chloride (saturated solution)
CAS Number: 7647-14-5
Primary Class: 340   Inorganic Salts

g NATURAL RUBBER	MARIGOLD BLACK HEVY	>480m	0	0	0.65	79
g NATURAL RUBBER	MARIGOLD FEATHERLIT	>480m	0	0	0.15	79
g NATURAL RUBBER	MARIGOLD MED WT EXT	>480m	0	0	0.45	79
g NATURAL RUBBER	MARIGOLD MEDICAL	>480m	0	0	0.28	79
g NATURAL RUBBER	MARIGOLD ORANGE SUP	>480m	0	0	0.73	79
g NATURAL RUBBER	MARIGOLD R SURGEONS	>480m	0	0	0.28	79
g NATURAL RUBBER	MARIGOLD RED LT WT	>480m	0	0	0.43	79

G A R M E Class & Number/ N test chemical/ T MATERIAL NAME	MANUFACTURER & PRODUCT IDENTIFICATION	Break- through Time in Minutes	Perm- eation Rate in mg/sq m /min.	I N D E X	Thick- ness in mm	degrad. and comments	R e f s
g NATURAL RUBBER	MARIGOLD SENSOTECH	>480m	0	0	0.28		79
g NATURAL RUBBER	MARIGOLD SUREGRIP	>480m	0	0	0.58		79
g NATURAL RUBBER	MARIGOLD YELLOW	>480m	0	0	0.38		79
g NATURAL+NEOPRENE	MARIGOLD FEATHERWT	>480m	0	0	0.35		79
g NITRILE	MARIGOLD BLUE	>480m	0	0	0.45		79
g NITRILE	MARIGOLD GREEN SUPA	>480m	0	0	0.30		79
g NITRILE	MARIGOLD NITROSOLVE	>480m	0	0	0.75		79
g UNKNOWN MATERIAL	MARIGOLD R MEDIGLOV	>480m	0	0	0.03		79

sodium cyanide (solid)
Primary Class: 345 Inorganic Cyano Compounds

g NATURAL RUBBER	ACKWELL 5-109	>480m	0	0	0.15		34
g NEOPRENE	ANSELLEDMONT 29-870	>480m	0	0	0.46		34
g NITRILE	ANSELLEDMONT 37-155	>480m	0	0	0.38		34
g PVC	ANSELLEDMONT 34-100	>480m	0	0	0.16		34
s SARANEX-23	DU PONT TYVEK SARAN	>480m	0	0	n.a.	95%	8
s UNKNOWN MATERIAL	DU PONT BARRICADE	>480m	0	0	n.a.	95%	8

sodium cyanide, 30-70% (salt solution)
CAS Number: 143-33-9
Primary Class: 345   Inorganic Cyano Compounds

c PE	DU PONT TYVEK QC	<240m	0.06	0	n.a.	70°C	8

sodium cyanide, <30% (salt solution)
Primary Class: 345   Inorganic Cyano Compounds

c PE	DU PONT TYVEK QC	<360m	0.09	0	n.a.	60°C	8

sodium cyanide, >70%
Primary Class: 345   Inorganic Cyano Compounds

s UNKNOWN MATERIAL	LIFE-GUARD RESPONDE	>180m			n.a.	solution	77

sodium dichromate, <30% (synonym: sodium bichromate)
CAS Number: 7789-12-0
Primary Class: 340   Inorganic Salts

c PE	DU PONT TYVEK QC	>480m	0	0	n.a.		8
s SARANEX-23	DU PONT TYVEK SARAN	>480m	0	0	n.a.		8

sodium fluoride
CAS Number: 7681-49-4
Primary Class: 340   Inorganic Salts

g NATURAL RUBBER	ACKWELL 5-109	>480m	0	0	0.18		34

G A R M E N T Class & Number/ test chemical/ MATERIAL NAME	MANUFACTURER & PRODUCT IDENTIFICATION	Break-through Time in Minutes	Perm-eation Rate in mg/sq m /min.	I N D E X	Thick-ness in mm	degrad. and comments	R e f s
g NEOPRENE	ANSELLEDMONT 29-870	>480m	0	0	0.50		34
g NITRILE	ANSELLEDMONT 37-155	>480m	0	0	0.38		34
g PVC	ANSELLEDMONT 34-100	>480m	0	0	0.18		34
s SARANEX-23	DU PONT TYVEK SARAN	>480m	0	0	n.a.		8

sodium hydroxide, 30-70%
CAS Number: 1310-73-2
Primary Class: 380   Inorganic Bases

g BUTYL	BEST 878	>480m	0	0	0.75		53
g BUTYL	BRUNSWICK BUTYL STD	>480m	0	0	0.63		89
g BUTYL	BRUNSWICK BUTYL-XTR	>480m	0	0	0.69		89
s BUTYL	LIFE-GUARD BUTYL	>480m	0	0	n.a.		77
s BUTYL	MSA CHEMPRUF	>480m	0	0	n.a.		48
s BUTYL	WHEELER ACID KING	>480m	0	0	n.a.		85
g BUTYL/ECO	BRUNSWICK BUTYL-POL	>480m	0	0	0.63		89
g BUTYL/NEOPRENE	COMASEC BUTYL PLUS	>480m	0	0	0.60		40
s BUTYL/NEOPRENE	MSA BETEX	>240m	0	0	n.a.		48
s CPE	STD. SAFETY	>180m			n.a.		75
g NAT+NEOPR+NITRILE	MAPA-PIONEER TRIONI	>780m	0	0	0.48		36
g NATURAL RUBBER	ANSELL CONFORM 4205	>360m		0	0.13		55
g NATURAL RUBBER	ANSELL FL 200 254	>360m		0	0.51		55
g NATURAL RUBBER	ANSELL ORANGE 208	>360m		0	0.76		55
g NATURAL RUBBER	ANSELL PVL 040	>360m		0	0.46		55
g NATURAL RUBBER	ANSELL STERILE 832	>360m		0	0.23		55
g NATURAL RUBBER	ANSELLEDMONT 36-124	>360m	0	0	0.46		6
g NATURAL RUBBER	ANSELLEDMONT 392	>360m		0	0.48		6
g NATURAL RUBBER	BEST 65NFW	>480m	0	0	n.a.		53
g NATURAL RUBBER	COMASEC FLEXIGUM	>480m	3000	4	0.95		40
g NATURAL RUBBER	MAPA-PIONEER L-118	>480m	0	0	0.46		36
g NATURAL RUBBER	MAPA-PIONEER L-118	>480m	0	0	0.46		36
g NATURAL RUBBER	MARIGOLD BLACK HEVY	>480m	0	0	0.65		79
g NATURAL RUBBER	MARIGOLD FEATHERLIT	>480m	0	0	0.15		79
g NATURAL RUBBER	MARIGOLD MED WT EXT	>480m	0	0	0.45		79
g NATURAL RUBBER	MARIGOLD MEDICAL	>480m	0	0	0.28		79
g NATURAL RUBBER	MARIGOLD ORANGE SUP	>480m	0	0	0.73		79
g NATURAL RUBBER	MARIGOLD R SURGEONS	>480m	0	0	0.28		79
g NATURAL RUBBER	MARIGOLD RED LT WT	>480m	0	0	0.43		79
g NATURAL RUBBER	MARIGOLD SENSOTECH	>480m	0	0	0.28		79
g NATURAL RUBBER	MARIGOLD SUREGRIP	>480m	0	0	0.58		79
g NATURAL RUBBER	MARIGOLD YELLOW	>480m	0	0	0.38		79
u NATURAL RUBBER	Unknown	>60m			n.a.		9
u NATURAL RUBBER	Unknown	>60m			0.43	degrad.	9
g NATURAL+NEOPRENE	ANSELL OMNI 276	>360m		0	0.56		55
g NATURAL+NEOPRENE	ANSELL TECHNICIANS	>360m		0	0.43		55
g NATURAL+NEOPRENE	MARIGOLD FEATHERWT	>480m	0	0	0.35		79
g NATURAL+NEOPRENE	PLAYTEX ARGUS 123	>480m	0	0	n.a.		72
u NATURAL+NEOPRENE	Unknown	>60m			n.a.		9

G A R M E Class & Number/ N test chemical/ T MATERIAL NAME	MANUFACTURER & PRODUCT IDENTIFICATION	Break- through Time in Minutes	Perm- eation Rate in mg/sq m /min.	I N D E X	Thick- ness in mm	degrad. and comments	R e f s
g NEOPRENE	ANSELL NEOPRENE 530	>360m		0	0.46		55
g NEOPRENE	ANSELLEDMONT 29-840	>360m	0	0	0.38		6
g NEOPRENE	ANSELLEDMONT NEOX	>360m	0	0	n.a.		6
g NEOPRENE	BEST 32	>480m	0	0	n.a.		53
g NEOPRENE	BEST 6780	>480m	0	0	n.a.		53
g NEOPRENE	BRUNSWICK NEOPRENE	>480m	0	0	0.88		89
g NEOPRENE	COMASEC COMAPRENE	>360m	0	0	n.a.		40
s NEOPRENE	LIFE-GUARD NEOPRENE	>480m	0	0	n.a.		77
g NEOPRENE	MAPA-PIONEER N-44	>480m	0	0	0.56		36
b NEOPRENE	RAINFAIR	>180m			n.a.		90
b NEOPRENE	RANGER	>180m			n.a.		90
b NEOPRENE	SERVUS NO 22204	>180m			n.a.		90
u NEOPRENE	Unknown	>60m			0.61		9
g NEOPRENE/NATURAL	ANSELL CHEMI-PRO	>360m		0	0.72		55
g NITRILE	ANSELL CHALLENGER	>360m		0	0.38		55
g NITRILE	ANSELLEDMONT 37-165	>360m	0	0	0.60		6
g NITRILE	BEST 22R	>480m	0	0	n.a.		53
g NITRILE	BEST 727	>480m	0	0	0.38		53
g NITRILE	COMASEC COMATRIL	>480m	0	0	0.55		40
g NITRILE	COMASEC COMATRIL SU	>480m	0	0	0.60		40
g NITRILE	COMASEC FLEXITRIL	150m	1800	4	n.a.		40
g NITRILE	MAPA-PIONEER A-14	>480m	0	0	0.56		36
g NITRILE	MARIGOLD BLUE	>480m	0	0	0.45		79
g NITRILE	MARIGOLD GREEN SUPA	>480m	0	0	0.30		79
g NITRILE	MARIGOLD NITROSOLVE	>480m	0	0	0.75		79
u NITRILE	Unknown	>60m			0.45	degrad.	9
b NITRILE+PUR+PVC	BATA HAZMAX	>180m			n.a.		86
g NITRILE+PVC	COMASEC MULTIPLUS	>480m		0	n.a.		40
b NITRILE+PVC	SERVUS NO 73101	>180m			n.a.		90
b NITRILE+PVC	TINGLEY	>180m			n.a.		90
c PE	DU PONT TYVEK QC	>480m	0	0	n.a.		8
g PE/EVAL/PE	SAFETY4 4H	>240m	0	0	0.07	35°C	60
g PE/EVAL/PE	SAFETY4 4H	>1440m	0	0	n.a.		60
g PVAL	ANSELLEDMONT PVA				n.a.	degrad.	6
g PVAL	EDMONT 15-552				n.a.	degrad.	6
g PVC	ANSELLEDMONT SNORKE	>360m	0	0	1.40		6
b PVC	BATA STANDARD	>580m	0	0	n.a.		86
g PVC	BEST 725R	>480m	0	0	n.a.		53
g PVC	COMASEC MULTIPOST	>480m	0	0	n.a.		40
g PVC	COMASEC MULTITOP	>480m	0	0	n.a.		40
g PVC	COMASEC NORMAL	>480m		0	n.a.		40
g PVC	COMASEC OMNI	>480m	0	0	n.a.		40
b PVC	JORDAN DAVID	>180m			n.a.		90
g PVC	MAPA-PIONEER V-20	>480m	0	0	0.51		36
s PVC	MSA UPC	>480m	0	0	0.20		48
b PVC	STD. SAFETY GL-20	>180m			n.a.		75
s PVC	STD. SAFETY WG-20	30m	<1	2	n.a.		75
u PVC	Unknown	>60m			0.22		9

G A R M E Class & Number/ N test chemical/ T MATERIAL NAME	MANUFACTURER & PRODUCT IDENTIFICATION	Break- through Time in Minutes	Perm- eation Rate in mg/sq m /min.	I N D E X	Thick- ness in mm	degrad. and comments	R e f s
u PVC	Unknown	>60m			n.a.		9
s PVC	WHEELER ACID KING	>480m	0	0	n.a.		85
b PVC+POLYURETHANE	BATA POLYBLEND	>580m	0	0	n.a.		86
b PVC+POLYURETHANE	BATA POLYBLEND	>180m			n.a.		90
b PVC+POLYURETHANE	BATA POLYMAX	>580m	0	0	n.a.		86
b PVC+POLYURETHANE	BATA SUPER POLY	>580m	0	0	n.a.		86
s SARANEX-23	DU PONT TYVEK SARAN	>480m	0	0	n.a.	40%	8
s SARANEX-23 2-PLY	DU PONT TYVEK SARAN	>480m	0	0	n.a.	50%	8
s TEFLON	CHEMFAB CHALL. 5100	>480m	0	0	n.a.		56
s TEFLON	CHEMFAB CHALL. 5100	>4260m	0	0	n.a.		65
s TEFLON	CHEMFAB CHALL. 5200	>300m	0	0	n.a.		80
s TEFLON	CHEMFAB CHALL. 6000	>180m			n.a.		80
s TEFLON	LIFE-GUARD TEFGUARD	>480m	0	0	n.a.		77
s TEFLON	WHEELER ACID KING	>480m	0	0	n.a.		85
s UNKNOWN MATERIAL	CHEMRON CHEMREL	>1440m	0	0	n.a.		67
s UNKNOWN MATERIAL	CHEMRON CHEMREL MAX	>1440m	0	0	n.a.		67
c UNKNOWN MATERIAL	CHEMRON CHEMTUFF	>480m	0	0	n.a.		67
s UNKNOWN MATERIAL	DU PONT BARRICADE	>480m	0	0	n.a.		8
s UNKNOWN MATERIAL	KAPPLER CPF III	>480m		0	n.a.		77
s UNKNOWN MATERIAL	LIFE-GUARD RESPONDE	>480m	0	0	n.a.		77
g UNKNOWN MATERIAL	NORTH SILVERSHIELD	>360m	0	0	0.08		7
g VITON	NORTH F-121	>60m			0.31		9
s VITON/BUTYL/UNKN.	TRELLEBORG HPS	>180m			n.a.		71
s VITON/CHLOROBUTYL	LIFE-GUARD VC-100	>480m	0	0	n.a.		77
s VITON/CHLOROBUTYL	LIFE-GUARD VNC 200	>480m	0	0	n.a.		77
s VITON/NEOPRENE	MSA VAUTEX	>240m	0	0	n.a.		48

sodium hydroxide, >70%
Primary Class: 380   Inorganic Bases

g NEOPRENE	GRANET 2714	>480m	0	0	n.a.		73
g PVC	GRANET 512-L	>480m	0	0	n.a.		73
s UNKNOWN MATERIAL	DU PONT BARRICADE	>480m	0	0	n.a.	99%	8

sodium hypochlorite (saturated solution)
Primary Class: 340   Inorganic Salts

g NATURAL RUBBER	MARIGOLD BLACK HEVY	>480m	0	0	0.65		79
g NATURAL RUBBER	MARIGOLD FEATHERLIT	>480m	0	0	0.15	degrad.	79
g NATURAL RUBBER	MARIGOLD MED WT EXT	>480m	0	0	0.45		79
g NATURAL RUBBER	MARIGOLD MEDICAL	>480m	0	0	0.28		79
g NATURAL RUBBER	MARIGOLD ORANGE SUP	>480m	0	0	0.73		79
g NATURAL RUBBER	MARIGOLD R SURGEONS	>480m	0	0	0.28		79
g NATURAL RUBBER	MARIGOLD RED LT WT	>480m	0	0	0.43		79
g NATURAL RUBBER	MARIGOLD SENSOTECH	>480m	0	0	0.28		79
g NATURAL RUBBER	MARIGOLD SUREGRIP	>480m	0	0	0.58		79
g NATURAL RUBBER	MARIGOLD YELLOW	>480m	0	0	0.38		79
g NATURAL+NEOPRENE	MARIGOLD FEATHERWT	>480m	0	0	0.35		79
g NITRILE	MARIGOLD BLUE	>480m	0	0	0.45		79

G A R M E Class & Number/ N test chemical/ T MATERIAL NAME	MANUFACTURER & PRODUCT IDENTIFICATION	Break- through Time in Minutes	Perm- eation Rate in mg/sq m /min.	I N D E X	Thick- ness in mm	degrad. and comments	R e f s
g NITRILE	MARIGOLD GREEN SUPA	>480m	0	0	0.30		79
g UNKNOWN MATERIAL	MARIGOLD R MEDIGLOV	>480m	0	0	0.03		79

sodium hypochlorite, 30-70%
CAS Number: 7681-52-9
Primary Class: 340   Inorganic Salts

g NATURAL RUBBER	COMASEC FLEXIGUM	>480m	0	0	0.95		40
g NATURAL+NEOPRENE	PLAYTEX ARGUS 123	>480m	0	0	n.a.		72
g NEOPRENE	COMASEC COMAPRENE	>480m	0	0	n.a.		40
g NITRILE	COMASEC COMATRIL	>480m	0	0	0.55		40
g NITRILE	COMASEC COMATRIL SU	>480m	0	0	0.60		40
g NITRILE	COMASEC FLEXITRIL	>480m	0	0	n.a.		40
g NITRILE	MARIGOLD NITROSOLVE	>480m	0	0	0.75		79
g NITRILE+PVC	COMASEC MULTIPLUS	>480m	0	0	n.a.		40
g PE/EVAL/PE	SAFETY4 4H	>240m	0	0	0.07	35°C	60
g PVC	COMASEC MULTIPOST	>480m	0	0	n.a.		40
g PVC	COMASEC MULTITOP	>480m	0	0	n.a.		40
g PVC	COMASEC NORMAL	>480m	0	0	n.a.		40
g PVC	COMASEC OMNI	>480m	0	0	n.a.		40
s SARANEX-23	DU PONT TYVEK SARAN	>480m	0	0	n.a.		8

sodium hypochlorite, < 30%
Primary Class: 340   Inorganic Salts

g BUTYL	BEST 878	>480m	0	0	0.75	4-6%	53
g NATURAL RUBBER	BEST 65NFW	>480m	0	0	n.a.	4-6%	53
g NEOPRENE	BEST 32	>480m	0	0	n.a.	4-6%	53
g NEOPRENE	BEST 6780	>480m	0	0	n.a.	4-6%	53
g NITRILE	BEST 22R	>480m	0	0	n.a.	4-6%	53
g NITRILE	BEST 727	>480m	0	0	0.38	4-6%	53
g PVC	BEST 725R	>480m	0	0	n.a.	4-6%	53

sodium pentachlorophenate, 4.2% in diesel oil
Primary Class: 860   Coals, Charcoals, Oils

g NATURAL RUBBER	BEST 65NFW	<1m	0.2	1	n.a.		64
g NATURAL+NEOPRENE	PLAYTEX 835	>470m	0	0	0.41		64
g NITRILE	ANSELLEDMONT 37-165	>900m	0	0	0.64		64
g PVC	DAYTON TRIFLEX	>300m	0	0	0.19		64
g PVC	GRANET SAFETY 1012	>900m	0	0	n.a.		64

sodium thiosulfate (synonym: sodium hyposulfate)
CAS Number: 7772-98-7
Primary Class: 340   Inorganic Salts

g NATURAL RUBBER	COMASEC FLEXIGUM	>480m	0	0	0.95		40
g NEOPRENE	COMASEC COMAPRENE	>480m	0	0	n.a.		40

G A R M E Class & Number/ N test chemical/ T MATERIAL NAME	MANUFACTURER & PRODUCT IDENTIFICATION	Break- through Time in Minutes	Perm- eation Rate in mg/sq m /min.	I N D E X	Thick- ness in mm	degrad. and comments	R e f s
g NITRILE	COMASEC COMATRIL	>480m	0	0	0.55		40
g NITRILE	COMASEC COMATRIL SU	>480m	0	0	0.60		40
g NITRILE	COMASEC FLEXITRIL	>480m	0	0	n.a.		40
g NITRILE	MARIGOLD NITROSOLVE	>480m	0	0	0.75		79
g NITRILE+PVC	COMASEC MULTIPLUS	>480m	0	0	n.a.		40
g PVC	COMASEC MULTIPOST	>480m	0	0	n.a.		40
g PVC	COMASEC MULTITOP	>480m	0	0	n.a.		40
g PVC	COMASEC NORMAL	>480m	0	0	n.a.		40
g PVC	COMASEC OMNI	>480m	0	0	n.a.		40

sodium thiosulfate (saturated solution)
Primary Class: 340 Inorganic Salts

g NATURAL RUBBER	MARIGOLD BLACK HEVY	>480m	0	0	0.65		79
g NATURAL RUBBER	MARIGOLD FEATHERLIT	>480m	0	0	0.28		79
g NATURAL RUBBER	MARIGOLD MED WT EXT	>480m	0	0	0.45		79
g NATURAL RUBBER	MARIGOLD MEDICAL	>480m	0	0	0.28		79
g NATURAL RUBBER	MARIGOLD ORANGE SUP	>480m	0	0	0.73		79
g NATURAL RUBBER	MARIGOLD R SURGEONS	>480m	0	0	0.28		79
g NATURAL RUBBER	MARIGOLD RED LT WT	>480m	0	0	0.43		79
g NATURAL RUBBER	MARIGOLD SENSOTECH	>480m	0	0	0.28		79
g NATURAL RUBBER	MARIGOLD SUREGRIP	>480m	0	0	0.58		79
g NATURAL RUBBER	MARIGOLD YELLOW	>480m	0	N	0.38		79
g NATURAL+NEOPRENE	MARIGOLD FEATHERWT	>480m	0	0	0.35		79
g NITRILE	MARIGOLD BLUE	>480m	0	0	0.45		79
g NITRILE	MARIGOLD GREEN SUPA	>480m	0	0	0.30		79
g UNKNOWN MATERIAL	MARIGOLD R MEDIGLOV	>480m	0	0	0.03		79

Solmaster (synonym: methylene chloride & ethyl alcohol)
Primary Class: 900   Miscellaneous Unclassified Chemicals

g PE/EVAL/PE	SAFETY4 4H	>240m	0		0.07	35°C	60

solvent 60
Primary Class: 900   Miscellaneous Unclassified Chemicals

g NITRILE	MARIGOLD NITROSOLVE	>480m	0	0	0.75		79

soybean oil
CAS Number: 8001-22-7
Primary Class: 860   Coals, Charcoals, Oils

s BUTYL	MSA CHEMPRUF	<60m			n.a.		48
s PVC	MSA UPC	<120m			0.20		48

G A R M E Class & Number/ N test chemical/ T MATERIAL NAME	MANUFACTURER & PRODUCT IDENTIFICATION	Break- through Time in Minutes	Perm- eation Rate in mg/sq m /min.	I N D E X	Thick- ness in mm	R degrad. e and f comments s

styrene (synonym: vinylbenzene)
CAS Number: 100-42-5
Primary Class: 292   Hydrocarbons, Aromatic
Related Class: 294   Hydrocarbons, Aliphatic and Alicyclic, Unsaturated

s BUTYL	MSA CHEMPRUF	<60m			n.a.	48
g BUTYL	NORTH B-174				0.65	degrad. 34
m BUTYL	PLYMOUTH RUBBER	<1m			n.a.	46
s CPE	ILC DOVER CLOROPEL	65m			n.a.	46
g NATURAL RUBBER	ACKWELL 5-109				0.15	degrad. 34
g NATURAL RUBBER	ANSELL PVL 040				n.a.	degrad. 55
g NATURAL RUBBER	ANSELLEDMONT 36-124				0.46	degrad. 6
g NATURAL RUBBER	ANSELLEDMONT 392				0.48	degrad. 6
g NATURAL RUBBER	COMASEC FLEXIGUM	24m	2880	4	0.95	40
g NATURAL RUBBER	MARIGOLD BLACK HEVY	4m	4200	5	0.65	degrad. 79
g NATURAL RUBBER	MARIGOLD FEATHERLIT	1m	69000	5	0.15	degrad. 79
g NATURAL RUBBER	MARIGOLD MED WT EXT	1m	4200	5	0.45	degrad. 79
g NATURAL RUBBER	MARIGOLD MEDICAL	1m	42000	5	0.28	degrad. 79
g NATURAL RUBBER	MARIGOLD ORANGE SUP	4m	4200	5	0.73	degrad. 79
g NATURAL RUBBER	MARIGOLD R SURGEONS	1m	42000	5	0.28	degrad. 79
g NATURAL RUBBER	MARIGOLD RED LT WT	1m	9600	5	0.43	degrad. 79
g NATURAL RUBBER	MARIGOLD SENSOTECH	1m	42000	5	0.28	degrad. 79
g NATURAL RUBBER	MARIGOLD SUREGRIP	3m	4200	5	0.58	degrad. 79
g NATURAL RUBBER	MARIGOLD YELLOW	1m	20400	5	0.38	degrad. 79
g NATURAL+NEOPRENE	ANSELL TECHNICIANS				n.a.	degrad. 55
g NATURAL+NEOPRENE	MARIGOLD FEATHERWT	1m	25800	5	0.35	degrad. 79
g NEOPRENE	ANSELLEDMONT 29-840				0.38	degrad. 6
g NEOPRENE	ANSELLEDMONT 29-870				0.48	degrad. 34
g NEOPRENE	BEST 6780		300	5	n.a.	53
g NEOPRENE	COMASEC COMAPRENE	12m	5160	5	n.a.	40
g NITRILE	ANSELLEDMONT 37-155				0.40	degrad. 34
g NITRILE	ANSELLEDMONT 37-165				0.60	degrad. 6
g NITRILE	BEST 22R		4560	5	n.a.	53
g NITRILE	COMASEC COMATRIL	25m	9000	4	0.55	40
g NITRILE	COMASEC COMATRIL SU	30m	7320	4	0.60	55
g NITRILE	COMASEC FLEXITRIL	18m	9600	4	n.a.	40
g NITRILE	MARIGOLD BLUE	33m	3212	4	0.45	degrad. 79
g NITRILE	MARIGOLD GREEN SUPA	13m	3100	5	0.30	degrad. 79
g NITRILE	MARIGOLD NITROSOLVE	37m	3234	4	0.75	degrad. 79
g NITRILE+PVC	COMASEC MULTIPLUS	40m	1860	4	n.a.	40
g PE/EVAL/PE	SAFETY4 4H	>240m	0	0	0.07	35°C 60
g PE/EVAL/PE	SAFETY4 4H	>1440m	0	0	0.07	60
g PVAL	ANSELLEDMONT PVA	>360m		0	n.a.	6
g PVAL	EDMONT 15-552	>360m	0	0	n.a.	6
g PVAL	EDMONT 25-545	>360m	0	0	n.a.	6
g PVC	ANSELLEDMONT SNORKE				n.a.	degrad. 6

G A R M E Class & Number/ N test chemical/ T MATERIAL NAME	MANUFACTURER & PRODUCT IDENTIFICATION	Break- through Time in Minutes	Perm- eation Rate in mg/sq m /min.	I N D E X	Thick- ness in mm	degrad. and comments	R e f s
g PVC	BEST 812		1560	5	n.a.		53
g PVC	COMASEC MULTIPOST	20m	2040	4	n.a.		40
g PVC	COMASEC MULTITOP	35m	1980	4	n.a.		40
g PVC	COMASEC NORMAL	20m	2160	4	n.a.		40
g PVC	COMASEC OMNI	14m	2280	5	n.a.		40
s SARANEX-23	DU PONT TYVEK SARAN	43m	699	3	n.a.		8
s TEFLON	CHEMFAB CHALL. 5100	>240m	0	0	n.a.		65
c UNKNOWN MATERIAL	CHEMRON CHEMTUFF	16m	142	3	n.a.		67
s UNKNOWN MATERIAL	DU PONT BARRICADE	>480m	0	0	n.a.		8
s UNKNOWN MATERIAL	KAPPLER CPF III	>480m	0	0	n.a.		77
s UNKNOWN MATERIAL	LIFE-GUARD RESPONDE	>180m			n.a.		77
u VITON/CHLOROBUTYL	Unknown	>180m			n.a.		46

styrene monomer resin
Primary Class: 810   Epoxy Products

g NATURAL RUBBER	MAPA-PIONEER LL-301	14m	407	4	n.a.		36
g NATURAL+NEOPRENE	MAPA-PIONEER NS-53	6m	453	4	0.81		36
g NEOPRENE	MAPA-PIONEER NS-35	28m	400	3	0.55		36
g PVC	MAPA-PIONEER V-20	27m	397	3	0.64		36

sulfallate (synonym: 2-chloro-2-propenyldiethyldithiocarbamate)
CAS Number: 95-06-7
Primary Class: 233   Esters, Non-Carboxylic, Carbamates and Others

g BUTYL	NORTH B-161	>480m	0	0	0.43		34
g NEOPRENE	ANSELLEDMONT 29-870	275m		0	0.46		34
g NITRILE	ANSELLEDMONT 37-155	>480m	0	0	0.38		34
g VITON	NORTH F-091	>480m	0	0	0.25		34

sulfur dichloride (synonym: chlorine sulfide)
CAS Number: 10545-99-0
Primary Class: 360   Inorganic Acid Halides

s UNKNOWN MATERIAL	LIFE-GUARD RESPONDE	448m	3	1	n.a.		77

sulfur dioxide
CAS Number: 7446-09-5
Primary Class: 350   Inorganic Gases and Vapors

c PE	DU PONT TYVEK QC	<1m	>1000	5	n.a.		8
s SARANEX-23	DU PONT TYVEK SARAN	>480m	0	0	n.a.		8
s UNKNOWN MATERIAL	DU PONT BARRICADE	>480m	0	0	n.a.		8
s UNKNOWN MATERIAL	LIFE-GUARD RESPONDE	>480m	0	0	n.a.		77
s VITON/BUTYL	DRAGER 500 OR 710	>60m			n.a.		91

G A R M E Class & Number/ N test chemical/ T MATERIAL NAME	MANUFACTURER & PRODUCT IDENTIFICATION	Break- through Time in Minutes	Perm- eation Rate in mg/sq m /min.	I N D E X	Thick- ness in mm	degrad. and comments	R e f s

sulfuric acid & potassium dichromate & water (synonym: dichromate etch)
Primary Class: 820   Etching Products

g NAT+NEOPR+NITRILE	MAPA-PIONEER TRIONI	>480m	0		0	0.46	36

sulfuric acid & sodium dichromate, 3% & water
Primary Class: 820   Etching Products

c PE	DU PONT TYVEK QC	>480m	0		0	n.a.	8
s SARANEX-23	DU PONT TYVEK SARAN	>480m	0		0	n.a.	8

sulfuric acid, 30-70%
CAS Number: 7664-93-9
Primary Class: 370   Inorganic Acids

g NATURAL RUBBER	ANSELL PVL 040	>360m	0		0	0.46	55
g NATURAL RUBBER	ANSELLEDMONT 392	>360m	0		0	0.48 47% acid	6
g NATURAL RUBBER	MAPA-PIONEER L-118	>480m	0		0	0.46	36
g NATURAL RUBBER	MARIGOLD BLACK HEVY	>480m	0		0	0.65	79
g NATURAL RUBBER	MARIGOLD FEATHERLIT	>480m	0		0	0.15	79
g NATURAL RUBBER	MARIGOLD MED WT EXT	>480m	0		0	0.45	79
g NATURAL RUBBER	MARIGOLD MEDICAL	>480m	0		0	0.28	79
g NATURAL RUBBER	MARIGOLD ORANGE SUP	>480m	0		0	0.73	79
g NATURAL RUBBER	MARIGOLD R SURGEONS	>480m	0		0	0.28	79
g NATURAL RUBBER	MARIGOLD RED LT WT	>480m	0		0	0.43	79
g NATURAL RUBBER	MARIGOLD SENSOTECH	>480m	0		0	0.28	79
g NATURAL RUBBER	MARIGOLD SUREGRIP	>480m	0		0	0.58	79
g NATURAL RUBBER	MARIGOLD YELLOW	>480m	0		0	0.38	79
g NATURAL+NEOPRENE	ANSELL OMNI 276	>360m	0		0	0.45	55
g NATURAL+NEOPRENE	ANSELL TECHNICIANS	>360m	0		0	n.a.	55
g NATURAL+NEOPRENE	MARIGOLD FEATHERWT	>480m	0		0	0.35	79
g NEOPRENE	ANSELL NEOPRENE 520	>360m	0		0	n.a.	55
g NEOPRENE	ANSELLEDMONT 29-840	>360m			0	0.38 47% battery	6
g NEOPRENE	ANSELLEDMONT NEOX	>360m			0	n.a. 47% battery	6
g NEOPRENE	MAPA-PIONEER N-44	>480m	0		0	0.56	36
g NEOPRENE/NATURAL	ANSELL CHEMI-PRO	>360m	0		0	n.a.	55
g NITRILE	ANSELLEDMONT 37-165	>360m			0	0.54 47% battery	6
g NITRILE	MAPA-PIONEER A-14	>480m	0		0	0.56	36
g NITRILE	MARIGOLD BLUE	180m	62686	5	0.45 degrad.		79
g NITRILE	MARIGOLD NITROSOLVE	215m	55223	5	0.75 degrad.		79
c PE	DU PONT TYVEK QC	>480m	0		0	n.a.	8
g PVC	ANSELLEDMONT M GRIP	>360m			0	n.a. 47% battery	6
g PVC	MAPA-PIONEER V-20	>480m	0		0	0.51	36
s SARANEX-23	DU PONT TYVEK SARAN	>480m	0		0	n.a.	8

G A R M E Class & Number/ N test chemical/ T MATERIAL NAME	MANUFACTURER & PRODUCT IDENTIFICATION	Break- through Time in Minutes	Perm- eation Rate in mg/sq m /min.	I N D E X	Thick- ness in mm	degrad. and comments	R e f s
sulfuric acid, 96% & sulphur dioxide, 65% in H2O Primary Class: 370 Inorganic Acids Related Class: 365 Inorganic Acid Oxides							
g PE/EVAL/PE	SAFETY4 4H	30m			0.07	35°C	60
g PE/EVAL/PE	SAFETY4 4H	120m			0.07		60
sulfuric acid, <30% Primary Class: 370 Inorganic Acids							
s BUTYL	MSA CHEMPRUF	>480m	0	0	n.a.		48
g NAT+NEOPR+NITRILE	MAPA-PIONEER TRIONI	>900m	0	0	0.48		36
g NATURAL RUBBER	MARIGOLD BLACK HEVY	>480m	0	0	0.65		79
g NATURAL RUBBER	MARIGOLD FEATHERLIT	>480m	0	0	0.15		79
g NATURAL RUBBER	MARIGOLD MED WT EXT	>480m	0	0	0.45		79
g NATURAL RUBBER	MARIGOLD MEDICAL	>480m	0	0	0.28		79
g NATURAL RUBBER	MARIGOLD ORANGE SUP	>480m	0	0	0.73		79
g NATURAL RUBBER	MARIGOLD R SURGEONS	>480m	0	0	0.28		79
g NATURAL RUBBER	MARIGOLD RED LT WT	>480m	0	0	0.43		79
g NATURAL RUBBER	MARIGOLD SENSOTECH	>480m	0	0	0.28		79
g NATURAL RUBBER	MARIGOLD SUREGRIP	>480m	0	0	0.58		79
g NATURAL RUBBER	MARIGOLD YELLOW	>480m	0	0	0.38		79
g NATURAL+NEOPRENE	MARIGOLD FEATHERWT	>480m	0	0	0.35		79
g NITRILE	MARIGOLD BLUE	>480m	0	0	0.45		79
g NITRILE	MARIGOLD GREEN SUPA	>480m	0	0	0.30		79
g NITRILE	MARIGOLD NITROSOLVE	480m	0	0	0.75		79
g NITRILE	NORTH LA-142G				0.38	degrad.	7
c PE	DU PONT TYVEK QC	>480m	0	0	n.a.		8
g PE/EVAL/PE	SAFETY4 4H	>240m	0	0	0.07	35°C	60
s PVC	MSA UPC	>480m	0	0	0.20		48
s SARANEX-23	DU PONT TYVEK SARAN	>480m	0	0	n.a.		8
g UNKNOWN MATERIAL	NORTH SILVERSHIELD	>360m	0	0	0.08		7
sulfuric acid, >70% Primary Class: 370 Inorganic Acids							
g BUTYL	BEST 878	>480m	0	0	0.75		53
g BUTYL	BRUNSWICK BUTYL STD	>480m	0	0	0.63		89
g BUTYL	BRUNSWICK BUTYL-XTR	>480m	0	0	0.69		89
s BUTYL	LIFE-GUARD BUTYL	>480m	0	0	n.a.		77
s BUTYL	MSA CHEMPRUF	>480m	0	0	n.a.		48
s BUTYL	WHEELER ACID KING	>480m	0	0	n.a.		85
g BUTYL/ECO	BRUNSWICK BUTYL-POL	>480m		0	0.63		89
g BUTYL/NEOPRENE	COMASEC BUTYL PLUS	>480m	0	0	0.60		40
s BUTYL/NEOPRENE	MSA BETEX	>480m	0	0	n.a.		48
u CPE	ILC DOVER	>180m			n.a.		29
s CPE	STD. SAFETY	>180m			n.a.		75

G A R M E N T	MANUFACTURER	Break-	Perm- eation	I N D	Thick-		R
Class & Number/	& PRODUCT	through	Rate in		ness	degrad.	e
N test chemical/	IDENTIFICATION	Time in	mg/sq m	E	in	and	f
T MATERIAL NAME		Minutes	/min.	X	mm	comments	s
g HYPALON	COMASEC DIPCO	>480m		0	0.51		40
g NAT+NEOPR+NITRILE	MAPA-PIONEER TRIONI	108m	1596	4	0.46	degrad.	36
g NATURAL RUBBER	ANSELL ORANGE 208	150m			0.76		55
g NATURAL RUBBER	ANSELL PVL 040				0.46	degrad.	55
g NATURAL RUBBER	ANSELLEDMONT 36-124				0.46	degrad.	6
g NATURAL RUBBER	ANSELLEDMONT 392				0.48	degrad.	6
g NATURAL RUBBER	COMASEC FLEXIGUM	150m	13200	5	0.95		40
u NATURAL RUBBER	Unknown	>60m			n.a.	degrad.	9
u NATURAL RUBBER	Unknown	>60m			0.43	degrad.	9
g NATURAL+NEOPRENE	ANSELL TECHNICIANS				0.43	degrad.	55
g NATURAL+NEOPRENE	PLAYTEX ARGUS 123	92m	4620	4	n.a.		72
u NATURAL+NEOPRENE	Unknown	>60m			n.a.	degrad.	9
g NEOPRENE	ANSELL NEOPRENE 530				0.46	degrad.	55
g NEOPRENE	ANSELLEDMONT 29-840	180m			0.38		6
g NEOPRENE	ANSELLEDMONT NEOX	>360m	0	0	n.a.		6
g NEOPRENE	BEST 32	480m	270	3	n.a.		53
g NEOPRENE	BEST 6780	>480m	0	0	n.a.		53
g NEOPRENE	BRUNSWICK NEOPRENE	>480m		0	0.88		89
g NEOPRENE	COMASEC COMAPRENE	150m			n.a.		40
g NEOPRENE	GRANET 2714	265m	47	2	n.a.		73
s NEOPRENE	LIFE-GUARD NEOPRENE	>480m	0	0	n.a.		77
b NEOPRENE	RAINFAIR	>180m			n.a.		90
b NEOPRENE	RANGER	>180m			n.a.		90
b NEOPRENE	SERVUS NO 22204	>180m			n.a.		90
u NEOPRENE	Unknown	>60m			0.61	degrad.	9
g NITRILE	ANSELL 650				n.a.	degrad.	55
g NITRILE	ANSELL CHALLENGER				0.38	degrad.	55
g NITRILE	ANSELLEDMONT 37-165				0.60	degrad.	6
g NITRILE	BEST 727	180m	480	3	0.38	degrad.	53
g NITRILE	COMASEC COMATRIL	150m	17040	5	0.55		40
g NITRILE	COMASEC COMATRIL SU	>480m	0	0	0.60		40
g NITRILE	COMASEC FLEXITRIL	20m	23000	5	n.a.		40
g NITRILE	MARIGOLD BLUE	180m	62686	5	0.45	degrad.	79
g NITRILE	MARIGOLD NITROSOLVE	215m	55223	5	0.75	degrad.	79
u NITRILE	Unknown	>60m			0.45	degrad.	9
b NITRILE+PUR+PVC	BATA HAZMAX	>180m			n.a.		86
g NITRILE+PVC	COMASEC MULTIPLUS	75m	85550	5	n.a.		40
b NITRILE+PVC	SERVUS NO 73101	>180m			n.a.		90
b NITRILE+PVC	TINGLEY	>180m			n.a.		90
c PE	DU PONT TYVEK QC	>480m	0	0	n.a.		8
c PE	DU PONT TYVEK QC	120m			n.a.	65°C	8
g PE/EVAL/PE	SAFETY4 4H	>240m	0	0	0.07	35°C	60
g PE/EVAL/PE	SAFETY4 4H	>1440m	0	0	0.07	93%	60
g PVAL	EDMONT 15-552				n.a.	degrad.	6
g PVC	ANSELLEDMONT M GRIP	220m			n.a.		6
b PVC	BATA STANDARD	180m	0.01	1	n.a.		86
g PVC	COMASEC MULTIPOST	75m	37200	5	n.a.		40
g PVC	COMASEC MULTITOP	80m	3480	4	n.a.		40

G A R M E Class & Number/ N test chemical/ T MATERIAL NAME	MANUFACTURER & PRODUCT IDENTIFICATION	Break- through Time in Minutes	Perm- eation Rate in mg/sq m /min.	I N D E X	Thick- ness in mm	degrad. and comments	R e f s
g PVC	COMASEC NORMAL	95m	34200	5	n.a.		40
g PVC	COMASEC OMNI	70m	37200	5	n.a.		40
g PVC	GRANET 512-L	143m	232	3	n.a.		73
b PVC	JORDAN DAVID	>180m			n.a.		90
s PVC	MSA UPC	<120m			0.20		48
b PVC	STD. SAFETY GL-20	>180m	0	1	n.a.		75
s PVC	STD. SAFETY WG-20	150m	<1	1	n.a.		75
u PVC	Unknown	9m			0.22	degrad.	9
s PVC	WHEELER ACID KING	24m	<900	3	n.a.		85
b PVC+POLYURETHANE	BATA POLYBLEND	580m	0.01	0	n.a.		86
b PVC+POLYURETHANE	BATA POLYBLEND	>180m			n.a.		90
b PVC+POLYURETHANE	BATA POLYMAX	>580m	0.01	0	n.a.		86
b PVC+POLYURETHANE	BATA SUPER POLY	590m	0.01	0	n.a.		86
s SARANEX-23	DU PONT TYVEK SARAN	>480m	0	0	n.a.	98%	8
s SARANEX-23	DU PONT TYVEK SARAN	330m	0	0	n.a.	65°C	8
s SARANEX-23 2-PLY	DU PONT TYVEK SARAN	>480m	0	0	n.a.	98%	8
s TEFLON	CHEMFAB CHALL. 5100	>4320m	0	0	n.a.	concentr.	65
s TEFLON	CHEMFAB CHALL. 5200	>300m	0	0	n.a.	90%	80
s TEFLON	CHEMFAB CHALL. 6000	>180m	n.a.		80		
s TEFLON	LIFE-GUARD TEFGUARD	>480m	0	0	n.a.		77
s TEFLON	WHEELER ACID KING	>480m	0	0	n.a.		85
s UNKNOWN MATERIAL	CHEMRON CHEMREL	>1440m	0	0	n.a.		48
s UNKNOWN MATERIAL	CHEMRON CHEMREL MAX	>1440m	0	0	n.a.		67
c UNKNOWN MATERIAL	CHEMRON CHEMTUFF	>480m	0	0	n.a.		67
s UNKNOWN MATERIAL	DU PONT BARRICADE	>480m	0	0	n.a.	95%	8
s UNKNOWN MATERIAL	KAPPLER CPF III	>480m		0	n.a.		77
s UNKNOWN MATERIAL	LIFE-GUARD RESPONDE	>480m	0	0	n.a.		77
g VITON	NORTH F-121	>60m			0.31		9
s VITON/BUTYL/UNKN.	TRELLEBORG HPS	>180m			n.a.		71
s VITON/CHLOROBUTYL	LIFE-GUARD VC-100	>480m	0	0	n.a.		77
s VITON/CHLOROBUTYL	LIFE-GUARD VNC 200	>480m	0	0	n.a.		77
u VITON/CHLOROBUTYL	Unknown	>180m			n.a.		29
s VITON/NEOPRENE	MSA VAUTEX	>480m	0	0	n.a.		48

sulfuric acid, fuming (synonym: oleum)
Primary Class: 370   Inorganic Acids

g BUTYL	BRUNSWICK BUTYL STD	270m	>5000	4	0.63		89
g BUTYL	BRUNSWICK BUTYL-XTR	440m	>5000	4	0.63		89
g BUTYL/ECO	BRUNSWICK BUTYL-POL	420m	320	3	0.63		89
s BUTYL/NEOPRENE	MSA BETEX	>480m	0	0	n.a.		48
g NEOPRENE	BRUNSWICK NEOPRENE	170m	>5000	4	0.63		89
c PE	DU PONT TYVEK QC	<1m			n.a.	65%	8
s SARANEX-23	DU PONT TYVEK SARAN	37m			n.a.	65%	8
s TEFLON	CHEMFAB CHALL. 5100	>480m	0	0	n.a.		56
s TEFLON	CHEMFAB CHALL. 5100	>180m	0	1	n.a.		65
s UNKNOWN MATERIAL	LIFE-GUARD RESPONDE	>480m	0	0	n.a.		77
s VITON/CHLOROBUTYL	LIFE-GUARD VC-100	>240m		0	n.a.		77
s VITON/NEOPRENE	MSA VAUTEX	>480m	0	0	n.a.		48

G A R M E Class & Number/ N test chemical/ T MATERIAL NAME	MANUFACTURER & PRODUCT IDENTIFICATION	Break- through Time in Minutes	Perm- eation Rate in mg/sq m /min.	I N D E X	Thick- ness in mm	degrad. and comments	R e f s

sulphur trioxide (synonym: sulphuric anhydride)
CAS Number: 7446-11-9
Primary Class: 365   Inorganic Acid Oxides

s BUTYL	TRELLEBORG TRELLCHE				n.a.	degrad.	71

sulphuryl chloride
CAS Number: 7791-25-5
Primary Class: 360   Inorganic Acid Halides

s BUTYL	TRELLEBORG TRELLCHE				n.a.	degrad.	71
s PVC	TRELLEBORG TRELLCHE				n.a.	degrad.	71

t-butyl peroxybenzoate
CAS Number: 614-45-9
Primary Class: 300   Peroxides

g BUTYL	NORTH B-161	>480m	0	0	0.43		34
g NEOPRENE	ANSELLEDMONT 29-870	240m	1.7	1	0.46		34

t-butylhydroperoxide
CAS Number: 75-91-2
Primary Class: 300   Peroxides

g PE/EVAL/PE	SAFETY4 4H	>240m	0	0	0.07		60

tannic acid (synonym: tannin)
CAS Number: 1401-55-4
Primary Class: 316   Hydroxy Compounds, Aromatic (Phenols)

s BUTYL	MSA CHEMPRUF	>480m	0	0	n.a.		48
g NATURAL RUBBER	ANSELL CONFORM 4205	>360m		0	0.13		55
g NATURAL RUBBER	ANSELL FL 200 254	>360m		0	0.51		55
g NATURAL RUBBER	ANSELL ORANGE 208	>360m		0	0.76		55
g NATURAL RUBBER	ANSELL PVL 040	>360m		0	0.46		55
g NATURAL RUBBER	ANSELL STERILE 832	>360m		0	0.23		55
g NATURAL RUBBER	ANSELLEDMONT 36-124	>360m	0	0	0.46		6
g NATURAL RUBBER	ANSELLEDMONT 392	>360m		0	0.48		6
g NATURAL+NEOPRENE	ANSELL OMNI 276	>360m	0	0	0.56		55
g NATURAL+NEOPRENE	ANSELL TECHNICIANS	>360m	0	0	0.43		55
g NEOPRENE	ANSELL NEOPRENE 530	>360m	0	0	0.46		55
g NEOPRENE	ANSELLEDMONT 29-840	>360m	0	0	0.38		6
g NEOPRENE	ANSELLEDMONT NEOX	>360m	0	0	n.a.		6
g NEOPRENE/NATURAL	ANSELL CHEMI-PRO	>360m	0	0	0.72		55
g NITRILE	ANSELL CHALLENGER	>360m	0	0	0.38		55
g NITRILE	ANSELLEDMONT 37-165	>360m	0	0	0.60		6
g PVAL	ANSELLEDMONT PVA				n.a.	degrad.	6

G A R M E Class & Number/ N test chemical/ T MATERIAL NAME	MANUFACTURER & PRODUCT IDENTIFICATION	Break- through Time in Minutes	Perm- eation Rate in mg/sq m /min.	I N D E X	Thick- ness in mm	degrad. and comments	R e f s
g PVAL	EDMONT 15-552				n.a.	degrad.	6
g PVC	ANSELLEDMONT M GRIP	>360m	0	0	n.a.		6
s PVC	MSA UPC	>480m	0	0	0.20		48

teak oil
Primary Class: 860   Coals, Charcoals, Oils

g PE/EVAL/PE	SAFETY4 4H	>240m	0	0	0.07	35°C	60

tert-butyl alcohol (synonyms: 2-methyl-2-propanol; tert-butanol)
CAS Number: 75-65-0
Primary Class: 313   Hydroxy Compounds, Aliphatic and Alicyclic, Tertiary

g BUTYL	NORTH B-161	>480m	0	0	0.41		7
g PE/EVAL/PE	SAFETY4 4H	>240m		0	0.07	35°C	60
s UNKNOWN MATERIAL	LIFE-GUARD RESPONDE	>480m	0	0	n.a.		77

tert-butyl methyl ether (synonym: 2-methoxy-2-methylpropane)
CAS Number: 1634-04-4
Primary Class: 241   Ethers, Aliphatic and Alicyclic

g NATURAL RUBBER	ANSELLEDMONT 36-124				0.46	degrad.	6
g NATURAL RUBBER	ANSELLEDMONT 392				0.48	degrad.	6
g NEOPRENE	ANSELLEDMONT 29-840				0.38	degrad.	6
g NEOPRENE	ANSELLEDMONT NEOX				n.a.	degrad.	6
g NITRILE	ANSELLEDMONT 37-165	>360m	0	0	0.60		6
g PVAL	ANSELLEDMONT PVA	>360m		0	n.a.		6
g PVAL	EDMONT 15-552	>360m	0	0	n.a.		6
g PVC	ANSELLEDMONT SNORKE				n.a.	degrad.	6
s UNKNOWN MATERIAL	LIFE-GUARD RESPONDE	>180m	0	1	n.a.		77

tert-butylamine
CAS Number: 75-64-9
Primary Class: 141   Amines, Aliphatic and Alicyclic, Primary

g BUTYL	NORTH B-174	>480m	0	0	0.68	degrad.	34
g NATURAL RUBBER	ACKWELL 5-109				0.15	degrad.	34
g NEOPRENE	ANSELLEDMONT 29-870	70m	2132	4	0.50	degrad.	34
g NITRILE	ANSELLEDMONT 37-155	84m	1440	4	0.40	degrad.	34
g PVC	ANSELLEDMONT 34-100	2m	18120	5	0.20	degrad.	34
g VITON	NORTH F-091				0.26	degrad.	34

tetrachloroethylene (synonym: perchloroethylene)
CAS Number: 127-18-4
Primary Class: 263   Halogen Compounds, Vinylic

g BUTYL	BEST 878	28m	6560	4	0.75	degrad.	53

G A R M E N T Class & Number/ test chemical/ MATERIAL NAME	MANUFACTURER & PRODUCT IDENTIFICATION	Break-through Time in Minutes	Perm-eation Rate in mg/sq m /min.	I N D E X	Thick-ness in mm	degrad. and comments	R e f s
g BUTYL	BRUNSWICK BUTYL STD	<4m	>5000	5	0.67		89
g BUTYL	BRUNSWICK BUTYL-XTR	4m	>5000	5	0.80		89
g BUTYL	COMASEC BUTYL	17m	198780	5	0.57		40
s BUTYL	LIFE-GUARD BUTYL	2m	1	3	n.a.		77
s BUTYL	MSA CHEMPRUF	<60m			n.a.		48
g BUTYL	NORTH B-131				0.35	degrad.	65
g BUTYL	NORTH B-161	10m	>7500	5	0.38		25
g BUTYL	NORTH B-161	8m	8940	5	0.40		35
g BUTYL	NORTH B-174				0.65	degrad.	34
s BUTYL	WHEELER ACID KING	14m	<9000	5	n.a.	degrad.	85
g BUTYL/ECO	BRUNSWICK BUTYL-POL	113m	>5000	4	0.63		89
s BUTYL/NEOPRENE	MSA BETEX	20m	1470	4	n.a.		48
s CPE	STD. SAFETY	65m	18	2	n.a.		75
g NAT+NEOPR+NITRILE	MAPA-PIONEER TRIONI	16m			0.48		36
g NAT+NEOPR+NITRILE	MAPA-PIONEER TRIONI	8m	16000	5	0.63		70
g NATURAL RUBBER	ACKWELL 5-109				0.15	degrad.	34
g NATURAL RUBBER	ANSELL PVL 040				n.a.	degrad.	55
g NATURAL RUBBER	ANSELLEDMONT 36-124				0.46	degrad.	6
g NATURAL RUBBER	ANSELLEDMONT 392				0.48	degrad.	6
g NATURAL RUBBER	BEST 65NFW				n.a.	degrad.	53
g NATURAL RUBBER	COMASEC FLEXIGUM	10m	5880	5	0.95		40
u NATURAL RUBBER	Unknown	<1m	>7500	5	0.23		25
g NATURAL+NEOPRENE	ANSELL OMNI 276				0.45	degrad.	55
g NATURAL+NEOPRENE	PLAYTEX ARGUS 123	3m	14760	5	n.a.		72
g NEOPRENE	ANSELLEDMONT 29-840				0.38	degrad.	6
g NEOPRENE	ANSELLEDMONT 29-870	7m	>6400	5	0.43		25
g NEOPRENE	ANSELLEDMONT 29-870				0.48	degrad.	34
g NEOPRENE	ANSELLEDMONT 29-870	12m	9780	5	0.48		35
g NEOPRENE	ANSELLEDMONT 29-870	15m	12000	5	0.53		70
g NEOPRENE	BEST 32	14m	8570	5	n.a.	degrad.	53
g NEOPRENE	BEST 6780	15m	3110	4	n.a.	degrad.	53
g NEOPRENE	BRUNSWICK NEOPRENE	48m	>5000	4	1.01		89
g NEOPRENE	COMASEC COMAPRENE	7m	5700	5	n.a.		40
s NEOPRENE	LIFE-GUARD NEOPRENE	17m	97	3	n.a.		77
b NEOPRENE	RAINFAIR	35m	>370	3	n.a.		90
b NEOPRENE	RANGER	32m	>320	3	n.a.		90
b NEOPRENE	SERVUS NO 22204	51m	>340	3	n.a.		90
g NITRILE	ANSELL 632	40m			0.36		58
g NITRILE	ANSELL CHALLENGER	130m	2	1	0.38		55
g NITRILE	ANSELLEDMONT 37-155	144m			0.32		58
g NITRILE	ANSELLEDMONT 37-165	300m	<90	2	0.60		6
g NITRILE	ANSELLEDMONT 37-165	>120m			0.55		70
g NITRILE	ANSELLEDMONT 37-175	210m	282	3	0.34		35
g NITRILE	BEST 22R	10m	50	4	n.a.		81
g NITRILE	BEST 22R	5m	50	4	n.a.	pre-exp 48h	81
g NITRILE	BEST 727	>480m	0	0	0.38		53
g NITRILE	BEST 727	138m			0.38		58
g NITRILE	COMASEC COMATRIL	78m	1080	4	0.55		40

G A R M E Class & Number/ N test chemical/ T MATERIAL NAME	MANUFACTURER & PRODUCT IDENTIFICATION	Break-through Time in Minutes	Perm-eation Rate in mg/sq m /min.	I N D E X	Thick-ness in mm	degrad. and comments	R e f s
g NITRILE	COMASEC COMATRIL SU	240m	60	2	0.60		40
g NITRILE	COMASEC FLEXITRIL	6m	6180	5	n.a.		40
g NITRILE	GRANET 490	92m			0.32		58
g NITRILE	MAPA-PIONEER A-15	290m	0	0	0.33		58
g NITRILE	MARIGOLD NITROSOLVE	273m	12	2	0.75		79
g NITRILE	NORTH LA-142G	77m	330	3	0.36		34
g NITRILE	NORTH LA-142G	145m			0.32		58
g NITRILE	SURETY	377m	440	3	0.38		25
g NITRILE	SURETY 315R	296m		0	0.32		58
g NITRILE	UNIROYAL 4-15	128m			0.32		58
b NITRILE+PUR+PVC	BATA HAZMAX	>180m			n.a.		86
g NITRILE+PVC	COMASEC MULTIMAX	61m	498	3	n.a.		40
g NITRILE+PVC	COMASEC MULTIPLUS	72m	900	3	n.a.		40
b NITRILE+PVC	SERVUS NO 73101	>180m			2.40		90
b NITRILE+PVC	TINGLEY	156m	4	1	3.30		90
g PE	ANSELLEDMONT 35-125	<1m	>6850	5	0.03		25
g PE	ANSELLEDMONT 35-125	<1m	7680	5	0.07		35
c PE	DU PONT TYVEK QC	1m	4100	5	n.a.		8
g PE/EVAL/PE	SAFETY4 4H	>240m	0	0	0.07	35°C	60
g PE/EVAL/PE	SAFETY4 4H	>1440m	0	0	0.07		60
g PVAL	ANSELLEDMONT PVA	>360m		0	n.a.		6
g PVAL	EDMONT 15-552	300m	<9	1	n.a.		6
g PVAL	EDMONT 15-552	36m	150	3	0.46		25
g PVAL	EDMONT 15-554	>120m			n.a.		70
g PVAL	EDMONT 25-545	>360m	0	0	n.a.		6
g PVAL	EDMONT 25-545	>480m	0	0	0.45		35
g PVAL	EDMONT 25-950	>960m	0	0	0.38		34
g PVC	ANSELLEDMONT 3-318	36m	2000	4	n.a.		70
g PVC	ANSELLEDMONT 34-100	<1m	1806	5	0.18		34
g PVC	ANSELLEDMONT SNORKE				n.a.	degrad.	6
b PVC	BATA STANDARD	52m	18	3	n.a.		86
g PVC	BEST 725R				n.a.	degrad.	53
g PVC	BEST 812	20m	3500	4	n.a.		70
g PVC	BEST 812	12m	50	4	n.a.	degrad.	81
g PVC	BEST 812	15m	30	3	n.a.	pre-exp 48h	81
g PVC	COMASEC MULTIPOST	47m	1020	4	n.a.		40
g PVC	COMASEC MULTITOP	52m	1080	4	n.a.		40
g PVC	COMASEC NORMAL	43m	1140	4	n.a.		40
g PVC	COMASEC OMNI	26m	840	3	n.a.		40
b PVC	JORDAN DAVID	145m	21	2	n.a.		90
g PVC	MAPA-PIONEER V-5	<1m	7440	5	0.10	degrad.	35
s PVC	MSA UPC	<3m			0.20		48
b PVC	STD. SAFETY GL-20	12m	>4000	5	n.a.		75
s PVC	STD. SAFETY WG-20	8m	>4000	5	n.a.		75
s PVC	WHEELER ACID KING	17m	<9000	4	n.a.	degrad.	85
b PVC+POLYURETHANE	BATA POLYBLEND	14m	8	3	n.a.		86
b PVC+POLYURETHANE	BATA POLYBLEND	>180m			n.a.		90

G A R M E Class & Number/ N test chemical/ T MATERIAL NAME	MANUFACTURER & PRODUCT IDENTIFICATION	Break-through Time in Minutes	Perm-eation Rate in mg/sq m /min.	I N D E X	Thick-ness in mm	degrad. and comments	R e f s
b PVC+POLYURETHANE	BATA POLYMAX	156m	12	2	n.a.		86
b PVC+POLYURETHANE	BATA SUPER POLY	120m	12	2	n.a.		86
s SARANEX-23	DU PONT TYVEK SARAN	13m	11	4	n.a.		8
s SARANEX-23	DU PONT TYVEK SARAN	3m	150	4	n.a.		25
s SARANEX-23 2-PLY	DU PONT TYVEK SARAN	303m	120	3	n.a.		8
u TEFLON	CHEMFAB	2m	178	4	n.a.		65
s TEFLON	CHEMFAB CHALL. 5100	108m			n.a.		65
s TEFLON	CHEMFAB CHALL. 5200	>300m	0	0	n.a.		80
s TEFLON	CHEMFAB CHALL. 6000	>180m			n.a.		80
g TEFLON	CLEAN ROOM PRODUCTS	>1440m	0	0	0.05		25
g TEFLON	CLEAN ROOM PRODUCTS	26m	23	3	0.05	crumpled	25
s TEFLON	LIFE-GUARD TEFGUARD	>480m	0	0	n.a.		77
s TEFLON	WHEELER ACID KING	>480m	0	0	n.a.		85
v TEFLON-FEP	CHEMFAB CHALL.	>180m			0.25		65
s UNKNOWN MATERIAL	CHEMRON CHEMREL	21m	6	2	n.a.		67
s UNKNOWN MATERIAL	CHEMRON CHEMREL MAX	>480m	0	0	n.a.		67
s UNKNOWN MATERIAL	DU PONT BARRICADE	>480m		0	n.a.		8
s UNKNOWN MATERIAL	KAPPLER CPF III	>480m		0	n.a.		77
s UNKNOWN MATERIAL	LIFE-GUARD RESPONDE	>480m	0	0	n.a.		77
g UNKNOWN MATERIAL	NORTH SILVERSHIELD	>360m	0	0	0.08		7
g VITON	NORTH F-091	>1440m	0	0	0.23		25
g VITON	NORTH F-091	>1440m	0	0	0.23	preexp. 24h	25
g VITON	NORTH F-091	>1440m	0	0	0.24	10°C	25
g VITON	NORTH F-091	180m	42	2	0.24	45°C	25
g VITON	NORTH F-091	1020m	0	0	0.27		34
g VITON	NORTH F-091	2040m	0.27	1	0.30		62
g VITON	NORTH F-121	>480m	0	0	0.30		35
g VITON	NORTH SF	>1440m	0	0	0.23		25
s VITON/BUTYL/UNKN.	TRELLEBORG HPS	>180m			n.a.		71
s VITON/CHLOROBUTYL	LIFE-GUARD VC-100	>480m	0	0	n.a.		77
s VITON/CHLOROBUTYL	LIFE-GUARD VNC 200	>480m	0	0	n.a.		77
s VITON/NEOPRENE	MSA VAUTEX	>240m	0	0	n.a.		48
g VITON/NITRILE	NORTH VITRILE	>1440m	0	0	0.20		25

tetraethylenepentamine
CAS Number: 112-57-2
Primary Class: 148   Amines, Poly, Aliphatic and Alicyclic
Related Class: 141   Amines, Aliphatic and Alicyclic, Primary
              142   Amines, Aliphatic and Alicyclic, Secondary

g BUTYL	NORTH B-174	>480m	0	0	0.63		34
g NATURAL RUBBER	ACKWELL 5-109	106m	110	3	0.15		34
g NEOPRENE	ANSELLEDMONT 29-870	>480m	0	0	0.48		34
g PVC	ANSELLEDMONT 34-100				0.20	degrad.	34
g VITON	NORTH F-091	>480m	0	0	0.35		34

G A R M E Class & Number/ N test chemical/ T MATERIAL NAME	MANUFACTURER & PRODUCT IDENTIFICATION	Break- through Time in Minutes	Perm- eation Rate in mg/sq m /min.	I N D E X	Thick- ness degrad. in and mm comments	R e f s
tetrafluoroboric acid (synonym: fluoroboric acid) CAS Number: 16872-11-0 Primary Class: 370 Inorganic Acids						
g PE/EVAL/PE	SAFETY4 4H	>240m		0	0.07	60
tetrafluoroethylene (synonym: perfluoroethylene) CAS Number: 116-14-3 Primary Class: 263 Halogen Compounds, Vinylic						
g BUTYL	NORTH B-174	>480m	0	0	0.61	34
g NEOPRENE	ANSELLEDMONT 29-870	>480m	0	0	0.64	34
g PVAL	EDMONT 25-950	>480m	0	0	0.25	34
g VITON	NORTH F-091	>480m	0	0	0.25	34
tetrahydrofuran (synonym: THF) CAS Number: 109-99-9 Primary Class: 241 Ethers, Aliphatic and Alicyclic Related Class: 277 Heterocyclic Compounds, Oxygen, Furans						
g BUTYL	BEST 878	24m	1060	4	0.75 degrad.	53
g BUTYL	BRUNSWICK BUTYL STD	25m	5000	4	0.66	89
g BUTYL	BRUNSWICK BUTYL-XTR	27m	>5000	4	0.96	89
g BUTYL	COMASEC BUTYL	12m			0.50	40
s BUTYL	LIFE-GUARD BUTYL	9m	33	4	n.a.	77
g BUTYL	NORTH B-131				n.a. degrad.	65
g BUTYL	NORTH B-131	31m			0.35	83
g BUTYL	NORTH B-174	27m	6702	4	0.71 degrad.	34
m BUTYL	PLYMOUTH RUBBER	10m			n.a.	46
s BUTYL	WHEELER ACID KING	11m	<9000	5	n.a.	85
g BUTYL/ECO	BRUNSWICK BUTYL-POL	12m	>5000	5	0.63	89
s BUTYL/NEOPRENE	MSA BETEX	<5m	1470	5	n.a.	48
u CPE	ILC DOVER	12m			n.a.	29
s CPE	ILC DOVER CLOROPEL	33m			0.75	46
s CPE	STD. SAFETY	16m	154	3	n.a.	75
g HYPALON	COMASEC DIPCO	16m	15864	5	0.60	40
g NATURAL RUBBER	ACKWELL 5-109				0.15 degrad.	34
g NATURAL RUBBER	ANSELL ORANGE 208				0.72 degrad.	55
g NATURAL RUBBER	ANSELL PVL 040				0.46 degrad.	55
g NATURAL RUBBER	ANSELLEDMONT 36-124				0.46 degrad.	6
g NATURAL RUBBER	ANSELLEDMONT 392				0.48 degrad.	6
g NATURAL RUBBER	ANSELLEDMONT 46-320	2m	250000	5	0.31	5
g NATURAL+NEOPRENE	ANSELL OMNI 276	3m	23000	5	0.45	5
g NATURAL+NEOPRENE	ANSELL TECHNICIANS				0.43 degrad.	55
g NEOPRENE	ANSELL NEOPRENE 520				0.38 degrad.	55
g NEOPRENE	ANSELLEDMONT 29-840				0.38 degrad.	6
g NEOPRENE	ANSELLEDMONT 29-870				0.48 degrad.	34

G				Perm-	I			R
A				eation	N	Thick-		
R			Break-	eation				R
M			through	Rate in	D	ness	degrad.	e
E Class & Number/	MANUFACTURER	Time in	mg/sq m	E	in	and	f
N test chemical/	& PRODUCT	Time in	mg/sq m	E	in	and	f
T MATERIAL NAME	IDENTIFICATION	Minutes	/min.	X	mm	comments	s

GARMENT	Class & Number/ test chemical/ MATERIAL NAME	MANUFACTURER & PRODUCT IDENTIFICATION	Break-through Time in Minutes	Permeation Rate in mg/sq m /min.	INDEX	Thickness in mm	degrad. and comments	Refs
g	NEOPRENE	ANSELLEDMONT NEOX	1m	80000	5	n.a.		5
g	NEOPRENE	BEST 32	6m	1730	5	n.a.	degrad.	53
g	NEOPRENE	BEST 6780	9m	4880	5	n.a.		53
g	NEOPRENE	BRUNSWICK NEOPRENE	12m	>5000	5	0.96		89
s	NEOPRENE	LIFE-GUARD NEOPRENE	11m	54	4	n.a.		77
g	NEOPRENE	MAPA-PIONEER GF-N-	5m	160000	5	0.47		5
g	NEOPRENE	MAPA-PIONEER N-44	11m	40260	5	0.74	degrad.	36
g	NEOPRENE	MAPA-PIONEER N-73	4m	140000	5	0.46		5
b	NEOPRENE	RAINFAIR	28m	>170	3	n.a.		90
b	NEOPRENE	RANGER	17m	>200	3	n.a.		90
b	NEOPRENE	SERVUS NO 22204	44m	>130	3	n.a.		90
g	NEOPRENE/NATURAL	ANSELL CHEMI-PRO				n.a.	degrad.	55
g	NITRILE	ANSELL CHALLENGER				0.38	degrad.	55
g	NITRILE	ANSELLEDMONT 37-165	6m	27000	5	0.55		5
g	NITRILE	ANSELLEDMONT 37-165				0.60	degrad.	6
g	NITRILE	ANSELLEDMONT 37-175	4m	43000	5	0.37		5
g	NITRILE	BEST 22R		9300	5	n.a.	degrad.	53
g	NITRILE	BEST 727	6m	1720	5	0.38	degrad.	53
g	NITRILE	MAPA-PIONEER A-14	17m	40260	5	0.54	degrad.	36
g	NITRILE	MARIGOLD NITROSOLVE				0.75	degrad.	79
g	NITRILE	NORTH LA-142G	<1m	10040	5	0.36	degrad.	34
b	NITRILE+PUR+PVC	BATA HAZMAX	152m	52	2	n.a.		86
g	NITRILE+PVC	COMASEC MULTIMAX	24m	96	3	n.a.		40
b	NITRILE+PVC	SERVUS NO 73101	148m	>170	3	n.a.		90
b	NITRILE+PVC	TINGLEY	132m	460	3	n.a.		90
g	PE	ANSELLEDMONT 35-125	1m	2000	5	0.03		5
c	PE	DU PONT TYVEK QC	<1m	1620	5	n.a.		8
g	PE/EVAL/PE	SAFETY4 4H	6m	0.05	1	0.07	35°C	60
g	PE/EVAL/PE	SAFETY4 4H	>1440m		0	0.07		60
g	PVAL	ANSELLEDMONT PVA	90m	<9	1	n.a.		6
g	PVAL	EDMONT 15-552				n.a.	degrad.	6
g	PVAL	EDMONT 25-950	282m	25	2	0.33		34
g	PVC	ANSELLEDMONT 34-100				0.20	degrad.	34
g	PVC	ANSELLEDMONT SNORKE				n.a.	degrad.	6
b	PVC	BATA STANDARD	77m	28	2	n.a.		86
g	PVC	BEST 725R				n.a.	degrad.	53
b	PVC	JORDAN DAVID	85m	>250	3	n.a.		90
g	PVC	MAPA-PIONEER V-20	<1m	>90000	5	0.31		5
b	PVC	STD. SAFETY GL-20	4m	>4000	5	n.a.		75
s	PVC	STD. SAFETY WG-20	4m	>4000	5	n.a.		75
s	PVC	WHEELER ACID KING	6m	<90000	5	n.a.	degrad.	85
b	PVC+POLYURETHANE	BATA POLYBLEND	15m	36	3	n.a.		86
b	PVC+POLYURETHANE	BATA POLYBLEND	164m	82	2	n.a.		90
b	PVC+POLYURETHANE	BATA POLYMAX	111m	42	2	n.a.		86
b	PVC+POLYURETHANE	BATA SUPER POLY	7m	25	4	n.a.		86
s	SARANEX-23	DU PONT TYVEK SARAN	<1m	>1000	5	n.a.		8
s	SARANEX-23 2-PLY	DU PONT TYVEK SARAN	12m	1260	5	n.a.		8
u	TEFLON	CHEMFAB	2m	30	4	0.10		65

G A R M E Class & Number/ N test chemical/ T MATERIAL NAME	MANUFACTURER & PRODUCT IDENTIFICATION	Break-through Time in Minutes	Perm-eation Rate in mg/sq m /min.	I N D E X	Thick-ness in mm	degrad. and comments	R e f s
s TEFLON	CHEMFAB CHALL. 5100	>480m	0	0	n.a.		56
s TEFLON	CHEMFAB CHALL. 5100	>330m	0	0	n.a.		65
s TEFLON	CHEMFAB CHALL. 5200	>300m	0	0	n.a.		80
s TEFLON	CHEMFAB CHALL. 6000	>180m			n.a.		80
s TEFLON	LIFE-GUARD TEFGUARD	>480m	0	0	n.a.		77
s TEFLON	WHEELER ACID KING	>480m	0	0	n.a.		85
v TEFLON-FEP	CHEMFAB CHALL.	>180m			0.25		65
s UNKNOWN MATERIAL	CHEMRON CHEMREL	7m	120	4	n.a.		67
s UNKNOWN MATERIAL	DU PONT BARRICADE	>480m	0	0	n.a.		8
s UNKNOWN MATERIAL	KAPPLER CPF III	>480m	0	0	n.a.		77
s UNKNOWN MATERIAL	LIFE-GUARD RESPONDE	>480m	0	0	n.a.		77
g UNKNOWN MATERIAL	NORTH SILVERSHIELD	16m	3	2	0.08		7
g UNKNOWN MATERIAL	NORTH SILVERSHIELD	16m			0.08		83
g VITON	NORTH F-091	<1m	19600	5	0.26	degrad.	34
g VITON	NORTH F-091	4m			0.25		83
s VITON/BUTYL/UNKN.	TRELLEBORG HPS	>180m			n.a.		71
s VITON/CHLOROBUTYL	LIFE-GUARD VC-100	14m			n.a.		77
s VITON/CHLOROBUTYL	LIFE-GUARD VNC 200	22m	11	3	n.a.		77
u VITON/CHLOROBUTYL	Unknown	8m			n.a.		29
u VITON/CHLOROBUTYL	Unknown	10m			n.a.		46
s VITON/NEOPRENE	MSA VAUTEX	<5m	2810	5	n.a.		48

tetramethylammonium hydroxide
CAS Number: 75-59-2
Primary Class: 550   Organic Salts (Solutions)

g PE/EVAL/PE	SAFETY4 4H	>240m	0	0	0.07	35°C	60

thiocarbamide, <30% in ETH
CAS Number: 62-56-6
Primary Class: 137   Amides, Carbamides, and Guanidines

g PE/EVAL/PE	SAFETY4 4H	>240m	0	0	0.07	35°C	60

thiophene
CAS Number: 110-02-1
Primary Class: 279   Heterocyclic Compounds, Sulfur
Related Class: 502   Sulfur Compounds, Sulfides and Disulfides

g BUTYL	NORTH B-174	108m	1027	4	0.66	degrad.	34
g NATURAL RUBBER	ACKWELL 5-109				0.15	degrad.	34
g NEOPRENE	ANSELLEDMONT 29-870				0.48	degrad.	34
g NITRILE	ANSELLEDMONT 37-155				0.40	degrad.	34
g PVAL	EDMONT 25-950	150m	5	1	0.36		34
g PVC	ANSELLEDMONT 34-100	1m	1540	5	0.20	degrad.	34
g VITON	NORTH F-091	>480m	0	0	0.28		34

G A R M E Class & Number/ N test chemical/ T MATERIAL NAME	MANUFACTURER & PRODUCT IDENTIFICATION	Break- through Time in Minutes	Perm- eation Rate in mg/sq m /min.	I N D E X	Thick- ness in mm	degrad. and comments	R e f s
titanium tetrachloride CAS Number: 7550-45-0 Primary Class: 360 Inorganic Acid Halides							
s PVC	TRELLEBORG TRELLCHE				n.a.	degrad.	71
s SARANEX-23	DU PONT TYVEK SARAN	>996m	0	0	n.a.		8
toluene (synonym: methylbenzene) CAS Number: 108-88-3 Primary Class: 292 Hydrocarbons, Aromatic							
g BUTYL	BEST 878	22m	5340	4	0.75	degrad.	53
g BUTYL	BRUNSWICK BUTYL STD	28m	>5000	4	0.68		89
g BUTYL	BRUNSWICK BUTYL-XTR	28m	>5000	4	0.68		89
g BUTYL	COMASEC BUTYL	13m	1260	5	n.a.		40
g BUTYL	ERISTA BX	9m			0.59	37°C TL	11
s BUTYL	LIFE-GUARD BUTYL	6m	77	4	n.a.		77
m BUTYL	NASA	25m	960	3	n.a.		66
m BUTYL	NASA	14m	1660	5	n.a.	45°C	66
g BUTYL	NORTH 8B 1532A	18m	1460	4	n.a.		66
g BUTYL	NORTH 8B 1532A	11m	2990	5	n.a.	45°C	66
g BUTYL	NORTH B-131				n.a.	degrad.	65
g BUTYL	NORTH B-161	17m	20200	5	0.49		11
g BUTYL	NORTH B-174	21m	1326	4	0.58		34
m BUTYL	PLYMOUTH RUBBER	3m			n.a.		46
u BUTYL	Unknown	28m	690	3	0.36		1
s BUTYL	WHEELER ACID KING	12m	<90000	5	0.30		85
g BUTYL/ECO	BRUNSWICK BUTYL-POL	8m	>5000	5	0.63		89
g BUTYL/NEOPRENE	COMASEC BUTYL PLUS	13m	1260	5	0.50		40
s BUTYL/NEOPRENE	MSA BETEX	<5m	1700	5	n.a.		48
s CPE	ILC DOVER CLOROPEL	72m			n.a.		46
s CPE	STD. SAFETY	28m	64	3	n.a.		75
g HYPALON	COMASEC DIPCO	24m	20778	5	0.60		40
g NATURAL RUBBER	ACKWELL 5-109				0.15	degrad.	34
g NATURAL RUBBER	ANSELL PVL 040				0.46	degrad.	55
g NATURAL RUBBER	ANSELLEDMONT 36-124				0.46	degrad.	6
g NATURAL RUBBER	ANSELLEDMONT 392				0.48	degrad.	6
g NATURAL RUBBER	ANSELLEDMONT 46-320	2m	47000	5	0.31		5
g NATURAL RUBBER	BEST 65NFW	26m	2550	4	n.a.	degrad.	53
g NATURAL RUBBER	COMASEC FLEXIGUM	9m	6360	5	0.95		40
g NATURAL RUBBER	MAPA-PIONEER 258	6m	850	4	0.64	30°C	87
g NATURAL RUBBER	MAPA-PIONEER 296	2m	500	4	0.51	30°C	87
g NATURAL RUBBER	MAPA-PIONEER 300	4m	200	4	n.a.	30°C	87
g NATURAL RUBBER	MAPA-PIONEER 730	14m	160	4	1.15	30°C	87
u NATURAL RUBBER	Unknown	1m	>5000	5	n.a.		10
g NATURAL+NEOPRENE	ANSELL TECHNICIANS				0.43	degrad.	55
g NATURAL+NEOPRENE	MAPA-PIONEER 424	6m	550	4	0.58	30°C	87

G A R M E N T Class & Number/ test chemical/ MATERIAL NAME	MANUFACTURER & PRODUCT IDENTIFICATION	Break-through Time in Minutes	Perm-eation Rate in mg/sq m /min.	I N D E X	Thick-ness in mm	degrad. and comments	R e f s
g NEOPRENE	ANSELL NEOPRENE 530	3m	0.7	3	0.46		55
g NEOPRENE	ANSELLEDMONT 29-840				0.38	degrad.	6
g NEOPRENE	ANSELLEDMONT 29-870	12m	1308	5	0.51	degrad.	34
g NEOPRENE	ANSELLEDMONT 29-870	10m	250	4	n.a.		81
g NEOPRENE	ANSELLEDMONT 29-870	8m	240	4	n.a.	pre-exp 48h	81
g NEOPRENE	BEST 32	6m	3130	5	n.a.	degrad.	53
g NEOPRENE	BEST 6780	22m	1800	4	n.a.	degrad.	53
g NEOPRENE	BRUNSWICK NEOPRENE	20m	>5000	4	0.93		89
g NEOPRENE	COMASEC COMAPRENE	9m	4980	5	n.a.		40
s NEOPRENE	LIFE-GUARD NEOPRENE	12m	92	4	n.a.		77
g NEOPRENE	MAPA-PIONEER 360	10m	80	4	n.a.	30°C	87
g NEOPRENE	MAPA-PIONEER 420	6m	350	4	0.75	30°C	87
g NEOPRENE	MAPA-PIONEER N-44	14m	22560	5	0.56		36
g NEOPRENE	MAPA-PIONEER N-73	17m	9000	4	0.46		5
g NEOPRENE	NOLATO 1505	31m	25000	5	0.85		11
b NEOPRENE	RAINFAIR	27m	>420	3	n.a.		90
b NEOPRENE	RANGER	20m	>700	3	n.a.		90
b NEOPRENE	SERVUS NO 22204	32m	>550	3	n.a.		90
u NEOPRENE	Unknown	7m	7660	5	0.38		1
u NEOPRENE	Unknown	18m			0.40	37°C	16
g NEOPRENE/NATURAL	ANSELL CHEMI-PRO	4m	16	4	0.72		55
g NITRILE	ANSELL CHALLENGER	30m	3	2	0.38		55
g NITRILE	ANSELLEDMONT 37-155	19m	7000	4	0.35		5
g NITRILE	ANSELLEDMONT 37-155	14m	2400	5	0.40	degrad.	34
g NITRILE	ANSELLEDMONT 37-155	33m	4080	4	0.37		41
g NITRILE	ANSELLEDMONT 37-165	>60m			0.55		5
g NITRILE	ANSELLEDMONT 37-165	10m	<9000	5	0.60		6
g NITRILE	ANSELLEDMONT 37-165	61m	4000	4	0.64		11
g NITRILE	ANSELLEDMONT 37-165	29m			0.60	37°C TL	11
g NITRILE	ANSELLEDMONT 37-175	36m	5000	4	0.37		5
g NITRILE	ANSELLEDMONT 37-175	23m	3300	4	0.37		11
g NITRILE	ANSELLEDMONT 37-175	17m	3300	4	0.37		11
g NITRILE	ANSELLEDMONT 37-175	14m	2200	5	0.37	34°C	11
g NITRILE	ANSELLEDMONT 37-175	13m			0.38	37°C TL	11
g NITRILE	ANSELLEDMONT 49-125	4m	<900	4	0.23		42
g NITRILE	ANSELLEDMONT 49-155	20m	<9000	4	0.38		42
g NITRILE	BEST 22R				n.a.	degrad.	53
g NITRILE	BEST 727				0.38	degrad.	53
g NITRILE	COMASEC COMATRIL	45m	4020	4	0.55		40
g NITRILE	COMASEC COMATRIL SU	60m	3300	4	0.60		40
g NITRILE	COMASEC FLEXITRIL	8m	5280	5	n.a.		40
g NITRILE	ERISTA P	41m	5100	4	0.90		11
g NITRILE	ERISTA SPECIAL	7m	6200	5	0.35		11
g NITRILE	ERISTA SPECIAL	6m	7100	5	0.35	34°C	11
g NITRILE	ERISTA SPECIAL	6m			0.40	37°C TL	11
g NITRILE	KURSAAL 85/62	10m	1500	5	0.41		15
g NITRILE	MAPA-PIONEER 370	10m	70	4	n.a.	30°C	87
g NITRILE	MAPA-PIONEER 490	9m	250	4	0.39	30°C	87

G A R M E N T	Class & Number/ test chemical/ MATERIAL NAME	MANUFACTURER & PRODUCT IDENTIFICATION	Break-through Time in Minutes	Perm-eation Rate in mg/sq m /min.	I N D E X	Thick-ness in mm	degrad. and comments	R e f s
g	NITRILE	MAPA-PIONEER 492	18m	262	3	0.42	30°C	87
g	NITRILE	MAPA-PIONEER 493	22m	135	3	0.55	30°C	87
g	NITRILE	MAPA-PIONEER A-14	28m	4500	4	0.56	degrad.	36
g	NITRILE	MARIGOLD BLUE	44m	507	3	0.43	degrad.	79
g	NITRILE	MARIGOLD GREEN SUPA	30m	653	3	0.40	degrad.	79
g	NITRILE	MARIGOLD NITROSOLVE	47m	477	3	0.75	degrad.	79
g	NITRILE	NORTH LA-102G	15m	2200	4	0.28		11
g	NITRILE	NORTH LA-132G17m		2200	4	0.36		11
g	NITRILE	NORTH LA-132G	9m	2000	5	0.36	34°C	11
g	NITRILE	NORTH LA-132G	8m			0.37	37°C TL	11
g	NITRILE	NORTH LA-142G	11m	4086	5	0.36		7
u	NITRILE	Unknown	13m			0.50	37°C	16
b	NITRILE+PUR+PVC	BATA HAZMAX	>180m			n.a.		86
g	NITRILE+PVC	COMASEC MULTIMAX	23m	6120	4	n.a.		40
g	NITRILE+PVC	COMASEC MULTIPLUS	40m	3660	4	n.a.		40
b	NITRILE+PVC	SERVUS NO 73101	111m	>200	3	n.a.		90
b	NITRILE+PVC	TINGLEY	64m	>5000	4	n.a.		90
g	PE	ANSELLEDMONT 35-125	<1m	22000	5	0.03		5
c	PE	DU PONT TYVEK QC	<1m	1650	5	n.a.		8
u	PE	Unknown	7m			0.12	37°C	16
g	PE/EVAL/PE	SAFETY4 4H	>240m	0	0	0.07	35°C	60
g	PE/EVAL/PE	SAFETY4 4H	>1440m	0	0	0.07		60
g	PVAL	ANSELLEDMONT PVA	>360m		0	n.a.		6
g	PVAL	COMASEC SOLVATRIL	>480m	0	0	n.a.		40
g	PVAL	EDMONT 15-552	15m	<90	3	n.a.		6
g	PVAL	EDMONT 15-552	>60m			n.a.		10
g	PVAL	EDMONT 15-552	>1500m	0	0	n.a.		11
g	PVAL	EDMONT 15-552	>1500m	0	0	n.a.	in use	7
Mo		11						
g	PVAL	EDMONT 25-545	>360m	0	0	n.a.		6
g	PVAL	EDMONT 25-950	>480m	0	0	0.61		30
g	PVC	ANSELLEDMONT 34-100	2m	1680	5	0.20		34
g	PVC	ANSELLEDMONT SNORKE				n.a.	degrad.	6
b	PVC	BATA STANDARD	138m	0.5	1	n.a.	degrad.	86
g	PVC	COMASEC MULTIPOST	20m	4200	4	n.a.		40
g	PVC	COMASEC MULTITOP	35m	3900	4	n.a.		40
g	PVC	COMASEC NORMAL	20m	4260	4	n.a.		40
g	PVC	COMASEC OMNI	12m	5280	5	n.a.		40
b	PVC	JORDAN DAVID	102m	312	3	n.a.		90
g	PVC	KID 490	28m	2000	4	n.a.		11
g	PVC	KID 490	21m			n.a.	37°C TL	11
g	PVC	KID 500	8m	3100	5	n.a.		11
g	PVC	KID 500	8m	3100	5	n.a.	34°C	11
g	PVC	KID VINYLPRODUKTER	8m	23000	5	0.55		11
g	PVC	MAPA-PIONEER V-20	3m	21000	5	0.31		5
g	PVC	NORTH 800	10m	10000	5	n.a.		15
b	PVC	STD. SAFETY GL-20	6m	>400	4	n.a.		75
s	PVC	STD. SAFETY WG-20	4m	>400	4	n.a.		75

G A R M E N T Class & Number/ test chemical/ MATERIAL NAME	MANUFACTURER & PRODUCT IDENTIFICATION	Break-through Time in Minutes	Perm-eation Rate in mg/sq m /min.	I N D E X	Thick-ness in mm	degrad. and comments	R e f s
s PVC	WHEELER ACID KING	9m	<9000	5	n.a.		85
b PVC+POLYURETHANE	BATA POLYBLEND	151m	0.4	1	n.a.	degrad.	86
b PVC+POLYURETHANE	BATA POLYBLEND	167m	7	1	n.a.		90
b PVC+POLYURETHANE	BATA POLYMAX	124m	42	2	n.a.		86
b PVC+POLYURETHANE	BATA SUPER POLY	175m	0.4	1	n.a.	degrad.	86
s SARANEX-23	DU PONT TYVEK SARAN	<5m	200	4	n.a.		8
s SARANEX-23 2-PLY	DU PONT TYVEK SARAN	82m	480	3	n.a.		8
u TEFLON	CHEMFAB	2m	>1000	5	0.10		65
s TEFLON	CHEMFAB CHALL. 5100	>480m	0	0	n.a.		56
s TEFLON	CHEMFAB CHALL. 5100	>180m	0	1	n.a.		65
s TEFLON	CHEMFAB CHALL. 5200	>300m	0	0	n.a.		80
s TEFLON	CHEMFAB CHALL. 6000	>180m			n.a.		80
s TEFLON	LIFE-GUARD TEFGUARD	>480m	0	0	n.a.		77
s TEFLON	WHEELER ACID KING	>480m	0	0	n.a.		85
v TEFLON-FEP	CHEMFAB CHALL.	>180m			0.25		65
s UNKNOWN MATERIAL	CHEMRON CHEMREL	142m	<1	1	n.a.		67
s UNKNOWN MATERIAL	CHEMRON CHEMREL MAX	>480m	0	0	n.a.		67
s UNKNOWN MATERIAL	DU PONT BARRICADE	>480m	0	0	n.a.		8
s UNKNOWN MATERIAL	KAPPLER CPF III	>480m		0	n.a.		77
s UNKNOWN MATERIAL	LIFE-GUARD RESPONDE	>480m	0	0	n.a.		77
g UNKNOWN MATERIAL	NORTH SILVERSHIELD	>360m	0	0	0.08		7
g VITON	NORTH F-091	>960m	0	0	0.24		7
g VITON	NORTH F-091	>270m	0	0	0.24		11
g VITON	NORTH F-091	71m			0.24	37°C	11
g VITON	NORTH F-091	>480m	0	0	0.30		34
g VITON	NORTH F-091	780m	0.4	1	0.30		62
g VITON	NORTH F-121	>420m	0	0	0.33	in use 1 Mo	11
s VITON/BUTYL/UNKN.	TRELLEBORG HPS	>180m			n.a.		71
s VITON/CHLOROBUTYL	LIFE-GUARD VC-100	398m		0	n.a.		77
s VITON/CHLOROBUTYL	LIFE-GUARD VNC 200	>480m	0	0	n.a.		77
u VITON/CHLOROBUTYL	Unknown	>180m			n.a.		43
g VITON/NEOPRENE	ERISTA VITRIC	252m	2000	4	0.60		11
g VITON/NEOPRENE	ERISTA VITRIC	162m			0.55	37°C	11
g VITON/NEOPRENE	ERISTA VITRIC	90m	80	2	0.45		15
s VITON/NEOPRENE	MSA VAUTEX	>240m	0	0	n.a.		48

toluene, 30-70% & butyl alcohol, butyl acetate, ethyl acetate and methanol
Primary Class: 800   Multicomponent Mixtures With > 2 Components

g PE/EVAL/PE	SAFETY4 4H	>240m	0	0	0.07	35°C	60
g PVAL	EDMONT 15-552	26m	420	3	n.a.		33
g VITON/NEOPRENE	ERISTA VITRIC	16m	3000	4	0.51	degrad.	33

toluene, 5-20% & butanol, butyl acetate, ethanol, methyl ethyl ketone, xylene
Primary Class: 800   Multicomponent Mixtures With >2 Components

g BUTYL	NORTH B-161	156m	66	2	0.43		33
g BUTYL	NORTH B-161	126m	73	2	0.43	preexp.	

G A R M E Class & Number/ N test chemical/ T MATERIAL NAME	MANUFACTURER & PRODUCT IDENTIFICATION	Break-through Time in Minutes	Perm-eation Rate in mg/sq m /min.	I N D E X	Thick-ness in mm	degrad. and comments	R e f s
						20h	33
g HYPALON	COMASEC DIPCO	26m	4100	4	0.60		33
g NITRILE	ANSELLEDMONT 37-175	6m	9150	5	0.40		33
g NITRILE	ANSELLEDMONT 37-175	6m	7500	5	0.40	preexp.	
						20h	33
g NITRILE	ERISTA SPECIAL	14m	8410	5	0.40		33
g NITRILE	ERISTA SPECIAL	12m	7800	5	0.40	preexp.	
						20h	33
g PE/EVAL/PE	SAFETY4 4H	>240m	0	0	0.07	35°C	60
g PVAL	EDMONT 15-552	>240m	0	0	n.a.		33
g PVAL	EDMONT 15-552	>240m	0	0	n.a.	preexp.	
						20h	33
g PVC	KID 490	9m	2900	5	n.a.		33
g PVC	KID VINYLPRODUKTER	4m	8540	5	0.53		33
g PVC	KID VINYLPRODUKTER	6m	5760	5	0.55		33
g VITON	NORTH F-091	5m	6700	5	0.25		33
g VITON	NORTH F-091	4m	7200	5	0.25	preexp.	
						20h	33

toluene, 50% & chloroform, 50%
Primary Class: 292   Hydrocarbons, Aromatic

g NEOPRENE	MAPA-PIONEER N-30	4/3m	45000	5	0.29		5

toluene, 50% & dimethylsulfoxide, 50%
Primary Class: 292   Hydrocarbons, Aromatic

toluene, 50% & hexane, 50%
Primary Class: 292   Hydrocarbons, Aromatic

toluene, 50% & isopropyl alcohol, 50%
Primary Class: 292   Hydrocarbons, Aromatic
Related Class: 312   Hydroxy Compounds, Aliphatic and Alicyclic, Secondary

g PE/EVAL/PE	SAFETY4 4H	>240m	0	0	0.07	35°C	60

toluene, 50% & MEK, 15% & methanol, 15% & butylglycol, 10% & ethyl acetate, 10%
Primary Class: 800   Multicomponent Mixtures With >2 Components

g PE/EVAL/PE	SAFETY4 4H	>240m		0	0.07		60

toluene, 50% & methyl alcohol, 50%
Primary Class: 292   Hydrocarbons, Aromatic

u NEOPRENE	Unknown	22/23m			0.40	37°C	TL 16

G A R M E Class & Number/ N test chemical/ T MATERIAL NAME	MANUFACTURER & PRODUCT IDENTIFICATION	Break-through Time in Minutes	Perm-eation Rate in mg/sq m /min.	I N D E X	Thick-ness in mm	R degrad. e and f comments s
u NITRILE	Unknown	19/18m			0.50	37°C TL 16
u PE	Unknown	15/19m			0.12	37°C TL 16
u PVC	Unknown	20/18m			0.30	37°C TL 16

toluene, 50% & methyl alcohol, 50%
Primary Class: 292   Hydrocarbons, Aromatic
Related Class: 292   Hydrocarbons, Aromatic

g PE/EVAL/PE	SAFETY4 4H	9m			0.07	35°C 60
g PE/EVAL/PE	SAFETY4 4H	114m			0.07	60
s UNKNOWN MATERIAL	CHEMRON CHEMREL	32m	1	2	n.a.	67

toluene, 50% & methyl isobutyl ketone, 50%
Primary Class: 292   Hydrocarbons, Aromatic

toluene, 50% & methylene chloride, 50%
Primary Class: 292   Hydrocarbons, Aromatic

toluene, 50% & p-xylene, 50%
Primary Class: 292   Hydrocarbons, Aromatic

toluene, 50% & water, 50%
Primary Class: 292   Hydrocarbons, Aromatic

u PVAL	Unknown	4/3m			0.07	37°C TL 16

toluene, 75% & naphthalene, 25%
Primary Class: 292 Hydrocarbons, Aromatic

g PE/EVAL/PE	SAFETY4 4H	>240m		0	0.07	60

toluene, 75% & p-xylene, 25%
Primary Class: 292 Hydrocarbons, Aromatic

g NITRILE	ANSELLEDMONT 37-155	26/29m	2760/	4	0.35	41

toluene-2,4-diisocyanate (synonym: TDI)
CAS Number: 584-84-9
Primary Class: 212   Isocyanates, Aromatic

g BUTYL	BEST 878	>480m	0	0	0.75	53
g BUTYL	BRUNSWICK BUTYL STD	>480m	0	0	0.63	89

G A R M E Class & Number/ N test chemical/ T MATERIAL NAME	MANUFACTURER & PRODUCT IDENTIFICATION	Break-through Time in Minutes	Perm-eation Rate in mg/sq m /min.	I N D E X	Thick-ness in mm	degrad. and comments	R e f s
g BUTYL	NORTH B-161	>480m	0	0	0.41		34
u CPE	ILC DOVER	>180m			n.a.		29
g NAT+NEOPR+NITRILE	MAPA-PIONEER TRIONI	27m	>5000	4	0.48	80%	36
g NATURAL RUBBER	ANSELLEDMONT 36-124	7m	<900	4	0.46		6
g NATURAL RUBBER	ANSELLEDMONT 392	7m	<9	3	0.48		6
g NATURAL RUBBER	MAPA-PIONEER L-118	>480m	0	0	0.46		36
g NATURAL RUBBER	MARIGOLD BLACK HEVY	29m	2	2	0.65	degrad.	79
g NATURAL RUBBER	MARIGOLD FEATHERLIT	15m	11	3	0.15	degrad.	79
g NATURAL RUBBER	MARIGOLD MED WT EXT	15m	3	2	0.73	degrad.	79
g NATURAL RUBBER	MARIGOLD MEDICAL	15m	8	2	0.28	degrad.	79
g NATURAL RUBBER	MARIGOLD ORANGE SUP	30m	2	2	0.73	degrad.	79
g NATURAL RUBBER	MARIGOLD R SURGEONS	15m	8	2	0.28	degrad.	79
g NATURAL RUBBER	MARIGOLD RED LT WT	15m	4	2	0.43	degrad.	79
g NATURAL RUBBER	MARIGOLD SENSOTECH	15m	8	2	0.28	degrad.	79
g NATURAL RUBBER	MARIGOLD SUREGRIP	25m	2	2	0.58	degrad.	79
g NATURAL RUBBER	MARIGOLD YELLOW	15m	5	2	0.38	degrad.	79
g NATURAL+NEOPRENE	MARIGOLD FEATHERWT	15m	6	2	0.35	degrad.	79
g NEOPRENE	ANSELLEDMONT 29-840				0.38	degrad.	6
g NEOPRENE	BRUNSWICK NEOPRENE	>480m	0	0	0.63		89
g NITRILE	ANSELLEDMONT 37-165				0.60	degrad.	6
g NITRILE	MAPA-PIONEER A-14	>480m	0	0	0.56	degrad.	36
g NITRILE	MARIGOLD BLUE	464m	1	1	0.45	degrad.	79
g NITRILE	MARIGOLD GREEN SUPA	455m	1	1	0.30	degrad.	79
g NITRILE	MARIGOLD NITROSOLVE	466m	0	0	0.75	degrad.	79
g NITRILE	NORTH LA-142G	222m	108	3	0.33		34
c PE	DU PONT TYVEK QC	<1m	420	4	n.a.		8
g PE/EVAL/PE	SAFETY4 4H	>240m		0	0.07	35°C	60
g PVAL	ANSELLEDMONT PVA	>360m		0	n.a.		6
g PVAL	EDMONT 15-552	>360m	0	0	n.a.		6
g PVAL	EDMONT 25-545	30m	<9	2	n.a.		6
g PVAL	EDMONT 25-950	>960m	0	0	0.25		34
g PVC	ANSELLEDMONT SNORKE				n.a.	degrad.	6
g PVC	MAPA-PIONEER V-20	>480m	0	0	0.51	degrad.	36
s SARANEX-23	DU PONT TYVEK SARAN	>480m	0	0	n.a.	80%	8
s TEFLON	CHEMFAB CHALL. 5100	>240m	0	0	n.a.		65
s UNKNOWN MATERIAL	CHEMRON CHEMREL	>480m	0	0	n.a.		67
s UNKNOWN MATERIAL	DU PONT BARRICADE	>480m	0	0	n.a.		8
s UNKNOWN MATERIAL	KAPPLER CPF III	>480m	0	0	n.a.		77
s UNKNOWN MATERIAL	LIFE-GUARD RESPONDE	>480m	0	0	n.a.		77
g UNKNOWN MATERIAL	NORTH SILVERSHIELD	>480m	0	0	0.10		7
g VITON	NORTH F-091	>960m	0	0	0.26		34
s VITON/BUTYL	DRAGER 500 OR 710	>30m			n.a.		91
u VITON/CHLOROBUTYL	Unknown	>180m			n.a.		29

toluene-2,4-diisocyanate, 40% & xylene, 60%
Primary Class: 212   Isocyanates, Aromatic

g PE/EVAL/PE	SAFETY4 4H	>240m	0	0	0.07	35°C	60
g PE/EVAL/PE	SAFETY4 4H	>480m	0	0	0.07		60

G A R M E Class & Number/ N test chemical/ T MATERIAL NAME	MANUFACTURER & PRODUCT IDENTIFICATION	Break- through Time in Minutes	Perm- eation Rate in mg/sq m /min.	I N D E X	Thick- ness in mm	degrad. and comments	R e f s
trans-1,2-dichloroethylene CAS Number: 156-60-5 Primary Class: 263 Halogen Compounds, Vinylic							
g BUTYL	NORTH B-174	8m	88260	5	0.63	degrad.	34
g NATURAL RUBBER	ACKWELL 5-109				0.15	degrad.	34
g NEOPRENE	ANSELLEDMONT 29-870				0.48	degrad.	34
g NITRILE	ANSELLEDMONT 37-155				0.40	degrad.	34
g PVAL	EDMONT 25-950	160m	6840	4	0.60	degrad.	34
g PVC	ANSELLEDMONT 34-100	1m	37500	5	0.20		34
g VITON	NORTH F-091	70m	135	3	0.25		34
transformer oil, Nytro 10X Primary Class: 860 Coals, Charcoals, Oils							
g PE/EVAL/PE	SAFETY4 4H	>240m	0	0	n.a.	35°C	60
transmission oil, opel Primary Class: 860 Coals, Charcoals, Oils							
g PE/EVAL/PE	SAFETY4 4H	>240m	0	0	0.07	35°C	60
treflan EC (synonym: trifluralin) CAS Number: 1582-09-8 Primary Class: 146 Amines, Aromatic, Secondary and Tertiary Related Class: 442 Nitro Compounds, Substituted 840 Pesticides, Mixtures and Formulations							
g NATURAL RUBBER	ANSELLEDMONT 36-124	<30m	420	3	0.51		74
g NEOPRENE	ANSELLEDMONT 29-865	<30m	420	3	0.51		74
g NITRILE	ANSELLEDMONT 37-175	90m	24	2	0.46		74
g PVC	EDMONT CANADA 14112	<30m	120	3	n.a.		74
tri-n-propylamine (synonym: tripropylamine) CAS Number: 102-69-2 Primary Class: 143 Amines, Aliphatic and Alicyclic, Tertiary							
g BUTYL	NORTH B-174				0.65	degrad.	34
g NATURAL RUBBER	ACKWELL 5-109				0.15	degrad.	34
g NEOPRENE	ANSELLEDMONT 29-870	>480m	0	0	0.48		34
g NITRILE	ANSELLEDMONT 37-155	>480m	0	0	0.40		34
g PVAL	EDMONT 25-950	>480m	0	0	0.63		34
g PVC	ANSELLEDMONT 34-100				0.15	degrad.	34
g VITON	NORTH F-091	>480m	0	0	0.30		34

G A R M E Class & Number/ N test chemical/ T MATERIAL NAME	MANUFACTURER & PRODUCT IDENTIFICATION	Break- through Time in Minutes	Perm- eation Rate in mg/sq m /min.	I N D E X	Thick- ness in mm	degrad. and comments	R e f s
triallylamine							
CAS Number: 102-70-5							
Primary Class: 143 Amines, Aliphatic and Alicyclic, Tertiary							
g BUTYL	NORTH B-174				0.66	degrad.	34
g NATURAL RUBBER	ACKWELL 5-109				0.15	degrad.	34
g NEOPRENE	ANSELLEDMONT 29-870	63m	3387	4	0.32	degrad.	34
g NITRILE	ANSELLEDMONT 37-155	>480m	0	0	0.32		34
g PVC	ANSELLEDMONT 34-100	5m	3725	5	0.20		34
g VITON	NORTH F-091	>480m	0	0	0.20		34
tribromomethane (synonym: bromoform)							
CAS Number: 75-25-2							
Primary Class: 261 Halogen Compounds, Aliphatic and Alicyclic							
g BUTYL	NORTH B-174	>480m	0	0	0.63	degrad.	34
g NATURAL RUBBER	ACKWELL 5-109				0.15	degrad.	34
g NEOPRENE	ANSELLEDMONT 29-870				0.48	degrad.	34
g NITRILE	ANSELLEDMONT 37-155	36m	1858	4	0.38		34
g PVAL	EDMONT 15-552	>480m	0	0	0.25		34
g PVC	ANSELLEDMONT 34-100				0.20	degrad.	34
g VITON	NORTH F-091	>480m	0	0	0.23		34
tributyl phosphate (synonym: TBP)							
CAS Number: 126-73-8							
Primary Class: 462 Organophosphorus Compounds and Derivatives of Phosphorus-based Acids							
g PE/EVAL/PE	SAFETY4 4H	>240m		0	0.07	35°C	60
tributyltin oxide (synonym: bis(tributyltin) oxide)							
CAS Number: 56-35-9							
Primary Class: 470 Organometallic Compounds							
s VITON/CHLOROBUTYL	LIFE-GUARD VC-100	>240m		0	n.a.		77
trichloroacetaldehyde (synonym: chloral)							
CAS Number: 75-87-6							
Primary Class: 121 Aldehydes, Aliphatic and Alicyclic							
g BUTYL	NORTH B-174	200m	294	3	0.71	degrad.	34
g NATURAL RUBBER	ACKWELL 5-109				0.15	degrad.	34
g NEOPRENE	ANSELLEDMONT 29-870				0.48	degrad.	34
g NITRILE	ANSELLEDMONT 37-155				0.40	degrad.	34
g PVAL	EDMONT 25-950	>480m	0	0	0.76		34
g PVC	ANSELLEDMONT 34-100	4m	16890	5	0.20	degrad.	34
g VITON	NORTH F-091	295m	0	0	0.25		34

G A R M E Class & Number/ N test chemical/ T MATERIAL NAME	MANUFACTURER & PRODUCT IDENTIFICATION	Break- through Time in Minutes	Perm- eation Rate in mg/sq m /min.	I N D E X	Thick- ness in mm	degrad. and comments	R e f s

trichloroacetic acid (synonym: trichloroethanoic acid)
CAS Number: 76-03-9
Primary Class: 103   Acids, Carboxylic, Aliphatic and Alicyclic, Substituted

g NITRILE	ANSELLEDMONT 37-145	>240m	0	0	0.29	35°C	57
c PE	DU PONT TYVEK QC	5m			n.a.	65°C	8
c PE	MOLNLYCKE H D	240m		0	n.a.	35°C	57
s SARANEX-23	DU PONT TYVEK SARAN	>120m			n.a.	65°C	8

trichloroacetonitrile (synonym: trichloromethyl cyanide)
CAS Number: 545-06-2
Primary Class: 431   Nitriles, Aliphatic and Alicyclic
Related Class: 261   Halogen Compounds, Aliphatic and Alicyclic

g BUTYL	NORTH B-174	120m	3156	4	0.61	degrad.	34
g NATURAL RUBBER	ACKWELL 5-109				0.15	degrad.	34
g NEOPRENE	ANSELLEDMONT 29-870	19m	9258	4	0.63	degrad.	34
g NITRILE	ANSELLEDMONT 37-155				0.40	degrad.	34
g PVAL	EDMONT 25-950	>480m	0	0	0.61		34
g PVC	ANSELLEDMONT 34-100				0.20	degrad.	34
g VITON	NORTH F-091	60m	1842	4	0.25	degrad.	34

trichloroethylene (synonyms: TCE; trichloroethene)
CAS Number: 79-01-6
Primary Class: 263   Halogen Compounds, Vinylic

g BUTYL	BEST 878	13m	1010	5	0.75	degrad.	53
g BUTYL	COMASEC BUTYL	7m	33000	5	0.57		40
s BUTYL	MSA CHEMPRUF	<60m			n.a.		48
m BUTYL	NASA	82m	16	2	0.53		81
m BUTYL	NASA	59m	17	3	0.53	pre-exp 48h	81
g BUTYL	NORTH B-161	5m	>8250	5	0.38		25
g BUTYL	NORTH B-161	5m	20400	5	0.40		35
g BUTYL	NORTH B-174	14m	33000	5	0.63	degrad.	34
g BUTYL/NEOPRENE	COMASEC BUTYL PLUS	7m	33000	5	n.a.		40
u CPE	ILC DOVER	12m			n.a.		29
g HYPALON	COMASEC DIPCO	17m	4560	4	0.60		40
g HYPALON/NEOPRENE	NORTH		10200	5	0.88		50
g NATURAL RUBBER	ACKWELL 5-109				0.15	degrad.	34
g NATURAL RUBBER	ANSELLEDMONT 36-124				0.46	degrad.	6
g NATURAL RUBBER	ANSELLEDMONT 392				n.a.	degrad.	6
g NATURAL RUBBER	ANSELLEDMONT 46-320	1m	94000	5	0.31		5
g NATURAL RUBBER	BEST 65NFW				n.a.	degrad.	53
g NATURAL RUBBER	COMASEC FLEXIGUM	22m	12600	5	0.95		40
g NATURAL+NEOPRENE	ANSELL OMNI 276	3m	73000	5	0.45		5
g NEOPRENE	ANSELLEDMONT 29-840				0.31	degrad.	6
g NEOPRENE	ANSELLEDMONT 29-870	3m	>5660	5	0.43		25

GARMENT Class & Number/ test chemical/ MATERIAL NAME	MANUFACTURER & PRODUCT IDENTIFICATION	Break-through Time in Minutes	Perm-eation Rate in mg/sq m /min.	INDEX	Thick-ness in mm	degrad. and comments	Refs
g NEOPRENE	ANSELLEDMONT 29-870				0.48	degrad.	34
g NEOPRENE	ANSELLEDMONT 29-870	5m	21900	5	0.48		35
g NEOPRENE	ANSELLEDMONT 29-870	10m	12	4	0.53		81
g NEOPRENE	ANSELLEDMONT 29-870	8m	27	4	0.53	pre-exp	
48h	81						
g NEOPRENE	BEST 32	5m	4070	5	n.a.	degrad.	53
g NEOPRENE	BEST 6780	6m	150	4	n.a.	degrad.	53
g NEOPRENE	COMASEC COMAPRENE	8m	11580	5	n.a.		40
g NEOPRENE	MAPA-PIONEER N-44	11m	39330	5	0.56		36
g NEOPRENE	MAPA-PIONEER N-73	12m	19000	5	0.46		5
g NEOPRENE	NORTH		7300	5	0.38		50
g NEOPRENE	NORTH		13700	5	0.76		50
g NEOPRENE	PIONEER		8800	5	0.84		50
u NEOPRENE	Unknown	10m	530	4	0.79		3
g NEOPRENE/NATURAL	ANSELL CHEMI-PRO				n.a.	degrad.	55
g NITRILE	ANSELLEDMONT 37-165	26m	9000	4	0.55		5
g NITRILE	ANSELLEDMONT 37-165				0.56	degrad.	6
g NITRILE	ANSELLEDMONT 37-165	110m	15	2	0.55		81
g NITRILE	ANSELLEDMONT 37-165	67m	16	2	0.55	pre-exp 48h	81
g NITRILE	ANSELLEDMONT 37-175	9m	21000	5	0.37		5
g NITRILE	ANSELLEDMONT 37-175	9m	16440	5	0.34	degrad.	35
g NITRILE	BEST 22R		12960	5	n.a.	degrad.	53
g NITRILE	BEST 727	9m	1170	5	0.38	degrad.	53
g NITRILE	COMASEC COMATRIL	15m	12000	5	0.55		40
g NITRILE	COMASEC COMATRIL SU	21m	11040	5	0.60		40
g NITRILE	COMASEC FLEXITRIL	6m	8700	5	n.a.		40
g NITRILE	KURSAAL 85/62	7m	11000	5	0.42		15
g NITRILE	MAPA-PIONEER A-14	9m	3744	5	0.56	degrad.	36
g NITRILE	MAPA-PIONEER A-15	12m	1767	5	0.33	degrad.	36
g NITRILE	MARIGOLD NITROSOLVE	20m	1164	4	0.75	degrad.	79
g NITRILE	NORTH LA-142G	4m	16980	5	0.38	degrad.	34
g NITRILE	SURETY	11m	>8250	5	0.38		25
u NITRILE	Unknown	10m	600	4	0.94		3
g NITRILE+PVC	COMASEC MULTIMAX	17m	5430	4	n.a.		40
g NITRILE+PVC	COMASEC MULTIPLUS	27m	12420	5	n.a.		40
g PE	ANSELLEDMONT 35-125	<1m	15000	5	0.03		5
g PE	ANSELLEDMONT 35-125	<1m	>6550	5	0.03		25
g PE	ANSELLEDMONT 35-125	<1m	13920	5	0.07		35
g PE/EVAL/PE	SAFETY4 4H	>240m	0	0	0.07	35°C	60
g PE/EVAL/PE	SAFETY4 4H	>1440m	0	0	0.07		60
g PVAL	ANSELLEDMONT PVA	>360m	0	0	n.a.		6
g PVAL	COMASEC SOLVATRIL	>480m	0	0	n.a.		40
g PVAL	EDMONT 15-552	30m	<9	2	n.a.		6
g PVAL	EDMONT 15-552	>1440m	0	0	n.a.		25
g PVAL	EDMONT 15-552	>1440m	0	0	n.a.	preexp 24h	25
g PVAL	EDMONT 15-552	>1440m	0	0	n.a.	10°C	25
g PVAL	EDMONT 15-552	>1440m	0	0	n.a.	45°C	25
g PVAL	EDMONT 25-545	>360m	0	0	n.a.		6

G A R M E Class & Number/ N test chemical/ T MATERIAL NAME	MANUFACTURER & PRODUCT IDENTIFICATION	Break- through Time in Minutes	Perm- eation Rate in mg/sq m /min.	I N D E X	Thick- ness in mm	degrad. and comments	R e f s
g PVAL	EDMONT 25-545	>480m	0	0	0.45		35
g PVAL	EDMONT 25-950	>960m	0	0	0.38		34
g PVC	ANSELLEDMONT SNORKE				n.a.	degrad.	6
g PVC	BEST 725R				n.a.	degrad.	53
g PVC	BEST 812		9000	5	n.a.		53
g PVC	COMASEC MULTIPOST	27m	12420	5	n.a.		40
g PVC	COMASEC MULTITOP	25m	12600	5	n.a.		40
g PVC	COMASEC NORMAL	20m	12540	5	n.a.		40
g PVC	COMASEC OMNI	12m	14980	5	n.a.		40
g PVC	MAPA-PIONEER V-5	<1m	7440	5	0.10	degrad.	35
s PVC	MSA UPC	<3m			0.20		48
g PVC	NORTH 800	4m	23000	5	n.a.		15
s SARANEX-23	DU PONT TYVEK SARAN	<1m	>3000	5	0.15		25
s TEFLON	CHEMFAB CHALL. 5100	144m	0.34	1	n.a.		65
g TEFLON	CLEAN ROOM PRODUCTS	>1440m	0	0	0.05		4
g TEFLON	CLEAN ROOM PRODUCTS	>1440m	0	0	0.05		25
v TEFLON-FEP	CHEMFAB CHALL.	>180m			0.25		65
c UNKNOWN MATERIAL	CHEMRON CHEMTUFF	7m	29	4	n.a.		67
s UNKNOWN MATERIAL	LIFE-GUARD RESPONDE	>240m	0	0	n.a.		77
g UNKNOWN MATERIAL	NORTH SILVERSHIELD	>360m	0	0	0.08		7
g VITON	NORTH F-091	660m	17	2	0.23		25
g VITON	NORTH F-091	330m	10	2	0.23	preexp 24h	25
g VITON	NORTH F-091	>1440m	0	0	0.23	10°C	25
g VITON	NORTH F-091	48m	220	3	0.23	45°C	25
g VITON	NORTH F-091	444m	14	2	0.25		34
g VITON	NORTH F-121	>480m	0	0	0.30		35
g VITON	NORTH SF	282m	33	2	0.20		25
s VITON/CHLOROBUTYL	LIFE-GUARD VC-100	>240m		0	n.a.		77
g VITON/NEOPRENE	ERISTA VITRIC	5m	50	4	0.45		15
g VITON/NEOPRENE	ERISTA VITRIC	540m	22	2	0.45		45
g VITON/NEOPRENE	ERISTA VITRIC	162m	250	3	0.45	28°C	45
g VITON/NEOPRENE	ERISTA VITRIC	100m	250	3	0.45	38°C	45
g VITON/NITRILE	NORTH VITRILE	491m	130	3	0.20		25

trichloroethylene, 50% & pentane, 50%
Primary Class: 263   Halogen Compounds, Vinylic

g PE	HANDGARDS	2m	4000	5	0.17		5

tricresyl phosphate (synonyms: TCP; tritolyl phosphate)
CAS Number: 78-30-8
Primary Class: 462   Organophosphorus Compounds and Derivatives of Phosphorus-based Acids
Related Class: 233   Esters, Non-Carboxylic, Carbamates and Others

g BUTYL	NORTH B-174	>480m	0	0	0.70		34
g NATURAL RUBBER	ANSELL CONFORM 4205	>360m		0	0.13		55

G A R M E Class & Number/ N test chemical/ T MATERIAL NAME	MANUFACTURER & PRODUCT IDENTIFICATION	Break- through Time in Minutes	Perm- eation Rate in mg/sq m /min.	I N D E X	Thick- ness in mm	degrad. and comments	R e f s
g NATURAL RUBBER	ANSELL FL 200 254	>360m		0	0.51		55
g NATURAL RUBBER	ANSELL ORANGE 208	>360m	0		0.76		55
g NATURAL RUBBER	ANSELL PVL 040	>360m		0	0.46		55
g NATURAL RUBBER	ANSELL STERILE 832	>360m	0		0.23		55
g NATURAL RUBBER	ANSELLEDMONT 36-124	45m	<9	2	0.46		6
g NATURAL RUBBER	ANSELLEDMONT 392	45m	<9	2	0.48		6
g NATURAL+NEOPRENE	ANSELL OMNI 276	>360m	0	0	0.56		55
g NATURAL+NEOPRENE	ANSELL TECHNICIANS	>360m	0	0	0.43		55
g NEOPRENE	ANSELL NEOPRENE 530	>360m	0	0	0.46		55
g NEOPRENE	ANSELLEDMONT 29-840	>360m	0	0	0.31		6
g NEOPRENE	ANSELLEDMONT NEOX	>360m	0	0	n.a.		6
g NEOPRENE/NATURAL	ANSELL CHEMI-PRO	>360m	0	0	0.72		55
g NITRILE	ANSELL CHALLENGER	>360m	0	0	0.38		55
g NITRILE	ANSELLEDMONT 37-165	>360m	0	0	0.60		6
g PVAL	ANSELLEDMONT PVA	>360m		0	n.a.		6
g PVAL	EDMONT 15-552	>360m	0	0	n.a.		6
g PVAL	EDMONT 25-545	>360m	0	0	n.a.		6
g PVAL	EDMONT 25-950	>480m	0	0	0.74		34
g PVC	ANSELLEDMONT 34-100	>480m	0	0	0.19		34
g PVC	ANSELLEDMONT M GRIP	>360m	0	0	n.a.		6
g VITON	NORTH F-091	>480m	0	0	0.38		34

triethanolamine (synonym: TEA)
CAS Number: 102-71-6
Primary Class: 143   Amines, Aliphatic and Alicyclic, Tertiary
Related Class: 311   Hydroxy Compounds, Aliphatic and Alicyclic, Primary

Class & Number/ test chemical/ MATERIAL NAME	MANUFACTURER & PRODUCT IDENTIFICATION	Break-through Time in Minutes	Permeation Rate in mg/sq m /min.	INDEX	Thickness in mm	degrad. and comments	Refs
g BUTYL	NORTH B-174	>480m	0	0	0.71		34
g NATURAL RUBBER	ACKWELL 5-109	>480m	0	0	0.15		34
g NATURAL RUBBER	ANSELLEDMONT 36-124	60m	<9	1	0.46	85%	6
g NATURAL RUBBER	ANSELLEDMONT 392	>360m		0	0.48	85%	6
g NATURAL RUBBER	MAPA-PIONEER L-118	>480m	0	0	0.46		36
g NATURAL+NEOPRENE	ANSELL OMNI 276	>360m	0	0	0.76		55
g NEOPRENE	ANSELL NEOPRENE 530	>360m	0	0	0.46		55
g NEOPRENE	ANSELLEDMONT 29-840	>360m	0	0	0.38	85%	6
g NEOPRENE	ANSELLEDMONT 29-870	>480m	0	0	0.50		34
g NEOPRENE	ANSELLEDMONT NEOX	>360m	0	0	n.a.	85%	6
g NEOPRENE	MAPA-PIONEER N-44	>480m	0	0	0.74		36
g NEOPRENE/NATURAL	ANSELL CHEMI-PRO	>360m	0	0	0.72		55
g NITRILE	ANSELL CHALLENGER	>360m	0	0	0.38		55
g NITRILE	ANSELLEDMONT 37-155	>480m	0	0	0.36		34
g NITRILE	ANSELLEDMONT 37-165	>360m	0	0	0.60	85%	6
g NITRILE	MAPA-PIONEER A-14	>480m	0	0	0.54		36
g PVAL	ANSELLEDMONT PVA	>360m		0	n.a.	85%	6
g PVAL	EDMONT 15-552	>360m	0	0	n.a.	85%	6
g PVAL	EDMONT 25-545	>360m	0	0	n.a.	85%	6
g PVC	ANSELLEDMONT M GRIP	>360m	0	0	n.a.	85%	6
g PVC	MAPA-PIONEER V-20	>480m	0	0	0.45		36

G A R M E Class & Number/ N test chemical/ T MATERIAL NAME	MANUFACTURER & PRODUCT IDENTIFICATION	Break- through Time in Minutes	Perm- eation Rate in mg/sq m /min.	I N D E X	Thick- ness in mm	degrad. and comments	R e f s
triethanolamine, 30-70% Primary Class: 143 Amines, Aliphatic and Alicyclic, Tertiary							
g PE/EVAL/PE	SAFETY4 4H	>240m	0	0	0.07	35°C	60
triethylamine CAS Number: 121-44-8 Primary Class: 143 Amines, Aliphatic and Alicyclic, Tertiary							
g BUTYL	NORTH B-174				0.65	degrad.	34
u CPE	ILC DOVER	>180m			0.05		29
g NATURAL RUBBER	ACKWELL 5-109				0.15	degrad.	34
g NEOPRENE	ANSELLEDMONT 29-870	37m	8769	4	0.48	degrad.	34
g NITRILE	ANSELLEDMONT 37-155	>480m	0	0	0.40		34
g NITRILE	ERISTA SPECIAL	>240m	0	0	0.40		33
g PVC	ANSELLEDMONT 34-100	4m	4854	5	0.20	degrad.	34
s SARANEX-23	DU PONT TYVEK SARAN	>480m	0	0	n.a.		8
s UNKNOWN MATERIAL	LIFE-GUARD RESPONDE	>480m	0	0	n.a.		77
g VITON	NORTH F-091	>480m	0	0	0.28		34
u VITON/CHLOROBUTYL	Unknown	9m			0.36		29
triethylenediamine, <30% CAS Number: 280-57-9 Primary Class: 143 Amines, Aliphatic and Alicyclic, Tertiary							
g PE/EVAL/PE	SAFETY4 4H	>240m	0	0	0.07	35°C	60
triethylenetetraamine (synonym: TETA) CAS Number: 112-24-3 Primary Class: 148 Amines, Poly, Aliphatic and Alicyclic Related Class: 141 Amines, Aliphatic and Alicyclic, Primary 142 Amines, Aliphatic and Alicyclic, Secondary							
g BUTYL	NORTH B-174	>480m	0	0	0.46		34
g NATURAL RUBBER	ACKWELL 5-109				0.15	degrad.	34
g NEOPRENE	ANSELLEDMONT 29-870	>480m	0	0	0.40		34
g NITRILE	ANSELLEDMONT 37-155	>480m	0	0	0.41		34
g PE/EVAL/PE	SAFETY4 4H	>240m		0	0.07	35°C	60
g PVC	ANSELLEDMONT 34-100				0.20	degrad.	34
g VITON	NORTH F-091	>480m	0	0	0.22		34
triethylenetetraamine, 50% & methyl ethyl ketone, 50% Primary Class: 148 Amines, Poly, Aliphatic and Alicyclic							

G A R M E N T										
Class & Number/ test chemical/ MATERIAL NAME	MANUFACTURER & PRODUCT IDENTIFICATION	Break- through Time in Minutes	Perm- eation Rate in mg/sq m /min.	I N D E X	Thick- ness in mm	degrad. and comments	R e f s

---

trifluoroacetic acid (synonym: perfluoroacetic acid)
CAS Number: 76-05-1
Primary Class: 103   Acids, Carboxylic, Aliphatic and Alicyclic, Substituted
Related Class: 261   Halogen Compounds, Aliphatic and Alicyclic

g PE/EVAL/PE	SAFETY4 4H	>240m	0	0	0.07	35°C	60

---

trimethyl phosphate
CAS Number: 512-56-1
Primary Class: 462   Organophosphorus Compounds and Derivatives of Phosphorus-based Acids
Related Class: 233   Esters, Non-Carboxylic, Carbamates and Others

g NITRILE	ANSELLEDMONT 37-155			0.40	degrad.	34
g PVC	ANSELLEDMONT 34-100			0.20	degrad.	34
g VITON	NORTH F-091			0.25	degrad.	34

---

trimethylamine
CAS Number: 75-50-3
Primary Class: 143   Amines, Aliphatic and Alicyclic, Tertiary

s BUTYL	TRELLEBORG TRELLCHE		n.a.	degrad.	71

---

trimethylolpropane triacrylate (synonym: TMPTA)
CAS Number: 15625-89-5
Primary Class: 223   Esters, Carboxylic, Acrylates and Methacrylates

g BUTYL	NORTH B-161	>480m	0	0	0.46	94
g NATURAL RUBBER	MAPA-PIONEER L-118	>360m	0	0	0.45	94
g NITRILE	ANSELLEDMONT 37-155	>480m	0	0	0.37	94

---

tripropyleneglycol diacrylate
CAS Number: 42978-66-5
Primary Class: 223   Esters, Carboxylic, Acrylates and Methacrylates

g PE/EVAL/PE	SAFETY4 4H	>240m	0	0	0.07	35°C	60

---

tris(1,3-dichloropropyl) phosphate
CAS Number: 40120-74-9
Primary Class: 462   Organophosphorus Compounds and Derivatives of Phosphorus-based Acids

g NITRILE	ANSELLEDMONT 37-155	>480m	0	0	0.38	34

Turco 5092 stripping agent
Primary Class: 900   Miscellaneous Unclassified Chemicals

g PE/EVAL/PE	SAFETY4 4H	132m		0.07	21°C	60

G A R M E Class & Number/ N test chemical/ T MATERIAL NAME	MANUFACTURER & PRODUCT IDENTIFICATION	Break- through Time in Minutes	Perm- eation Rate in mg/sq m /min.	I N D E X	Thick- ness in mm	degrad. and comments	R e f s
g PE/EVAL/PE	SAFETY4 4H	28m			0.07	35°C	60

turpentine
CAS Number: 8006-64-2
Primary Class: 294   Hydrocarbons, Aliphatic and Alicyclic, Unsaturated

g BUTYL	BEST 878				0.75	degrad.	53
g NATURAL RUBBER	ANSELLEDMONT 36-124				0.46	degrad.	6
g NATURAL RUBBER	ANSELLEDMONT 392				0.48	degrad.	6
g NATURAL RUBBER	BEST 65NFW				n.a.	degrad.	53
g NATURAL+NEOPRENE	PLAYTEX ARGUS 123	4m	2640	5	n.a.		72
g NEOPRENE	ANSELL NEOPRENE 530	13m	14	4	0.46		55
g NEOPRENE	ANSELLEDMONT 29-840				0.38	degrad.	6
g NEOPRENE	MAPA-PIONEER N-44	>480m	0	0	0.56		36
g NEOPRENE/NATURAL	ANSELL CHEMI-PRO	11m	36	4	0.72		55
g NITRILE	ANSELL CHALLENGER	30m	2	2	0.38		55
g NITRILE	ANSELLEDMONT 37-165	30m	<9	2	0.60		6
g NITRILE	MAPA-PIONEER A-14	>480m	0	0	0.56		36
g PE/EVAL/PE	SAFETY4 4H	>480m	0	0	0.07		60
g PVAL	ANSELLEDMONT PVA	>360m			0	n.a.	6
g PVAL	EDMONT 15-552	360m	<9	1	n.a.		6
g PVAL	EDMONT 25-545	>360m	0	0	n.a.		6
g PVC	ANSELLEDMONT SNORKE				n.a.	degrad.	6

U-V resin 20074
Primary Class: 900   Miscellaneous Unclassified Chemicals

g PE/EVAL/PE	SAFETY4 4H	>240m	0	0	0.07	35°C	60

valeronitrile (synonym: 1-pentanenitrile)
CAS Number: 110-59-8
Primary Class: 431   Nitriles, Aliphatic and Alicyclic

g BUTYL	NORTH B-174	>480m	0	0	0.63		34
g NATURAL RUBBER	ACKWELL 5-109	2m	1260	5	0.15		34
g NEOPRENE	ANSELLEDMONT 29-870	40m	1260	4	0.46	degrad.	34
g NITRILE	ANSELLEDMONT 37-155				0.40	degrad.	34
g PVAL	EDMONT 25-950	>480m	0	0	0.64		34
g PVC	ANSELLEDMONT 34-100				0.20	degrad.	34
g VITON	NORTH F-091				0.26	degrad.	34

vegetable oil
CAS Number: 68956-68-3
Primary Class: 860   Coals, Charcoals, Oils

s BUTYL	MSA CHEMPRUF	<120m			n.a.		48

G A R M E Class & Number/ N test chemical/ T MATERIAL NAME	MANUFACTURER & PRODUCT IDENTIFICATION	Break- through Time in Minutes	Perm- eation Rate in mg/sq m /min.	I N D E X	Thick- ness in mm	degrad. and comments	R e f s
s PVC	MSA UPC	>480m	0	0	0.20		48

vincristine
CAS Number: 57-22-7
Primary Class: 870   Antineoplastic Drugs and Other Pharmaceuticals

g EMA	Unknown	90m			0.07		59
u NATURAL RUBBER	Unknown	>360m	0	0	0.18		59
g NEOPRENE	ANSELL NEOPRENE 530	>360m	0	0	0.28		59
u PVC	Unknown	>360m	0	0	0.19		59

vinyl acetate
CAS Number: 108-05-4
Primary Class: 222   Esters, Carboxylic, Acetates
Related Class: 294   Hydrocarbons, Aliphatic and Alicyclic, Unsaturated

g NEOPRENE	MAPA-PIONEER N-44	30m	1980	4	0.74		36
g NITRILE	MAPA-PIONEER A-14	30m	4020	4	0.54		36
g PE/EVAL/PE	SAFETY4 4H	>240m	0	0	0.07	35°C	60
s TEFLON	CHEMFAB CHALL. 5100	74m	0.6	1	n.a.		65
v TEFLON-FEP	CHEMFAB CHALL.	>180m	<1	1	0.25		65
s UNKNOWN MATERIAL	LIFE-GUARD RESPONDE	>180m	0	1	n.a.		77

vinyl chloride (synonym: chloroethene)
CAS Number: 75-01-4
Primary Class: 263   Halogen Compounds, Vinylic
Related Class: 261   Halogen Compounds, Aliphatic and Alicyclic

g BUTYL	NORTH B-161				0.43	degrad.	34
u CPE	ILC DOVER	>180m			n.a.		29
g NITRILE	BEST 22R	<1	4		n.a.		53
g NITRILE	NORTH LA-142G	342m	8	1	0.36		7
s UNKNOWN MATERIAL	LIFE-GUARD RESPONDE	>180m			n.a.		77
g UNKNOWN MATERIAL	NORTH SILVERSHIELD	>480m	0	0	0.10		7
g VITON	NORTH F-091	264m	6	1	0.24		7
s VITON/BUTYL	DRAGER 500 OR 710	>60m			n.a.		91
u VITON/CHLOROBUTYL	Unknown	>180m			n.a.		29

vinyl fluoride (synonym: fluoroethene)
CAS Number: 75-02-5
Primary Class: 263   Halogen Compounds, Vinylic

s BUTYL	TRELLEBORG TRELLCHE				n.a.	degrad.	71

G							
A							
R			Perm-		I		
M		Break-	eation	N	Thick-		R
E Class & Number/	MANUFACTURER	through	Rate in	D	ness	degrad.	e
N test chemical/	& PRODUCT	Time in	mg/sq m	E	in	and	f
T MATERIAL NAME	IDENTIFICATION	Minutes	/min.	X	mm	comments	s

vinylidene chloride (synonym: 1,1-dichloroethene)
CAS Number: 75-35-4
Primary Class: 263   Halogen Compounds, Vinylic
Related Class: 261   Halogen Compounds, Aliphatic and Alicyclic

g BUTYL	NORTH B-174				0.66	degrad.	34
g NATURAL RUBBER	ACKWELL 5-109				0.15	degrad.	34
g NEOPRENE	ANSELLEDMONT 29-870				0.48	degrad.	34
g NITRILE	ANSELLEDMONT 37-155	6m	7080	5	0.43	degrad.	34
g PVAL	EDMONT 25-950	360m	0	0	0.60		34
g PVC	ANSELLEDMONT 34-100	1m	13320	5	0.20	degrad.	34
s UNKNOWN MATERIAL	LIFE-GUARD RESPONDE	>180m	0	1	n.a.		77
g VITON	NORTH F-091	94m	480	3	0.30		34

vinylidene fluoride (synonym: 1,1-difluoroethene)
CAS Number: 75-38-7
Primary Class: 263   Halogen Compounds, Vinylic
Related Class: 261   Halogen Compounds, Aliphatic and Alicyclic

g BUTYL	NORTH B-174	>480m	0	0	0.72		34
g NATURAL RUBBER	ACKWELL 5-109	1m	60	4	0.15		34
g NEOPRENE	ANSELLEDMONT 29-870	1m	18	4	0.45		34
g VITON	NORTH F-091	>480m	0	0	0.24		34

water
CAS Number: 7732-18-5
Primary Class: 900   Miscellaneous Unclassified Chemicals

g BUTYL	NORTH B-174	>480m	0	0	0.43	deionized	34
g NATURAL RUBBER	ACKWELL 5-109	<30m	1038	4	0.15	deionized	34
g NEOPRENE	ANSELLEDMONT 29-870	>480m	0	0	0.46	deionized	34
g PVAL	EDMONT 25-545	<15m	3120	4	0.40	deionized	34
g UNKNOWN MATERIAL	NORTH SILVERSHIELD	>480m	0	0	0.10	deionized	34
g VITON	NORTH F-091	>480m	0	0	0.23	deionized	34

witch hazel
CAS Number: 68916-39-2
Primary Class: 900   Miscellaneous Unclassified Chemicals

g BUTYL	NORTH B-174	>480m	0	0	0.54		34
g NATURAL RUBBER	ACKWELL 5-109	2m	16200	5	0.15		34
g PVAL	EDMONT 25-950				0.50	degrad.	34
g PVC	ANSELLEDMONT 34-100	3m	2	3	0.20		34
g VITON	NORTH F-091	>480m	0	0	0.25		34

G A R M E Class & Number/ N test chemical/ T MATERIAL NAME	MANUFACTURER & PRODUCT IDENTIFICATION	Break- through Time in Minutes	Perm- eation Rate in mg/sq m /min.	I N D E X	Thick- ness in mm	degrad. and comments	R e f s
wood creosote CAS Number: 8021-39-4 Primary Class: 316 Hydroxy Compounds, Aromatic (Phenols)							
g NEOPRENE	ANSELLEDMONT 29-840	>240m	0	0	0.45	degrad.	33
g PE/EVAL/PE	SAFETY4 4H	>240m	0	0	0.07	35°C	60
g VITON	NORTH F-091	>1140m	0	0	0.25		33
Xylamon Primary Class: 900 Miscellaneous Unclassified Chemicals							
g PE/EVAL/PE	SAFETY4 4H	>240m	0	0	0.07	35°C	60
xylene (undefined mixture) CAS Number: 1330-20-7 Primary Class: 292 Hydrocarbons, Aromatic							
g BUTYL	ANSELL LAMPRECHT	39m	6550	4	0.70		38
g BUTYL	BEST 878				0.75	degrad.	53
g BUTYL	COMASEC BUTYL	40m	4080	4	0.50		40
g BUTYL	NORTH B-161				0.65	degrad.	34
g BUTYL	NORTH B-161	9m	7600	5	0.38		93
g BUTYL/NEOPRENE	COMASEC BUTYL PLUS	40m	4080	4	0.50		40
g HYPALON	COMASEC DIPCO	35m	5040	4	0.60		40
g NAT+NEOPR+NITRILE	MAPA-PIONEER TRIONI	4m	600	4	0.48		36
g NAT+NEOPR+NITRILE	MAPA-PIONEER TRIONI	2m	8370	5	0.46		93
g NATURAL RUBBER	ACKWELL 5-109				0.15	degrad.	34
g NATURAL RUBBER	ANSELLEDMONT 30-139				0.25	degrad.	42
g NATURAL RUBBER	ANSELLEDMONT 30-139	4m	8860	5	0.51		93
g NATURAL RUBBER	ANSELLEDMONT 36-124				0.46	degrad.	6
g NATURAL RUBBER	ANSELLEDMONT 392				0.48	degrad.	6
g NATURAL RUBBER	ANSELLEDMONT 46-320	2m	34000	5	0.31		5
g NATURAL RUBBER	BEST 65NFW				n.a.	degrad.	53
g NATURAL RUBBER	COMASEC FLEXIGUM	7m	4440	5	0.95		40
g NATURAL+NEOPRENE	ANSELL OMNI 276	4m	3	3	0.56		55
g NATURAL+NEOPRENE	PLAYTEX 827	3m	10310	5	0.38		93
g NEOPRENE	ANSELL NEOPRENE 520	4m	3	3	0.46		55
g NEOPRENE	ANSELLEDMONT 29-840				0.38	degrad.	6
g NEOPRENE	ANSELLEDMONT 29-870				0.48	degrad.	34
g NEOPRENE	BEST 32				n.a.	degrad.	53
g NEOPRENE	BEST 6780		300	5	n.a.	degrad.	53
g NEOPRENE	COMASEC COMAPRENE	8m	4080	5	n.a.		40
g NEOPRENE	MAPA-PIONEER N-73	23m	8000	4	0.46		5
g NEOPRENE/NATURAL	ANSELL CHEMI-PRO	5m	8	3	0.72		55
g NITRILE	ANSELL CHALLENGER	34m	2	2	0.38		55
g NITRILE	ANSELLEDMONT 37-155	27m	1704	4	0.40	degrad.	34

G A R M E Class & Number/ N test chemical/ T MATERIAL NAME	MANUFACTURER & PRODUCT IDENTIFICATION	Break-through Time in Minutes	Perm-eation Rate in mg/sq m /min.	I N D E X	Thick-ness in mm	degrad. and comments	R e f s
g NITRILE	ANSELLEDMONT 37-165	>60m			0.55		5
g NITRILE	ANSELLEDMONT 37-165	75m	<9000	4	0.60		6
g NITRILE	ANSELLEDMONT 37-165	75m	1530	4	0.60		31
g NITRILE	ANSELLEDMONT 37-165	4m	23	4	0.60	preexp. 15h	31
g NITRILE	ANSELLEDMONT 37-175	57m	1000	4	0.37		5
g NITRILE	ANSELLEDMONT 37-175	41m	2790	4	0.42		38
g NITRILE	ANSELLEDMONT 49-125	6m	<900	4	0.25		42
g NITRILE	ANSELLEDMONT 49-155	40m	<9000	4	0.38		42
g NITRILE	ANSELLEDMONT 49-155	62m	1960	4	0.43		93
g NITRILE	BEST 22R		840	5	n.a.	degrad.	53
g NITRILE	BEST 727	41m	1220	4	0.38		53
g NITRILE	COMASEC COMATRIL	60m	4560	4	0.60		40
g NITRILE	COMASEC COMATRIL SU	100m	3000	4	0.60		40
g NITRILE	COMASEC FLEXITRIL	5m	4380	5	n.a.		40
g NITRILE	MAPA-PIONEER A-14	92m	216	3	0.56	degrad.	36
g NITRILE	MAPA-PIONEER A-14	92m	240	3	0.54	degrad.	36
g NITRILE	MARIGOLD NITROSOLVE	79m	96	2	0.75	degrad.	79
g NITRILE	NORTH LA-132G	42m	1700	4	0.35		30
g NITRILE+PVC	COMASEC MULTIMAX	36m	5160	4	n.a.		40
g NITRILE+PVC	COMASEC MULTIPLUS	45m	3300	4	n.a.		40
g PE	HANDGARDS	4m	1000	5	0.07		5
u PE	Unknown	1m	1075	5	0.04		30
g PE/EVAL/PE	SAFETY4 4H	>240m	0	0	0.07	35°C	60
g PE/EVAL/PE	SAFETY4 4H	>1440m	0	0	0.07		60
g PVAL	ANSELLEDMONT PVA	>360m			n.a.		6
g PVAL	EDMONT 15-552	>360m	0	0	n.a.		6
g PVAL	EDMONT 25-545	>360m	0	0	n.a.		6
g PVAL	EDMONT 25-950	>480m	0	0	0.60		34
g PVC	ANSELLEDMONT 34-100	1m	1920	5	0.20		34
g PVC	ANSELLEDMONT SNORKE				n.a.	degrad.	6
g PVC	BEST 812		720	5	n.a.		53
g PVC	COMASEC MULTIPOST	26m	3420	4	n.a.		40
g PVC	COMASEC MULTITOP	40m	3600	4	n.a.		40
g PVC	COMASEC NORMAL	39m	3720	4	n.a.		40
g PVC	COMASEC OMNI	8m	3720	5	n.a.		40
g PVC	MAPA-PIONEER V-10	4m	17000	5	0.20		6
u PVDC/PE/PVDC	MOLNLYCKE	23m	160	3	0.07		54
s TEFLON	CHEMFAB CHALL. 5100	>180m	0	1	n.a.		65
s UNKNOWN MATERIAL	DU PONT BARRICADE	>480m	0	0	n.a.		8
s UNKNOWN MATERIAL	KAPPLER CPF III	>480m	0	0	n.a.		77
s UNKNOWN MATERIAL	LIFE-GUARD RESPONDE	>180m			n.a.		77
g VITON	NORTH F-091	>160m			0.25		30
g VITON	NORTH F-091	>480m	0	0	0.26		34
g VITON	NORTH F-091	>60m			0.26		38
g VITON/NEOPRENE	ERISTA VITRIC	>60m			0.60		38

xylene (undefined mixture), 50% & methyl isobutyl ketone, 50%
Primary Class: 292   Hydrocarbons, Aromatic

G A R M E Class & Number/ N test chemical/ T MATERIAL NAME	MANUFACTURER & PRODUCT IDENTIFICATION	Break- through Time in Minutes	Perm- eation Rate in mg/sq m /min.	I N D E X	Thick- ness in mm	degrad. and comments	R e f s

xylene sulfonic acid sodium salt (synonym: sodium dimethylbenzene sulfonate)
CAS Number: 1300-72-7
Primary Class: 550   Organic Salts (Solutions)
Related Class: 504   Sulfur Compounds, Sulfonic Acids
              508   Sulfur Compounds, Thiones

g NATURAL RUBBER	ANSELL 9070-5	>480m	0	0	0.13		34
g NEOPRENE	ANSELLEDMONT 29-870	>480m	0	0	0.51		34
g PVC	ANSELLEDMONT 34-100	>480m	0	0	0.19		34

xylene, 50% & 2-ethoxyethanol, 50%
Primary Class: 292   Hydrocarbons, Aromatic
Related Class: 245   Ethers, Glycols

g PE/EVAL/PE	SAFETY4 4H	>240m	0	0	0.07	35°C	60

xylene, 60% & toluene-2,4-diisocyanate, 40%
Primary Class: 292   Hydrocarbons, Aromatic

g PE/EVAL/PE	SAFETY4 4H	>480m	0	0	0.07		60
g PE/EVAL/PE	SAFETY4 4H	>240m	0	0	0.07	35°C	60

xylene, 75% & 2-butoxyethanol, 25%
Primary Class: 292   Hydrocarbons, Aromatic
Related Class: 245   Ethers, Glycols

g PE/EVAL/PE	SAFETY4 4H	>240m	0	0	0.07	35°C	60

xylene, 80-92% & 2-methoxyethanol, < 6% & cyclized polyisoprene, 5-15%
Primary Class: 800   Multicomponent Mixtures With > 2 Components

g BUTYL	NORTH B-161	12m	5540	5	0.38		93
g BUTYL	NORTH B-161	8m	8300	5	0.38	37°C	93
g NAT+NEOPR+NITRILE	MAPA-PIONEER TRIONI	4m	6040	5	0.46		93
g NATURAL RUBBER	ANSELLEDMONT 30-139	6m	6690	5	0.51		93
g NATURAL+NEOPRENE	PLAYTEX 827	4m	5690	5	0.38		93
g NITRILE	ANSELLEDMONT 49-155	72m	120	3	0.36		93
g NITRILE	ANSELLEDMONT 49-155	49m	700	3	0.36	37°C	93

xylenol
CAS Number: 1300-71-6
Primary Class: 316   Hydroxy Compounds, Aromatic (Phenols)

s TEFLON	CHEMFAB CHALL. 5100	>198m	0	1	n.a.		65